Contemporary Trends and Issues in Science Education

Volume 61

The book series Contemporary Trends and Issues in Science Education provides a forum for innovative trends and issues impacting science education. Scholarship that focuses on advancing new visions, understanding, and is at the forefront of the field is found in this series. Authoritative works based on empirical research and/or conceptual theory from disciplines including historical, philosophical, psychological and sociological traditions are represented here. Our goal is to advance the field of science education by testing and pushing the prevailing sociocultural norms about teaching, learning, research and policy. Book proposals for this series may be submitted to the Publishing Editor: Claudia Acuna E-mail: Claudia.Acuna@springer.com

Dana L. Zeidler
Editor

A Moral Inquiry into Epistemic Insights in Science Education

Personal and Global Perspectives of Socioscientific Issues

With Contrib. by Ly Do

Springer

Editor
Dana L. Zeidler
University of South Florida
Tampa, FL, USA

With Contrib. by
Ly Do
Department of Teaching and Learning
College of Education
University of South Florida
Tampa, FL, USA

ISSN 1878-0482 ISSN 1878-0784 (electronic)
Contemporary Trends and Issues in Science Education
ISBN 978-3-031-63381-2 ISBN 978-3-031-63382-9 (eBook)
https://doi.org/10.1007/978-3-031-63382-9

This Springer imprint is published by the registered company Springer Nature Switzerland AG
The registered company address is: Gewerbestrasse 11, 6330 Cham, Switzerland

This volume is dedicated to the scholarly family of colleagues who supported this venture and continue to inspire present and future generations of educators in science education, as well as educators in related disciplines around the globe. It is also dedicated to those past, present, and future generations of scholars that we are fortunate to work with. Further, this work is certainly dedicated to each of our own individual families who constantly support and guide us. Finally, on a personal level, I would like to dedicate this work to Annie, Judy, John, Christopher, Matthew, and Sophie—who brought me into the fold to become a part of their family. (D.L.Z.)

Foreword

If you are familiar with the label "socioscientific issues" (SSI), which denotes a major domain of research and practice in the field of science education, and/or you know (or know of) Dana Zeidler, a science education scholar whose name is synonymous with SSI, then you have ample reason to get and read this book.

Or so you think!

I venture to say that you might think you know Dana Zeidler or fully grasp SSI theory, research, and practice, as well as its global reach. I thought I knew the person and the work. I was *mistaken*.

If you made it to this paragraph and are reading these words, then you likely are still deciding whether to get your hands on this unique volume and read through its content. Maybe you picked up the volume off the publisher's table at conference or a store shelf somewhere or, more likely, are scrolling down some limited preview of the book on the publisher's or a vendor website. You likely skimmed the table of contents and were struck by the uniqueness (or oddity) of the book's structure. The book is unique in that it is neither sole authored nor an edited work—at least, not in the traditional sense of authored and edited books. It is rare for a book editor to write several opening chapters in an edited volume just to set the stage for invited authors to make their contributions. You might also have realized that the book is, at once, inward looking and deeply personal, as well as a collective, multi-lensed, global effort. This book is unique.

So, should you get the book and read it? My answer is an unequivocal yes. At the same time, I have reasons to doubt this answer. You still need to decide.

For full disclosure, I have known Dana Zeidler for close to three decades now, since I first met him in the company of my doctoral adviser, the late Norman Lederman. Dana and Norm went through their doctoral programs in science education together at Syracuse University. Over the years, I got to know Dana and work with him professionally, having contributed chapters to two of his edited books. We also participated together in more than a dozen conference symposia and presentations over the years. From 2016 to 2020, Dana and I co-edited the *Journal of Research in Science Teaching* (JRST).

Dana is a valued colleague and close friend. I thought I knew him well!

No doubt, Dana is the authority in the domain of SSI in science education. He is among a few colleagues in the field who can lay claim to having founded a whole domain of research and practice that has gained a wide global following and that continues to be vigorously adopted, researched, practiced, expanded, and transformed. Dana is widely published in the premier science education research journals and has helped shape the field through his co-editorship of *JRST* and the most recent *Handbook of Research in Science Education*, as well as several book series. His scholarly work is widely used, having been cited by more than 2,500 documents in SCOPUS to date, and garnering more than 17,000 citations on *Google Scholar*, and counting. Dana's contributions, and the rigor and impact of his work and leadership in science education, have landed him multiple distinctions and accolades—too many to list here. These include receiving the *Distinguished Contributions to Science Education Through Research Award* from NARST in 2016. So, if you want to take a deep and current dive into SSI, this is the book to read. You are getting the story from the horse's mouth.

I vouch for the author. I know Dana. Or so I thought!

Would you still want to get and read this book if I were to share with you some personal "brute facts" about Dana that paint a completely different—even shocking—picture of the moral character of the person he really is? Intellectual integrity dictates that I must, especially given the central role that moral character and ethical conduct play in SSI theory and practice. Brute facts, I should note, is a phrase that Dana himself uses repeatedly in this volume to articulate his SSI theory.

Here are the brute facts:

As a high school student, Dana Zeidler was handcuffed and arrested for allegedly assaulting police officers in an attempt to prevent them from injuring a helpless fellow student who was protesting in the midst of a riot. His first year in college, Dana wrote academic papers, as well as letters to his probation officer to demonstrate that he was on good behavior. It would take an expensive lawyer to eventually get charges against Dana to be dropped by a "friendly" judge.

This was not the end of it.

During his college years, Dana Zeidler spent time in a maximum-security prison in upstate New York in the company of hardened criminals and dangerous inmates. He socialized with these individuals and was on a "first-name-basis" sort of relationship with many of them. In transacting their mutual business, Dana managed to establish rules of engagement with this cohort of characters to ensure they achieved their common interests without disturbing the peace. Dana somehow managed to ensure the sharp implements he and his associates used did not seep into the prison's general population. Even more interesting, is the brute fact that he was a repeat visitor with this "community." Dana came back again inside the prison walls, albeit his later associations were less peaceful and involved heated arguments, among other interesting forms of socialization.

I am not betraying Dana by divulging his secrets. I was as shocked as you might be right now by these revelations. These brute facts are laid out in all their glory and intrigue in the introductory chapters of this book for you to read and, in some cases, see in explicit visual documentation of events.

Would you still want to get and read this book? My colorful skirting with the truth of the aforementioned "brute facts" aside, my advice is an unequivocal yes and an even stronger endorsement of the book!

The book makes two major contributions. First, it expounds SSI theory and practice from the perspective of its founder. Dana and his collaborators and colleagues have written much about SSI over the past two decades as they worked to birth, develop, synthesize, and elaborate SSI theory and its associated elements. The book opening chapters offer a powerful synthesis and elaborate primer into this domain of research and practice, which would have utility for both novice and veteran scholars and practitioners interested in SSI. Here, Dana explicates a powerful argument for how SSI theory and practice extend the often expounded, important contributions of precollege science education to learners' understanding of science content and processes in preparation for more advanced studies, and development of scientific ways of thinking and scientific literacies in preparation for citizenry in our scientifically laden world. Specifically, Dana argues that science education also could and should contribute, and demonstrates how SSI enables this contribution, to the character, moral, and ethical development of students in our precollege schools, indeed to the cultivation virtue, character, and independence among students.

Next, the book takes the reader on a *tour de force* of global perspectives on, and issues related to, SSI. Chapters, representing 16 nations and 40 scholars, explore a broad array of SSI-related educational issues from around the globe, ranging from the metalevel political lens of national emancipation and justice, to ontological and epistemological questions, to various syntheses of sub-domains of SSI research. Other chapters address more specific questions about the place and role of SSI in national science education policies and curricula, examining student learning of various SSI topics and acquisition of relevant skills, and investigating teachers' instructional practice and professional development related to SSI, among other interesting and leading-edge topics.

The second major contribution of the book is rather unique and equally, if not more, important. It offers a *rare, rich, and nuanced* rendering of the often deployed and true statement that "every scholarly journey is deeply personal." The statement should not be taken to mean that scholarly journeys are undertaken alone or are lonely endeavors. Far from it, scholars are members of their professional communities. Successful scholars understand and take full advantage of what their colleagues and communities offer in terms of socialization, distributed intelligence and insights, and work and supports to be able to expand our understanding, build new knowledge, and extend positive impacts in the sphere of influence of our chosen domain of scholarship and practice.

The statement also bears the meaning that scholarship is "deeply personal" in the sense that scholars often spend lengthy stretches of time alone and expend enormous personal effort in the seclusion of their offices or preferred places of work. In many instances, they are on their own: They read and work hard to synthesize ideas, struggle with analyzing and interpreting their data, write and rewrite manuscripts, and find their own colorful way through frustrations of receiving, initially rejecting wholesale, and eventually reconciling themselves to dealing with (often anonymous and, to them, unjustifiably harsh) critiques of their work as they revise their way once and again into publication and communal acknowledgment. This, however, is not the sense in which I mean to invoke this statement.

The sense of "every scholarly journey is deeply personal" I am after here goes a level deeper. It speaks to how each of us journeys into the world of ideas from within the inescapable window of our *individual, personal histories* and, particularly, those *critical junctures* in our personal and societal lives that shape or transform our growing up. These unique personal junctures equip us with what Dana calls "a predisposition to inquiry in unfamiliar territory" and often enable us to make novel, and hitherto unmade, connections among seemingly unrelated ideas in supposedly disparate scholarly fields. The forming of these new connections is the *single* most important factor in advancing human knowledge and understanding.

In this book, Dana takes us on a deeply personal journey and explores those critical junctures in his personal life that jumpstarted a journey of discovery during his doctoral studies and invited him to explore and, more importantly, make novel connections among an array of disciplines. The latter included science education, sociology, moral philosophy, developmental psychology, social justice, and various subfields that ranged from moral reasoning to ethical conduct, moral judgment, and moral decision-making. Concepts in these scholarly fields were the ingredients that eventually gave birth to the field we now identify as SSI, where the (once very novel) connections made (by Dana) are now so obvious to us and seem to would have come naturally to any of us who might have read across these fields—assuming that reading across these varying fields would have happened as a chance occurrence to start with. To my mind, this is the beauty and power of what Dana achieves in this book, namely, the rich and nuanced way he interlaces his personal and scholarly journeys to unearth hidden connections between the two. Suddenly, my colorful twisting of the brute facts of Dana having been arrested, corresponded with a parole officer during college, lawyered by his family out of a trial, and ventured to interact with prison inmates; all these take on a rather different and deeply satisfying turn into a narrative that is at once insightful and educative.

It goes without saying that Dana was not aware of these connections as they occurred, that the book is an intentional, introspective, and reflective journey few of us undertake. Dana took the journey, and did that convincingly and masterfully, as well as with integrity and authenticity. As an empirically textured case study in how scholarly journeys are deeply personal, the book is uniquely instructive to novice scholars and inviting of the more seasoned ones to appreciate their uniqueness and the most important asset they bring to their scholarship journey and professional communities, namely, *themselves*. Dana's book is a testament to the crucial importance that our diversity, in *all* its myriad forms, is constitutive for advancing our collective scholarship and advancing our goal to do good in the world—and SSI, literally, aims to do good in the world.

Finally, and this is personal. You could tell from my aforementioned accolades that I am a fan of Dana as a phenomenal scholar, generous colleague, and close friend. I thought I knew him very well. Reading this work made me realize that he is all that and *more*. Dana is deeply ethical, unwaveringly moral, and, simply put, a *superbly good* human being. I am fortunate to have him as a friend.

Enjoy!

University of North Carolina
Chapel Hill, NC, USA

Fouad Abd-El-Khalick

Preface

The doctrine of scientific disconnection from social morals goes all the way back to the ancient Greek belief that thought is independent of society, that it stands alone, born without parents. Ancient Greeks such as Socrates and Pythagoras paved the way for the fundamental principle behind science: that truth stands independently of social opinion. It is to be determined by direct observation and experiment, not by hearsay. Religious authority always has attacked this principle as heresy. For its early believers, the idea of a science independent of society was a very dangerous notion to hold. People died for it.

The defenders who fought to protect science from church control argued that science is not concerned with morals. Intellectuals would leave morals for the church to decide. But ... this political battle of science to free itself from domination by social moral codes was in fact a moral battle! It was the battle of a higher, intellectual level of evolution to keep itself from being devoured by a lower, social level of evolution. .... Once this political battle is resolved, … re-ask the question, "Just exactly how independent is science, in *fact,* from society?" The answer … is, "not at all." A science in which social patterns are of no account is as unreal and absurd as a society in which biological patterns are of no account. It's an impossibility (Pirsig, 1991, p. 140).

In Part I of this book, the serendipitous personal experiences described in Chaps. 1, 2, and 3, from being arrested to teaching in a maximum-security prison, served as the impetus for developing the underlying presuppositions embedded in *The Construction of a Framework for Socioscientific Issues* described in Chap. 4. That framework serves as a roadmap to think about SSI as a kind of moral inquiry into epistemic insight, presented in Chap. 5, that is informed by recent philosophical, conceptual, and empirical intersections of scholarship connected to SSI.

But what does a moral inquiry into SSI look like beyond a personal foundational formulation of this framework? What impact has the SSI framework had on the perceptions of scholars around the globe? What are their perceptions of the "State of the Art" connected to SSI-related topics, moral inquiry, and epistemic insight? It is understood that no one person, or even a team of scholars, represents the sum total of diverse perspective related to SSI. However, having representation from differing geographical and sociocultural perspectives about research, policy, and practice of SSI-related work, remained an open question. I was particularly interested in what other scholars thought ought to be advocated, in relation to those areas, as well as what were possible "blind spots," sticky wickets, problematic or overlooked areas, and what fruitful/emerging directions might the SSI framework be directed towards in future research?

To that end, Part II of this book aims at presenting perspectives that converge on those questions. I am humbled that 40 scholars representing sixteen countries agreed to contribute to this volume. These countries include Australia, Brazil, Canada, Cyprus, and Spain (together sharing concerns about Ocean Literacy impacting both countries), Greece, China, Lebanon, Portugal, Spain, South Africa, South Korea, Sweden, Taiwan, Turkey, United Kingdom, and the United States. Collectively, these contributors representing global perspectives about SSI-related themes are collectively better situated to address those questions than my singular personal prospective. However, I argue in this volume that the experiences which left their mark on my slate have managed to intersect and resonate with others around the globe who hold in common a sociocultural view toward the role of human flourishing in science education. I am indebted to these scholars for their insights and dedicated efforts to advance moral inquiry and epistemic insight as it related to functional scientific literacy. They have contributed significantly to the field of science education.

These dedicated efforts are also bracketed first by a Foreword from Dr. Fouad Abd-El-Khalick, and second by an Afterword from Dr. Benjamin C. Herman. I asked these particular individuals to serve in those "bookend" roles because of their expertise in research and scholarship in the field. They too have been deeply invested and immersed in advancing science education toward fruitful and progressive directions. But that expertise aside, both have been part of a journey connected to my personal well-being as a colleague and a friend. Some things are just brute facts.

Tampa, FL, USA Dana L. Zeidler

Contents

Part I Personal Perspectives of the Socioscientific Issues Framework

1 **Introduction: Daring to Be Wise About Serendipity** 3
Dana L. Zeidler

2 **Arrested Development** 11
Dana L. Zeidler

3 **Prison, Morality, and Nascent Socioscientific Issues** 19
Dana L. Zeidler

4 **Brute Facts About the Construction of a Framework for Socioscientific Issues** 25
Dana L. Zeidler

5 **Moral Inquiry in the Practice of Socioscientific Issues** 39
Dana L. Zeidler

Part II Global Perspectives of the Socioscientific Issues Framework

6 **Is There a Place for Socioscientific Issues in Australian Secondary School Science?** 67
Vaille Dawson

7 **Politicization of Socioscientific Issues as an Inalienable Condition for Scientific Education in the Brazilian Context** 77
Paulo Gabriel Franco dos Santos, Polliane Santos de Sousa, Michel Pisa Carnio, and André Vitor Fernandes dos Santos

8 **Encouraging and Enabling Student Socio-political Actions Based on Longitudinal Dispositif Analyses** 107
Lawrence Bencze and Chantal Pouliot

9 **Ocean Literacy as a Socioscientific Issue for Hope in the Anthropocene** 129
Blanca Puig, Isabel María Cruz-Lorite, and Maria Evagorou

10 **Widening Students' World Views via the Implementation of Socioscientific Issues in Educational Practice** 151
Martha Georgiou

11 **"Content-Teaching First, Socioscientific Issue Next" as Mutant SSI Instruction: Making Sense of Hong Kong Preservice Science Teachers' Pedagogical Decisions** 181
Kennedy Kam Ho Chan and Maurice Man Wai Cheng

12 **The State of the Art of Socioscientific Issues in the Arab World** 197
Rola Khishfe

13 **Approaches to Socioscientific Issues in Portugal: A Scoping Literature Review** 215
Patrícia Pessoa, Maria João Fonseca, Bruno Sousa, António Pedro Costa, and Xana Sá-Pinto

14 **SSI as An Educational Innovation in Spain: Easier Said Than Done** 245
Silvia Alcaraz-Dominguez

15 **The Socioscientific Issues Framework as a Tool to Enhance Health Education Within the Life Sciences Curriculum in South Africa** 269
Lindelani Mnguni and Velisiwe Gasa

16 **Evolution and Future Directions of SSI Education in South Korea: A 2-Decade Review and Beyond** 287
Hyunju Lee, Hyunok Lee, and Yeonjoo Ko

17 **Science and Moral Inquiry as the Yin and Yang of SSI Education: Two Examples of SSI Research from Sweden** 311
Carl-Johan Rundgren and Shu-Nu Chang Rundgren

18 **Comprehensive Exploration of Science Education Research within Socioscientific Contexts in Taiwan: Examining Curriculum Reform, Research Status, Reflections and Implications** 333
Shu-Sheng Lin, Shiang-Yao Liu, and Ying-Shao Hsu

19 **A Moral Inquiry into Epistemic Insight through Socioscientific Issues: A Turkish Perspective** 355
Özgül Yılmaz-Tüzün and Mustafa Sami Topçu

20 The Contribution of Socioscientific Issues to Science Education: A United Kingdom Perspective 387
Alex Craig Jeffries, Amanda McCrory, and Michael J. Reiss

21 Next Generation Socioscientific Issues Education: Incorporating Practices in SSI Learning Experiences 409
Troy D. Sadler, Li Ke, and Amanda Peel

Afterward. 427

Editor and Contributors

About the Editor

Dana L. Zeidler (AAS, BS, MS, PhD) is a distinguished university professor and the department chair of Teaching and Learning in the College of Education at the University of South Florida, Tampa, Florida. He has developed a global research program in the field of science education centered on socioscientific issues—taking a sociocultural approach to teaching and learning about how moral and ethical issues can be a means to foster the formation of epistemological sophistication and character in the pursuit of scientific literacy. Accordingly, his research incorporates aspects of argumentation and discourse, moral reasoning and character formation, epistemology, place-based immersive scenarios, and the nature of science. Dr. Zeidler's passion is sharing his pursuit of moral inquiry in science education with students and colleagues around the world helping to progress the field in fruitful directions.

Contributors

Fouad Abd-El-Khalick is Provost and Senior Vice Chancellor for Academic Affairs at the University of Massachusetts Amherst (UMass) and Professor of Science of Education in its College of Education. Prior to joining UMass, he was Dean of the School of Education and Alumni Distinguished Professor at the University of North Carolina at Chapel Hill. His research focuses on the teaching and learning about nature of science (NOS) in precollege grades, and in preservice and inservice science teacher education settings.

Silvia Alcaraz-Dominguez (PhD) is an associate lecturer in the Department of Didactics and Educational Organisation at the Faculty of Education of the University of Barcelona, Spain. During the last decade, Dr. Alcaraz-Dominguez has researched pre-college teachers' experiences with educational changes, especially in science education. Accordingly, her research covers aspects such as in-service teacher professional development, innovative teaching practices, STEM/STEAM/STEAME education, and educators' pedagogical competences. Dr. Alcaraz-Dominguez has developed her research in the context of competitive grants from the European Union and the Spanish Ministry of Science, Innovation and Universities, involving collaboration with several Spanish and international institutions. As a result, she is the author of multiple research reports, guides, and materials for educators and policy makers, as well as publications in prestigious journals such as the *Journal of Research in Science Teaching* and the *International Journal of Science Education*.

Lawrence Bencze (PhD, MSc, BSc, BEd) is an associate professor (Emeritus) in Science Education at the University of Toronto (1998–present). Prior to this role, he worked for fifteen years as a science teacher and as a science education consultant in Ontario, Canada. His research program emphasizes critical analyses—drawing on history, philosophy, sociology, etc.—of science and technology, along with student-led research-informed and negotiated socio-political actions to address personal, social, and environmental harms associated with fields of science and technology. Recent publications include two edited books about proactive citizenship. He also is co-editor of an open-source activist journal (goo.gl/ir7YRj).

Michel Pisa Carnio is a Brazilian professor and researcher at the Federal University of São Carlos, in the city of São Carlos, Brazil. He has a master's and doctorate in Science Education from the São Paulo State University and a postdoctorate with a focus on public educational policies from the Maringá State University. His work is permeated by references from critical theory and historical-dialectical materialism, in dialogue with Brazilian and Latin American authors who defend a humanized and emancipatory scientific education. He works in the areas of science and biology teaching and teacher training, focusing on CTSA education and socioscientific issues. He is currently developing critical studies on the relationship between human beings and nature in this world in crisis, on the possibilities of citizenship training in a social context averse to democracy, and on the construction of possible paths for training science and biology teachers at the university-school interface.

Kennedy Kam Ho Chan (BSc, PGDE (distinction), M.Phil, PhD) is an associate professor at the Faculty of Education, the University of Hong Kong. He is passionate about enhancing student learning experiences in science by improving the teaching quality of science teachers. His research interests include pedagogical content knowledge, teacher noticing, and the use of video in teacher education. He is particularly interested in the situated and dynamic aspects of teacher expertise and has conducted a number of studies to characterize, assess, and develop these aspects of teacher expertise in Hong Kong.

Maurice Man Wai Cheng (BPharm, PCEd, MEd, PhD, FRSB) is an associate professor in the Faculty of Education and Social Work, the University of Auckland. His research interest is student affective and cognitive aspects of learning science, in particular, in the areas of socioscientific issues, the intersections of science and culture, and the roles of visual representations in science education. He is also interested in science teachers' knowledge and capacities through the lens of funds of knowledge and learning ecologies. He has been involved in Trends in International Mathematics and Science Study (TIMSS) for more than two decades, and as the National Research Coordinator (Science) of TIMSS Hong Kong study 2019. He also serves in the Extended Science Expert Group for OECD's PISA (Programme for International Student Assessment) 2024.

António Pedro Costa is a principal researcher at the Research Centre on Didactics and Technology in the Education of Trainers (CIDTFF). He also collaborates with the Artificial Intelligence and Computer Science Laboratory at the University of Porto, Portugal. He is one of the authors of the qualitative data analysis *software* webQDA (www.webqda.net). He coordinates two qualitative research conferences (www.ciaiq.org; www.wcqr.info). He is the author of more than 300 publications in international conferences and journals.

Vaille Dawson is Professor of Science Education at the University of Western Australia. Originally a medical researcher and then a science teacher, she has conducted educational research with science teachers and their students at the secondary and tertiary level in Australia, Indonesia, and India for over 30 years. Her research expertise spans climate change education, genetics education, scientific literacy, initial teacher education, critical thinking, and teaching in disadvantaged schools. The key outcome of this research is to ensure that all young people receive a quality science education that enables them to make informed decisions about issues that impact their current and future lives. She has been awarded external research funding from various organizations, including the Australian Research Council. Vaille has co-edited five highly respected and widely used initial teacher education textbooks for prospective school science teachers that celebrate the role of science inquiry and science for all.

Polliane Santos de Sousa is a Brazilian researcher, with a degree in Physics and a master's degree in Science Education from the State University of Santa Cruz, and a PhD in Scientific and Technological Education from the Federal University of Santa Catarina. Currently, she serves as a faculty member at the Teacher Training Center of the Federal University of Recôncavo of Bahia and has dedicated herself to deepening studies on an ethical-critical scientific and technological education, focusing on overcoming situations of social injustice experienced by school communities. Drawing on the educational principles of Paulo Freire, she has been furthering studies on curriculum organization based on thematic approaches. Moreover, she investigates the shaping of research on socioscientific issues in Brazil, establishing connections of this field with Latin American theoretical frameworks such as those of Paulo Freire, Milton Santos, and Enrique Dussel.

André Vitor Fernandes dos Santos is a Brazilian researcher holding a degree in Biology and with master's and doctoral degrees in Education from the Federal University of Rio de Janeiro. He serves as a professor at the University of Brasília, campus of Planaltina. With a focus on science education, policy education, and curriculum, in his works we can find an integration between critical and post-critical references, focusing on the analysis of educational policies, especially curriculum and assessment policies, with the aim of constructing "a" curriculum history. His studies have been focusing on how neoliberal ideology has been articulating with disciplinary traditions, reshaping what we conceive as school knowledge and, therefore, on how and who we should be as teachers and students. By focusing on these themes, his work has been making visible the systems of reasoning that underlie the formulation of public policies, opening spaces for us to reimagine new futures through engagement in disputes over the meaning of the world.

Paulo Gabriel Franco dos Santos is a Physics graduate with a master's and doctorate in science education from São Paulo State University, and postdoctoral research at the University of Havana, serves as a professor at the University of Brasília, campus of Planaltina. With a focus on teacher education, socioscientific issues, and critical perspectives in scientific education, he has conducted researches at regional, national, and international levels, particularly in collaboration with Latin American and Caribbean partners. Currently, his academic endeavors encompass teaching, research, and university extension, with a special emphasis on socioscientific issues, science, technology, society, and environment (STSE) education, as well as in-depth exploration of peripheral, racial, and class-related challenges. His work is directed toward the transformation and nurturing of dreams and strategies for the creation of another possible reality through Brazilian and Latin American scientific education.

María Evagorou is an associate professor at the Department of Education, University of Nicosia, Cyprus. Her scholarly activity focuses on exploring and enhancing young students' and pre-service teachers' argumentation skills within socioscientific issues (SSI). More specifically, the emphasis of her work is on students' and teachers' talk when they engage in the discussion of SSI and aims to explore ways in which teachers can be supported in their effort to include SSI in their teaching. Through the engagement with SSI, Maria aims to promote responsible citizenship. Maria has worked as a principal investigator and senior researcher on various EU and local projects, served as a member on the JRST editorial board, and as a strand coordinator for ESERA. She has published widely in international referred journals and recently co-edited a book titled *Science Teacher Education for Responsible Citizenship: Towards a pedagogy of Relevance Through Socioscientific Issues*.

Maria João Fonseca is Head of Communication at the Natural History and Science Museum of the University of Porto and Director of the Ciência Viva Center Hall of Biodiversity, Portugal. She is also an invited auxiliary professor at the Faculty of Sciences of the University of Porto. With a background in animal biology and biology teaching and a PhD in science education and communication, she is in charge of general administration and coordinating and promoting science communication and outreach activities in the museum. She has been involved in the design and evaluation of formal and informal educational activities, teacher training, and outreach initiatives for all sorts of audiences. Among her interests are topics such as art and science in the promotion of scientific culture and well-being, science education in informal environments, especially museums and science centers, active learning methodologies, and evaluation of the impact of educational experiences.

Velisiwe Gasa (PhD, MEd) is a professor at the University of South Africa. She leads various research and community engagement projects. Gasa is the editor-in-chief of *Africa Education Review* (AER) journal, chairperson of the Editorial Board for the *International Journal of Educational Development in Africa* (IJEDA), *Dilla Journal of Education* (DJE) Editorial Board member as well as Chair of Teacher Education and Interdisciplinary Research (TEIR) Conference. Her research interests include postgraduate supervision support, research administration and management, decoloniality in higher education, distance learning, learners' aggressive/challenging behavior, family dynamics, inclusive education, and gender studies.

Martha Georgiou (BSc, MEd, PhD) is a member of the teaching staff of the Department of Biology of the National and Kapodistrian University of Athens (NKUA), specialized in Education. Since her doctoral studies, she has worked extensively on socioscientific issues in science education, especially in biology education. She has combined the study of socioscientific issues with argumentation and informal reasoning, thus deepening the respective abilities of Greek students. Recognizing and supporting socioscientific issues as a framework to promoting scientific literacy, she recommended their inclusion in the new biology curricula for upper secondary education (2022) as a member of the author team. Moreover, she was recently a visiting professor at the Université Paris-Cité, France (ex ParisV,VII Sorbonne) to lecture to Master 2 students on socioscientific issues and argumentation. Latterly, being responsible for the educational programs for the Museum of Zoology (NKUA), she works on combining socioscientific issues with non-formal education.

Benjamin C. Herman occupies a joint faculty appointment across the Department of Teaching, Learning and Culture and the Department of Biology at Texas A&M University. He received his PhD in Science Education with a minor in Statistics from Iowa State University. He also possesses a BS in Animal Ecology, an MS in Wildlife Resources, an MAT in Science Education and extensive experiences teaching science and researching science teaching and learning in formal and informal settings. These experiences include working as a certified secondary biology and chemistry teacher and helping pre-service and in-service science teachers implement research-based practices. Dr. Herman's research focuses on promoting informed civic engagement with socioscientific issues (SSI). To this end, he investigates factors related to SSI decision-making such as science knowledge, sociocultural affiliation, and perspective taking. His research also focuses on improving nature of science and SSI pedagogical practices.

Ying-Shao Hsu (MS, PhD) is a chair professor at the Graduate Institute of Science Education at National Taiwan Normal University (NTNU), Taiwan. Her primary research interests include socioscientific issues (SSI) learning, technology-infused science education, curriculum design, metacognition, and STEM education. She specializes in developing SSI-based curricula and investigating scaffolds that improve students' decision-making and metacognition in SSI contexts. She also leads projects that explore STEM teachers' professional development and the use of educational technology to enhance inquiry-based learning. Her international collaborations with leading researchers worldwide have established her as a global scholar in science education.

Alex Craig Jeffries is a senior lecturer in the Department of Life Sciences, University of Bath, United Kingdom. Housed in the Milner Centre for Evolution, his interests include bacterial genomes and use bioinformatics approaches. His interests also extend to expertise related to United Nations Sustainable Development Goals with respect to good health, well-being, and quality education. He also works with progressive scholars in the field of science education.

Li Ke is an assistant professor of Science Education in the College of Education and Human Development at the University of Nevada, Reno. His research centers around fostering meaningful science teaching and learning from elementary through high school classrooms. He is interested in how students negotiate socioscientific issues such as climate change and how issue-based learning experiences support students' development in scientific literacy. His work also explores co-design as an effective way of teacher professional development to promote teacher agency.

Rola Khishfe is an associate professor at the Department of Education and faculty of Arts and Sciences of the American University of Beirut (AUB), Lebanon. She received a PhD in science education from Illinois Institute of Technology; MA in science education from AUB; Certification in teaching secondary science from AUB; and a BS in biology from AUB. Her research interests focus on the teaching and learning of nature of science and argumentation skills in the context of socioscientific and scientific issues. She has published articles in international journals such as the *International Journal of Science Education* and the *Journal of Research in Science Teaching*. She has been a guest speaker at international conferences in science education.

Yeonjoo Ko is an assistant professor in the Department of Science Education at Jeju National University, located in the Jeju Special Self-Governing Province, South Korea. Holding an MEd in Learning, Design, and Technology and a PhD in Science Education, she has adeptly integrated these two disciplines by focusing on socioscientific issues education and technology-based instructional design. Recently, she has extended her research to explore the application of artificial intelligence in science learning and provided education for future STEM professionals to promote their competencies and character through cutting-edge educational methods.

Hyunju Lee is a professor in the Department of Science Education at Ewha Womans University in South Korea. Her research centers on teaching socioscientific issues (SSI), especially the development of SSI curricula and programs, teacher professional development for teaching SSI, and analysis of students' informal reasoning patterns on SSI. She has designed and implemented various types of SSI curricula while collaborating with school teachers. Recently, she conducted a 3-year research project on SSIs aimed at enhancing students' social responsibility and willingness to act toward SSIs. Currently, she is expanding her research and expertise in SSI education by integrating it with the concept of risk literacy.

Hyunok Lee is an assistant professor at Kangwon National University in South Korea. She holds a PhD in Science Education and an MS in the History and Philosophy of Science. Her research focuses on socioscientific issues and the nature of science and technology. She has broadened her research to encompass education on risk, particularly in the context of scientific uncertainty and future forecasting. Dr. Lee is committed to incorporating historical and philosophical perspectives into the field of SSI, not only to improve conceptual frameworks but also to develop educational programs and discuss strategic directions.

Shu-Sheng Lin (MS, PhD) is a professor of the Master's program in Mathematics and Science Education at the Department of Education, National Chiayi University, Taiwan. His primary research focuses on enhancing higher-order thinking skills, such as argumentation, decision-making, and critical thinking, among students and science teachers through socioscientific instruction. Additionally, he is dedicated to promoting the professional development of science teachers by helping them acquire knowledge and skills through socioscientific instruction. His current research endeavors include exploring university students' understanding of the nature of science in socioscientific contexts, as well as analyzing the news framing of socioscientific issues and understanding their implications for science education and communication.

Shiang-Yao Liu (MS, PhD) is a professor of the Graduate Institute of Science Education at the National Taiwan Normal University, Taiwan. Her original research is on students' and teachers' conceptions of nature of science (NOS) and how their worldviews and epistemological beliefs influence their science learning and teaching. Having previously engaged in environmental education research, her subsequent research interests have been focused on integrating socioscientific and environmental issues in science curricula and designing teaching activities to improve students' scientific literacy. Her current research delves into the characteristics of scientific knowledge in local social controversies, using tailored teaching materials to enhance learners' argumentation, decision-making, and civic engagement.

Isabel María Cruz-Lorite is a researcher at the Department of Education of the University of Nicosia (UNIC), Cyprus, and is currently working on the European Union funded project "ICSE Science Factory: Open schooling for science education and a learning continuum for all." Her background is in Environmental and Marine Sciences (bachelor's degrees) and Environmental Education (Master's degree). She holds a PhD in Science Education with a thesis on the analysis of environmental awareness and argumentation skills of preservice primary teachers in the context of a role-playing game on the use of nuclear energy. Her research interests focus on the use of environmental socioscientific issues/socially acute questions as a context for science education.

Amanda McCrory (Ph.D, FCCT, SFHEA) is an associate professor of Science Education at Institute of Education, University College London. She is an academic and researcher of ethics in science education. She leads the Science Special Interest Research Group at UCL IOE and is a co-convenor of the British Education Research Association STEAM Special Interest Research Group. She strongly believes that teaching is the sine qua none of research. Her publications have been built around (a) ethics, socioscientific issues and argumentation in science education; (b) primary science pedagogy and teacher education; and (c) intersectionality/equality, diversity, and inclusion in science education. She is interested in how teachers and parents/guardians can support children in their understanding of topical, yet relevant, socioscientific issues while developing and promoting ethical and scientific thinking in children in both formal and informal learning environments. She is passionate about promoting the agency and voice of children via her research.

Lindelani Mnguni (PhD, MSc, BSc Hons, BSc) is an associate professor and deputy dean in the Faculty of Health Sciences at the University of the Witwatersrand and is a research fellow at the University of South Africa. He holds a C2 NRF rating in biology education and PhD in Curriculum and Instructional Design from the University of Pretoria. With over 15 years in higher education, Mnguni has accrued extensive teaching and supervisory experience, working across various academic roles and institutions. His research interest includes visual literacy in molecular biology and the interplay between science education, health education, and curriculum ideologies.

Amanda Peel (BS, MS, PhD) is an assistant professor of STEM Education in the Teacher Preparation, Administration, and Leadership Program in the College of Health, Education, and Social Transformation at New Mexico State University in Las Cruces, NM. She has worked to integrate computational thinking into science education by supporting both teacher and student learning. She also researches science learning and engagement through socioscientific issues, such as antibiotic resistance and climate change. Her research blends computer science education, science education, and relevant issues to better align science education with modern STEM practice and societal concerns.

Patrícia Pessoa is a primary school teacher and a PhD student in Science and Technology Teaching at the University of Trás-os-Montes e Alto Douro, with the Centre for Research in Didactics and Technology in the Training of Trainers at the University of Aveiro, Portugal, as her host institution. She has held a PhD scholarship from the Portuguese National Foundation for Science and Technology since 2020 and is researching how addressing socioscientific issues informed by biological evolution contributes to primary school students' sustainability literacy.

Chantal Pouliot (BS, MS, PhD) is a full professor in the Department of Teaching and Learning Studies at Laval University*, Quebec, Canada. She holds a bachelor's degree in biology, a master's degree in literature, a teaching certificate, and a doctorate in science education. Her research interests include the teaching of socially relevant environmental and health issues, civic capacity, and the challenges of engaging researchers in socio-political conversations. She is co-director of the collection Apprendre de la controverse [Learning from controversies] (PUQ). She was a commissioner on the Independent Scientific and Technical Commission on the Recognition of Academic Freedom in the University setting (also known as the Cloutier Commission, 2021) whose work led to the creation of Quebec's Bill 32 (June 2022). Chantal takes every opportunity to teach through comics, podcasts, films, museum exhibitions, and literary works.

*Département d'études sur l'enseignement et l'apprentissage, Faculté des sciences de l'éducation, Canada.

Blanca Puig is an associate professor at the Faculty of Education at the University of Santiago de Compostela (USC), Galicia, Spain. She coordinates the distinguished research group RODA (Reasoning, Argumentation, and Discourse) (www.rodausc.gal) in which she has developed her research on argumentation and critical thinking on socioscientific issues. Her research interest is to explore the practice of critical thinking in socioscientific contexts related to the post-truth era challenges. She is involved as a principal investigator and as a senior researcher in diverse national, local, and EU projects toward this goal. She works together with science teachers and science educators in the design of learning environments to promote and study critical thinking in science education. Understanding the ocean is one of her passions, and the main topic she currently addresses in her research on socioscientific issues.

Michael J. Reiss is Professor of Science Education at University College London, UK, a member of the Nuffield Council on Bioethics, a Fellow of the Academy of Social Sciences, and President of the Association for Science Education. The former Director of Education at the Royal Society and the former President of the International Society for Science and Religion, his academic interests are in science education, bioethics, sex education, and science and religion. Recent books of his include: McCrory, A. & Reiss, M. J. (2023) *The Place of Ethics in Science Education: Implications for Practice*, Bloomsbury; Reiss, M. J. & Ruse, M. (2023)

The New Biology: A Battle between Mechanism and Organicism, Harvard University Press; Briggs, A. & Reiss, M. J. (2021) *Human Flourishing: Scientific Insight and Spiritual Wisdom in Uncertain Times*, Oxford University Press; and Barmania, S. & Reiss, M. J. (2018) *Islam and Health Policies Related to HIV Prevention in Malaysia*, Springer.

Carl-Johan Rundgren (MA, MS, PhD) is a professor in Science Education at the Department of Teaching and Learning, Stockholm University, Sweden. His main research interest relates to development of scientific literacy and different ways to represent science content. In his research, he has studied how science content is presented in media, and how science education can contribute to the formation of a critical awareness of science-related claims presented in media among students. He has worked with socioscientific issues (SSI) in research during several years. His research covers SSI-based teaching and learning, inquiry-based science education as well as language, and other forms of representations in science education.

Shu-Nu Chang Rundgren (BS, MS, PhD) is a professor in Education and the research leader of the research group, Educational Research in Citizenship Competences (ERiCC), in the Department of Education at Stockholm University, Sweden. She has worked with socioscientific issues (SSI) in her research during the past decades, taking perspectives on exploring and developing students' skills on SSI informal argumentation and teachers' teaching competence on engaging SSI to achieve critical scientific literacy and Education for Sustainable Development (ESD). Her research covers SSI-based teaching and learning, inquiry- and context-based science education, and teacher professional development for ESD.

Troy D. Sadler (BS, MEd, PhD) is a professor of science education at the University of North Carolina at Chapel Hill School of Education and serves at the Thomas James Distinguished Professor of Experiential Learning. Sadler's research focuses on how students negotiate complex socioscientific issues and how these issues may be used as contexts for science learning. He is interested in how issues-based learning experiences can support student learning of science and development of practices essential for full participation in modern societies. He is co-editor of the *Journal of Research in Science Teaching* and has served in multiple leadership roles for NARST: A Global Organization for Improving Science Education through Research. Prior to his work at UNC, Sadler was a science education faculty member at the University of Florida and the University of Missouri and served as the director of a STEM education research center.

Xana Sá-Pinto is an assistant researcher at CIDTFF, University of Aveiro, Portugal. She is interested in understanding how to profit from transdisciplinary research networks to develop effective solutions for sustainability problems and to promote participants' key competencies in sustainability. For that she integrated, supported, and developed transdisciplinary research networks, including BiblioLab (https://www.bibliolab.pt/), EvoKE (https://evokeproject.org), and EuroScitizen COST Action (http://www.euroscitizen.eu/). Her interest in the SSI approach comes from its potential to promote education for sustainability and students' key competencies in sustainability.

Bruno Sousa is a Biology and Geology high school teacher, school manager, and science schoolbook author. He has a Masters' in Educational sciences and a post-graduation on school management and administration and is a PhD student in Didactics and Curricular development at the University of Aveiro, Portugal, triangulating sustainability, co-constructing practices and open schools. He was part of EuroScitizen COST action (http://www.euroscitizen.eu/) and is a board member of EvoKE association (https://evokeproject.org).

Mustafa Sami Topçu is a professor of Science Education in the Department of Mathematics and Science Education at Yıldız Technical University, Türkiye. As a managing director, he has established STEM Education Center at Yıldız Technical University since 2016. His research interests are socioscientific issues, informal, computational and mechanistic reasoning, and engineering education in K-12.

Özgül Yılmaz-Tüzün is a professor of Science Education in the Department of Mathematics and Science Education at Middle East Technical University in Türkiye. She completed her undergraduate education and master degree in biology education at Middle East Technical University and received her doctorate in science and environmental education at Indiana University in the USA. Her research interests are socioscientific issues, environmental education, epistemological beliefs, argumentation, discourse analysis, and misconceptions.

Part I
Personal Perspectives of the Socioscientific Issues Framework

Chapter 1
Introduction: Daring to Be Wise About Serendipity

Dana L. Zeidler

1.1 Sapere Aude

A good number of decades ago, when I was in graduate school, the uber goal of science education was the pursuit of science literacy. It is important to note that the emphasis at that time, was on *science* literacy, in contrast to how we tend to frame its analogue contemporaneously as *scientific* literacy. The point is, if you approached a science educator in the 1970s and 1980s to describe the major goal of science education, that person (likely a male) would offer a description of science literacy that was technocratic in nature. The space-race following the Soviets launch of Sputnik, and other residual artifacts from the "Cold War" had been won by the United States, and many of the "alphabet soup" curriculums (e.g., BSCS, IPS, PSSC, SAPA, SCIS) tended to still focus on content-related skills aimed at producing the next generation of engineers and scientists. The intent was also to include more "activity-based" curricula that stood in contrast to "traditional" lecture-based courses that required students to be passive note-takers and receivers of information. The idea that science could be better connected to other social factors began to make their appearance with the emphasis on "Science for All Americans" (AAAS, 1990). DeBoer (1991, 2014) and Kyle et al. (1983) provide good overviews of this time period. I realize I am presenting an "American" perspective of science literacy here, and I am by no means suggesting that the US should have been the "gold standard," then or now, in the development, implementation and impact of science curriculum. But I do know, based on my own understanding of the global literature, that the US did tend to set the tone for much science education research agendas in countries around the world.

D. L. Zeidler (✉)
University of South Florida, Tampa, FL, USA
e-mail: zeidler@usf.edu

D. L. Zeidler (ed.), *A Moral Inquiry into Epistemic Insights in Science Education*, Contemporary Trends and Issues in Science Education 61,
https://doi.org/10.1007/978-3-031-63382-9_1

It was during that time, for reasons explained elsewhere (e.g., Zeidler, 2014; Zeidler & Sadler, 2011), that the conversation shifted from speaking about science literacy to scientific literacy. That shift, while subtle, implied that the focus was no longer relegated to science-content understanding but moved in a direction toward understanding the *character* of science, or nature of science (NOS) (Lederman & Lederman, 2014). This was important because the emphasis was no longer solely on content-understanding, but rather on conceptual understanding of the activity of science, important to both science majors as well as non-science individuals. Those decades of past, I was in graduate school at Syracuse University (New York) with my friend and colleague, Dr. Norman Lederman. They were both fun and heady days, to be sure. Those days were *our* Enlightenment—as least it seemed that way for us. We often spoke of the distinction between science and scientific literacy, and the implications of what that distinction held for the field. At the same time, I was beginning to suggest that those mainstream conceptions of science, and even scientific literacy, were short-sighted, and not inclusive enough to meet the problems and issues that confronted science education (Rudolph, 2019), both from a US-perspective as well as a global perspective. Norm's ideas and mine were synergistic and we were both excited to do our part to shift the conversation, for those who would listen, and for those who we could convince, to the notion of functional scientific literacy (FSL). I began to argue the point (Zeidler, 1984; Zeidler & Schafer, 1984) that being scientifically literate necessarily entailed the condition of a moral context. Achieving FSL, in other words, would mean being able to make decisions about scientific issues that were embedded in the social fabric of our society as well as the organic and physical world in which we dwell in a manner that produced eudaimonia, best translated from the Greek as "human flourishing."

The arguments I have made over the years boil down to this: In order to reach an operational threshold of FSL, one must necessarily engage in the realm of ethical conduct and moral reasoning. This view suggests that notions about conventional inquiry in science education (e.g., process skills such as controlling variables, making observations and predictions, using the 5-E model, etc.) must also include another domain of inquiry skills. Perhaps we can refer to this as a domain of *moral inquiry*. Therefore, FSL will need to attend to what we will refer to later (Chap. 4), as brute facts about the prudent nature of human beings, and how moral education may elevate those brute facts to allow passage through a threshold of normative expectations as one enters the house of reflective consciousness.

The research program that I was developing then was the beginning of the Socioscientific Issues (SSI) framework, and I have been pleasantly surprised at its genesis and global impact since those early days to the present. More recently, another colleague, Dr. Troy Sadler, and I have written a review of literature of SSI from 2014 to 2022 for the third edition of the *Handbook of Research in Science Education*, and we have identified well over 400 scholarly research articles, 68 book chapters, and 42 dissertations directly connected to the SSI research program during that time (Zeidler & Sadler, 2023). I offer this as one form of empirical evidence, that the topic of SSI is omnipresent and relevant to science education, as well as to fields external to the field, at this point in time and for the foreseeable future. It is also important to recognize that the scholarship embedded within the *SSI framework* is not simply an ideology in search of a theory. It has been crafted from theory, conceptual frameworks, empirical studies (quantitative, qualitative, mixed methods), sociology, developmental psychology, and conceptual analyses that are found in the research base. Accordingly, the common threads of this book situate SSI as a *Science-in-Context* (Bencze et al., 2020) approach to science education.

If one adopts a philosophy of teaching that views the etymology of education (i.e., educere, educare) as meaning "to lead out, bring forth," then the practice and teaching of SSI are understood as providing the conditions necessary to allow the mind to be nurtured and cultivated to excel. Kant's (1785/2012) Categorial Imperative and Age of Enlightenment (1784/2017) touch on this as representative of the banner for the Enlightenment Period in Europe which was "Sapere Aude" or "Dare to Know," or equally applicably translated as "Dare to be Wise." It is noteworthy that Kant was well aware of the almost ageless origin of the phrase written by the Roman poet Horace in his First Book of Letters in 20 BCE "Why so quick to remove a speck from your eye, when if it's your mind, you put off the cure till next year? Who's started has half finished: dare to be wise: begin!" (Book I, Epistle II 32–54). Every child, every student ought to have opportunities to become immersed in their own Zeitgeist of Enlightenment, a period to cultivate their own manner of knowing in the world. This is what SSI, at its best, strives to do—allow children the opportunity to dare to know.

Of course, the teacher is central in providing the conditions under which students may flourish. This is precisely what Kant meant when he stated that a teacher should act as a "midwife" of pupil's thoughts. More specifically, Kant states: "For if the teacher wants to question his pupil's reason, he must do this in a dialogue in which teacher and pupil reciprocally question and answer each other. The teacher, by his questions, guides the pupil's thinking merely by presenting him with situations in which his disposition for certain concepts will develop" (cited by Peck, 2007). In doing so, students may develop a sense of agency not in only their own sensibilities, but in the context of SSI, about activity of science as well. Students are not merely the passive receivers of knowledge, but active participants in the co-construction of both scientific and socio-moral knowledge. Since SSI pedagogy will compel students to practice reflective reasoning, socioscientific reasoning, argumentation, discourse, and generally exercise good judgment in rendering decisions, they are led to cultivate their moral reasoning and apply it in ways that will render them better citizens more likely to be engaged in societal issues and practices. This, of course, is the citizenship view of moral education practice, which is also a part of the SSI framework.

My own small excursion into my corner of Enlightenment allowed me to cross paths with an eminent philosopher at Syracuse University, Professor Thomas F. Green, who was also a mentor to me. We studied moral philosophy and moral education, where he stressed the work of Aristotle's *Nichomachean Ethics* in the pursuit of conceptualizing what an exploration into moral education should, at a minimum, entail. In that work, Aristotle emphasized two important intertwined concepts in the pursuit of moral virtue. One was the notion of "Phronesis," usually translated as a type of wisdom and intelligence gleaned from experience in the world. So, it is not surprising that the etymology of phronesis is also aligned with Prudence (Prudentia). The second concept was that of "Sophia" which described a type of wisdom that is derived from theoretical knowledge which is often equated with the logic of science and Epistemology (Episteme). It is the marriage of both Phronesis and Sophia that results in the virtuous pursuit of wisdom—the kind of wisdom that leads to eudaimonia. This would certainly play a role in framing part of journey in discussing moral education in the context of science education.

At the same time, while I was consuming all that I could about moral philosophy, I also took courses in developmental psychology, where I found the work of Jean Piaget and Lawrence Kohlberg to be most insightful and interesting in terms of their implications for teaching and learning. It also complemented much of what I was learning from my courses in the Cultural Foundations Department at Syracuse, which subsumed History, Philosophy and Sociology of Education. I was fortunate that the Science Education Program at Syracuse was quite liberal in their thinking, and as long as I could make connections to science education, the program allowed me to pursue such interests. They also provided the means for me to attend an intense week-long workshop at Harvard University, led by Kohlberg, to study his writings in depth, test the presuppositions of his theoretical ideas and learn to assess stages and sub-stages of moral reasoning and development. This also impacted how I began to instantiate work cognitive and moral development into science education as well. The subject of my dissertation was *Identifying Mediating Factors of Moral Reasoning in Science Education*. But to understand the passion behind this work, you must allow me to indulge you with some formative events that directed me to that point and beyond, such as the spread of virtue, the stance that development is an aim of education, and how certain events that were masked in the moment, could turn out to be both fortuitous and serendipitous. These events form the genesis and subsequent development of the SSI framework. It is also the foundation of a plethora of scholarship found connected to such a sociocultural view of education.

1.2 Serendipity of Socioscientific Issues

Serendipity is a word that has endured throughout the centuries partly because of its rich, melodious tone, and because its definition or interpretation has varied slightly with its application by different individuals under different circumstances. Horace Walpole (1754) is credited with having coined the term in a letter to his uncle, Sir Horace Mann, explaining to him:

> I once read a silly fairy tale, called "The Three Princes of Serendip;" as their Highnesses travelled, they were always making discoveries, by accidents and sagacity, of things which they were not in quest of: for instance, one of them discovered that a mule, blind of the right eye, had travelled the same road lately, because the grass was eaten only on the left side, where it was worse than on the right --- now do you understand Serendipity?" (Rosenau, 1935, p. 91).

Examining an array of English dictionaries, you will come to find definitions of serendipity that tend to center around something like “the happy knack of making unexpected and delightful discoveries by accident.” I have a hankering for this conglomerate of descriptions because it implies the intercourse of two things: (1) a knack; and (2) a discovery. Hence, serendipity may be thought of as the gift of finding, by the combination of chance and wisdom, certain valuable or agreeable things not sought, while in pursuit of something else. “Accidental discovery” is not a synonym for serendipity. Serendipity is, rather, a discovery made because the individual is mentally equipped not to disregard or overlook something new or strange while on the quest for another goal. The fact that such discoveries come with them a sense of happiness and delight both touch on and highlight the emotive nature of act.

I view my own research program on socioscientific issues (SSI) as a journey that has come about through a kind of serendipity (Zeidler, 1984, 2014; Zeidler et al., 2019; Zeidler & Schafer, 1984). I wasn’t looking to develop a particular research program on SSI, and I surely was not any more intelligent than the other graduate students in the program at Syracuse University. However, in retrospect, I think I was simply mentally equipped for the excursion. I was pursuing a fundamental question of how to better educate in the sciences in a manner that developed more than the cognitive functions of the student, but in a manner that also developed the moral faculties of the individual. The journey took me from the domain of the “hard sciences” to other domains of inquiry that tapped psychology, sociology, and philosophy. So being mentally equipped was no more than having a kind of *predisposition to wonder* beyond my immediate comfort zone and taking advantage of the flexibility afforded to me by my Ph.D. program, which provided opportunities to explore beyond my immediate program of study.

This is precisely what SSI enable both science educators and their students to be able to do. SSI provides a context for the act of teaching to afford the conditions for both students and their teachers to experience serendipity. As students search for answers to scientific issues, they find moral and ethical principles embedded in sociocultural concerns—not prescribed but thrust upon them in surprisingly interesting ways. SSI is the arena where students find the crossroads of science, morality, care, discourse and action. Teachers will find, quite often, that unanticipated questions that require different forms of inquiry and usurping teachable moments as they present themselves may be creatively leveraged. SSI, by its very nature, favors those the most, in possession of a prepared mind, one that possess a predisposition to inquiry in unfamiliar territory. It allows even laggardly souls to find interest and purpose, because educators have encouraged discoveries that are not merely fortuitous but destined to be unveiled in different ways by different students with different perspectives about how to construct meaning, relevance, and shared understanding.

Earlier, I mentioned that serendipity is a discovery made because the individual is mentally equipped not to disregard or overlook something new or strange while on the quest for another goal. The extant research base in this field of study is filled with voluminous examples of teachers and students surprising and engaging discoveries while implementing SSI (e.g., Zeidler et al., 2011). For teachers, their journey typically is a quest to improve their pedagogical science education practices. For students, their journey is one that leads them to make personally relevant connections to subject matter that might otherwise seem distant and remote. The goal in both cases has always been for the SSI framework to promote an open mind through the exploration of perspectives, positions and orientations where moral and ethical issues intersect science (Kahn & Zeidler, 2017). This *is* the serendipity of socioscientific issues. This is also the enactment of a *predisposition to wonder.* In this regard, I am reminded of a famous quote by Louis Pasteur while he presented a lecture at the University of Lille in 1854:

"Dans les champs de l'observation le hazard ne favorise que les esprits prepares"—which translates to something close to "Where observation is concerned, chance favors only the prepared mind."

1.3 Proxies

Like me, you may have noticed a historic trend in contemporary mainstream (read traditional) educative practices pursuant to schooling, seems to be one of teaching by the use of rather conventional, and often uninspiring, pedagogical practices. I am not talking about the exceptions, the teaches that could mesmerize the class with anecdotes and brilliant lectures. I am more concerned about how subject matter has typically been taught. I base this generalization of my collective observations over decades in both public and private, K-12, and higher education settings. My question to you is, why do we tend to teach proxies of critical learning rather than focusing on authentic acts of learning itself? Why do we teach mathematics as a proxy for symbolic logic or abstract thought? Why do we teach history and social science as a proxy for conflict and social change? Why is the teaching of literature a proxy for the exploration of human behavior and self-expression? And most germane to science education, why do we teach science as a proxy for inquiry and skepticism?

The notion of proxies might even be generous in that proxies, at least in statistical speak, need to correlate well with the dependent variable. It may be argued that the manner in which schooling has traditionally occurred, subject matter may at best be regarded as poor substitutes for the cultivation of virtue, character, independence, and the like. My argument for an imperative in the development of the Socioscientific Issues (SSI) Framework has been to raise counter questions that stand in contrast to tradition. Why not immerse students in meaningful and authentic learning experiences that provide an engaging context for learning subject matter? Such experiences are likely to better equip students to engage in actual inquiry, exhibit well-placed skepticism, explore varied perspectives of human behavior, engage in moral decision making and develop scientific literacy. In this sense, scientific literacy is at once humanistic and functional in that it necessarily requires the wedding of scientific understanding with a kind of moral inquiry. The preferred term, at least for me, that is ascribed to this idea is functional scientific literacy (FSL).

The way I frame FSL in the context of SSI is an attempt to render traditional proxies impotent. SSI as an immersive academic experience is one that deliberately moves away from compartmentalized academic content, fosters students' creativity, instills wonder and scientific habits of mind in children, and emphasizes a sociocultural approach to learning. But I may be getting ahead of myself. Allow me to unpack several important caveats about the SSI framework in the following section, before we return to virtue, character, and other related elements of FSL.

Throughout this book are essays that provide some personal and historical context for the development of the SSI framework. I may not have been as lucky as Galilei to find many shoulders of giants to stand on, at least in the classical sense, but there have certainly been a fruitful assortment of impressive scholars, colleagues, students and teachers that bent the arc of inquiry my way. I have also rubbed elbows with many academicians, colleagues, friends and students who have pressed my thinking and encouraged me in ways far too numerous and detailed to recount with perfect pitch. (Some of these important global contributions will be discussed in Part II of this book.) But I do strive to achieve a high degree of fidelity in my recollections, reflections and sharing of that journey with you. This much I promise you: these anecdotes will have meaning and provide a foreshadowing for the development of the SSI framework.

References

American Association for the Advancement of Science (AAAS). (1990). *Science for all Americans*. Oxford University Press. (Published in 1989 by AAAS).

Bencze, J. L., Pouliot, C., Pedretti, E., Simonneaux, L., Simonneaux, J., & Zeidler, D. L. (2020). SAQ, SSI & STSE education: Defending and extending 'Science-in-Context'. *Cultural Studies in Science Education, 15*, 825–851. https://doi.org/10.1007/s11422-019-09962-7

DeBoer, G. E. (1991). *A history of ideas in science education: Implications for practice*. Teachers College Press.

DeBoer, G. E. (2014). The history of science curriculum reform in the United States. In N. G. Lederman & S. K. Abell (Eds.), *Handbook of research on science education* (Vol. II, pp. 559–578). Routledge.

Kahn, S., & Zeidler, D. L. (2017). A case for the use of conceptual analysis in science education research. *Journal of Research in Science Teaching, 54*(4), 538–551.

Kant, I. (1784/2017). *What is enlightenment?* Translated by C. Warman (Ed.) Open Book Publishers.

Kant, I. (1785/2012). *Groundwork of the metaphysic of morals*, 3rd edition. Translated by M. Gregor Cambridge University Press.

Kyle, W., Shymansky, J., & Alport, J. (1983). The effects of new science curricula on student performance. *Journal of Research in Science Teaching, 20*(5), 387–404.

Lederman, N. G., & Lederman, J. S. (2014). Research on teaching and learning of nature of science. In N. G. Lederman & S. K. Abell (Eds.), *Handbook of research on science education* (Vol. II, pp. 600–620). Routledge.

Peck, L. A. (2007). *Sapere Aude!* The Importance of a Moral Education in Kant's *Doctrine of Virtue*. *Journal of Mass Media Ethics, 22*(2–3), 208–214. https://doi.org/10.1080/08900520701315731

Rosenau, M. J. (1935). Serendipity. *Journal of Bacteriology, 29*(2), 91–98.

Rudolph, J. L. (2019). *How we teach science—What's changed, and why*. Harvard University Press.

Zeidler, D. L. (1984). Moral issues and social policy in science education: Closing the literacy gap. *Science Education, 68*(4), 411–419.

Zeidler, D. L. (2014). Socioscientific issues as a curriculum emphasis: Theory, research and practice. In N. G. Lederman & S. K. Abell (Eds.), *Handbook of research on science education* (Vol. II, pp. 697–726). Routledge.

Zeidler, D. L., Applebaum, S. M., & Sadler, T. D. (2011). Enacting a socioscientific issues classroom: Transformative transformations. In T. D. Sadler (Ed.), *Socio-scientific issues in science classrooms: Teaching, learning and research* (pp. 277–306). Springer.

Zeidler, D. L., Herman, B. C., & Sadler, T. D. (2019). New directions in socioscientific issues research. *Disciplinary and Interdisciplinary Science Education Research, 1*(11), 1–9. https://doi.org/10.1186/s43031-019-0008-7

Zeidler, D. L., & Sadler, T. D. (2011). An inclusive view of scientific literacy: Core issues and future directions. In C. Linder, L. Ostman, D. A. Roberts, P. O. Wickman, G. Erickson, & A. MacKinnon (Eds.), *Exploring the landscape of scientific literacy* (pp. 176–192). Routledge.

Zeidler, D. L., & Sadler, T. D. (2023). Exploring and expanding the frontiers of socioscientific issues: Crossroads and future directions. In N. G. Lederman, D. L. Zeidler, & J. S. Lederman (Eds.), *Handbook of research on science education* (Vol. III, pp. 899–929). Routledge.

Zeidler, D. L., & Schafer, L. E. (1984). Identifying mediating factors of moral reasoning in science education. *Journal of Research in Science Teaching, 21*(1), 1–15.

Chapter 2
Arrested Development

Dana L. Zeidler

2.1 High School: Righteous Indignation

I can remember with mixed degrees of fondness my High School years. The consensus was, by most observers' accounts (my parents, teachers, guidance counselors), that I was an average student with my share of below average grades. I was inquisitive, showed fleeting flashes of creativity and interest, could advance a line of argumentative reasoning quite well, but academically, tended to fall short of the mark. If I "liked" a teacher and the topic at-hand, I could even excel. Mostly, I simply got by.

The early 1970s was punctuated by the various shattering of social norms, the natural culmination of the counter-culture of the late 1960s in the United States. Institutions and policies were questioned, the Vietnam War was taking its toll on disillusioned youth. I had just turned 18 years old, a full-fledged legal adult during spring of my senior year of High School, when I found myself on the outskirts of a political demonstration during a bomb-scare in the main traffic circle of school. There were hundreds of students and dozens of police that had been called in from multiple precincts to restore order, as a police helicopter circled overhead to survey the scene. I did not know it at the time, but serendipity would be impressed upon me, although I would not realize it for many years to come.

I noticed along the edge of the crowd, that three policemen (they were all men at that time) had handcuffed a student, were removing him from the crowd, when one of them struck him in the stomach with his baton while the other two were holding him. He was not a friend of mine, but I knew his name. He was slight of stature, already handcuffed, and doubled over. I had an immediate sense of righteous indignation (although I am sure I did not know what the term meant then), and I did not

D. L. Zeidler (✉)
University of South Florida, Tampa, FL, USA
e-mail: zeidler@usf.edu

D. L. Zeidler (ed.), *A Moral Inquiry into Epistemic Insights in Science Education*, Contemporary Trends and Issues in Science Education 61,
https://doi.org/10.1007/978-3-031-63382-9_2

give prolonged consideration to what I did next, I simply knew I needed to "act" because what had just transpired did not seem to serve any purpose but to brutally inflict pain on a helpless person.

I ran about 25 yards to attempt to intervene and while I knew never to "hit" a policeman, I immediately grabbed the student by his shirt and the policeman who hit him by his shirt, pulled them apart and stood there between both of them. As I said, this action was not well thought out on my part as I yelled loudly and incredulously, "*What* are you doing?" Everyone stood frozen for a split second, me included, not sure what to make of this new turn of events, when I heard, quite loudly, the very distinct words "Get the *fuck* outta here!" immediately followed by "Let's get *this* fucker!" It was at precisely that moment when I realized policemen do not like to be touched, questioned, or interfered with during the course of enacting their charge. While I had the outward appearance of a "hippy" complete with long hair, OshKosh Overalls, and a Carpenters Coat, I was also athletic having played ice hockey, lacrosse, wrestled, and partook in judo. In a nano-second, I had already plotted out my entire escape route: sprint across the traffic circle, around the side of the school, over the back athletic fields, climbing a fence between the school property and the surrounding neighborhood, cutting through several backyards, hopping the fence in a different location and making back into the safe confines of the school, and take refuge in possibly a Restroom or the cafeteria (as I was sure to work up an appetite). I was as fast as the wind back in the day, and I knew very few could likely keep up with me.

That plan lasted about 2.5 seconds in time and 10 feet in distance, as two pairs of hands on each side of my shoulders were attempting to drag me down somewhere between the concrete and the grass. Not wanting to faceplant into the ground, I arched my back, grabbed behind each side of my head onto some part of the foreign bodies that were latching on to me, shot both my legs out and using Newton's three laws of motion to keep the inertia of the all-in-one forward motion as I tossed those two policemen off of me, and did a rather impressive roll on the ground. (I later thought how proud my science teachers would be of me having applied conceptual physics to this problem!) A third policeman dove at me attempting a tackle, but I was able to maneuver into a hip throw to lessen my burden. I was about to continue to execute my escape route when one of the policemen on the ground reached up and grabbed a piece of my baggy carpenter's jacket. I let it slide off of one arm and had intended on letting it slide down and off my other arm, when the entire jacket simply bunched up and got stuck on my remaining limb, just long enough for a fourth policeman to sufficiently tackle me while the others that had been scattered around me, jumped on top of me for extra measure. As my "escape plan" evaporated, a brilliant thought crossed my mind, and I uttered the words, "OK, just relax, I'm not going anywhere," and if they did miraculously release me from their chokehold, I would breathe new life into my now dormant escape plan. I had often wondered until that point in time if it was permissible to tell a falsehood under the circumstances, as there seemed in my mind, a greater moral imperative to be had. Instead, much to my profound disappointment, I was promptly handcuffed and escorted away into a waiting patrol car. It was at this point I had the thought that had

I known I would have been in the midst of a riot when I left for school that morning, I surely would have dressed more appropriately for the occasion—sans baggy clothing. Unbeknownst to me at that time, the first encounter with moral indignation, social justice and serendipity that would, in less than a decade, be revealed in the nascent formation of the SSI framework, had transpired (see Fig. 2.1).

There is more to this story, but out of regard to you, the reader, I will only highlight certain aspects of the aftermath as I think these points add to the foreshadowing of aspects for the conceptual framework associated with SSI. After being threatened with a beating, I was taken to the local police precinct, and remained handcuffed at all times behind my back, pushed down a steep flight of stairs, (I had

Fig. 2.1 The commencement of moral indignation, social justice, and latent serendipity

pretty good coordination and balance, and may have only set a foot down on two or three steps) slamming into the wall at the bottom of the staircase. I turned, facing upwards towards the policeman at the top of the steps and smugly said with a wry grin, “Didn’t fall!” That rather displeased the gaggle of police around me who continued to threaten me and always had me secured to chairs, overhead pipes, or bars.

Finally, after some time, a detective wearing a white collared shirt, black tie, and suit jacket slung over his chair had requested my presence in his office. He re-handcuffed me with my hands now in front (a nice gesture, I thought), took a long drag from his cigarette, and said, “Now Dana, you’re in a lot of serious trouble and I think you ought tell me your side of what happened today so the judge will have your statement to look at later when you appear in court.” He took another long drag of his cigarette, put it out, positioned the paper in the typewriter and began typing down whatever was coming out of my mouth. Now, this was pre-CSI and courtroom dramas that are so common today, and I had no inkling of the “good cop/bad cop” playbook, nor that my words could actually be twisted and used against me. The detective just seemed so nice, rational, calm, reasonable and assuring, and I was happy to converse with him after being treated rather coarsely by others throughout that morning.

After only about 2 or 3 minutes, his phone rang and he said, “Excuse me, Dana, while I take this.” Impeccable manners, I thought. He is definitely on my side! His mood and tone changed quite abruptly as I heard him say into the phone—“Oh…yeah…yeah, right…OK!” and promptly slammed the receiver down on his desk. His next words created one of my first epiphanies of this whole situation—and ranked right up there in its impact along with other epiphanies of my life, like finding out that Santa Clause was not real, when he said “That was your lawyer on the phone, he said not to say anything!” (I had a lawyer?) As he pulled the paper out of the typewriter and ripped it up throwing in the trash basket, he yelled to his colleagues “Get him the fuck outta here!” That is when the proverbial light bulb over my head—my newest epiphany—set in. I thought to myself that none of these cops appear to be playing on my team and I can’t trust them. Just shut up and don’t say anything else.

I was transported in the back of a paddy wagon (a charming if not out-of-date term) to the central county courthouse and jail complex in Nassau County, New York, to await in a cell and hopefully face a judge later by the end of the day. Eventually, after many hours of sitting on a wooden bench in a jail cell, musing about the day’s events, I was led into a courtroom, where my parents, who were contacted by school officials (and subsequently contacted a lawyer), were seated. My dear mother, laying eyes upon me, immediately burst into tears. I was, by all appearances, disheveled with torn clothes and wild hair pointing in many directions at once. The judge told me I could only set foot off my property to attend school until my next court appearance. I meekly protested and said, “But your honor, I play an afternoon sport after school” thinking that was a point in my favor that he would approve of. The judge’s response was, “You *used* to play an afternoon sport after school!” I was released in my parents’ custody until my next court date where my long-term fate would be decided.

Having just turned 18 years old a week before this happened, I was a full-fledged adult and charged with two felonies (Assault Second Degree of a Police Officer, Resisting Arrest) and one misdemeanor (Obstructing Governmental Administration). Happy Birthday to me!

2.2 Post-Riot Indignation

I arrived for my court date several weeks later. I watched as the arresting officer, who took the stand and told, what we in the U.S. would now call, a story of alternative facts. Once again, I experienced a strong sense of righteous indignation. I could not fathom that a police officer would blatantly lie under oath. My lawyer, who was hired by my parents at considerable cost, calmed me down and moved on to cross-examine the officer's version of events, which went something like this: "The defendant ran up to us and yelled, 'You fuck'en pigs' and then punched me in the face and tried to run away. We said, 'Halt! You're under arrest!' He kept running and we apprehended him." Comparing it to my rendition above, the officer providing testimony and my rendition only converged correctly on the "running away" part. Long story short, my lawyer caught him in a number of contradictions, so much so that even the judge was suspect of his story. My lawyer argued that having never been in legal trouble before, and being from a "good home," I should be eligible for a probation diversion program where after demonstrating that I could be a good citizen and stay out of trouble for 1 year, my case could be revisited and at that time a final judgment could be rendered. The interesting backstory is that this was actually a program that existed for first-time drug offenders—but as I said my lawyer was both expensive and persuasive. The judge granted that request. I realized life is full of ironies and then graduated from High School.

2.3 Higher Education and White Privilege Reckoning

I now had a probation officer assigned to me that would visit me in my home every month to make sure I was progressing appropriately on my path to redemption. I was going off to college in the fall, and since Zoom would not exist in any functional form for another 45 years or so, I wrote letters longhand and sent them to my probation officer once per month from college—regaling him with tails of serious study and future plans. I failed to mention anything else about my personal college experiences figuring that some things were best to remain private. He was sure to grace me with his visits during holidays and spring break when I would return home. Now, 1 year later, I had proven myself worthy to possibly live as a free person and, not having failed out of college nor participated in any more riots, was ready to potentially be a productive member of society. I readied myself for my final court appearance as I finished my first year of higher education.

My case was on the docket for early in the morning. My parents and I were waiting in the back of the courtroom while my attorney discussed something in private with the judge. There were many cases on the docket and anxious families awaited their turn. Suddenly, my lawyer promptly ushered us out of the courtroom to have lunch, and to return later, much later, in the afternoon. It tuned out that there was a case before mine which entailed that of a young black man who faced very similar charges, was not a beneficiary of a diversion program, did not have an expensive lawyer, and would likely be sent to prison for some time. It just wouldn't be good optics to adjudicate my case so markedly different right before his, and so we returned to court much later that afternoon when those previous defendants and their families had already left the courtroom.

My lawyer assured my family that this was a preferable logistical maneuver. Finally, it was my turn before the judge and after reviewing my case, he asked me if I had anything to say for myself. Of course, this provided me the opportunity to launch into a short speech on finding redemption in higher education and lessons learned. The judge summarily said, "Case Dismissed" and brought his gavel down for good measure. I found out later that day, that my case was dismissed, in part, not because the original charges against me were trumped up and supported with lies, but because it was the judge's last day on the bench before he would ascend from county court to the New York State Supreme Court. He also knew my lawyer well. Something was mentioned later to me about the judge owing my lawyer a favor, but I was never quite clear on that.

What I do know was that the verdict was a *fait accompli*. It was at that moment that I understood what White Privilege entailed many decades before that term, or ideas that subsumed its meaning such as critical race theory (Delgado & Stefancic, 2011; Ladson-Billings & Tate IV, 1995) made it into the public consciousness. This moment of self-awareness in relation to others would also prove to be an instance of serendipity in terms of developing a broader perspective of injustice and compassion. Serendipity had allowed a culture of righteous indignation to take hold in me, beginning with my own sense of the injustice aimed at me, and expanding outward seeing what misfortunes others experience.

I went on to finish an Associate Degree in Applied Science in the Natural Sciences and Conservation, followed by a Bachelor of Science in Education with a major in Biology and minor in Earth Science. I then turned my attention to graduate school and earned a Master of Science in Science Education, followed by a Doctor of Philosophy in the same area. It was during my graduate school days where I began to rethink about many of those formative serendipitous experiences that were never quite forgotten, but now made their way back from dormancy to conscious thought. This was partly fueled by the catalytic experiences I had spending time in a maximum-security prison.

References

Delgado, R., & Stefancic, J. (2011). *Critical race theory: An introduction*. New York University Press.

Ladson-Billings, G., & Tate, W. F., IV. (1995). Toward a critical race theory of education. *Teachers College Record, 97*(1), 47–68.

Chapter 3
Prison, Morality, and Nascent Socioscientific Issues

Dana L. Zeidler

I can remember it quite distinctly. The commemorative plaque outside of the main door to the prison boldly displaying the words: "*Auburn Prison: Erection Commenced 1816. First Prisoners 1817 Assisted in Construction. First Electrocution in the World 1890.*" I would be patted down (frisked) and have my books, folders, and papers searched. I heard the sounds of a succession of steel gates each being locked behind me before the next one in sequence became unlocked for me to pass through. Needless to say, I was well aware of the high-powered rifles that scanned the courtyard as well as the stares and catcalls from inmates pressing weights as I made my way to the "school," a structure which served as the education building deep inside this maximum-security prison. The guards addressed the convicts as "Inmate 20065" or "Convict 14680." I addressed each by their name. There were Jack, Jesús, Robert, Gerardo, among 46 others. I addressed each of them as though they were my students. In fact, there *were* my students.

The year was 1978 and I found myself teaching courses for Syracuse University in General Biology and Bioethics at Auburn Correctional Facility in upstate New York. Syracuse offered courses through a program called the Independent Study Degree Program (ISDP), which was a bit of a misnomer because the courses for the prison were actually offered on-site and face-to-face. This was one of two assistantships I had to pay my way through graduate school. I taught Genetics, Human Genetics, and Science Methods courses on the main campus. However, this opportunity was certainly a first for several reasons: it was the first time that I had ever set foot inside of a maximum-security prison, the first time that my students would be taking a college level biology course, and the first time that a course in

D. L. Zeidler (✉)
University of South Florida, Tampa, FL, USA
e-mail: zeidler@usf.edu

D. L. Zeidler (ed.), *A Moral Inquiry into Epistemic Insights in Science Education*, Contemporary Trends and Issues in Science Education 61,
https://doi.org/10.1007/978-3-031-63382-9_3

Bioethics would ever have been offered at Syracuse University—either on the main campus, or in any other alternative setting.

It sounds quaint, doesn't it? An alternative setting. This was indeed an *alternative* setting. The charges among my students ranged from murder and armed robbery to possession of narcotics and breaking and entering. It was also my first attempt at developing a responsible pedagogy, contextualized in science, aimed at developing *citizenship*. For such crimes, one could argue that a punitive, retributive, or reformative approach to incarceration was each appropriate. Perhaps some combination of those might bring about some measure of justice. My personal choice, both as a citizen and as an instructor, was to focus on the latter, which is one of the reasons I accepted this position. If and when any of my students be released from prison, they become members of a pluralistic society, but would they function as "productive" and "responsible" members of the human condition? Would they be reformed? Educated?

Of course, the students could not have the luxury of selecting from a platter of concentrations to major in, but the ISDP provided a pathway to obtain a general liberal arts degree. The first semester I was assigned to teach a Biology 101 course. I argued to the powers that be, both at the university and the administration of the prison, that since this was a Syracuse University course, it had to be the equivalent course in breadth, depth, and laboratory experiences as what was offered on the main campus. Accordingly, I further argued that that experience needed to include several opportunities to perform dissections. (Keep in mind this was before the days of computer-assisted instruction where animation could substitute for actual dissections.) This meant that I needed to convince my prison and university superiors that having 50 students enrolled in the class meant that I would need to bring 50 dissecting kits, complete with scalpels, scissors, probes, tweezers, (etc.) into the prison on the days that class would include such laboratory experiences. To this day, I am astounded that my rudimentary powers of persuasion were enough for them to sanction my request. Maybe it was also partly due to the vigor, passion, and authenticity that I coupled with whatever powers of rational discourse I could muster at the time. It was determined that I was to be locked in the "school" building where our classrooms were located and further locked within my classroom itself. Every scalpel, scissor, probe and tweezer would have to be accounted for when I entered and before I left the prison for that day. Pairs of students were assigned a numbered dissection kit that correlated to their prison ID number.

I had prefaced my initial lectures with discussions of what a "*Just Community Approach*" to education, and corrections, might look like. This approach represented the embryonic stages of that movement, which was founded on Lawrence Kohlberg's theory of moral development and moral education theory at that time (Kohlberg et al., 1975). My students understood the importance of having a classroom culture of trust and respect for our journey, and we never came up short at the end of those classes. We collectively agreed on rules of conduct and discourse, socializing those rules by living them and self-enforcing them through weekly meetings in class to solve and resolve new issues as they revealed themselves. We had had discussions about what we owe each other to partake in this social experiment. To the extent possible, students were extended responsibility for deciding how to best implement rules, how to handle violation of those rules, in addition to mediating conflicts that might arise in a just and democratic manner. In a way, it was a foreshadowing of the *communities of practice movement* in science education (e.g., Akerson et al., 2012; Kinskey & Zeidler, 2020) that became popular decades later. The sense of *agency* nurtured by the just community in this academic experiment resulted in every scalpel, scissor, probe and tweezer being fully accounted for on lab days and students being thoroughly prepared for each class.

We extended the Just Community Approach the second semester, when I proposed teaching a new course on Bioethics. This provided a more *authentic context* for the development of moral reasoning skills, various forms of discourse, dialogue and quite enthusiastic debate. In my mind, the Just Community Approach helped to bridge the gap between theory and practice in the realm of education in general, and science education in particular in at least four ways: First, it was focused on democratic principles of justice and fairness. Second, it fostered participatory contributions beginning from the classroom and extending to microcosms of community and society. Third, it required variant forms of argumentation and discourse. Fourth, it required the support of evidence, whether that entailed making clear the presuppositions of one's epistemological orientations or providing empirical support and/or other forms of evidential forms of substantiation for one's position.

From a science education instructional paradigm perspective, the Just Community Approach was also congruent with a *progressive* viewpoint of schooling where the focus of education was on developing as much individual responsibility as it was in cultivating a sense of a shared community. In this sense, students take credit for their own actions whether those actions lead to success or failure. The progressive approach provided the opportunity to create a learning environment that is student-centered. Under this learning condition, the teacher's role is to provide support, facilitation of argument and discourse, challenge students, encourage questioning of sources and appeals to authority, provide clarification of concepts, provide inquiry contexts for the co-construction of meaning, and take advantage of serendipitous moments. Here, students' personal epistemologies are built through actions and deeds as they become more cognizant of their beliefs and the justifications they hold for those beliefs. The learning outcomes under the progressive paradigm were, therefore, derived from a sense of autonomy, where students understood that learning does not require blind dependence of authority or faith, but that "daring to know" can be derived from a sense of scientific identity where openness of mind is valued more than orthodoxy. This is, after all, how epistemic insight is nurtured and developed.

In this course, the first order of business was reading essays by selected ethical theorists, to cover topics such as metaethics, normative ethics, utilitarianism, deontology and the messiness of the distinction between act and rule principles when applied to topics like active versus passive euthanasia. Of course, those topics inevitably lead to finer grain distinctions among related topics like duty, beneficence, the bifurcation of facts and values, rights, justice, nonmaleficence, paternalism, and the like. Next, we considered some particularly sensitive controversial questions and issues including: "*Can prisoners' consent to chemical castration for time shaved off their sentence in the case of sex-crimes? Should therapeutic trials in medicine be conducted on prisoner populations? Can there ever be free and informed consent in a system that is inherently restrictive or coercive? Who, if anyone, should be able to make consent decisions for minors, or people with cognitive disabilities? Are there circumstances where active or passive euthanasia should be acceptable?*"

The day finally arrived where, after a good deal of background readings (remember, the internet did not exist and the prison library did not typically have books on these topics, so thoughts and ideas were shaped by the selected readings and the experiences of the students themselves) we would engage in a "spirited" debate. Students were chosen to represent different sides of the debate topic for that day, along with a group that was to evaluate the quality of arguments and select a "winning team." I expected there would be a high degree of chaos and indeed that expectation was quickly fulfilled! To say that things got heated would be an understatement as these topics were not only *personally relevant* to the students, but for many, *personally lived.*

With much shouting, yelling, interrupting, pointing fingers (middle digits seemed to be favored), the students were quite engaged, animated, with the focus on a zero-sum end result for their sides. We did not come to any resolution that week, but I did suggest that they keep their thoughts and points in mind and think about how that class went. Being on the "outside"—I sometimes slipped up when I said phrases that I would say without a second's thought on the main campus of the university, like "When you go home tonight, I want you to think about x…" Of course, my students would promptly remind me that they were going back to their respective cells and cell blocks—a long way from "home." They did not say that in so many words, but simply would echo the phrase, "That's cold, Zeidler!" I now fully understand the position of an etic perspective. The students received a fair degree of satisfaction by monitoring my reaction to see what degree of embarrassment I might register. I can say, with satisfaction, that our sense of community allowed us all to laugh at this raw human condition. Calling me by my Surname, sans Mr., or my first name, in the absence of any other "title" (e.g., Mr., Sir, etc.) was, oddly enough, evidence for the mutual respect of shared community norms established in our classroom. (This point is teased out later in broader context under the "*Brute Facts*" section of Chap. 4.)

I share the story above to emphasize that the pedagogical goal for SSI is to create an intellectual, social, and, if possible, a physically immersive and authentic academic experience that provides a context for learning subject matter (science, social science, art, language arts, music, etc.). What is both important and central to remember, however, is that *relationships* lie at the heart of the learning process and are fundamental to improving developmental and sociomoral outcomes for all students, whether they be in an elementary, middle, high school, or a college level. It is a necessary prerequisite to the *formation of virtue*, whether learning takes place in a classroom, on school grounds or any kind of informal learning experience. This is because the cultivation of virtue must be practiced, since we are not all born saints. But even the most brutish among us can be transformed by the exercise of virtuous acts. The SSI framework promotes the development of morality by providing the kinds of experiences to practice civil discourse and the exercise of what is referred to in the literature as *socioscientific perspective taking* (SSPT) (Herman et al., 2021; Kahn & Zeidler, 2017, 2019; Newton & Zeidler, 2020; Zeidler & Sadler, 2023). Hence, part of the sustained practice of SSI may be thought of as providing the conditions under which moral education may occur—for the aim of moral education can be nothing short of becoming moral.

A week later after our chaotic debate, I returned to our classroom within the deep confines of the prison walls. I was fully cognizant that the prior class would likely to go sideways—and I would have been surprised had it not done so. I solicited my students' reactions to what had transpired, and what suggestions they had to move forward with such discourse, since everyone agreed that it was an important, engaging and educationally-instructive experience. The students embarked on developing a set of rules, norms, and procedures for the conduct of future discussions and debates. The specifics of those rules were less important than the fact that their development was framed by a shared commitment to seeking a better moral atmosphere and sense of community within our classroom. That these articles of conduct were organically developed, discussed, and organized by the students themselves, and seen as fair and just, helped to foster group solidarity as well. This experience was consistent with Kohlberg's notion that an undertaking of developing a just community would require a radical restructuring of schools, classroom settings, and conventional pedagogy (Kohlberg et al., 1975).

An educational setting, in order to be transformed, required a greater commitment to justice in the form of participatory democracy where the citizens discuss and create group norms to promote the common good (see Althof, 2015; Oser et al., 2008; Power & Higgins-D'Alessandro, 2008 for further work). I was pleased to see that in subsequent discussions and debates, students, still brimming with passion, refrain from interrupting one another, impose restrictions on the allotted time different groups had to present their positions, engaged in rebuttal, and question one another. This also allowed the "elected judges" of the class to render a more informed verdict, one that attempted to resolve some of the differences among various perspectives and positions. And while the students knew it might not be the ideal solution for all involved, it offered the best resolution for a path forward to subsequent discussions and deeper understandings. These were formative opportunities afforded to me as yet another opportunity for serendipity broadened my thinking about the role of moral education in science education and further coalescence of what was then a nascent SSI framework.

The irony of attending school within the confines of prison walls was lost on me in 1978. It was not until a few years later that I learned the etymology of the word "school." It is derived from the Latin "schola" as well from the Greek "skhole." Both imply a sense of leisure and idleness (in the sense that one is not bound to have to work). In ancient times, if one was not compelled to work constantly in order to survive, then that person was certainly a member of the leisure class—a person of privilege who could spend their free time learning the seven liberal arts (i.e., quadrivium and trivium), including mathematics, music, rhetoric and the like. Having leisure of learning meant that one could participate in learned discussions, the challenges of argumentation, and become an active participant in leading oneself out of ignorance. The irony, of course, was that those serving "hard time" had plenty of "time," as it were, on their hands. Those who were selected to enroll in college coursework to be "schooled" had the penal analogue of leisure time so they might focus on moving from the house of brutes, if they so choose, into a home of higher consciousness. They could be led from ignorance and enter into a personal age of Enlightenment. (To this day, I take a certain measure of delight in letting my university students know that they are *not* working while they attend classes to be "schooled," as that time is, in fact, their leisure time.)

References

Akerson, V., Donnelly, L., Riggs, M., & Eastwood, J. L. (2012). Developing a community of practice to support preservice elementary teachers' nature of science instruction. *International Journal of Science Education, 34*(9), 1317–1392. https://doi.org/10.1080/09500693.2011.639100

Althof, W. (2015). Just community sources and transformations: A conceptual archeology of Kohlberg's approach to moral and democratic schooling. In B. Zizek, D. Garz, & E. Nowak (Eds.), *Kohlberg revisited* (pp. 51–89). Sense Publishers.

Herman, B. C., Newton, M. H., & Zeidler, D. L. (2021). Impact of place-based socioscientific issues instruction on students' contextualization of socioscientific orientations. *Science Education, 105*(4), 585–627. https://doi.org/10.1002/sce.21618

Kahn, S., & Zeidler, D. L. (2017). A case for the use of conceptual analysis in science education research. *Journal of Research in Science Teaching, 54*(4), 538–551.

Kahn, S., & Zeidler, D. L. (2019). A conceptual analysis of perspective taking: Positioning a tangled construct within science education and beyond. *Science & Education, 28*, 605–638.

Kinskey, M., & Zeidler, D. L. (2020). Elementary preservice teachers' challenges in designing and implementing socioscientific issues-based lessons. *Journal of Science Teacher Education, 32*(3), 350–372. https://doi.org/10.1080/1046560X.2020.1826079

Kohlberg, L., Kauffman, K., Scharf, P., & Hickey, J. (1975). The just community approach to corrections: A theory. *Journal of Moral Education, 4*(3), 243–260. https://doi.org/10.1080/0305724750040307

Newton, M., & Zeidler, D. L. (2020). Developing socioscientific perspective taking. *International Journal of Science Education, 42*(8), 1302–1319. https://doi.org/10.1080/09500693.2020.1756515

Oser, F. K., Althof, W., & Higgins-D'Alessandro, A. (2008). The just community approach to moral education: System change or individual change? *Journal of Moral Education, 37*(3), 395–415.

Power, F. C., & Higgins-D'Alessandro, A. (2008). The Just Community approach to moral education and the moral atmosphere of the school. In L. P. Nucci & D. Narváez (Eds.), *Handbook of moral and character education* (pp. 230–247). Routledge.

Zeidler, D. L., & Sadler, T. D. (2023). Exploring and expanding the frontiers of socioscientific issues: Crossroads and future directions. In N. G. Lederman, D. L. Zeidler, & J. S. Lederman (Eds.), *Handbook of research on science education* (Vol. III, pp. 899–929). Routledge.

Chapter 4
Brute Facts About the Construction of a Framework for Socioscientific Issues

Dana L. Zeidler

The serendipitous experiences described above led me to think more deeply about my dissertation work titled "Identifying Mediating Factors of Moral Reasoning in Science Education." I still owe a debt of gratitude to my advisor at the time, Dr. Larry Schafer, who's methodological acuity pushed me to clarify, again and again, the meaning and significance of my topic. He was, after all, the one who put the kibosh on the title of my first draft, "Why Are There So Many Assholes in the World?" He let me know that while the title was indeed an "eyecatcher," it was not sufficiently specific in scope or clarity and questioned the relevance to the field of science education. I think he was guardedly pleased at my attempt at sophomoric academic humor. He did have a point, however, because his questioning about the topic of moral reasoning and its connection to science education was something I would need to articulate with better clarity and argue to advance its cause, not just with Syracuse University, but to other gatekeepers in the science education field.

I eventually arrived at a moment of clarity when I viewed the capacity and enactment of reasoning, both in the cognitive domain, as well as the moral/affective domain, as an area worthy of pursuit in the field of science education. If students (or anyone) had the developmental capacity to reason at a given stage of cognition (e.g., formal operational reasoning), why would they not enact the corresponding stage of moral reasoning (i.e., postconventional reasoning)? Why did there often seem to be a chasm between these two domains of reasoning structures?

Assuming that higher levels of logical thinking would better resolve issues requiring scientific reasoning, wouldn't higher levels of moral reasoning better resolve issues of ethical matters? Is there not a role for both kinds of thinking to be used concurrently in science, and in everyday matters? Are there normative factors

D. L. Zeidler (✉)
University of South Florida, Tampa, FL, USA
e-mail: zeidler@usf.edu

D. L. Zeidler (ed.), *A Moral Inquiry into Epistemic Insights in Science Education*, Contemporary Trends and Issues in Science Education 61,
https://doi.org/10.1007/978-3-031-63382-9_4

that impact both kinds of thinking, which may either muddle the waters or add clarity to the formation and enactment of higher-order reasoning skills? (See Zeidler & Schafer, 1984 for a synthesis of this dissertation research.) The serendipitous experiences that made their mark on me in high school, were reawakened when I was working with my students in the maximum-security prison, and further aroused while I was teaching middle school students. I observed first-hand how my students' emergent reasoning structures revealed themselves as we discussed age-appropriate topics in science class that had moral and ethical implications, convincing me that the pursuit of such topics constituted overlooked pedagogical territory in science education, sitting dormant, and waiting to be explored. I pulled synergistic writings together from developmental psychology, sociology, moral philosophy, as well as the field of science education, in my attempt to create a conceptual framework that united these areas, and building a theoretical framework that was empirically predictive and provided conceptual links to the teaching and learning of science. This was my opportunity to tap these experiences and attempt to convey to the gatekeepers of the field that the constructs derived from such an array of academic disciplines were necessary to develop a meaningful framework and would challenge the field in novel ways.

I saw this challenge as one in which my pedagogical decisions would need to be embedded in a sense of *responsible pedagogy* if my distal aim was the promotion of responsible citizenship. This was my opening to delve into areas of normative ethics, meta-ethics, moral reasoning, moral development, character development, a *Just Community Approach* to education, argumentation, and related areas in the context of science, which were to become the precursors of the socioscientific issues (SSI) framework some years later (Zeidler & Keefer, 2003).

A central component in conceptualizing the SSI framework was recognizing that what it means to be scientifically literate is necessarily bound up in a sense of civic virtue and connects practical wisdom (phronesis) to moral intelligence (arête). Moral intelligence, at least in the classical sense, requires virtue—a kind of fittingness of one's potential to excellence and purpose. It is in this space that the exploration of sociocultural approaches to the development of character, conscience and care are intricately tied to a sense of responsibility, civic obligation and even human and ecological activism. An important aspect of developing a sense of civic virtue is also linked to the ability to take on various perspectives of those with competing interests. The question of how features of prudence, conscience, perspective taking, and virtue can purposefully spread relative to the SSI framework, has been a focal point of interest in the current literature. Embracing this area of scholarship also suggests a value-added approach to cultivating scientific literacy in that students not only become engaged in the activity of science but become normed into a democratic collective consciousness (Durkheim, 1893/2019) and collective moral responsibility (Isaacs, 2011).

So, I view science teacher education as a means to bring about responsible teachers who, in turn, bring about responsible students seeking to become virtuous in their words and deeds. This is nothing short of an educative exercise in participatory citizenship, and this is precisely what SSI and those who advocate for the practice of SSI recognize in both science teacher education, as well as the teaching of science more broadly.

4.1 Presuppositions of the Socioscientific Issues Framework: Brute Facts about Moral Education

The discussion above raises questions about how one becomes normed into a democratic collective consciousness, and for that matter, what sociological structures need to be in place for the process of normation to effectively occur. This point also has implications for the distinction between the aim of moral philosophy or moral theory and moral education, for the aim of the former is to describe different ways to think about what it means to adopt different philosophical positions or to empirically measure how human beings progress through structural stages of moral development, while the latter aims to guide us into a world that is nothing short of being virtuous—being moral. Much has been written about moral theory in previous work related to the SSI framework (see Zeidler, 2014; Zeidler & Keefer, 2003). However, there are certain presuppositions of the SSI framework that have been solidity entrenched in pragmatic sociological claims of what is necessary to actually *become* moral.

To explore and unpack a series of propositions related to the necessary conditions to becoming moral, I rely heavily on several works of Thomas F. Green (i.e., Green, 1985, 1988, 1996, 1999, 2005). It is important to understand that the sociological conditions which underlie moral education are directly tied to the formation of conscience. For Green, the formation of conscience necessitates the practice of reflexive judgement about "things that matter." In the context of science education, responsible scientific decision-making, then, requires the formation of conscience though the development of virtue and the practice of reflexive judgment (Zeidler & Sadler, 2008). But before the journey into the house of enlightenment can take place, we have to begin with what Green liked to call "brute facts." Brute facts may be thought of as fundamental attributes about our nonmoral selves—our primary nature as brutish humans fueled by a type of calculative rationality. However, attention to these attributes is necessary for the transformation of our brutish selves so that we might pass through a threshold upon entering the house of moral conscience where moral education can have its day. So, brute facts serve as precursors, as well as catalysts out of which our secondary moral natures may develop. Zeidler & Sadler (2008) touched on these, but it deserves further explanation to appreciate their importance in terms of how they are instantiated into a sociocultural normative view of SSI. I emphasize five brute facts, italicized here, that demand our attention and are accompanied by their corresponding educational implications, in order for any kind of moral education to be realized:

1. *We cannot have moral education without the functional presence of the sacred or holy.*

In a world where nothing is sacred there can be no sense of righteous indignation because if *nothing* is sacred then there is literally nothing left that can possibly offend. In such a world, no transgression against human nature would be noticed. It would be a Hobbesian state of nature with no form of social contract to bind humans together. In short, it would be brutish, solitary, fearful, and devoid of hope. There would also be no sense of outrage or moral indignation because there are no recognized moral boundaries. To entertain the idea of a world without the sacred, "… is to conceive a condition in which nothing excites horror. And in such a world, moral education cannot gain a foothold. The possibility of moral horror signals not only the presence of the sacred, but an essential prerequisite for moral education itself." (Green, 1999, p. 113).

To be clear, the presence of the sacred does not require any commitment to any particular religious doctrine, nor acceptance of God, Zeus, Krishna, Thor, or the like (although it does not exclude such commitments or communities). What it does demand, and here is the key point, is at least a *functional proxy of something that one considers sacred*—be it a symbol, a personal code or creed that can serve as an entry point into an educative normative process for the transformation of brutes into moral beings.

That symbol or personal code might constitute part of their cultural identity, their sense of belongingness, a sense of personal indignation, or in another sense, a degree of personal moral horror. Short of being a sociopath, even the most seemingly disengaged, unaffected student will bristle and take notice if the right button is pushed, or if the wrong line is crossed or potentially violated. The presence of guilt, shame, embarrassment, pride, joy, happiness, and like human emotions means that one can experience the functional analogue to the holy, a kind of moral law that has a degree of moral authority over the individual. In these cases, their functional equivalent of the scared is challenged and it is at this point where we can begin to attend to the educative cultivation of conscience, and moral reasoning—where brutes can enter into constructive conversations about "things that matter."

This is what engagement in SSI seeks to achieve. Conflicting perspectives about moral and ethical issues invite challenges to one's epistemic worldviews by introducing cognitive and moral dissonance into the mix. Those conflicting views may be found from the input of other class members, the opinions of others, or even carefully crafted challenges brought about from their teacher. Such challenges are not directed to a particular point of view—but to *other* points of view that may have been outside of their everyday purview, for further exploration, discussion, and/or debate. Good teaching under the SSI framework, is tantamount to good facilitation of these kinds of discourse to engage the otherwise disengaged.

2. *Moral argument is ineffective in establishing conscious because engaging in moral argument presupposes the existence of conscious.*

The argument here is that there is no point in beating a dead horse. You end up, eventually, beating the bare ground with a stick. The horse remains unaffected, your stick only gets shorter, and your clothes get dirtier as you kick up the dust. Thus, moral arguments will remain lost on amoral persons. For moral arguments to reach their mark, the person they are aimed at must already reside in the house of enlightenment. It seems then, in regard to any type of moral education and its cousin, character education, that it presupposes the formation of conscience. This seems like a tautology: if moral arguments appeal only to those who have already developed a conscious, then moral education is relegated to merely a well-intended exercise that falls short of its mark if the formation of conscience has yet to be fully hatched and we are no worse or better for the wear.

This is an important brute fact to consider so we don't waste our time developing educational approaches that fail to impact the development of higher-order reasoning—for reasoning needs something to act upon. St. Augustine's notion "Credo ut intellegas" (I believe in order to understand) suggests that belief is prior to understanding (Jacobs, 2000; McCloskey, 2008). While St. Augustine advocated the teaching and learning of dialectical discourse (what we might call critical thinking skills), it is necessary for a student to first believe in its usefulness and see how it advantages his, her, or their position. In the pursuit of SSI, we need to first appeal to the prudential nature of human beings—their belief that a particular action or deed serves them well. Hence, beliefs are more fundamental than reason, but may be acted upon by reason.

This leads us to consider the next two brute facts, in tantum, about prudence.

3. *Prudence is more primitive and natural, than morality, and;*
4. *If the primacy of prudence is the foundation for the formulation of conscience, then an education in prudence must be part of moral education.*

I find it difficult to untangle number 3 from number 4. While they are different claims, they are interdependent in the conduct of human affairs and therefore its treatment requires joint consideration. The eventual attainment of conscience as an educational aim is made possible by the recognition that people are first prudential by nature. This is a claim of self-interest—that people tend to act in ways that they believe will promote their own advantages, benefits, comforts, the lower parts of the conceptual totem pole representing Maslow's hierarchy of needs (Maslow, 1943, 1987). (Note: I am aware of criticisms of cultural bias and lack of empirical support for Maslow's hierarchy, but I reference it here not as an endorsement, but as a metaphor for people seeking what is in their own interests.) I can think of no better way to express this perspective than following quote by Green (1988):

> Being prudent, in the sense of looking after one's own interest, is not something that needs to be taught at all. Persons may need to be taught *what* is in their own interests, and they may even need to be taught *how* to pursue their own interests. But they do not need to be taught *to* pursue their own interests. Left alone, they will do that. Sometimes they will do it ineptly, sometimes shortsightedly, and sometimes with little self-knowledge. And, thus, they will make mistakes … Being moral, seen in this way, requires education, but being prudent does not. This is one sense in which prudence is prior to morality. (p. 138, italics added).

This encapsulates the argument of the primacy of prudence, and therefore, provides a space to begin lessons in prudence as a gateway to lessons in moral reasoning and moral education. Securing conditions and goods of basic needs may allow one to skillfully navigate the foundation of a good life, though not necessarily a moral life. The former secures conditions for what people think is important, and this is necessary as *a part* of moral education, i.e., how can one achieve goals with competence. Certainly, a life filled with regrets and misgivings because of one's incompetence cannot be part of what we may call a virtuous life. But the rational calculative drive to secure (ineptly or skillfully) what is in one's interests is certainly an important part of a moral education in the exercise of prudence. The notion of prudential behavior, in the sense of self interest, is also consistent with research done in the field of moral reasoning in that individuals will tend to make moral judgements and partake in perceived moral behaviors to engender themselves and those around them to uphold systems of cooperative exchange in order to effectively secure their own goals and advance their own interests (Krebs & Denton, 2005). However, the exercise of skillful prudential behavior is also concerned with the necessary conditions that will lead towards *epistemic insight* about the nature of virtue, eudaimonia, and how that is spread throughout the world.

Morality in this latter case stands in contrast to prudence as it is enmeshed with moral law and moral dictums that impact how we conduct ourselves as responsible and caring members of communities and social institutions (e.g., marriage, schools, friendship, democracy, etc.). In this sense, morality transcends prudence, and while it stands in contrast to prudence, any exercise in moral education must acknowledge the brute fact that it is the consistent exercise of prudence that provides a platform from which morality can be built.

Under the SSI framework, this presupposition underscores the critical factors required for the spread of human flourishing relative to topics that crosscut multiple disciplines. It includes topics like what it means to resolve issues of equity, duties, obligations, rights, and the like, relative to short- and long-term impacts, fairly and justly, between and among friends, family, and global communities.

Of equal importance, at least from a science education position, SSI also requires attention to possible ecological and ecojustice considerations (Burek & Zeidler, 2015; Herman et al., 2018; Mueller & Zeidler, 2010; Mueller et al., 2011; Zeidler et al., 2019; Zeidler & Newton, 2017). The marriage of prudence, with its self-interest drivers, to human-centric and eco-centric views enlarges the conceptualization of what is good for the planet and likely good for the individual. To ask teachers to be value-neutral appears to contradict those aspects of SSI that make it an appropriate and impactful context for developing moral character and functional scientific literacy. Functional scientific literacy strives to facilitate movement from self-centeredness towards collective consciousness through participatory experiences where teachers and students collaborate using a full spectrum of knowledge skills, and learning experiences which are inseparable from community. Teaching within an ethical framework requires awareness and understanding of students' interests as well as the larger communities that surrounds them. This is fully consistent with Dewey's (1916/1966) advocacy of *progressive pragmatism*.A final important pedagogical point about prudence may be found in its dualistic nature. On the one hand, the first sense of prudence is connected with foresight. It entails thinking ahead, planning and strategizing, whether for the short term or the long term. One must scheme, plot, and plan ahead to skillfully implement a next move, or sequence of moves, to fulfill one's desires, wishes or goals. Doing so with fidelity to goals requires the formation of practical judgements about human nature and public affairs. This is also one sense of what we mean by phronesis—or intelligence applied to action. On the other hand, prudence reveals itself in a second sense when one takes the time (and has the inclination) to look backwards. Doing so requires examining one's proximate acts, as well as distal acts that have contributed to getting you where you have arrived. This second sense of prudence compels one to seek understanding by examining acts in contextual hindsight—which is critical to the contribution of a collective, socially-shared ethics of memory embedded in community (Margalit, 2002). Asking "Did I hit the mark? Did I do that well? Did I achieve X with due regard to those around me?" and the like, acknowledges that one lives in a pluralistic world where actions are evaluated against community norms. *Understanding* how those norms came to be requires a sense of history by way of a rootedness in a shared social memory. (This is the reason Aristotle thought it a challenge to teach ethics to the young for they did not have adequate experience constructing a sense of communal history in relation to the genesis and subsequent development of culturally instantiated normative expectations.) Hence, the dualistic nature of prudence necessitates the ability to engage in reflective foresight as well as reflective hindsight. Taken together, looking forward and looking backward make up the yin-yang of prudence. Not acknowledging both senses of prudence under the SSI framework puts in danger the ability to engage in substantive discussions about moral and character education.

Prudence, therefore, necessitates participation in the affairs of others, be it family, friendship, community, society, or global affairs. Aristotle noted that the polis (Greek City State—or community) ministers to a wider range of needs than is can possibly be procured by the individual. "What a man needs to stay alive is much less than what he needs to live well. Thus, a community that is 'self-sufficient for good life' is more self-sufficient than one that is merely 'self-sufficient in necessaries.'" (Nicomachean Ethics translated by Keyt, 1987). It is clear that Aristotle believed that whatever the "good life" consists of, it certainly requires the capacity to share in community to not only advance the collective good, but to secure individual goods as well. It is interesting to note that the etymology for the word "individual" in Greek is "idiotes" which shares the same bloodline as the word "idiot." Green (1999) brings this point to the forefront: "He who lives outside the polis, says Aristotle, must be either beast or God—a beast because beyond the laws (nomoi) or a God because self-sufficient; but in neither case can such a creature be a human being. Human beings are neither self-sufficient nor are they creatures who can forsake the maternal and pedagogical service rendered by the law." (p. 74). This is yet another lesson in the primacy of prudence as a prior commitment to the education of moral beings.

5. *Moral education must include an education of emotions.*

One can conceptually deconstruct the main elements of a symphony and the instrumental families that typically perform it. We can study the four movements, and examine each role the strings, woodwinds, brass and percussion play. Clearly, the whole concoction becomes greater than the sum of its parts when orchestrated into one magnificent performance.

By way of analogy, we can see that it is necessary to deconstruct various domains of education (i.e., cognitive, affective, and behavioral). This kind of reductionism is necessary, at least from a scientific viewpoint, to delve deeper into the function of each domain. It allows for finer grain distinctions at a micro level of understanding. But at the macro level, it makes little sense to teach in a manner that prioritizes one over another, or worse still, teach in a manner where one domain is virtually excised. Children, as human beings, need the ability to think, feel, and act on things. While there are conceptual distinctions to be had in learning about these domains, their distinction becomes less important in the pedagogy of moral education. But teachers tend to shy away from tapping the range of human emotions in classroom teaching. While teachers generally desire their students to be "happy" and enjoy their classroom experience, dealing with other "messy" emotions is not part of the game plan.

Consider, for example, the case of keeping an oath or promise. In this case, I know what I should do, ought to do, obligated to do, but am tempted not to do it (for a plethora of reasons), or further, choose not to do it. Simply put, I may think to myself, "Perhaps I should have acted otherwise!" Ayer (1954) points out that if my action was voluntary (not coerced), to say I *could* have acted otherwise is equivalent to saying I *should* have acted otherwise. There are many reasons not to have done so. Maybe it was a function of sloth, deceitfulness or vanity. In any case, I may experience sadness, disappointment, or even anger at myself. If on the receiving end, my friend or spouse violated an oath or promise, then my spirit might be broken. That broken spirit is a function of the emotive nature of my conscience. Plato referred to this as thumos—that was distinct from reason and our appetites (the behavioral component of acting on desire). Having wronged someone else, or having been wronged, invokes a sense of moral indignation because our sense of worth has been called into question. (Consider that indignation and dignity have evolved from the same root word as "dignus" which connotes worth.) Expressing moral or righteous indignation becomes a function of my sense of worth being violated—either by having wronged someone else or having been wronged by the trust I placed in another. This is precisely why some functional equivalent of the sacred or holy is necessary in the conduct of both moral education and SSI (which is embedded in moral education). In this case, what was sacred was a sense of self-respect, of worth, or more precisely, a sense of dignity.

In a way then, we might think of SSI practice as a symphony in which all the elements of the cognitive, affective, and behavioral domains are orchestrated together to create a harmony of epistemic insight. An eastern Zen philosophy might liken this to a fusion of mind, body and spirit. Plato thought of it as the tripartite soul consisting of logos (reason), thumos (spirit), and eros (appetites). The point is, that in each of these cases, the whole becomes greater than the sum of its parts. It is not unlike the practice of educational research where we liken this to an interaction effect.

Regardless of how we view it, conscience in the absence of emotion is no better than a mind in a perpetual state of *tabula rasa* (Locke, 1689/1996). It is the practice of overt behavioral acts, the epistemic weighing of likely outcomes, arguments, logical consistency, and emotive sensibilities that leave its mark on the slate. Locke saw this as an act in the formation of a personal identity, which entails a conscious awareness of past actions (or thought) in relation to the present. Interestingly, that notion also foreshadowed what is essentially a contemporary holistic argument about the complex nature of character and moral reasoning. Berkowitz (2002) suggests it is instructive to conceptualize the moral nature of individuals as a kind of moral anatomy that includes moral values, moral reasoning, moral emotion, moral identity, and meta-moral characteristics (attributes that are not moral in and of themselves but support or add technical competence to moral functioning) that collectively comprise moral character. Hence, moral character "…entails the capacity to think about right and wrong, experience moral emotions (guilt, empathy, compassion), engage in moral behaviors (sharing, donating to charity, telling the truth), believe in moral goods, demonstrate an enduring tendency to act with honesty, altruism, responsibility, and other characteristics that support moral functions." (Berkowitz, 2002, p. 48).

The emphasis on such a holistic approach to pedagogy that supports SSI instruction reinforces the claims that our educational aims cannot be directed towards isolated parts of a student's moral anatomy nor prioritize one domain (cognitive, affective, behavioral) over another. Zeidler & Sadler (2008) have stated, "Argumentation, without context relevant to students' lives, is meaningless. Moral reasoning that ignores real-world evidence is fundamentally flawed. Science classrooms that deny emotive venues of discourse in the discussion of socioscientific issues curtail students' personal development." (p. 207).

These five brute facts make up the main presuppositions underlying the SSI framework. They speak to the sociocultural nature of SSI, and the primacy of attending to the holistic nature of human beings to spread human flourishing within the individual and within our school communities. Whether the SSI we engage in are global, local, scientific, political, personal, or environmental in nature, embedded in all such scenarios are elements of cognitive, affective, and behavioral conflicts which contribute to moral dissonance. This is the rendezvous where scientific evidence, real-time data collection, analysis, interpretation, and investigation become enmeshed with sociopolitical decisions, and where public understanding of science becomes tantamount to not only a life lived well, but may impact life or death decisions for us, others, non-human species and the planet. For science education in general, and moral education in particular, the attention to brute facts is generally contained within four elements that constitute an SSI-oriented curriculum:

1. *Utilize personally relevant, controversial, and ill-structured problems that require scientific, evidence-based reasoning to inform decisions about such topics*. By this we mean that while there certainly are many issues with important implications to the world around us that are relevant to the scientific community proper, as educators, we need to find a means to connect features of scientific

importance, that might seem distant and alien to students, to features of personal meaningfulness in their lived experiences.
2. *Employ the use of scientific topics with social ramifications that require students to engage in dialogue, discussion, debate, and argumentation.* This reflects our pedagogical view that the progression of science, as well as our understanding of the science of education, is fundamentally a sociocultural process connected to the teaching and learning of science necessitating shifts between emic and etic perspectives of issues—something that requires what we refer to as socioscientific perspective taking.
3. *Integrate implicit and/or explicit ethical components that require some degree of moral reasoning.* The emphasis here is our recognition that a fully functional view of scientific literacy must necessarily subsume the notion that the quality of the human condition depends on blending elements of the better angles found within our cognitive and moral talents.
4. *Emphasize the formation of virtue and character as long-range pedagogical goals.* In providing the kinds of experiences through SSI that will live fruitfully in educationally- significant ways, we aim to wed practical wisdom with moral virtue to provide the conditions under which eudaimonia may take root.

The four elements of the SSI framework are certainly aligned with Roberts (2007) Vision II of scientific literacy and extends it to subsume Vision III as well (Haglund & Hultén, 2017; McCrory & Reiss, 2023; Sjöström et al., 2017). Whereas Vision II makes a concerted effort to situate science in everyday sociocultural contexts, Vision III is aimed at problematizing the issues to explicitly focus on sociopolitical concerns, social justice and activism, and an ethic of care that allows one to enter the architecture of a moral life. Certainly, Vision II and III are in the SSI wheelhouse.

I have argued the case for the elevation of Functional Scientific Literacy (FSL) over Science Literacy (Vision I) which is primarily focused on those who will enter scientific fields (e.g., STEM), thereby making science education more inclusive by emphasizing Scientific Literacy (Vision II) that considers the "non-science" major in becoming scientifically literate in the decisions and choices about everyday affairs. FSL is also an organic bridge to Vision III Scientific Literacy (with its emphasis on advocacy and participatory democratic citizenship) as has been emphasized elsewhere (Bennett, 2020; Bennett & Zeidler, 2023; Sadler, 2009; Zeidler, 2014; Zeidler & Sadler, 2023). It is important to recognize, as Sadler (2009, 2011) has made clear, that SSI in general, and FSL in particular, have procedural and/or conceptual connections to the disciplinary nature of science. To be clear, decisions about science can be made in the absence moral and ethical understandings, because those decisions need only to be informed by some measure of scientific understanding. But the converse is also true—sort of. While decisions about science can certainly be made in the absence of moral or ethical considerations, (i.e., Vision I science literacy), FSL necessarily is, by definition, informed by such sociocultural considerations (i.e., Vision II). My colleagues and I have argue further that FSL may be leveraged to tap Vision III as well—perhaps this may mean what it is to be Socioscientific Literate (SSL) by developing socioscientific reasoning and epistemic insight (Bennett, 2020; Bennett & Zeidler, 2023; Kahn & Zeidler, 2019;

Zeidler et al., 2019; Zeidler & Sadler, 2023). That may be parsing constructs as this point, but it is a point worth noting. Table 4.1 presents, in broad brushstrokes, the features of each Vision of science education-related literacy relative to its ends.

While the SSI framework is centered around what we might arguably call, by Vision II and III, a progressive philosophy of education in general, it also has dendrites that touch upon classical views of moral philosophy and moral education. The study of moral philosophy may result in having a domain of expertise in that area, it is hardly different than having a domain of expertise in any other subject area. Such experts may be well-informed, but does it mean that one is more inclined to live a life of moral excellence over others? Knowledge of moral philosophy is one thing. Knowing how to *enact* a life of moral excellence is quite another. This is where moral education, which is a distinctly different beast, becomes critical in the cultivation and transformation of brutes into human beings who are concerned with the collective welfare of society. Green (1999) echoes this point well:

> If moral philosophy is concerned with the architecture of the moral institution of life, then the philosophy of moral education deals with how we enter it; if moral philosophy studies the shape and content of a mature conscience, then a philosophy of moral education asks how we come to own that kind of conscience; if moral philosophy addresses the nature of virtue, then a philosophy of moral education is concerned with how virtue is spread and the worldly conditions of its spread. It is one think to know what courage is, another to know

Table 4.1 Visions of literacy and their features relative to SSI in science education

Vision I Science Literacy	Vision II Scientific Literacy/ Functional Scientific Literacy	Vision III Functional Scientific Literacy/Socioscientific Literacy
Features of SL	*Features of FSL*	*Features of FSL/SSL*
SL is a "colony" of concepts (poly-conceptual) that are closed and emic in nature	A functional view of SL extends from the "visions" of SL that open and etic in nature	Emphasis on moral growth, reflective reasoning and the formation of character that require etic-emic/emic-etic shifts
SL is dehumanized to particular interests, purposes, and biases SL has no outside referents	A functional view of SL, humanizes science, not through colonization, but through interaction, transaction, and awareness.	Pedagogy is deliberately directed toward issues embedded in the crossroads of cultural, ethical, and personal norms
SL is insular to particular interests, purposes, and biases SL carries influence from science and other westernized systems	A functional view of SL attempts to connect directly with globalized and localized ethical/ moral realms	Sociocultural identity formation via enculturation in communities of practice
SL open to mythologizing (scientism)	Socioscientific Reasoning used to expose affordances and limitations of science	Contextually and culturally embedded emotive reasoning particularly elevating perspective taking
SL is imbued with *power* SL has no obligation or responsibilities	A functional view of SL allows for the possibility of the impossible or attempts to do justice for the *other*	Sensitive to both dominant and alternative normative views of SL

> how it is acquired. It is one thing to ask on what rational grounds our moral principles can be understood to rest; another to ask by what path we acquire such principles. The first in each of these pairs poses a problem of moral philosophy, the second, a problem for the philosophy of moral education. (p. 16).

It is, therefore, important to understand that a functional view of scientific literacy (FSL) seen through the lens of the SSI framework necessitates understanding what moral learning entails. Brute facts must be taken into consideration not only in the development of curricula that touches on SSI, but in providing opportunities for students to practice and enact behaviors congruent with seeking a life guided by moral intuitions, consciousness, and agency. In short, a FSL must lead to establishing the conditions under which *moral inquiry* may take place. Our attention now turns to examining examples of that venture within several examples in the extant literature.

References

Ayer, A. J. (1954). *Philosophical essays: Freedom and necessity*. Macmillan, and Co Ltd.

Bennett, K. (2020). *Advancing functional scientific literacy to socioscientific literacy as a cross disciplinary educational goal: A philosophical analysis*. USF Tampa Graduate Theses and Dissertations. http://digitalcommons.usf.edu/etd/8515

Bennett, K., & Zeidler, D. L. (2023). Socioscientific Literacy: An Emancipatory Goal for Science Education. Paper Presented at the *96th Annual Meeting of NARST: A worldwide organization for improving science teaching and learning through research.* Chicago, IL.

Berkowitz, M. W. (2002). The science of character. In W. Damon (Ed.), *Bringing in a new era in character education* (pp. 43–63). Hoover Institution Press.

Burek, K., & Zeidler, D. L. (2015). Seeing the forest for the trees! Conservation and activism through socioscientific issues. In M. P. Mueller & D. J. Tippins (Eds.), *EcoJustice, citizen science and youth activism: Situated tensions for science education* (pp. 425–442). Springer International Press.

Dewey, J. (1916/1966). *Democracy and education*. Macmillan.

Durkheim, E. (1893). *The division of labor in society* (G. Simpson, Trans.). Digireads. (2019).

Green, T. F. (1985). The formation of conscience in an age of technology. *American Journal of Education, 94*(1), 1–32.

Green, T. F. (1988). The economy of virtue and the primacy of prudence. *American Journal of Education, 96*(2), 127–142.

Green, T. F. (1996). *Education and pluralism: Ideal and reality*. Syracuse University Press.

Green, T. F. (1999). *Voices: The educational formation of conscience*. University of Notre Dame Press.

Green, T. F. (2005). *Can there be a secular moral education in the public schools of a pluralistic society?* Personal Communication/Unpublished Paper.

Haglund, J., & Hultén, M. (2017). Tension between visions of science education: The case of energy quality in Swedish secondary science curricula. *Science & Education, 26*, 323–344. https://doi.org/10.1007/s11191-017-9895-1

Herman, B. C., Sadler, T. D., Zeidler, D. L., & Newton, M. (2018). A socioscientific issues approach to environmental education. In G. Reis & J. Scott (Eds.), *International perspectives on the theory and practice of environmental education: A reader* (pp. 145–161). Springer International Publishers.

Isaacs, T. L. (2011). *Moral responsibility in collective contexts*. Oxford University Press.

Jacobs, R. M. (2000). Augustine's pedagogy of intellectual liberation: Turning students from the "truth of authority" and returning them to the "authority of truth". In K. Paffenroth & K. Hughes (Eds.), *Augustine and liberal education* (pp. 111–123). Ashgate Press.

Kahn, S., & Zeidler, D. L. (2019). A conceptual analysis of perspective taking: Positioning a tangled construct within science education and beyond. *Science & Education, 28*, 605–638.

Keyt, D. (1987). Three fundamental theorems in Aristotle's politics. *Phronesis, 32*(1–3), 54–79. https://doi.org/10.1163/156852887X00037

Krebs, D. L., & Denton, K. (2005). Toward a more pragmatic approach to morality: A critical evaluation of Kohlberg's Model. *Psychological Review, 112*(3), 629–649.

Locke, J. (1689/1996). *An essay concerning human understanding*. II, edited by K.P. Winkler. Hackett Publishing Company.

Margalit, A. (2002). *The ethics of memory*. Harvard University Press.

Maslow, A. H. (1943). A theory of human motivation. *Psychological Review, 50*(4), 370–396.

Maslow, A. H. (1987). *Motivation and personality* (3rd ed.). Pearson Education.

McCloskey, G. N. (2008). *Encounters of learning: Saint Augustine on education*. Saint Augustine Institute for Learning and Teaching, Merrimack College. Retrieved 6 June 2023.

McCrory, A., & Reiss, M. J. (2023). *The place of ethics in science education: Implications for practice*. Bloomsbury Academic.

Mueller, M. P., & Zeidler, D. L. (2010). Moral-ethical character and science education: Ecojustice ethics through socioscientific issues (SSI). In D. Tippins, M. Mueller, M. van Eijck, & J. Adams (Eds.), *Cultural studies and environmentalism: The confluence of ecojustice, place-based (science) education, and indigenous knowledge systems* (pp. 105–128). Springer.

Mueller, M. P., Zeidler, D. L., & Jenkins, L. L. (2011). Earth's role in moral reasoning and functional scientific literacy. In J. L. DeVitis & T. Yu (Eds.), *Character and moral education: A reader* (pp. 382–391). Peter Lang.

Roberts, D. A. (2007). Scientific literacy/science literacy. In S. K. Abell & N. G. Lederman (Eds.), *Handbook of research on science education* (pp. 729–780). Lawrence Erlbaum Associates.

Sadler, T. D. (2009). Situated learning in science education: socio-scientific issues as contexts for practice. *Studies in Science Education, 45*(1), 1–42.

Sadler, T. D. (2011). *Socio-scientific issues in science classroom: Teaching, learning and research*. Springer.

Sjöström, J., Frerichs, N., Zin, V. G., & Eilks, I. (2017). Use of the concept of Bildung in the international science education literature, its potential, and implications for teaching and learning. *Studies in Science Education, 53*, 165–192. https://doi.org/10.1080/03057267.2017.1384649

Zeidler, D. L. (2014). Socioscientific issues as a curriculum emphasis: Theory, research and practice. In N. G. Lederman & S. K. Abell (Eds.), *Handbook of research on science education* (Vol. II, pp. 697–726). Routledge.

Zeidler, D. L., Herman, B. C., & Sadler, T. D. (2019). New directions in socioscientific issues research. *Disciplinary and Interdisciplinary Science Education Research, 1*(11), 1–9. https://doi.org/10.1186/s43031-019-0008-7

Zeidler, D. L., & Keefer, M. (2003). The role of moral reasoning and the status of socioscientific issues in science education: Philosophical, psychological and pedagogical considerations. In D. L. Zeidler (Ed.), *The role of moral reasoning on socioscientific issues and discourse in science education* (pp. 7–38). Kluwer Academic Press.

Zeidler, D. L., & Newton, M. (2017). Using a socioscientific issues framework for climate change education: An ecojustice approach. In D. Shepardson & R. Roychoudhury (Eds.), *Teaching and learning about climate change: A framework of educators* (pp. 56–65). Routledge.

Zeidler, D. L., & Sadler, T. D. (2008). The role of moral reasoning in argumentation: Conscience, character and care. In S. Erduran & M. Pilar Jimenez-Aleixandre (Eds.), *Argumentation in science education: Perspectives from classroom-based research* (pp. 201–216). Springer Press.

Zeidler, D. L., & Sadler, T. D. (2023). Exploring and expanding the frontiers of socioscientific issues: Crossroads and future directions. In N. G. Lederman, D. L. Zeidler, & J. S. Lederman (Eds.), *Handbook of research on science education* (Vol. III, pp. 899–929). Routledge.

Zeidler, D. L., & Schafer, L. E. (1984). Identifying mediating factors of moral reasoning in science education. *Journal of Research in Science Teaching, 21*(1), 1–15.

Chapter 5
Moral Inquiry in the Practice of Socioscientific Issues

Dana L. Zeidler

5.1 The Case for Warmly Analytic Moral Inquiry

The undertaking of a moral inquiry is never aimed at just anything, but rather at some particular thing. Hitting the mark on virtuous moral inquiry will require doing so with excellence, because one cannot rely on the generosity of gods any more than thinking the earth is a worldly paradise. If we should find ourselves in paradise, then there would be no need for moral inquiry because all human needs would be met. So, we must pursue a moral inquiry into the affairs of humans regarding the decisions made about the world we inhabit. The Greeks referred to worldly excellence as virtue, but they did so by evoking a more encompassing notion of arête which denoted the quality of excellence as 'functional goodness.' Interestingly, arête is also closely connected to the Greek word *ergon*, which connotes 'function.' In a sense, we could speak of a virtuous object, say a house, because it performed its function well. We could imagine such a house that was built solidly and securely, was comfortable, aesthetically pleasing, in short, possessed excellence to the extent that it functioned well. It embodied, both, form and function. The same may be said of a virtuous horse, that it be in harmony with its rider's intent, or might even refuse to obey a command because of a danger unforeseen to its mount.

Virtuous inquiry also requires a kind of fittingness of form to function. Again, it is aimed at something in particular such as, for example, resolving competing conflicts or claims, but doing so with compassion, while exploring the parameters of a given situation, examining potentially fallacious arguments, determining what data is relevant to the dilemma, thinking about the short-term and long-term consequences of potential decisions to be had, considering historic factors that may have

D. L. Zeidler (✉)
University of South Florida, Tampa, FL, USA
e-mail: zeidler@usf.edu

D. L. Zeidler (ed.), *A Moral Inquiry into Epistemic Insights in Science Education*, Contemporary Trends and Issues in Science Education 61,
https://doi.org/10.1007/978-3-031-63382-9_5

disadvantaged particular actors, and making just and fair decisions that are recognized by those involved as the best case that could be secured at that time. Cultivating the kind of skills that enable one to partake in such a venture is what is needed to lead a moral, good life. The practice of such skills requires a kind of *moral inquiry*—fittingness of intent and action to achieve desirable outcomes in the spread of eudaimonia. It is not unintentional that I refer to functional scientific literacy in the framing of the SSI framework, for FSL enacted in the sense that I propose can be nothing less than the coordination of deliberate intent with action in the distribution of human flourishing.

Since we dwell in the arena of science education, the context of moral inquiry is aimed at what the socioscientific issues framework provides as stated in the four elements of the SSI curriculum of Chap. 4. I am not suggesting that those elements are exhaustive—merely that they provide the pedagogical backbone for moral inquiry to take place. To do so, means that we have to understand the emotive piece of the educational framework in better scope and detail. This is not to undermine the rationalistic-cognitive pieces of scientific thinking such as the role of critical thinking, the ability to evaluate the worth, accuracy, or authenticity of various propositions, and possessing the inclination to consider questions that arise by way of healthy (but not frivolous) skepticism (Ennis, 1996, 2018; Kuhn, 2019; Zeidler et al., 1992).

The importance of distinguishing between verifiable facts and value claims, relevant from irrelevant information, recognizing unstated assumptions, credibility of sources, and the like is certainly central to carefully crafting an analysis of a problem. It is important to note that critical thinking or reasoning is also tied to the concept of thoughtful reflective thinking mentioned earlier. While much in the literature connected to science education focuses on that cognitive domain, it is arguably equally important, if not more so because of its inherent nature to promoting just and equitable environments, as well as being understudied in science education, to parse out the impact of emotive reasoning particularly as it applies to resolving issues that entail moral conflict. While it is easy to think of critical reasoning being aligned in the positivist tradition as being "coldly analytic"—we might think it odd to think about emotive reasoning as residing in the "warmly analytic" realm (thanks to Professor Tom Green for that analogy). In applying moral inquiry to SSI-based issues, we need to give thoughtful and due consideration to traits and constructs like empathy, sympathy, and compassion, to ensure practice in the formation of the kind of Aristotelian virtues (e.g., moral courage, truthfulness, righteous indignation / justice) that align with a life well lived. This begs for the kind of inquiry that Hume suggested in *An Enquiry Concerning the Principles of Morals*, which aimed at a logical scientific study of morality to empirically establish virtues (and vices) of the mind. His affinity toward the emotive elements can be discerned from Hume's analysis:

> The social virtues must, therefore, be allowed to have a natural beauty and amiableness, which, at first, antecedent to all precept or education, recommends them to the esteem of uninstructed mankind, and engages their affections. And as the public utility of these virtues is the chief circumstance, whence they derive their merit, it follows, that the end, which they

> have a tendency to promote, must be some way agreeable to us, and take hold of some natural affection. It must please, either from considerations of self-interest, or from more generous motives and regards. (Section V, Part I) Animals are found susceptible of kindness, both to their own species and to ours; nor is there, in this case, the least suspicion of disguise or artifice. Shall we account for all THEIR sentiments, too, from refined deductions of self-interest? Or if we admit a disinterested benevolence in the inferior species, by what rule of analogy can we refuse it in the superior? (Appendix II) (Hume, 1777/1912).

Moral inquiry, then, is an activity aimed at advancing virtue when confronted with conflicting or unclear moral judgements. Moral inquiry aims at illuminating moral questions as we engage in reflective equilibrium (Rawls, 1971). But it is generally not correct to think of reflective equilibrium as only lessening dissonance in my own mind. Reflective equilibrium is a pluralistic experience—something that we do together as one tries to align their worldview with that of others. Moral inquiry "… attempt(s) to reach and sustain certain kinds of agreement, shared understanding of what should be the moral terms of our life in community together … to build and sustain the moral commonalities that make human community and human relations possible" (Lenman, 2007, p. 76). While logical sensibilities like careful reflection, internal consistency, and coherence are certainly employed (Timmons, 2002), my central claim is such critical (cognitive) thinking traits that factor into moral judgments can be enhanced, relative to the spread of virtue, by the emotive traits associated with passion and care. The fusion of cognitive and moral structures plays out in the formation of virtues and psychological structures tied to practical wisdom—phronesis. And if warmly analytic moral inquiry is to have its day, then a further understanding of phronesis and, in particular, emotive factors that help to humanize it, are deserving of due consideration.

5.2 Phronesis in the Practice of Moral Inquiry

Phronesis is an intellectual virtue that lies between moral instincts and practical wisdom. In doing so, it serves as an anchor point that connects moral identity, moral emotions, and moral reasoning. Moral emotions, both conceptually and empirically, are understood to provide the affective factors that supply the underlying motivational impetus to connect reasoning with action (Kristjansson et al., 2020). One way to conceptualize how these factors interact is to consider a neo-Aristotelian model of phonetic (practical wisdom) moral decision-making suggested by Kristjansson and Pollard (2021) (Fig. 5.1). More specifically, moving from a translation of Greek to contemporary terms, "phronesis refers to the capacity of knowing and enacting the right course of (moral) action through a process of identifying and deliberating between competing values, emotions and alternatives" (Kristjansson et al., 2020, p. 8). In short, this conceptualization, on the one hand, is consistent with Aristotle's philosophical treatment of phronesis, while on the other hand, is couched in terms (constructs) consistent with current social scientists' conceptualizations of moral functions.

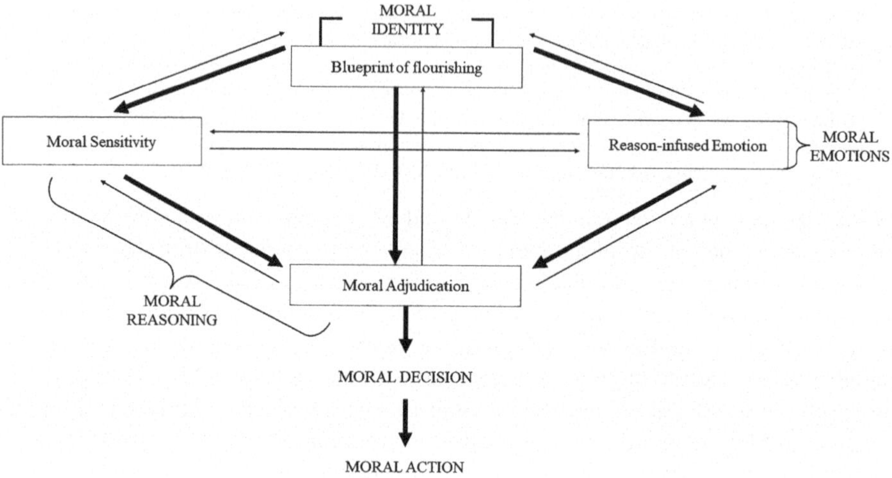

Fig. 5.1 A neo-aristotelian model of phonetic moral decision-making. (Adapted from Kristjansson & Pollard, 2021, p. 9)

What interests me are the connections among moral inquiry and moral action. It shows that moral decision-making precedes moral action, and that developing moral sensitivity through reason-induced emotions in the process of evaluating (moral adjudication) varied perspectives prior to making and enacting that decision, contributes to the development of moral identity. While studies exist aimed at advancing emotional intelligence (e.g., Durlak et al., 2011) or focus on isolated components contained within Fig. 5.1, there is scant research on how these components play out and work in an orchestrated manner within a holistic view of phronesis. The "Blueprint of flourishing" depicted below functions as a kind of meta-construction providing an overall sense of the kinds of things that matter for the practice of eudaimonia—connected the various constructs together. One can think of this meta-construction as the formation of a *moral identity* which ultimately serves as a blueprint that is connected to moral action. The SSI framework is sensitive to these factors as necessary conditions for the growth, practice and spread of human flourishing. This is because the SSI framework provides the intellectual space to experience the kinds of issues that require practical wisdom, practice, formulation of decisions, and the opportunity to potentially act on those decisions (the practice of Vision III scientific literacy). Afterall, Aristotle (Nicomachean Ethics, 1145a) noted that "Virtue makes us reach the end in our action, while phronesis makes us reach what promotes the end." (Aristotle, 2002).

5.3 Moral Inquiry in the Context of Socioscientific Issues: Place-Based Emphasis

While there is evidence to support the idea that moral emotions are a predictor of moral behavior, it is not clear how these affective qualities reveal themselves in thinking about SSI. One exploratory study by Herman et al. (2020) is particularly salient here as it fleshes out how emotive factors may mediate moral inquiry in the context of environmental SSI problems. We believe this work illustrates, from a psychological perspective, phronesis in action.

This study was the result of a deliberately designed place-based immersive context with college students who spent 10 days in the greater Yellowstone Area—a national park in the United States. There has been research reported that suggests place-based experiential environmental education helps students to develop compelling emotive/affective connections to nature, in addition to supporting situated learning approaches that focus on authentic engagement with sociocultural, moral, and affective perspectives related to SSI (Herman, 2018; Kurdyavtsev et al., 2012; Sadler, 2009). The purposeful connection to these sociocultural elements connected to emotive reasoning / experiences was amplified in pre-immersive and post field experiences. Students' experiences entailed exploring the reintroduction of wolves in the national park, and interviewing citizens and many stakeholder groups (e.g., Wildlife Biologists, Environmental Scientists, Tour Guides, Native Americans, Livestock Ranchers, Local Business Owners, etc.) impacted by the controversy to gain more insight into the multiple perspectives related to this environmentally-contextualized SSI.

In this research, we were interested in exploring and examining the nuanced ways students tapped emotive / affective reasoning structures when asked to engage with environmental SSI, and how these emotive responses shift form and function through dissonance and conflict as a result of their immersive personal experiences and interactions among ecocentric and anthropocentric perspectives. The problematic context was the reintroduction of wolves into Yellowstone National Park in the US, and its impact on multiple groups within the surrounding communities having competing claims. It is important to note that multiple emotive factors identified were the result of empirically analysis of the students thinking, not the result of a-priori categorization. Table 5.1 summarizes this portion of the research. The analysis revealed four overarching taxonomies of emotive reasoning: (1) Apathy (which is really the absence of emotive considerations); (2) Passive Care (kind of a general not firmly committed concern); (3) Moderated Concern (revealing emotive awareness and concern for the welfare of ecocentric and/or anthropocentric factors, but may include certain caveats to protect a degree of psychological distancing when vicariously experiencing others' potential suffering; and (4) Empathetic Dissonance (which showed strong visceral emotive responses for the well-being of nature and people.

By unpacking Moderated Concern and Empathetic Dissonance, the research reveals some interesting pathways a student might find to instantiate their

Table 5.1 Empirically derived categories and subcategories of emotive reasoning about place-based SSI

Categories of emotive reasoning	Subtypes of emotive reasoning	Operational definition
Apathy		An explicit emotive disregard for the suffering or ill treatment of people and nature
Passive care		A blanket generic caring about the welfare of people and nature
Moderated concern		A statement of emotive awareness and concern for people and nature containing caveats or stipulations
	Value judgment	A value statement identifying that those (i.e., people and nature) impacted were somehow responsible for their misfortune
	Helplessness	A statement indicating a lack of insight or influence regarding the resolution of the issue
	Diffusion of responsibility	A statement shifting responsibility away from oneself toward some larger component of community or nature
Empathetic dissonance		A visceral emotive response identifying with people and nature that was proportionally similar to the negative impacts incurred
	Compassion	A strong feeling of concern for people and nature who suffer without concern for the cause of the suffering
	Guilt	Feeling badly or shameful for not engaging in sustainable resolution that would relieve the negative impacts experienced by people and nature
	Anger	Feelings of empathy toward people and nature who suffer, while demonstrating overt ire toward an explicitly identified cause
	Righteous indignation	A visceral emotive response toward people and nature who suffer due to perceived inequities or injustices

reasoning. In particular, for Moderated Concern, we noted three subtypes of emotive reasoning including (1) value judgment, (2) helplessness, and (3) diffusion of responsibility; while for Empathetic Dissonance, four additional subcategories were reported including (1) Compassion, (2) Guilt, (3) Anger, and (4) Righteous Indignation. Operational definitions based on the data were formulated for each of these subcategories as well.

While Herman et al. (2020) provide explicit examples of each of these subcategories, suffice it to say that the subcategories demonstrated a degree of emotive sophistication that accumulates with practice and experience. It is as if experience via active participation (function) contributes to form (affective structure), while form synergistically impacts function. Figure 5.2 is illustrates a range of emotive states that students may purposefully or unwittingly move through as they experience and reason about SSI. It does not mean that every student passes through this sequence, nor does it imply a *fixed* hierarchy. What it does imply, however, is that empathetic dissonance does tend to reveal more pluralistic worldviews and ecojustice orientations. In this research, we were not suggesting that forms of empathetic

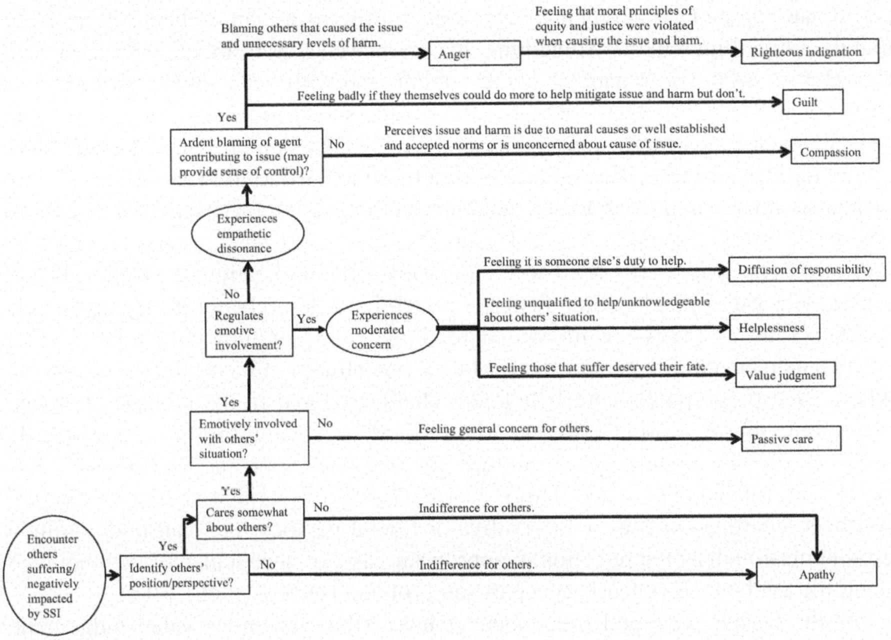

Fig. 5.2 Hypothetical pathways of student' emotive reasoning impacted by place-based SSI

dissonance were better or more effectively resolved an SSI than other emotive responses. Rather, we did imply that educators can broaden the perspective taking faculties of the inquirer. This aligns with the kind of virtues that Aristotle thought central (e.g., thumos (anger); righteous indignation (social justice and/or ecojustice) in the pursuit of moral inquiry. One can infer here that the role of the teacher is one that may facilitate strategic opportunities for meta-reflection so students can better navigate through a range of emotive perspectives.

I think it noteworthy to zero in on the categories of emotive reasoning (1- compassion, 2- guilt, 3- anger, 4- righteous indignation) that align with empathetic dissonance, which may be thought of as a subset of moral dissonance. It has been recognized that dissonance, in general, is a precursor to more advanced states of reasoning and conceptual understanding both in the realms of cognitive and moral development because the kind of disequilibrium experienced in either domain, serves as a driver to reduce or eliminate itself (Blake & Pope, 2008; Lee & Francesca, 2018).

The magnitude of dissonance will factor in on the degree of cognitive or moral tension present where different actions or pathways forward in the resolution of a given scenario led to conflicts with competing values or preferences. These factors, these tensions, are *not* something that should be avoided by teachers. On the contrary, tapping such tensions crafted in skillful ways can be used as a means to compel students to seek out new information and/or perspectives in resolving these internal conflicts. In Piagetian (cognitive development) terms, I may need to coordinate my understanding and viewpoint with some aspect of the external world that does not align with it—something that science educators are quite familiar with though the use of discrepant events in conjunction with conceptual change theory (diSessa, 2006; Kural & Kocakulah, 2016; Nadelson et al., 2018). In Kohlbergian (moral development) terms, moral dissonance occurs when other moral stances are incongruent with mine. This creates a kind of ethical disequilibrium as conflicting principles now become uncovered and laid out in the open for inspection. If I am to engage in further discourse, the task requires me to coordinate my worldview with that of others. It is as if I need to lay bare my structural rationale of choices and preferences side-by-side with others as I search for ways to better justify my beliefs to others (and vice versa), while with defending against, or allowing for, the counter arguments from others' perspectives. Mutual bootstrapping is certainly in play here, where each participant's core beliefs are challenged and in the process of shared resolution, moral tensions become diminished, and equilibrium is temporarily restored, thereby creating opportunities for our reasoning and worldviews have become elevated and enriched. This is a case where adversarial dialogue over shared conflicts regardless of, say, conservative or liberal perspectives, can find common ground inasmuch as the rejection, inappropriate use, or acceptance of well-sourced scientific evidence is not strictly a partisan problem (Owens et al., 2017).

While passive care and moderated concern (that subsumes value judgement, helplessness and diffusion of responsibility) represent various psychological expressions of emotive reasoning, it is the categories of empathetic dissonance that, at least on a conceptual level, that are more likely to lead toward interventive actions because of its impactful connections to visceral affective responses. Herman et al. (2020) suggest an explanatory link that exists between the forms of empathetic dissonance, and the motivational elements contained therein to enact pro-social behaviors that may lead to the resolution of an issue(s). In particular, these conceptual categories (along with examples offered by participants)of empathetic dissonance include:

1. **Compassion** was directed toward entities suffering as a result of being impacted by environmental SSI hardships (e.g., "Taking the wolves from Canada and just putting them in Yellowstone. It throws them off. You feel bad. They have to relearn where everything is. They are not in a familiar place anymore.")
2. **Guilt** described affective responses where students expressed feeling sad, ashamed, or remorseful for not enacting behaviors congruent with proenvironmental and prosocial moral agency such as donating time, making reparations, protecting, and stewardship kinds of actions (e.g., "Me not recycling I feel guilty because I say we should but then I guess it's an inconvenience to me, so I don't. I do not know. It's bad. I feel like a bad person for not (recycling).")
3. **Anger** revealed an intense sense of moral agency (e.g., a sense of duty) and empathetic feelings for entities that suffer and was accompanied with an overt ire toward those who were seen as accountable (e.g., "With the animals and the environment, I am actually really upset that humans are as devastating as we are to the environment. If we cannot respect how things work in the natural world

then shame on us, there is no need for individuals to gorge themselves with massive amounts of food and materialistic items that cause more harm to nature.")

4. **Righteous Indignation** was characterized by an emotive response that included anger (like in the defining characteristics above), but in addition, coupled with the perception or recognition that distinct moral principles of equity and justice were violated (e.g., "Humans think of themselves first before the effects it could have on the environment …The bison were here before any residential areas were in place, but humans consider themselves superior to all things, and now killing bison is a suggested way to manage that issue. I think these issues, such as reintroducing wolves into Yellowstone Park that raise a lot of social and ethical concerns. This concern could be ethical on the side of who are WE to play God? It could also be social/ethical on the side of the impact it may have on the residential locals, the animals within the park, and the tourists that come to see Yellowstone.")

Of particular importance is that these forms of empathetic dissonance were directed toward both ecocentric and anthropocentric orientations, and the SSI intervention showed a higher presence of both orientations post intervention. This was encouraging because showing compassion toward humans seems more apprehensible than showing compassion toward other organisms, and certainly toward abiotic factors in the environment. We attribute this to the nature of a place-based immersive experiences where students made connections between the inherent virtues of a massive, yet delicate, integration of ecosystems contained within the Greater Yellowstone National Park area (balance of form and function as found in the notion of arête), and the virtues contained within the human condition (Arendt, 1958; Hominh, 2021; Raz, 1988) with respect to what it means to live a good and virtuous life in a global society and embedded in an ecological context. I describe such a state as the result of virtuous, moral inquiry.

Related research has reconfirmed and reinforced the importance and impact of authentic immersive sociocultural experiences and affective elements in the pursuit of functional scientific literacy. Herman et al. (2021) further demonstrated that SSI instruction embedded in a place-based setting, where students were engaged in detailed discussions with various stakeholders impacted by these environmental SSI, helped students better articulate contextualized perspectives, emotive concerns and feeling of accountability in the resolution of environmental problems. Again, as in our prior work, SSI instruction with students from diverse academic majors were immersed in the Greater Yellowstone National Park and assessed on four broad dimensions of socioscientific orientations. These socioscientific orientations consisted of: (1) *Ecological Worldviews* (which subsumed Interconnectedness and Sustainable Development); (2) *Social and Moral Compassion* (which subsumed Moral and Ethical Sensitivity, Perspective Taking, and Empathetic Concern); and (3) *Socioscientific Accountability* (that subsumed Feeling of Responsibility and Willingness to Act); and (4) Scientific Evidence Views (that subsumed Affordances and Constraints of Scientific Evidence). It is important to credit that this framework was premised on the seminal research by Lee et al. (2012) and Lee et al. (2013) who posited that the emotive/affective traits of character and values were central to not only linking social values to science education, but also for the need of students to be educated as whole human beings in relation to global issues impacting the world they inhabit. These four orientations, along with their corresponding sub-categories, reflect the multidimensional nature of socioscientific reasoning about SSI. Indeed, individuals may position themselves in diverse ways and will need to understand how others conceptualize and perceive these interconnected dimensions. (This notion is further expanded below and later in Table 5.2).

While these four themes are fully operationalized in the research, the essential take-away was that students revealed more robust and better contextualized claims (both statistically significant and qualitatively meaningful) across these four orientations (and the sub-categories subsumed in each one) in a manner that reflected elements of social and moral compassion. The other focal point to take notice of in this research was the construction of standardized qualitative measures (rubrics) that entailed a rigorous coding and rating process whereby qualitative data was transformed and quantified for nonparametric statistical analyses. The rubrics permitted the quantitative evaluation of the four socioscientific orientations dimensions and was, at the same time, sensitive to SSI contextualizing links proffered by the students. Transformation of qualitative to quantitative scales were performed for Interconnectedness, Sustainable Development, Moral and Ethical Sensitivity, Perspective Taking, Empathetic Concern, Feelings of Responsibility, Willingness to Act, and Affordances and Constraints of Scientific Evidence. To select one rubric, *Moral and Ethical Sensitivity*, as a case in point, the progressions for each factor followed this simplified prototypical pathway, with .5 added to levels 1–4 if a student qualifies / contextualizes their statement with an SSI connection (note that number 5 necessarily subsumes contextualized links to SSI):

Table 5.2 Interaction of cognitive and affective factors directed toward moral sensitivity

	Orientation of moral inquiry	
Reasoning function ⬇	Orientation directed toward internal and self-directed toward personal growth	Orientation directed toward external / community, societal and/or global considerations
Rationalistic reasoning judgements	Moral sensitivity is focused on rational control. Rationalistic abilities help to determine the ethical problem and define the perspectives of the stakeholders relative to prudent interests	Moral sensitivity includes the ability to evaluate using metrics like justice, fairness, consequences, etc. other stakeholder's positions, orientations and perspectives
Emotive reasoning judgements	Moral sensitivity is aimed at control precedence. Emotive experiences include an evaluation of the ethical situation in terms positive or negative emotive states for self (e.g., reducing empathetic dissonance)	Moral sensitivity concerns the well-being of the group(s). The counterpart of community living is that the individual must be aware or the values and norms of the group and act in concert with those or persuade others toward alternative norms out of an abundance of care and concern

Modified from Kraaijeveld et al. (2021)

1. Lacks awareness of and/or rejects moral and ethical sensitivity.
2. Mention of moral and ethical sensitivity appears shallowly tacit or with pretense (i.e., posturing).
3. Inconsistently provides reason(s) how/why moral and ethical considerations are important (e.g., tentatively states morals should be considered with weak justification.
4. Consistently provides reason(s) how/why moral and ethical considerations are important.
5. Consistently and explicitly provides reasons how and why moral and ethical considerations are important with clear contextualized links to SSI.

Two important themes have been deduced from this research. Herman et al. (2021) point out: "First, the hope is that effectively engaging SSI becomes a *reflective and transcending act*, where SSI engagement at one scale (e.g., local and immediate) is abstracted to another scale (e.g., globally and extended) in contextually appropriate ways. Second, effectively engaging SSI requires *explicit negotiation* of multiple complex and often disconcerting dimensions that include scientific evidence, compassion, ethical sensitivity, NOS, willingness to act, and the unequal impacts on people and the environment" (p. 3—italics added). This implies that SSI as a *reflective* and *transcending* act, provides the experiences and opportunities for students to see science as one valuable way of knowing, but cannot be divorced from other ways of constructing knowledge that ought to be factored into decisions that are informed by sociocultural, ethical, economic, personal and community consequences typically impacting the resolution of SSI-type problems. Accordingly, the framework guiding our approach is congruent with others who value and view epistemic insight as the ability to reflect on one's own thinking about their own construction of knowledge from a variety of perspectives and tap multiple disciplines that can inform the resolution of issues while adding diversity to epistemic knowledge (Billingsley, 2017; Billingsley et al., 2018; Green et al., 2016). In this way, the pervasive compartmentalization of knowledge may begin to evaporate, giving way to more fruitful and productive means of thinking and communicating (Zeidler, 2016; Zeidler et al., 2016).

Additional place-based and related free choice scholarship connected to SSI shows promise in the engagement of students impacted by authentic contexts. While school-based (classroom) settings can provide a productive means for the discussion of scientific concepts that underlie actual or hypothetical SSI (see next section on School-Based SSI), classroom discussions of SSI can also serve as an advanced organizer for place-based settings, immersive learning experiences in museums, nature, community, or even the school yard, free-choice experiences can certainly enrich and reinforce those classroom experiences. Further, authentic experiences embedded in real-world-problems may lead to a sense of moral identity concerning environmental stewardship by offering students a perspective of how today's decisions may impact their future. Falk and Dierking (2018) posit in their contextual model of learning that all learning is situated in overlapping contexts that include personal, sociocultural and physical contexts where meaning-making is through interactive experiences that are constructed over time and in constant flux as new experiences become integrated with prior understandings.

5.4 Moral Inquiry in the Context of Socioscientific Issues: Blended Emphasis

Some research has indeed focused on the integration of formal classroom instruction with some degree of connectedness to real-world settings or community issues. One such emphasis was aimed at developing a sense of place (SOP). Kim et al. (2020) brought together the contextual elements that SSI can tap with community-based programs (referred to as the SSI-COMM model) with the aim to connect students to an SOP. In this research, Korean students were engaged in an eight-week course that resonated with students' concerns such as ongoing pollution disputes that impacts several Asian countries which can lead to respiratory and cancer risks, as well as more local issues such as abandoned pets and recycling. The strategy entailed a four-part scaffolded sequence: (1) *recognition* of the issue by visiting sites in the community, (2) *exploration* of the issue by visiting local organizations and talking with experts to explore the issues, (3) *sharing* information and possible solutions with community members, and (4) *action-taking* in the community. It is essential to understand that connections within the local community were not limited to taking advantage of local resources through simple field trips, but more importantly students were provided the opportunity to engage and communicate with local experts and resident with an aim to formulate plans of action. The result of interest here is that assessments of students' SOP and character improved after participating in the SSI-COMM experience.

Other scholars have also used more formal classroom settings to serve as inquiry springboards to citizenship activism relative to science issues. Doing so helps to address the problem of science and technology issues being addressed in a manner devoid of their full societal context (Levinson, 2018). Hodson (2011, 2014, 2021) directly emphasizes that in addition to addressing the values and developing concern for the perspectives of others, a kind of critically responsive pedagogy should also allow for learning about possible sociopolitical actions, enacting feasible actions, and evaluating those actions in the context of SSI and related frameworks (e.g., Science, Technology, Society, Environment (STSE), and Socially Acute Questions (SAQ)). He stresses that students can learn from, and through, actions, and provides examples of teachers serving as a type of mentor as students apprentice themselves into activism; he also provides pedagogical models to illustrate how that might be accomplished through reflective practice. In particular, in addition to appreciating the impact of science and technology, understanding that decisions about science and technology are linked with the distribution of wealth and power, justifying one's views about SSI through discussion and argumentation, learning about action through doing action, afforded in projects that extend beyond the classroom can be impactful in preparation of informed citizenry. Additionally, Hodson (2021, p. 1) states that for teachers, "An understanding of emotional intelligence, emotional literacy and emotional competence is essential, and building students' capacities in these matters is crucial to helping them deal with potentially stressful and disconcerting learning situations that will inevitably arise."

Bencze (2017) and colleagues (e.g., Bencze & Alsop, 2014; Bencze & Krstovic, 2017; Bencze et al., 2012, 2020a) have also advocated for scientific activism through an ongoing research program referred to as STEPWISE framework (Science & Technology Education Promoting Wellbeing for Individuals, Societies & Environment). STEPWISE has distinct parallels to the SSI framework, but with explicit attention to advocacy. STEPWISE is directed at engaging students to act (and become) citizen scientists that need to navigate sociopolitical actions as they discuss SSI-related content. Many of the student projects begin as local issues that impact them in personal ways, and then become amplified to broader environmental contexts. What is quite interesting in this work, is the emphasis on problematizing instruction within SSI contexts in a manner that students begin to understand and conceptualize the products and goods created by science and technology as "…elements of complex global networks of living, nonliving and symbolic entities largely serving interests of relatively few pro-capitalist individuals and groups while creating various problems for individuals (e.g., poor labour conditions), societies (e.g., wealth disparities) and environments (e.g., species losses)." (Bencze et al., 2020b, p. 68). It is not difficult to surmise that both the STEPWISE and SSI frameworks are synergistic in advocating for developing and understanding social and environmental justice.

Additional research have also endorsed Vision III-type goals connected to SSI to better position science education in having a transformative effect for the students we teach. The transformative power of science education is situated in the secular concept of "Bildung" which originates from a German tradition of constructing one's character and has spread to other European nations. Bildung may be best thought of as building a capacity for self-determination through the cultivation of epistemic knowledge and moral maturity (Herdt, 2019). It begins within the self and moves outward towards family, community, humanity and universal principles owed to future generations. Achieving those universal ends (i.e., what *ought* to be) requires the kind of socio-political activism congruent with Vision III. Active citizenship, critical thinking and empowerment are essential to a healthy democracy. Our educational institutions have a prima facia directive to ensure that students are able to explore such options (Alsop & Bencze, 2014; Malmberg & Urbas, 2019; Sjöström & Eilks, 2016; Sjöström et al., 2016, 2017; Sjöström & Rydberg, 2018).

5.5 Moral Inquiry in the Context of Socioscientific Issues: School-Based Emphasis

Much of the research on place-based SSI was built on studies in formal school settings that sought, through an SSI responsive curriculum, to advance students reasoning by tapping emotive constructs. For example, Fowler et al. (2009) were able to demonstrate the impact of an SSI-focused course in Anatomy and Physiology over an academic year that showed significant gains in moral sensitivity in SSI-treatment groups over comparison groups. Moral sensitivity entails the ability to identify with a given moral issue, derive the possible moral consequences contained within different decisions, and recognize how one's actions may impact others. Moral sensitivity is also related to an individual's propensity to engage with moral elements contained within an SSI and aligns with the work of moral psychologists who have posited moral emotions as a center-piece to moral reasoning and moral education (Belenky et al., 1986; Berkowitz, 1997; Eisenberg, 2000; Sadler & Zeidler, 2005). It is also intricately connected to the ability to humanize (empathize) with others who are not directly connected to us (Suazo et al., 2020). Hence, rather than being distant, removed, amoral and remote from a situation (i.e., coldly analytic), we advance the notion of being "warmly analytic" which, enacted, entails becoming morally sensitive to the care and well-being of others in our decision-making process. In this case of the Fowler et al. (2009) study, students' identification and articulation of moral concerns in the context of SSI became more sophisticated over the academic year. (It is important to note that ANCOVA was used to control for any initial differences between treatment and comparison groups).

It is also important to recognize that the cognitive elements of reasoning can work in tandem with affective elements (e.g., moral sensitivity) in the context of moral issues, to humanize what might otherwise be non-personal scenarios and produce prosocial behaviors resulting in moral actions (Eisenberg et al., 2015; Hoffman, 2008; Kraaijeveld et al., 2021; Suazo et al., 2020). Table 5.2 depicts the interaction of rationalistic and emotive reasoning directed towards both personal and societal orientations. Indeed, in the context of education settings, it strains credulity in our pedagogical practice to think that emotive reasoning, particularly in the context of moral inquiry, would not be considered with respect to the role that moral sensitivity and moral compassion play in our students' epistemic beliefs.

The interaction of cognitive and affective factors is also embedded in the socio-scientific perspective taking (SSPT) framework developed by Kahn and Zeidler (2019) to augment socioscientific reasoning (SSR). Implementing a conceptual analysis, we identified a construct (SSPT) as one that requires engagement with other persons, groups, communities' perspectives or circumstances, an etic/emic shift in positionality (or viewpoints) and embedding its analysis in a moral context guided by conscience. Figure 5.3 illustrates the interaction among cognitive and affective factors when applied to the kind of reasoning expected to occur within classrooms, or life experiences in naturalistic settings, while engaging in SSR under

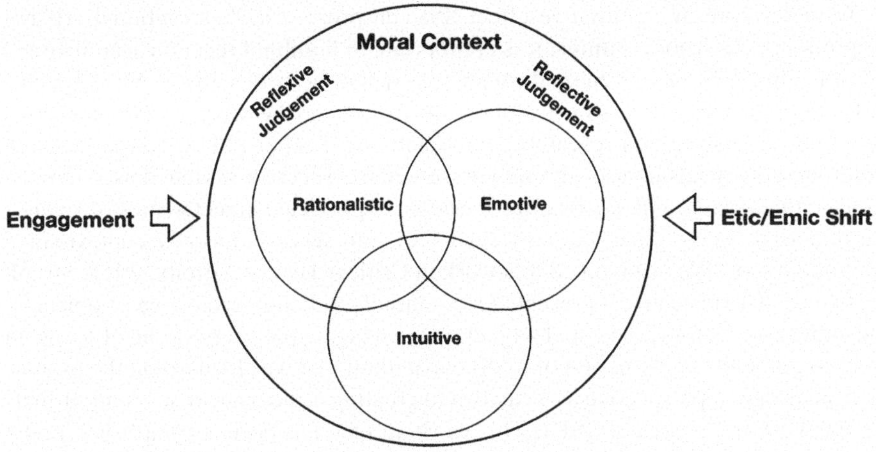

Fig. 5.3 Conceptual framework of socioscientific perspective taking

the SSI framework. From a practical standpoint, it matters not whether the source of reasoning resides within a particular domain or might be more fluid crossing membranes between domains. What matters more is that the exercise of *reflexive judgement* (evaluating one's own actions/decisions according to an interior bar of integrity) and *reflective judgement* (evaluating one's own actions / decisions according to social norms) may have its day because of one's personal engagement with the issue at-hand, and the inclination to a shift perspectives from that of an "outsider" to that of an "insider" and vice versa, as many times as it takes to formulate a resolution to the issue.

Newton and Zeidler (2020) provided empirical support for the effective use of SSPT in fostering social perspectives and facilitating dialogic engagement and reasoning. Whereas Kahn and Zeidler (2016) have supported the proposed the HARTSS (**H**umanities, **Art**s, and **S**ocial **S**ciences) model to promotes perspective taking from different disciplines which is congruent with an SSI approach. The findings of Newton and Zeidler also provide support for the efficacy of the underlying approach as well. For example, students in a place-based setting were expected to read background material related to wolf management provided by various stakeholders, such as livestock ranchers, environmental advocates, indigenous people, eco-tourism businesspeople, hunter/sportsmen, wildlife biologists, and ecologists as they examined varied claims and evidence underlying these perspectives before they argued the issues. (Note: While other work by Herman et al. (2020, 2021) has examined this topic in place-based settings (see above), this research was carried out in the classroom with the contribution of guest speakers in a carefully designed scaffolded curriculum.) Results indicated that students were able to change, shift, and represent multiple perspectives ranging from superficial changes to deep structural changes, which were attributed to the main parameters of the Kahn and Zeidler model (engagement, moral context, and etic / emic shifts).

To underscore the facilitative effect SSI can provide in "conventional" (if not progressive) classroom settings, it is appropriate to highlight recent research aimed at cultivating critical awareness through socioscientific perspectives. More specifically, this current scholarship results from extending the work of Nolan (2021) based on his dissertation research, and Nolan and Zeidler (2024) where research fostering critical awareness through informed perspective taking, based on the premise that sociomoral, sociocultural, and sociopolitical characteristics of science are important to integrate into curricular foci, can serve to morally contextualize scientific issues. We instantiated this work in Critical Theory (Giroux, 2011; Strunk & Betties, 2019) because it presented a potentially holistic perspective of objective and subjective factors associated with climate change—the course topic of focus, in order to guide the construction of a SSI curriculum that was focused on the promotion and usage of perspective as a catalyst to creating critical awareness and furthering the work in the area of SSPT. This work, in fact, has been accomplished in the context of a laboratory-focused setting where post-secondary students were examining and investigating issues related to climate change. Students were exposed to data-based reasoning through field methods like using geospatial technology to create heat maps from infrared data they collected on outdoor surfaces of the campus, critical reflection by conducting climate surveys with peers from natural science and social-science majors, weekly classroom discussions and writing assignments to voice and synthesize their evolving perspectives over the 16-week course.

We blended the conceptual framework of the Four Quadrants model (Marquis, 2008; Stein, 2019; Wilber, 2000), which is derived from Integral Theory (Esbjörn-Hargens, 2006; Wilber, 1995) with the framework of SSPT (describe above) to guide our pedagogical approach to instruction. This approach afforded the opportunity to cross discipline boundaries and be adaptive to multiple perspectives in educational contexts. The Four Quadrant model (Fig. 5.4) depicts a top row representing

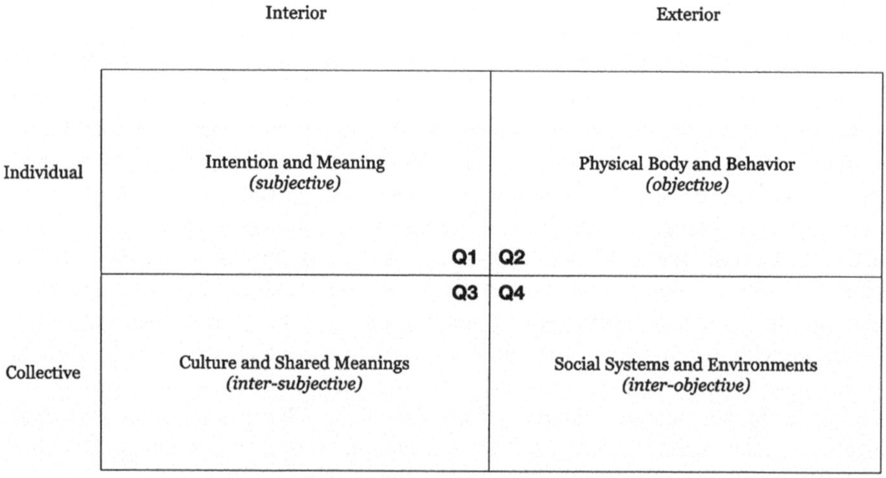

Fig. 5.4 Four quadrants model

interior (internal) and exterior (external) domains of consciousness of an individual, while the bottom row represents the interior and exterior domains of any given collective (group).

Each of the Quadrants in the model provide for educational experiences that intersect an individual's emic/etic shifts relative to internal and external domains of consciousness. Quadrant 1 is focused on an individual's interior and subject self that is aligned with contemplative, critical thinking—in short, developing a sense of self and self-awareness. Quadrant 2 represents an individual's exterior and objective self where one "acts" in the world and the nature of interactions with others. The focus is on being skillful and practical where conceptual beliefs, affective and behavioral intentions are realized in overt behaviors and actions with others or things, and understanding how the nature of those interactions affect a sense of self and their impact on others or parts of the environment. Quadrant 3 depicts how inter-subjective exchanges between the self in relation to others relative to the generation of shared understandings of meanings, ethics, values and worldviews. In an educational sense, its aim is the immersion into an educational culture of co-constructing knowledge through joint participation. It is systems thinking in nature but contained and directed within one's interpretation of their role or place within social systems. It requires an educational culture where ethical participation and perspectival connections with others are reflected upon. Finally, in Quadrant 4, represents the inter-objective area where social systems cross paths or interact with external environmental systems. It is this space where systems thinking is required to gauge collective impacts on the physical and biological world. From an educational perspective, experiences that develop a shared understanding of global dynamism, social sustainability, and ecological flourishing would fit here.

The integral nature of this blended approach (Four Quadrant Model with SSPT) can be illustrated in Fig. 5.5. This figure demonstrates how pedagogical themes could be (and were) developed over a semester by providing opportunities in the laboratory-based course on Climate Change to be examined and explored through investigative opportunities. Most importantly, it provided a holistic approach to moral inquiry through consideration of socioscientific issues. Nolan (2021) points out that this approach is not only consistent with a growing number of scholars who advocate for more holistic approaches to science education (e.g., Atabey & Topçu, 2017; Bencze et al., 2012; Herman, 2014; Liu et al., 2010; Rundgren & Rundgren, 2010; Zeidler, 2016; Zouda, 2018), but that it also aims at an equitable approach to fostering scientific identity.

Something significant occurred when the blended approach of the Four Quadrants Model was integrated with the SSPT model as students investigated factors related to climate change. It allowed for the attributes of SSPT (as noted in Fig. 5.5) to be expressed through authentic experiences elicited by each Quadrant as a kind of epistemic insight. Table 5.3 clearly conveys how the experience of sensemaking through reflective and reflexive moral reasoning created a holistic understanding of epistemic awareness.

A typical "push-back" to engage in moral inquiry via an SSI approach is the fact that teachers may question whether partaking in such practice can be effectively

	Interior	Exterior
Individual	• What does the thought of climate change mean to me? • Do I feel confident I understand the language of climate change? • What sense of morality do I hold toward climate change? • Do I feel a sense of responsibility? • Do I naturally want to engage or avoid discussing climate change? Q1	• What behaviors of mine contribute to climate change? • What behaviors of mine alleviate climate change? • Do I talk about climate change with others? • What is my carbon footprint? • How do I travel? What is the carbon footprint of my various modes of transportation? Q2
Collective	Q3 • What values do my family and community hold toward climate change? • How is climate change communicated within my direct surroundings? • How were the laws and policies on climate change determined by my community/city/state/nation? • Do you and your culture acknowledge what scientists say about climate change?	Q4 • What is the carbon footprint of other countries? • How is climate change affecting other parts of the world? • Which laws and policies govern climate change around the world?

Fig. 5.5 Climate change educational experiences conceptualized with the four quadrants and SSPT models

implemented in an age of accountability. Put differently, is it possible to attend to standards-based objectives aimed towards science achievement while teaching with a focus on SSI-related pedagogy? While SSI research has demonstrated gains in scientific content understanding via a sociocultural approach to contextualized instruction (Sadler et al., 2016), such research does little to assuage the fears of teachers who need to balance the mandates of US state or national accountability movements, since high-stakes achievement tests tend to be a reference point as the main metric for student success (Anderson, 2012; DeBoer et al., 2008). Even some international movements, such as the Programme for International Student Assessment (PISA) (OECD, 2019), while overtly rejecting a Vision I conceptualization of scientific literacy, falls short of overtly supporting the full aims of SSI in its assessment of supporting the goals of the SSI framework (Sadler & Zeidler, 2009). Research by Wikman (2023) was aimed to specifically address this issue, investigating the effect of subsuming standards-based objectives within the SSI framework on science achievement. Table 5.4 shows how the 12 units that spanned the academic year were guided by overarching questions that connected to moral inquiry elements of the curriculum. While some of these questions are global in nature, it is important to recognize the deliberate attempt to engage students in issues that were local in proximity (in this case, Florida) to the direct impact on students' lives (e.g., items 8 and nine).

Additional significant factors that added to the efficacy of this study included:

- Fourteen traditional high schools, 44 environmental science teachers, 5000 students, diverse racial population from which data was collected.

Table 5.3 Relationships between socioscientific perspective taking attributes and quadrant context

SSPT attribute	Quote exemplars	Quadrant context
Moral context		
Rationalistic (*role taking*)	I feel that I personally cannot lessen the effects of climate change, that's just ridiculous. I would need to be able to have a very wide-reach and influence to be able to lessen the effects of climate change. I feel that no matter what I do, I am contributing towards the effects of climate change, just because I am not "living off the land". I would love to be able to have this alleged influence, but that's what I am in college for, I guess	Q1, Q2 Individual reflection on interiority of self-thinking, and exteriority of actions
Emotive (*empathy*)	I think that I am rather moral with climate change. I think that we should be able to acknowledge that the effects of climate change are not equitable, meaning that the people who contribute to it the least are going to feel the effects of it the most, which is terrible	Q1, Q4 Individual interior reflection on moral judgement and exterior conditions impacting others
Intuitive (*"gut" feeling* or *deep sense*)	This class was not what I expected it to be for I thought it was going to be more science-oriented, but I feel what I gained from this class is better than what I could have gotten from it being strictly science-oriented	Q1 Individual interior reflection on initial feeling or perception of the issue
Etic/emic shift	Whenever I used to think about climate change, I always thought that it was strictly a scientific issue for I only learned about climate change in science-related classes. However, climate change is more than that, it is truly an issue that effects quite literally everything and requires people of all educational and professional backgrounds in order to help mitigate the effects of it	Q1, Q3, Q4 Individual interior reflection on initial understanding of issue with a shift towards new social perspectives and cultural alignments (i.e., shift from scientific communities to multi-disciplinary communities)
Engagement with others	I would say that my general mood toward climate change is hopeful. I think that it's terrible and all that, but it's sort of like the thing that could unite the world. If we are able to see climate change as the biggest threat to our livelihood, then we could all band together to help mitigate it	Q1, Q3, Q4 Individual interior reflection on hope, collective exterior recognition of potential to unite the world, and cultivation of collective interior values

– Target-task alignment of State Benchmarks for specific scientific concepts.
– District-wide training of the SSI framework.
– Multi-level assessment using proximal and distal measures, classroom observations, standardized interview rubrics, and semi-structured interviews.

Wikman found that science achievement for students was significantly positively affected when standardized objectives were subsumed with the SSI curricula. She was also able to parse out pedagogical differences among the fidelity of higher, middle, and lower SSI implementation teachers. Additional analysis also supported

Table 5.4 Environmental science SSI units, science content, and overarching questions (Wikman, 2023, p. 90)

Unit content	Overarching questions
1. Critical thinking and discourse about the environment	Is human self-interest the driving force behind all environmental problems?
2. Biodiversity and evolution	Should we eradicate invasive species?
3. Ecosystem dynamics and interconnectedness of living things	Should we save endangered species, whatever the cost?
4. Sustaining biodiversity, an ecosystem approach	Should we sustain Earth's biodiversity, whatever the cost?
5. Human Population Dynamics and Urban Sprawl? (human population dynamics and growth)	Should family size be limited?
6. Water resources and pollution)	Is access to clean water a human right? If so, who is responsible for assuring that it is available to all?
7. Food, soil, and pest management)	Should we all be vegetarians, at least part-time?
8. Nonrenewable mineral resources	Should phosphate mining be expanded in Florida?
9. Renewable and nonrenewable energy sources	Should hydraulic fracturing be allowed in the Florida Everglades?
10. Air pollution, climate change, and ozone depletion	Do my actions affect climate change?
11. Solid and hazardous waste	Should we be charged, based on the weight and content of our garbage, for its disposal?
12. Sustainability, economics, and equity	Can I really make a difference?

the claim that the resultant SSI curricula significantly impacted the development of students' character and values as global citizens, consistent with previous research (e.g., Herman et al., 2021; Lee et al., 2012, 2013). This work supports the idea that moral inquiry through the SSI framework can have its day (or in this case, have its year) even in the context of standards-driven school settings.

5.6 Conclusion

A moral inquiry into epistemic insights through socioscientific issues is not an overly idealistic notion. The extant literature, both what I have presented through my own experiences with research and practice, as well as the perspectives of scholars from around the globe, demonstrate that a contemporary vision of scientific literacy can be realized through the deliberate decisions of science educators who wish to spread eudaimonia. This is not to say that such a commitment is not without its difficulties. Challenges faced by teachers (Bossér et al., 2015; Kinskey & Zeidler, 2020; Lee & Yang, 2019; Lindahl et al., 2019) are recognized in the literature, but not insurmountable particularly in light of the potential impact on the quality of students' educative experiences, and on the lives of students beyond those experiences. Some authors (Bayram-Jacobs et al., 2018), for example, have offered ideas as to build and increase science teachers pedagogical content knowledge (PCK) relative to SSI instruction through the enactment of SSI curriculum. Their study included 30 teachers from four countries (Cyprus, Israel, Norway, and Spain), and demonstrated that while subject matter knowledge is necessary for the strong development of PCK, it is interaction of subject matter knowledge and SSI goals that are important for the robust development of PCK in the context of SSI teaching. Their findings also reaffirm the claim that increased emphasis in subject matter knowledge does not necessarily affect the quality of science instruction; rather, it is more likely that the *act* of teaching influences both subject matter knowledge as well as genesis of PCK. (Zeidler, 2002). Others (Bossér, 2023) have pointed out that tensions may arise between students' dependency on conventional pedagogy and teachers' desire to promote an exploratory classroom, but that could be overcome by supporting students with mutual trust and clear expectations thereby supporting functional scientific literacy.

As the world becomes more interconnected, preparing citizens to become engaged citizens having the ability to locate and evaluate evidence, make informed decisions while considering diverse perspectives and tempered by compassion about the world they inhabit, becomes tantamount to promoting Vision III scientific literacy. One example, consistent with the aims of this book, is the advancement of science education among numerous European nations that is focused on science education for responsible citizenship (European Commission, 2015). It is heartening to see the progress made in different countries around the globe toward that end, as evidenced by the scholars that contributed to this book. Science education in the collective global educational community must move toward a collective vision of cultivating epistemic insight, while understanding the sociocultural factors that marks off the uniqueness of the students we teach.

A sociocultural understanding of science education requires that we do not become entrenched, siloed and compartmentalized in our quest for engaging students with the aim of achieving functional scientific literacy. For science education to be transformative, pedagogical practices must allow for educationally-rich experiences that bring about a deeper shift in epistemic beliefs. The sociocultural view of science education requires that we elevate any form of knowledge if doing so can prudentially impact our students. This means we need not restrict our teaching to matters of the quadrivium, but we open our minds to tap matters of the trivium in creating a value-added approach to science education (Zeidler, 2016). This echoes the same appeal that Greene & Green (1982) advanced decades ago (the same year I finished my dissertation), when they entertained the question "Can we be scientific about policy research?" At that meeting of the American Association for Educational Research, they stated:

> Our belief is that education, educational research, and educational practice, however technically successful in forming a public, however guided by scientific insight, is dangerous and puts us all at risk if it is uninformed by the arts, by literature, and by attentiveness to the

> kind of social memory that is being framed. Polic research, for example, uninformed by The Federalist Papers, by the words of Shakespeare and Abraham Lincoln, is rootless research whatever its excellence by scientific standards. And research guided only by fact—by a knowledge of what already is—instead of by an image of what might be, but never was, is barren and dangerous. The arts, the texts we share, the criticism of our common memory—these are central to the many publics of educational practice and research. (1982, pp. 19–20)

My argument is that at the very *least*, we have a *pro tanto* duty to exercise prudential teaching in ways that we know can engage students in moral inquiry about the social and natural world that they inhabit. This is a matter of providing a kind of rootedness in scientific identity for students. At the very *most*, my claim is that if we wish to cultivate students who will become engaged citizens willing to act in varied capacities consistent with Vision III of scientific literacy, and are imbued with a sense of care, conscious and critical thinking, then we have a moral imperative to design coursework, policy and pedagogical practices in a manner congruent with the facilitative nature of the SSI framework.

References

Alsop, S., & Bencze, L. (2014). Activism! Toward a more radical science and technology education. In L. Bencze & S. Alsop (Eds.), *Activist science and technology education* (pp. 1–19). Springer.

Anderson, K. J. B. (2012). Science education and test-based accountability: Reviewing their relationship and exploring implications for future policy. *Science Education, 96*(1), 104–129.

Arendt, H. (1958). *The human condition*. Chicago University of Chicago Press.

Aristotle (2002). *Nicomachean ethics*. Rowe, C. (trans.) and Broadie, S. (Ed.), Oxford University Press.

Atabey, N., & Topçu, M. S. (2017). The development of a socioscientific issues-based curriculum unit for middle school students: Global warming issue. *International Journal of Education in Mathematics, Science and Technology, 5*(3), 153–153.

Bayram-Jacobs, D., Henze, I., Evagorou, M., Shwartz, Y., Aschim, E. L., Alcaraz-Dominguez, B., & M., & Dagan, E. (2018). Science teachers' pedagogical content knowledge development during enactment of socioscientific curriculum materials. *Journal of Research in Science Teaching, 56*, 1207–1233.

Belenky, M. F., Clinchy, B. M., Goldberger, N. R., & Tarule, J. M. (1986). *Women's ways of knowing: The development of self, voice, and mind*. Basic Books.

Bencze, J. L. (2017). *Science & technology education promoting wellbeing for individuals, societies, & environments*. Springer.

Bencze, J. L., & Alsop, S. (Eds.). (2014). *Activist science & technology education*. Springer.

Bencze, L., & Krstovic, M. (2017). Science students' ethical technology designs as solutions to socio-scientific problems. In *Science and technology education promoting wellbeing for individuals, societies and environments* (pp. 201–226). Springer.

Bencze, L., Sperling, E., & Carter, L. (2012). Students' research-informed socioscientific activism: Re/Visions for a sustainable future. *Research in Science Education, 42*(1), 129–148.

Bencze, J. L., El Hawany, S., & Zouda, M. (2020a). Critical and active public engagement in addressing socioscientific problems through science teacher education. In M. Evagorou, J. A. Alexis, & J. Dillon (Eds.), *Science teacher education for responsible citizenship: Toward a pedagogy for relevance through socioscientific issues* (pp. 62–83). Springer.

Bencze, J. L., Pouliot, C., Pedretti, E., Simonneaux, L., Simonneaux, J., & Zeidler, D. L. (2020b). SAQ, SSI & STSE education: Defending and extending 'Science-in-Context'. *Cultural Studies in Science Education, 15*, 825–851. https://doi.org/10.1007/s11422-019-09962-7

Berkowitz, M. W. (1997). The complete moral person: Anatomy and formation. In J. M. DuBois (Ed.), *Moral issues in psychology: Personalist contributions to selected problems* (pp. 11–41). University Press of America.

Billingsley, B. (2017). Teaching and learning about epistemic insight. *School Science Review, 98*(365), 59–64.

Billingsley, B., Nassaji, M., Fraser, S., & Lawson, F. (2018). A framework for teaching epistemic insight in schools. *Research in Science Education, 48*, 1115–1131. https://doi.org/10.1007/s11165-018-9788-6

Blake, B., & Pope, T. (2008). Developmental psychology: Incorporating Piaget's and Vygotsky's theories in classrooms. *Journal of Cross-Disciplinary Perspectives in Education, 1*(1), 59–67.

Bossér, U. (2023). Transformation of school science practices to promote functional scientific literacy. *Research in Science Education*. https://doi.org/10.1007/s11165-023-10138-1

Bossér, U., Lundin, M., Lindahl, M., & Linder, C. (2015). Challenges faced by teachers implementing socio-scientific issues as core elements in their classroom practices. *European Journal of Science and Mathematics Education, 3*(2), 159–176. https://doi.org/10.1007/s11165-017-9627-1

DeBoer, G. E., Lee, H. S., & Husic, F. (2008). *Designing coherent science education: Implications for curriculum, instruction, and policy*. Columbia University Teachers College Press.

diSessa, A. A. (2006). A history of conceptual change research: Threads and fault lines. In K. Sawyer (Ed.), *Cambridge handbook of the learning sciences* (pp. 265–281). Cambridge University Press.

Durlak, J. A., Weissberg, R. P., Gymnicki, A. B., Taylor, R. D., & Schellinger, K. B. (2011). The impact of enhancing students' social and emotional learning: A meta-analysis of school-based universal interventions. *Child Development, 82*, 405–432.

Eisenberg, N. (2000). Emotion, regulation, and moral development. *Annual Review of Psychology, 51*, 665–697.

Eisenberg, N., Spinrad, T. L., & Knafo-Noam, A. (2015). Prosocial development. In M. E. Lamb & C. Garcia-Coll (Eds.), *Handbook of child psychology and developmental science* (pp. 610–658). John Wiley & Sons.

Ennis, R. (1996). *Critical thinking*. Prentice-Hall.

Ennis, R. H. (2018). Critical thinking across the curriculum: A vision. *Topoi, 37*(1), 165–184. https://doi.org/10.1007/s11245-016-9401-4

Esbjörn-Hargens, S. (2006). Integral teacher, integral students, integral classroom: Applying integral theory to education. *Journal of Integral Theory and Practice, 2*(2), 1–42.

European Commission, (2015). *Directorate-General for research and innovation, science education for responsible citizenship—Report to the European Commission of the expert group on science education*, Publications Office. https://data.europa.eu/doi/10.2777/12626

Falk, J. H., & Dierking, L. D. (2018). *Learning from museums* (2nd ed.). Rowman & Littlefield Publishing Group, Inc.

Fowler, S. R., Zeidler, D. L., & Sadler, T. D. (2009). Moral sensitivity in the context of socioscientific issues in high school science students. *International Journal of Science Teacher Education, 31*(2), 279–296.

Giroux, H. A. (2011). *On critical pedagogy*. Bloomsbury.

Green, J. A., Sandoval, W. A., & Bråten, I. (2016). *Handbook of epistemic cognition*. Routledge.

Greene, M., & Green, T. F. (1982). 1982 annual meeting program. *Educational Researcher, 11*(2), 4–27.

Herdt, J. A. (2019). *Forming humanity: Redeeming the German Bildung tradition*. University of Chicago Press.

Herman, B. C. (2014). The influence of global warming science views and sociocultural factors on willingness to mitigate global warming. *Science Education, 99*(1), 1–38.

Herman, B. C. (2018). Students' environmental NOS views, compassion, intent, and action: Impact of place-based socioscientific issues instruction. *Journal of Research in Science Teaching, 55*(4), 600–638.

Herman, B. C., Zeidler, D. L., & Newton, M. (2020). Students' emotive reasoning through place-based environmental socioscientific issues. *Research in Science Education, 50*(3), 2081–2109. https://doi.org/10.1007/s11165-018-9764-1

Herman, B. C., Newton, M. H., & Zeidler, D. L. (2021). Impact of place-based socioscientific issues instruction on students' contextualization of socioscientific orientations. *Science Education, 105*(4), 585–627. https://doi.org/10.1002/sce.21618

Hodson, D. (2011). *Looking to the future: Building a curriculum for social activism.* Sense Publishers.

Hodson, D. (2014). Becoming part of the solution: Learning about activism, learning through activism, learning from activism. In L. Bencze (Ed.), *Activist science and technology education* (pp. 67–98). Springer.

Hodson, D. (2021). Going beyond STS education: Building a curriculum for sociopolitical activism. *Canadian Journal of Science Mathematics and Technology Education.* (Online) https://doi.org/10.1007/s42330-020-00114-6.

Hoffman, M. L. (2008). Empathy and prosocial behavior. In M. Lewis, J. M. Haviland-Jones, & L. Feldman (Eds.), *Handbook of emotions* (pp. 440–454). The Guilford Press.

Hominh, Y. (2021). Dewey and the tragedy of the human condition. *The Pluralist, 16*(1), 26–40. https://www.muse.jhu.edu/article/784086

Hume, D. (1777/1912). *An enquiry concerning the principles of morals.* E-text of a 1912 reprint of the 1777 edition. Download from: https://www.gutenberg.org/files/4320/4320-h/4320-h.htm

Kahn, S., & Zeidler, D. L. (2019). A conceptual analysis of perspective taking: Positioning a tangled construct within science education and beyond. *Science & Education, 28*, 605–638.

Kim, G., Ko, Y., & Lee, H. (2020). The effects of community-based socioscientific issues program (SSI-COMM) on promoting students' sense of place and character as citizens. *International Journal of Science and Mathematics Education, 18*(3), 399–418.

Kinskey, M., & Zeidler, D. L. (2020). Elementary preservice teachers' challenges in designing and implementing socioscientific issues-based lessons. *Journal of Science Teacher Education, 32*(3), 350–372. https://doi.org/10.1080/1046560X.2020.1826079

Kraaijeveld, M. I., Schilderman, J., & (Hans), van Leeuwen E. (2021). Moral sensitivity revisited. *Nursing Ethics, 28*(2), 179–189. https://doi.org/10.1177/0969733020930407

Kristjansson, K., & Pollard, D. (2021). *Phronesis: Using an Aristotelian model as a research tool.* Research report. University of Birmingham: The Jubilee Centre for Character & Virtues. ISBN: 9780704429729.

Kristjansson, K., Darnell, C. Fowers, B., Moller, F., & Pollard, D. (2020). *Phronesis: Developing a conceptualization and an instrument.* Research report. University of Birmingham: The Jubilee Centre for Character & Virtues. ISBN: 9780704429680.

Kuhn, D. (2019). Critical thinking as discourse. *Human Development, 62*(3), 146–164. https://doi.org/10.1159/000500171

Kural, M., & Kocakulah, M. S. (2016). Teaching for hot conceptual change: Towards a new model, beyond the cold and warm ones. *European Journal of Education Studies, 2*(8), 1–40.

Kurdyavtsev, A., Stedman, R. C., & Krasny, M. E. (2012). Sense of place in education. *Environmental Education Research, 18*(2), 229–250.

Lee, J. J., & Francesca, G. (2018). In search of moral equilibrium: Person, situation, and their interplay in behavioral ethics. In K. Gray & J. Graham (Eds.), *Atlas of moral psychology* (pp. 475–484). The Guilford Press.

Lee, H., & Yang, J. (2019). Science teachers taking their first steps toward teaching socioscientific issues through collaborative action research. *Research in Science Education, 49*(1), 51–71. https://doi.org/10.1007/s11165-017-9614-6

Lee, H., Chang, H., Choi, K., Kim, S. W., & Zeidler, D. L. (2012). Developing character and values for global citizens: Analysis of pre-service science teachers' moral reasoning on socioscientific issues. *International Journal of Science Education, 34*(6), 925–953.

Lee, H., Yoo, J., Choi, K., Kim, S., Krajcik, J., Herman, B., & Zeidler, D. L. (2013). Socioscientific issues as a vehicle for promoting character and values for global citizens. *International Journal of Science Education, 35*(12), 2079–2113.

Lenman, J. (2007). What is moral inquiry? *Aristotelian Society Supplementary Volume, 81*(1), 63–81. https://doi.org/10.1111/j.1467-8349.2007.00151.x

Levinson, R. (2018). Realizing the school science curriculum. *The Curriculum Journal, 29*(4), 522–537.

Lindahl, M. G., Folkesson, A.-M., & Zeidler, D. L. (2019). Students' recognition of educational demands in the context of a socioscientific issues curriculum. *Journal of Research in Science Teaching, 56*(9), 1155–1182. https://doi.org/10.1002/tea.21548

Liu, S. Y., Lin, C. S., & Tsai, C. C. (2010). College students' scientific epistemological views and thinking patterns in socioscientific decision making. *Science Education, 95*(3), 497–517.

Malmberg, C., & Urbas, A. (2019). Health in school—Stress, individual responsibility and democratic politics. *Cultural Studies of Science Education, 14*, 863–878. https://doi.org/10.1007/s11422-018-9882-0

Marquis, A. (2008). *The Integral intake: A guide to comprehensive idiographic assessment in Integral Psychotherapy*. Routledge.

Nadelson, L. S., Heddy, B. C., Jones, S., Taasoobshirazi, G., & Johnson, M. (2018). Conceptual change in science teaching and learning: Introducing the dynamic model of conceptual change. *International Journal of Educational Psychology, 7*(2), 151–195. https://doi.org/10.17583/ijep.2018.3249

Newton, M., & Zeidler, D. L. (2020). Developing socioscientific perspective taking. *International Journal of Science Education, 42*(8), 1302–1319. https://doi.org/10.1080/09500693.2020.1756515

Nolan, E. T. (2021). *Socioscientific perspective taking in a four quadrant-based curriculum on climate change*. Doctoral dissertation, Northern Arizona University.

Nolan, E.T., & Zeidler, D.L. (2024). Cultivating critical awareness through socioscientific perspectives. *Disciplinary and Interdisciplinary Science Education Research*. (Under Review).

Owens, D. C., Sadler, T. D., & Zeidler, D. L. (2017). Controversial issues in the science classroom. *Phi Delta Kappan, 99*(4), 45–49.

Rawls, J. (1971). *A theory of justice*. Belknap Press of Harvard University Press. https://doi.org/10.4159/9780674042605

Raz, J. (1988). *The Morality of Freedom*. Oxford University Press.

Rundgren, S.-N., & Rundgren, C.-J. (2010). SEE-SEP: From a separate to a holistic view of socioscientific issues. *Asia-Pacific Forum on Science Learning and Teaching, 11*(1), 1–24.

Sadler, T. D. (2009). Situated learning in science education: Socio-scientific issues as contexts for practice. *Studies in Science Education, 45*(1), 1–42.

Sadler, T. D., & Zeidler, D. L. (2005). Patterns of informal reasoning in the context of socioscientific decision-making. *Journal of Research in Science Education, 42*, 112–138.

Sadler, T. D., & Zeidler, D. L. (2009). Scientific literacy, PISA, and socioscientific discourse: Assessment for progressive aims of science education. *Journal of Research in Science Teaching, 46*, 909–921.

Sadler, T. D., Romine, W. L., & Topçu, M. S. (2016). Learning science content through socioscientific issues-based instruction: A multi-level assessment study. *International Journal of Science Education, 38*(10), 1622–1635.

Sjöström, J., & Eilks, I. (2016). Reconsidering different visions of scientific literacy and science education based on the concept of Bildung. In J. Dori, Z. Mevarech, & D. Bake (Eds.), *Cognition, metacognition, and culture in STEM education*. Springer.

Sjöström, J., & Rydberg, C. (2018). Towards transdisciplinary Didaktik: Didactic modelling of complex controversial issues teaching for reflexive Bildung and sustainability. In *Building bridges across disciplines for transformative education and a sustainable future*. Shaker Verlag.
Sjöström, J., Eilks, I., & Zuin, V. G. (2016). Towards eco-reflexive science education: A critical reflection about educational implications of green chemistry. *Science & Education, 25*, 321–341. https://doi.org/10.1007/s11191-016-9818-6
Sjöström, J., Frerichs, N., Zin, V. G., & Eilks, I. (2017). Use of the concept of Bildung in the international science education literature, its potential, and implications for teaching and learning. *Studies in Science Education, 53*, 165–192. https://doi.org/10.1080/03057267.2017.1384649
Stein, Z. (2019). Education in a time between worlds: essays on the future of schools, technology, and society. .
Strunk, K. K., & Betties, J. S. (2019). Using critical theory in educational research. In K. K. Strunk & L. A. Locke (Eds.), *Research methods for social justice and equity in education*. Palgrave Macmillan. https://doi.org/10.1007/978-3-030-05900-2_6
Suazo, I., Pérez-Fuentes, M., Jurado, M., Martínez, A. M., Márquez, M., Martín, A. B., Sisto, M., & Linares, J. J. G. (2020). Moral sensitivity, empathy and prosocial behavior: Implications for humanization of nursing care. *International Journal of Environmental Research and Public Health, 17*(23), 8914. https://doi.org/10.3390/ijerph17238914
Timmons, M. (2002). *Moral theory: An introduction*. Rowman & Littlefield.
Wikman, K. (2023). *The contextualization of socioscientific issues in an age of accountability*. Doctoral dissertation, University of South Florida.
Wilber, K. (1995). *Sex, ecology, spirituality: The spirit of evolution*. Shambhala.
Wilber, K. (2000). *A brief history of everything* (2nd ed.). Shambhala Publications.
Zeidler, D. L. (2002). Dancing with Maggots and Saints: Past and future visions for subject matter knowledge, pedagogical knowledge, and pedagogical content knowledge in reform and science teacher education. *Journal of Science Teacher Education, 13*(1), 27–42.
Zeidler, D. L. (2016). STEM education: A deficit framework for the 21st century? A sociocultural socioscientific response. *Cultural Studies of Science Education, 11*(1), 11–26.
Zeidler, D. L., Lederman, N. G., & Taylor, S. C. (1992). Fallacies and student discourse: Conceptualizing the role of critical thinking in science education. *Science Education, 75*(4), 437–450.
Zeidler, D. L., Herman, B. C., Clough, M. P., Olson, J. K., Kahn, S., & Newton, M. (2016). Humanitas Emptor: Reconsidering recent trends and policy in science teacher education. *Journal of Science Teacher Education, 25*(5), 465–476.
Zouda, M. (2018). Issues of power and control in STEM education: A reading through the postmodern condition. *Cultural Studies of Science Education, 27*, 1–20.

Part II
Global Perspectives of the Socioscientific Issues Framework

Chapter 6
Is There a Place for Socioscientific Issues in Australian Secondary School Science?

Vaille Dawson

6.1 Introduction

Like many Western countries, the post Second World War scientific advances in space exploration led to calls for a scientifically educated workforce in Australia. Policy changes meant that schools were supported financially and materially to introduce a science curriculum to 'train' future scientists. This view of school science education as a means to prepare a future scientific workforce emphasised science content and processes with scant attention paid to the nature and role of science in society. Roberts (2007) described this type of science education as Vision I scientific literacy. SSI were absent from the curriculum.

Three decades later, in the 1990s, the 'science for all' movement emerged, based on an awareness of the need for a whole of society approach to address global issues such as population growth and pollution. This movement led to school science curricula that promoted science in society and real-life relevance. There were opportunities to include SSI. Roberts (2007) termed this approach Vision II scientific literacy. While it was recognised that everyone in society needed to be scientifically literate, the status of science content in schools remained strong. Student-centred and collaborative inquiry-based pedagogies underpinned by socio-constructivist theory were recommended but not widely implemented (Rennie et al., 2001). In 2024, it would be hoped that secondary school science was moving towards Vision III scientific literacy encompassing global citizenship, socio-ecojustice and activism, where SSI are the starting points for teaching (Siarova et al., 2019; Sjöström & Eilks, 2020).

V. Dawson (✉)
Graduate School of Education, University of Western Australia, Perth, Australia
e-mail: vaille.dawson@uwa.edu.au

D. L. Zeidler (ed.), *A Moral Inquiry into Epistemic Insights in Science Education*, Contemporary Trends and Issues in Science Education 61,
https://doi.org/10.1007/978-3-031-63382-9_6

My own interest in the teaching of SSI arose from my professional experience some 35 years ago. I was part of a medical research team researching bone marrow transplantation as a treatment for childhood leukaemia. In that role, I became acutely aware of the types of decisions parents had to make about their child's treatment options. The parents had limited understandings of the underlying science and how to make decisions. These observations, in part, led to my retraining as a science teacher, then continuing on to my Masters and PhD on introducing bioethics into school science. As a science teacher/researcher, I incorporated the four bioethical principles of autonomy, beneficence, non-maleficence and justice (Beauchamp & Childress, 1994) into decision-making about controversial issues in biology. My research and teaching was initially informed by Cross and Price (1992) through their book, *Teaching science for social responsibility.*

In the late 1990s, at a National Association for Research in Science Teaching (NARST) conference in the US, I attended a session by a PhD student who had used Toulmin's argumentation framework in her study (Toulmin, 1958). Toulmin's framework seemed to be a useful scaffold to teach decision-making about SSI in science. This realisation informed my subsequent research career on SSI, argumentation and decision-making. I continue to be inspired by contemporary colleagues (Dana Zeidler, Anat Zohar, Troy Sadler, Michael Reiss, and Jonathan Osborne) and the future generation (Jan Nielson, Sakari Tolppanen, and Peta White) of SSI researchers.

6.1.1 Policy, Curriculum and SSI

Australia has a Federal government that delegates and partially funds the six states and two territories to provide compulsory schooling for children aged 5–17. In the 1970s and 80s, education authorities in each of the states and territories developed and implemented their own curriculum. This led to problems when children moved from one state to another or wished to attend university in a different state. There were concerns that some children did not have equitable access to a quality education. In 1989, the Education Council which comprised the state, territory and federal education ministers agreed to develop a national curriculum for years 1–10 in which science was one of eight curriculum areas (Curriculum Corporation, 1994). There was a move away from a rigid science syllabus to a more flexible curriculum of science understandings, skills and values that acknowledged prior learning and catered for students of varying abilities. The emphasis was on what students learn rather than what they should be taught. However, when the Education Council met to approve the national curriculum, due to political reasons (there had been changes of state and federal governments) and public pressure (some commentators refuted the role of values in science) it was rejected and each state continued with its own curriculum implementation.

A decade later, a national curriculum authority, now called the Australian Curriculum Assessment and Reporting Authority (ACARA), was set up and a national science curriculum was developed informed by national reports by Rennie et al. (2001) and Tytler (2007). These reports had demonstrated that despite international and national educational research evidence about student-centred socio-constructivist approaches, the science taught in most Australian schools was content heavy, teacher-centred and focussed on 'future scientists'. The state-based curricula aligned with Roberts (2007) Vision 1 scientific literacy. In addition, decreasing numbers of students were selecting science subjects in the final years of schooling.

In 2008, the Melbourne Declaration on Educational Goals for Young Australians (MCEETYA, 2008) stated that "complex environmental, social, and economic pressures such as climate change that extend beyond national borders pose unprecedented challenges ... to meet these challenges, Australians must be able to engage with scientific concepts and principles, and approach problem-solving in new and creative ways" (p. 5). In response to the Melbourne Declaration, ACARA released a new Australian curriculum in science in which the rationale stated that the curriculum must develop "scientific knowledge, understandings and skills to make informed decisions about local, national and global issues" (ACARA, 2011, p. 1).

The national curriculum authority was tasked with developing a science curriculum for all young Australians embodying a Vision II scientific literacy where the aim of school science education was to enable all students to develop a deep understanding of the world around them, and to use their understanding of science to contribute to public debate and make balanced decisions about scientific issues that impact their lives. Students would have an appreciation of the nature, practice and limitations of science, along with sufficient knowledge of science content to make informed evidence-based decisions.

The new science curriculum has three strands (*science understanding*, *science inquiry* and *science as a human endeavour*) which are further divided into sub-strands. *Science as a human endeavour* highlights the role of decision-making and problem-solving in science in ways that take account of ethical and other processes. Each sub-strand has elaborations or examples of the content to be addressed. However, when this science curriculum was released in 2012, it was subject to criticism for including values and ethics in science and 'watering down' science content. There were political calls for a back-to-basics approach of content knowledge only. Some states decided to adapt rather than adopt the curriculum while others delayed implementation. Fensham (2013) is recommended for further reading about the history of the science curricula in Australia.

Against a backdrop of neoliberalism in politics, the science curriculum has been reviewed every couple of years (Savage, 2017). Increasing neoliberalism has resulted in teacher registration and monitoring, a mandated curriculum, and high stakes testing at the end of secondary school (Year 12). There is national testing (NAPLAN) in literacy and numeracy of all children at the end of years 3, 5, 7 and 9 plus international testing in PISA and TIMMS in science and maths. As school level results are published, the testing requirements resulted in a recalibration of teaching time to prepare for the tests.

Despite the testing regime, one of the seven aims of the current Version 8.4 of the science curriculum is "an ability to solve problems and make informed, evidence-based decisions about current and future applications of science while taking into account ethical and social implications of decisions" (ACARA, 2024a). It would thus be expected that there would be explicit opportunities for students to achieve this aim in the compulsory years of schooling (up to year 10) through learning about SSI.

Yet there is no mention of SSI in the science curriculum. There is a single reference to contemporary issues in the content description of the *science as a human endeavour* strand in year 7/8 stating 'Solutions to contemporary issues that are found using science and technology may impact on other areas of society and may involve ethical considerations' (ACARA, 2024b). There are elaborations such as 'examine how proposed scientific responses to contemporary issues may impact on society and explore ethical, environmental, social and economic considerations' that imply the possible teaching of SSI (ACARA, 2024b). However, elaborations are not mandated and are examples only.

What can we conclude from this? There is a potential role to teach about SSI through *science as a human endeavour* but it is implicit. In Australia, what is taught in schools is dictated by the Australian and state curriculum authority's, science textbooks, industry resources and teacher developed resources. In Western Australia, the School Curriculum and Standards Authority (SCSA) (2020) provides teachers with detailed programs and lesson plans with teaching and assessment activities that are predominantly for the *science understanding* strand.

Educational policy, national and state curriculum authorities and education sectors have not enabled the teaching of SSI. This is due, in part, to a content heavy science curriculum especially in the final years of schooling and large scale summative assessment practices (e.g., pre-university exams) that measure what is easily assessed. Furthermore, school science is taught as discrete subjects of biology, chemistry, physics and earth science. Within each year, each of the subjects are further divided into discrete topics such as atoms, forces and genetics. This segmentation of science content is at odds with the multidisciplinary underlying science of many SSI which may also incorporate cultural and Indigenous knowledges.

6.2 My SSI Journey

In this section I focus on four of my research findings on using SSI for teaching and learning science. They are: (1) introducing ethics into secondary school science classrooms, (2) decision-making about ethical issues, (3) improving argumentation and decision-making about SSI; and (4) the crucial role of the teacher.

As indicated in the introduction, my first career was in medical science where I worked on the use of bone marrow transplantation as a treatment for childhood leukaemia. A parent myself, I observed first-hand the difficult decisions that parents had to make regarding treatment options for their young children. The underlying science and treatment outcomes were uncertain. I wondered about whether school science might be able to help people to make decisions. I retrained as a secondary science teacher. I noted that science was often viewed as objective, analytical, rational and unaffected by morals and values. I was concerned that if science teachers ignored ethical dilemmas then students may receive the implicit message that there are no ethics in science (Dawson, 1996). As a result, I embarked on a Masters study to develop a transplantation technology unit on bioethics for high school students. The unit was student-centred using role plays, structured discussions and other co-operative activities that enabled students to reflect on their ethical values and those of others. I shared the resources I developed with other teachers at teacher conferences and subsequently developed similar resources on forensic science, genetically modified organisms and genetic disease testing.

The starting point for resource development was determining students' prior knowledge and attitudes (Dawson & Schibeci, 2003). My aim was to improve students' understandings and attitudes and develop their ethical decision-making skills about these new technologies. I compared the decision-making of secondary school students to those of expert bioethicists (Dawson & Taylor, 2000). I found that while experts used the four bioethical principles of autonomy, beneficence, non maleficence and justice, students made decisions that were naïve and idealistic. The students gave undue weighting to the principle of autonomy or the right to choose and tended to only consider the short-term consequences of decisions. I shared the findings with teachers at teacher education conferences and incorporated the findings into future resource development.

My colleague, Grady Venville, and I conducted a comprehensive case study in a year 10 genetics classroom where the teacher used genetic testing SSI to improve students' decision-making skills (Venville & Dawson, 2010). Decision-making was defined as the ability to make a decision and construct an argument. Students' understandings of genetics concepts were also measured. Using a quasi-experimental design, the students' genetics understandings improved significantly more than a comparison group of students taught the same genetics content without the SSI and argumentation lessons. The genetics study was replicated with other biology teachers and their students (Dawson & Venville, 2013). Using a similar pattern of first establishing prior knowledge, curriculum resources for other SSI such as climate change (Dawson, 2015) and water use (Dawson & Venville, 2022) were also successfully developed and implemented. Improvements in argumentation about SSI were demonstrated with low achieving students (Dawson & Carson 2020) and younger students (Dawson, 2023).

However, a significant improvement in students' argumentation skills was not seen in all classrooms. While possible explanations included the length of the intervention, type of SSI, activities and student characteristics, as a science teacher educator, I began to wonder whether the behaviours and strategies used by the teachers may be important. This realisation led to a renewed focus on teacher behaviour and dialogue in SSI and argumentation lessons. My most recent published research suggests there are key teacher behaviours associated with improved argumentation using SSI (Dawson, 2023). They are: (1) developing a safe classroom environment; (2) providing clear instructions about the structure of an argument; (3) providing opportunities for all students to contribute and practice argument construction; and, (4) gradually removing scaffolds that support argument construction.

Finally, I have been part of a six country consortium of ten science education researchers comparing the presence of climate change topics in national science and geography curricula (Dawson et al., 2022). When compared to other countries, in Australia, opportunities to teach climate change are implied and depend on the willingness of teachers. Topics taught tend to relate to the science such as the greenhouse effect, greenhouses gases, radiation, weather and climate rather than the political, ethical and economic aspects. Education stakeholders (teachers, professional development providers, and stakeholders) from five countries were interviewed about what should be taught and the inhibitors and enablers (Ben Zvi Assaraf et al., 2024). We found that there was strong support for a cross-curriculum approach to teaching climate change. It was recommended the governments provide leadership, professional development and resources to support the explicit teaching of climate change as an SSI.

6.3 Australian Science Education Research on SSI

The Australian government provides limited funding for educational research with much of it restricted to priority areas of literacy, numeracy and neuroscience about how children learn. The Australian Education Research Organisation formed by the previous conservative government sets rather than facilitates the research agenda. Nevertheless, there are pockets of Australia-based research on the teaching of SSI as the following examples illustrate. Note that this section reports on papers using the term SSI, rather than education for sustainability, sustainable development, or environmental education.

White and Tytler (2022) in Victoria report on the outcomes of their preservice teacher (PST)/scientist partnerships where they facilitated local scientists to work with PST to communicate their scientific practices and enabled the PST to produce learning resources on contemporary issues and scientific practices. The authors argued that SSI, while shown to improve scientific literacy, had not been embedded in all science classrooms due to assessment practices, lack of resources and an overcrowded curriculum. They envisaged that the learning resources produced by the PST could enable the linking of SSI to "mainstream science" through "resources focused on contemporary science practice in societal contexts, in which scientists themselves are engaged in socioscientific challenges and issues". (p. 18). The resources incorporated 'socially acute questions' as a way of drawing students into the SSI of the context (Morin et al., 2014). Examples of SSI included stem cell research and local issues such as the culling of wild horses in alpine areas. When implemented by teachers, the learning materials engaged secondary school students in science inquiry practices and SSI and informed them about the work of local scientists (Vamvakis et al., 2021). Davis and Bellocchi (2020) in Queensland also used socially acute questions as a starting point for introducing SSI to their PST through gamification. The authors drew on their passion for gaming in introducing their PST to a game design framework in the context of skin cancer prevention. The use of gaming as a pedagogy was intended to introduce PST to SSI, as well as improve their critical thinking and promote activism.

Observing that a great deal of SSI research focused on school students, Saunders and Rennie (2013) embarked on an international literature review in order to develop a theoretical pedagogical model for ethical inquiry about SSI. Their model built on the ethical framework suggested by Reiss (2008) comprising consequences, harms and benefits, rights and duties, virtues, and right to choose by adding pluralism as another dimension for ethical decision-making about SSI. Through professional development, the model was introduced to four New Zealand secondary science teachers who subsequently implemented it using a variety of co-operative student activities. The teachers' confidence in teaching about SSI increased and they observed that their students' science understandings and ethical thinking improved. Working in a Western Australian Christian school, Yap (2014) in Western Australia also modified the ethical framework of Reiss (2008) through the addition of Christian ethics and used it to introduce her students to decision-making about gene technology and transplantation SSI.

6.4 Current Inclusion of SSI in Preservice Science Teacher Education

In Jan 2024, I contacted preservice teacher educators who are members of the Australasian Science Education Research Association (analogous to NARST and ESERA) about whether or not they included SSI teaching in their science teacher education units. The respondents (from across all states) included SSI teaching and learning in their teacher education programs. The reasons for teaching about SSI included media literacy, navigating mis- and dis-information, critical thinking, systems thinking and ethical thinking, integrated approaches to teaching science, perspective-taking and argumentation, and models of SSI teaching. One example from Victoria (Marangio & Cooper, 2022) combines values and knowledge education (VaKE) using dilemma stories and decision-making to help their PST become aware that science has a values dimension and is connected to society. What is not known though is whether the PST go on to effectively use SSI in their teaching.

6.5 Conclusions and Recommendations

Despite research findings in Australia and internationally demonstrating the positive outcomes of teachers using SSI with their students there are structural issues with curriculum content, implementation and assessment practices. The inclusion of SSI in secondary school science is not mandated although the national curriculum does suggest opportunities (in the elaborations) for teachers to include SSI. The way that SSI are represented in resources and text books tends to suggest that SSI are related to science in daily life and are used to increase interest in science or provide relevance for science concepts. The teaching of SSI is frequently an optional add-on not a starting point.

Teacher beliefs and expertise mean that not all students have the opportunity to become aware of the complexity of science and develop the skills required to become global citizens. There is a hesitancy by some teachers to introduce controversial issues into their science teaching. SSI are perceived as complex, difficult and nebulous. Some teachers do not want to alarm their students and worry about how to manage students' views. The problems are compounded by a lack of funding for teacher professional development, resources and sustainable large scale research about what works in science classrooms.

However, there is some cause for optimism in the status of SSI learning in Australian schools. White et al. (2022) "*dare us to think differently about education*", (p. 27, italics in original). The paper reports on the perspectives of two education researchers (Peta and Joseph) and two school aged Student Strike for Climate—Australia leaders (Niamh and Harriet). The students mobilised and encouraged other school students to engage in civil disobedience and activism about the SSI of climate change. This fusion of school and community led to the students engaging in climate activism and developing a sense of agency in their political power to bring about change. The authors call for "reimagining education systems to change so that they might become places of important and meaningful learning, preparing future leaders and citizens for engaging with global issues, impacts and crises" (p. 37).

In conclusion don't be disheartened. There are positive changes occurring in some schools through student activism and agency. As science education researchers we need to continue to work within existing systems and networks, support our PST and practising science teachers, disseminate our research findings and learning resources to teachers, and promote the advantages of SSI teaching and learning. In Australia, we are moving (albeit slowly) towards Vision III scientific literacy.

References

Australian Curriculum Assessment and Reporting Authority (ACARA). (2024a). *Science: Aims*. Retrieved Jan 23, 2024 https://www.australiancurriculum.edu.au/f-10-curriculum/science/aims/

ACARA. (2024b). *Science* (Version 8.4) Retrieved Jan 23, 2024 from https://www.australiancurriculum.edu.au/

ACARA. (2011). *The Australian Curriculum: Science*. Retrieved 30/12/11 from http://www.acara.edu.au

Beauchamp, T. L., & Childress, J. F. (1994). *Principles of biomedical ethics* (4th ed.). Oxford University Press.

Ben Zvi Assaraf, O., Dawson, V., Eilam, E., Gokpinar, T., Goldman, D., Naugauker, N., Putri, G.A.P., Subiantoro, A.W., Tolppanen, S., White, P., Widdop Quinton, H., & Dillon, J. (2024). Climate change education implementation: The voices of policymakers, professional development providers, and teachers in five countries. *International Journal of Science Education, [Online].*

Cross, R. T., & Price, R. F. (1992). *Teaching science for social responsibility*. St. Louis Press.

Curriculum Corporation. (1994). *A statement on science for Australian schools*. Curriculum Corporation.

Davis, J. P., & Bellocchi, A. (2020). Gamification of SSI's as a science pedagogy: Towards a critical rationality in teaching science. In M. Evagorou, J. Neilson, & J. Dillon (Eds.), *Science teacher education for responsible citizenship* (pp. 101–116). Springer.

Dawson, V. (1996). A constructivist approach to teaching transplantation technology in science. *Australian Science Teachers' Journal, 42*(4), 15–20.

Dawson, V. (2015). Western Australian high school students' understandings about the socioscientific issue of climate change. *International Journal of Science Education, 37*(7), 1024–1043.

Dawson, V. (2023). Teachers' support in developing year 7 students' argumentation skills about water-based socioscientific issues. *International Journal of Science Education, Online*, 1–14.

Dawson, V., & Carson, K. (2020). Introducing argumentation about climate change socioscientific issues in a disadvantaged school. *Research in Science Education, 50*(3), 863–883.

Dawson, V., Eilam, E., Tolpannen, S., et al. (2022). A multi-country comparison of climate change curricula in secondary schools. *International Journal of Science Education, 44*(9), 1379–1398.

Dawson, V., & Schibeci, R. (2003). Western Australian school students' understanding of biotechnology. *International Journal of Science Education, 25*(1), 57–69.

Dawson, V., & Taylor, P. C. (2000). Do adolescents' bioethical decisions differ from those of experts? *Journal of Biological Education, 34*(4), 1–5.

Dawson, V., & Venville, G. (2010). Teaching strategies for developing students' argumentation skills about socioscientific issues in high school genetics. *Research in Science Education, 40*(2), 133–148.

Dawson, V., & Venville, G. (2013). Introducing high school biology students to argumentation about socioscientific issues. *Canadian Journal of Science, Mathematics and Technology Education, 13*(4), 356–372.

Dawson, V., & Venville, G. (2022). Testing a methodology for the development of socioscientific issues to enhance middle school students' argumentation and reasoning. *Research in Science & Technological Education, 40*(4), 499–514.

Education Council. (2015). *National STEM school education strategy*. http://www.educationcouncil.edu.au

Fensham, P. J. (2013). The science curriculum; the decline of expertise and the rise of bureaucratise. *Journal of Curriculum Studies, 45*(2), 152–168.

Marangio, K., & Cooper, R. (2022). Challenging preservice teachers thinking about school science education with a values and knowledge education (V*a*KE) experience. *School Science and Mathematics*, 358–370.

Ministerial Council on Education, Employment, Training and Youth Affairs (MCEETYA). (2008). *Melbourne Declaration on Educational Goals for Young Australians*. Author.

Morin, O., Simonneaux, L., Simonneaux, L., & Tytler, R. (2014). Developing and using an S3R model to analyse reasoning in web-based cross-national exchanges on sustainability. *Science Education, 98*(3), 157–165.

Reiss, M. (2008). The use of ethical frameworks by students following a new science course for 16-18 year olds. *Science & Education, 17*(8–9), 889–902.

Rennie, L., Goodrum, D., & Hackling, M. (2001). Science teaching and learning in Australian schools: Results of a national study. *Research in Science Education, 31*, 455–498.

Roberts, D. A. (2007). Scientific literacy/science literacy. In S. K. Abell & N. G. Lederman (Eds.), *Handbook of research on science education* (pp. 729–780). Lawrence Erlbaum.

Saunders, K., & Rennie, L. (2013). A pedagogical model for ethical inquiry into socioscinetific issues. *Research in Science Education, 43*(1), 253–274.
Savage, G. (2017). Neoliberalism, education and curriculum. In *Powers of curriculum: Sociological perspectives on education* (pp. 143–165). Oxford University Press.
School Curriculum and Standards Authority. (2020). *Curriculum support materials*. Accessed Jan 23, 2024 from https://k10outline.scsa.wa.edu.au/home/year-7-to-year-10/science#cpsm
Siarova, H., Sternadel, D., & Szőnyi, E. (2019). *Research for CULT Committee – Science and Scientific Literacy as an Educational Challenge. European Parliament, Policy Department for Structural and Cohesion Policies*. European Union.
Sjöström, J., & Eilks, I. (2020). Correction to: Reconsidering different visions of scientific literacy and science education based on the concept of *Bildung*. In Y. J. Dori, Z. R. Mevarech, & D. R. Baker (Eds.), *Cognition, metacognition, and culture in STEM education* (Innovations in science education and technology) (Vol. 24). Springer.
Tomas, L., & Ritchie, S. M. (2012). Positive emotional responses to hybridised writing about a socio-scientific issue. *Research in Science Education, 42*(1), 25–49.
Toulmin, S. (1958). *The uses of argument*. Cambridge University Press.
Tytler, R. (2007). *Re-imagining science education: Engaging students in science for Australia's future*. Australian Council for Educational Research Press.
Vamvakis, M., White, P., & Tytler, R. (2021). Contemporary science practice in the classroom: A phenomenological exploration into how online curriculum resources can facilitate learning. *International Journal of Science Education, 43*(13), 2087–2107.
Venville, G. J., & Dawson, V. M. (2010). The impact of a classroom intervention on Grade 10 students' argumentation skills, informal reasoning, and conceptual understanding of science. *Journal of Research in Science Teaching, 47*(8), 952–977.
White, P., Ferguson, J., O'Connor Smith, N., & O'Shea Carre, H. (2022). School strikers enacting politics for climate justice: Daring to think differently about education. *Australian Journal of Environmental Education, 38*, 26–39.
White, P., & Tytler, R. (2022). Pre-service teachers representing socioscientific aspects of scientists work. In Y.-S. Hsu, R. Tytler, & P. White (Eds.), *Innovative approaches to socioscientific issues and sustainability education linking research to practice* (pp. 15–31). Springer.
Yap, S. (2014). Beliefs, values, ethics and moral reasoning in socio-scientific education. *Issues in Education Research, 24*(3), 299–319.

Chapter 7
Politicization of Socioscientific Issues as an Inalienable Condition for Scientific Education in the Brazilian Context

Paulo Gabriel Franco dos Santos, Polliane Santos de Sousa, Michel Pisa Carnio, and André Vitor Fernandes dos Santos

> Unemployment in the world is not a fatality, as I said and have repeated. Rather, it results from the globalization of the economy and technological advances that lack ethics truly at the service of human beings—not for profit nor unrestrained greed of the minorities that control the world. From my point of view, scientific and technological progress that does not fundamentally respond to human interests, that does not respond to the needs of our existence, loses its significance. Every technological advance would have to be matched by a real commitment to respond immediately to any challenge that would put men's and women's joy in life at risk. A technological advance that threatens thousands of women and men with losing their jobs should be matched by another technological advance that would serve the victims of previous progress. As you can see, this is an ethical and political issue, not a technological one. (Paulo Freire, A Pedagogia da Autonomia [The Pedagogy of Autonomy], 2021, p. 127)

P. G. F. dos Santos (✉) · A. V. F. dos Santos
University of Brasília, Brasília, Distrito Federal, Brazil
e-mail: paulosantos@unb.br; andre.santos@unb.br

P. S. de Sousa
Recôncavo of Bahia Federal University, Amargosa, Bahia, Brazil
e-mail: polliane.sousa@ufrb.edu.br

M. P. Carnio
Federal University of São Carlos, São Carlos, São Paulo, Brazil
e-mail: michelcarnio@ufscar.br

D. L. Zeidler (ed.), *A Moral Inquiry into Epistemic Insights in Science Education*, Contemporary Trends and Issues in Science Education 61,
https://doi.org/10.1007/978-3-031-63382-9_7

7.1 Introduction

Sometime after we lived through the terrible experience of the COVID-19 pandemic,[1] we still feel the impact on our science education spaces of teachers and researchers of questions and doubts about our role in the scientific education of the population and in combating the pressing problems of fake news, the political and opportunistic use of the pandemic, denialism, and the recognition of the lack of safe and culturally widespread scientific references to support the explanations that society required. Indeed, one of the historical lessons reinforced during the pandemic experience was the importance of politically oriented scientific education, whose meanings, usages, and deepening of related specific knowledge are tensioned, disputed, and defined in a border field. The nature of the border is linked to the nature of the problems addressed by scientific education research in Brazil, whose epistemic field, like what occurs in the broader educational field, is multidisciplinary and contributed by political, sociological, anthropological, psychological, and philosophical ideas of different tendencies that create the appropriate background for the political project in which the research or educational practice in question will be guided and from which its coherence will be determined.

For a Latin American country located on the periphery of capitalism, it is not enough to mindlessly follow global educational trends or engage in slogans and fads in an area of investigation without taking due contextual consideration, causing tensions based on concrete reality and adjusting the political and practices to the relevant demands. Our political commitment must be focused primarily on the objective transformation of our reality to overcome the conditions of exploitation and oppression that still make up our social fabric and our material formation. Furthermore, we must create conditions to include our creative and productive potential and those arising from our historical lessons of resistance, which implies us undertaking the potential of scientific education in emancipatory and popular projects.

Our global challenges stem from our geographical and geopolitical positioning and other particular issues rooted in our material and cultural makeup, both dimensions of historical origins. Among those problems are algorithmic racism and virtual vulnerability, relaxation of health standards that facilitate the entry and use of globally rejected pesticides, poisoning of breast milk due to the spread of these same chemicals in monoculture rural areas, impactful energy and production matrix and supply of international markets, environmental and labor impact resulting from the production of large international companies on Brazilian soil, planned obsolescence of electronic products, piracy, and biopiracy and their impacts on the human body and the environment, deforestation and effects of industry or human mining in different regions of the country, among many other problems that populate our reality.

[1] We refer here to the most acute periods of the beginning of contamination and the incessant fight against the virus, its political use, and the inequalities in the conditions of prevention, treatment, and care. We recognize that, to a lesser extent, we still suffer from COVID-19 sequels, especially in underdeveloped countries.

Teachers and researchers have assumed those and other issues in the context of science teaching from different perspectives, levels of depth, interests, and educational aspirations. In more or less critical, more or less well-founded spectrums, we find the application of problems in the form of themes, appropriations to illustrate reality or contextualization, education through thematic projects, approaches to the relationships between science, technology, society, and the environment (STSE) and socioscientific issues (SSI).

Theoretical appropriations and Brazilian scientific productions on SSI vary according to educational levels (elementary, secondary, or higher education, etc.), the formative context (students and teachers education, formal and non-formal spaces), and the fundamental theoretical basis (critical theories, Freirean perspective, science teaching perspectives).

In the field of STS education, the debate about the politicization of science education is not new, taking us back to precursory work from the 1990s (Hodson, 1994; Pedretti & Hodson, 1995). Given the two major trends in scientific education in force in the 1990s, namely, educating young people for scientific and technical careers and ensuring scientific and technological literacy for all, Hodson (1994) emphasizes learning personalization and the politicization of scientific education as a way of making them compatible and attainable. According to the author, personalization would be important because it foresees the "transformation of society, on which the solution of current environmental crises depends, can only be achieved by individuals who are critically literate, politicised, confident and assertive" (p. 75). By underscoring a curriculum that focuses more emphatically on each student's understandings and experiences, personalization would require scientific and technological education to generate learning at a conceptual, theoretical, methodological, relational, qualification, and problem-solving level. Although personalization aligns with the political and educational trends of the time, it is crucial to bring out the significance and duration of specific education demands and formative perspectives that we are still pursuing today.

The politicization of scientific education, a runup to what we now call socioscientific issues, shows how relevant it is to face topics with scientific, technological, and environmental dimensions, recognizing that those elements are a product of their time and space. Such formative perspective aims to generate contexts of motivation, overcome the abstraction of content and concepts and decontextualization, and construct significant perspectives. It is organized into four levels of sophistication:

Level 1. Appreciating the societal impact of scientific and technological change, and recognising that science and technology are, to some extent, culturally determined.

Level 2. Recognising that decisions about scientific and technological development are taken in pursuit of particular interests, and that benefits accruing to some may be at the expense of others. Recognising that scientific and technological development is inextricably linked with the distribution of wealth and power.

Level 3. Developing one's own views and establishing one's own underlying value positions.
Level 4. Preparing for and taking action. (Hodson, 1994, p. 85)

One of the crucial aspects of politicization is the rejection of technological determinism, proposing, to overcome it, the control of technology controllers or adequate education for choice and decision-making. We must mention that the action factor foregrounds an overcoming of the individualization and adaptation tendencies that the "decision-making" agenda usually points to, indicating concrete, political, well-founded actions as possibilities with a view to objective changes, even if the dimension of collectivity and the radical transformation of capital's mode of production is not on the agenda of the formative program in question.

Just over 20 years later, alluding particularly to the formative potential of socio-scientific issues, Hodson (2018) reaffirms the importance of scientific education oriented towards socio-political action, also requiring learning *about*, *through*, and *from* action, putting in evidence a formative-political agenda that seems convenient and quite promising to what we intend to address in this text.

From this context, this essay discloses an abridged overview of research on SSI developed in Brazil and signals the "viable unprecedented" (Freire, 2014) for the Brazilian educational context. This is not intended to be a state-of-the-art study, but we are anchored in research that has investigated the composition of research on socioscientific issues in Brazil to discuss some trends and defend a political synthesis focused on the relevance and need for metabolization of such discussions in the Brazilian educational scenario.

The first part of this chapter develops a historical contextualization of scientific education and the appropriation of SSI in Brazilian research, considering scenarios of educational reform, intensification of neoliberal agendas, and those unfavorable to the working classes. Next, it presents the political-theoretical perspectives with which we align and which allow us to develop theoretical categories for undertaking critical analyses of the current appropriations of SSI and possible emancipatory agendas for their treatment, given the concrete Brazilian reality and the status of science and technology resulting from historical processes. The theoretical design presented highlights our political positioning to announce perspectives for scientific education, the treatment of socioscientific issues from formative, moral, and ethical perspectives that agree with an emancipatory project for Brazil.

7.2 Contextualization and Historical Aspects of Socioscientific Issues in the Brazilian Scenario

According to Santos et al. (2018), the first studies on the approach to socioscientific issues in Brazil emerged in the second half of the 2000s. By mapping research focusing on SSI published at the National Meeting of Research in Science

Education,[2] from its first edition in 1997 until 2013, Sousa and Gehlen (2017) point to the first national publication related to the topic in 2001. This work refers to a study by Santos et al. (2001) that discusses how argumentation can contribute to the discussion of what the authors call socioscientific aspects (SSAs) and the importance of educators' pedagogical interventions in this process. This study is related to the doctoral thesis by Brazilian researcher Wildson Santos (2002), in which he investigated school chemistry teachers' approach to environmental, political, economic, ethical, social, and cultural issues related to science and technology at the Federal District of Brazil.

According to K. Silva (2016), the socioscientific aspects proposed by Santos (2002) are directly related to STS education, presenting historical, political, economic, cultural, environmental, ethical, and moral aspects, among others, which highlight the relationships and interrelationships between science, technology, society, and the environment in search of a contextualized approach. In the author's view, SSAs are drivers of critical discussions involving STS interrelations, which is why they are also evidenced in an SSI. Therefore, in the analysis carried out by da Silva (2016), the socioscientific issues approach involves the socioscientific aspects.[3]

Despite being relatively recent, the approach to SSI is a growing and consolidating field of study in Brazil (Sousa & Gehlen, 2017; Sousa, 2021). For the same reason, it has been the object of study in research concerned with understanding its conformation and insertion in the Brazilian educational context (Bezerra & Amaral, 2015; Duso & Hoffmann, 2013; da Silva, 2016; Sousa, 2021; Sousa & Gehlen, 2017). Below, we present some of the results of these investigations to highlight some trends and outline the scenario from which we defend the political-epistemological need to contextualize this theoretical framework for the Brazilian reality.

7.2.1 *An Overview of the SSI in Brazilian Research*

One of the challenges initially identified by research that has investigated productions on SSI in Brazil concerns the diversity of nominal expressions used in the country, such as socioscientific discussions, socioscientific issues, controversial themes, socioscientific controversies, controversial socioscientific issues, socially controversial issues, controversial topics, controversial socioscientific topics, etc. (Bezerra & Amaral, 2015; K. Silva, 2016; Sousa, 2021; Sousa & Gehlen, 2017).

[2] Important Brazilian congress in scientific and technological education, organized by the Brazilian Association for Research in Science Education (Associação Brasileira de Pesquisa em Educação em Ciências - Abrapec).

[3] This study initiated the debate on the different dimensions present in issues or themes involving scientific-technological development. We intend to highlight Wildson Santos's contribution to the area and pay him a posthumous tribute due to the important work he developed on STS and SSI education in the country.

Although this profusion of terminologies initially seemed uninteresting and neutral, it was associated, at first, with few discussions regarding the theoretical elements that supported the research (Bezerra & Amaral, 2015; Sousa, 2021; da Silva, 2016). Therefore, researchers in the area made an effort to analyze such productions and understand the movement toward contextualizing this theoretical perspective in Brazil. After all, the approach to SSI involves a network of concepts about which there are broad discussions and different positions, such as education for citizenship, scientific literacy, decision-making, critical thinking, moral reasoning, and the nature of science, among others (Sousa, 2021).

Among such studies, we will emphasize Sousa's (2021) mapping, which analyzes works on scientific and technological education published in important educational congresses in Brazil between 2001 and 2015. The congresses mentioned are the Encontro Nacional de Pesquisa em Educação em Ciências [National Meeting on Research in Science Education], Encontro Nacional de Ensino de Biologia [National Meeting on Biology Teaching], Encontro de Pesquisa em Ensino de Física [Research Meeting on Physics Teaching], and Encontro Nacional de Ensino de Química [National Meeting on Chemistry Teaching]. Although the mapping does not involve more recent studies, it provides a good overview of production in the area, especially considering the biennial nature of the congresses.

In total, 161 studies were analyzed. Among the results, we initially highlight the consolidation of the translation of the term socioscientific issue from 2016 and the creation of research groups in the country dedicated to its study. Regarding this last aspect, when analyzing the fundamental references used by Brazilian studies, Sousa (2021) points to the growth of research subsidized by national references. According to the author, we have some institutions with some tradition of research on SSI in Brazil, namely the Universidade de Brasília, the Universidade Estadual Paulista, and the Universidade Federal da Bahia.

Regarding the theoretical and methodological aspects of the research, this field has been established in Brazil based on empirical investigations related to interventions in basic education and spaces for initial and continuing teacher education (Sousa, 2021). Therefore, among the research problems, there is a greater emphasis on the analysis of aspects related to the involvement and/or perception of subjects in activities aimed at addressing socioscientific issues, such as performance, perceptions, discursive interactions, argumentation, mobilized knowledge, etc. (Sousa, 2021). Priority is also given to studies relating to the challenges inherent to teaching practice for elaborating and developing pedagogical proposals aimed at the SSI.

It is also necessary to emphasize the educational levels at which the studies are carried out. Although there is a predominance of investigations within the scope of secondary and higher education (Duso & Hoffmann, 2013; Sousa & Gehlen, 2017), Sousa (2021) identified the development of actions in more diverse contexts, albeit in an embryonic form, such as activities in the initial grades of elementary school and vocational courses. Furthermore, the author stresses the existence of studies combining different levels, such as basic education and initial and continuing teacher education. This connection has been made possible through the Institutional Teaching Initiation Scholarship Program (Programa Institucional de Bolsas de

Iniciação à Docência—PIBID), a public policy aimed at encouraging initiation into teaching, inserting undergraduates into the daily life of public primary and secondary education schools.

Considering the discussions about the structuring of school curricula based on themes (Delizoicov et al., 2011) existing in scientific and technological education in Brazil, Sousa (2021) also analyzes the criteria used to select socioscientific issues, their relationship with the school curriculum, and the methods and resources used in their approach. In this sense, even though some studies employ a set of characteristics to outline an SSI, such as those presented by Ratcliffe & Grace (2003), Sousa (2021) identified, among the main criteria for approaching socioscientific issues, their relationship with conceptual and normative aspects of school curricula, such as the list of contents provided for in the school institution, official documents, the textbook adopted by the school, etc. With this, it is possible to infer rigidity in Brazilian school curricula, which can constitute a challenge in addressing the SSI related to the demands of the school community or even distinguishing between conceptual and socioscientific discussions in the classroom.

According to the author, another result possibly related to the curriculum structure of Brazilian schools is that most classroom interventions occur through specific actions in the format of short courses and workshops. With this, Sousa (2021) highlights studies that have reinforced the importance of fostering and strengthening partnerships between universities and schools, such as research by Santos and Carvalho (2012).

Regarding the relationship of socioscientific issues with the local and school community, Sousa (2021) also points to the presence, albeit in smaller quantities, of works concerned with addressing issues in the students' experiential world. As an example, we can mention some themes the author highlighted that are related to local or national scientific and technological elements: "Problemas e injustiças ambientais relacionadas à Represa de São Pedro" [Environmental problems and injustices related to the São Pedro Dam] (de Araújo et al., 2016), "Ligações elétricas irregulars" [Irregular electrical connections] (Bernardo, 2012), "Agrotóxicos: toxidade versus custos" [Pesticides: toxicity versus costs] (Lopes et al., 2013), "Dengue Zika e febre Chikungunya" [Dengue Zika and Chikungunya fever] (Martins et al., 2016), "Utilização de agrotóxicos na produção de soja" [The use of pesticides in soybean production] (Ribeiro, 2016), "Produção de soja no norte do Mato Grosso" [Soybean production in northern Mato Grosso] (Ribeiro & Darsie, 2013), and "Implantação das usinas termonucleares no Nordeste do Brasil" [The implementation of thermonuclear plants in the Northeast of Brazil] (Souza & Medeiros, 2010).

Another noteworthy piece of information related to interventions in the classroom, which signals the maturation of research on this topic in Brazil, concerns the role of SSI in educational processes. Although Brazilian research tends to understand the socioscientific issues as a methodological perspective (Bezerra & Amaral, 2015; K. Silva, 2016; Sousa & Gehlen, 2017), Sousa (2021) argues that a diversity of didactic-pedagogical methods and resources in its approach suggests they must be understood as an object of knowledge in the classroom. The author even mentions studies that indicate some procedures for discussing SSI in class, such as

Conrado et al.'s (2016), which suggest using Hodson's (2011) three phases of the development process and Ferreira et al.'s (2016), who suggest using the pedagogical moments (Delizoicov et al., 2011).

In summary, although research on socioscientific issues in Brazil still needs further development, presenting a broad spectrum of possibilities (Bezerra & Amaral, 2015; K. Silva, 2016; Sousa, 2021; Sousa & Gehlen, 2017), there are indications of the growth of these studies in the country and the proposition of theoretical-methodological advances. Some discussions, for example, address the epistemic and didactic-pedagogical dimensions of the SSI in the light of different theoretical references (Lopes, 2013; K. Silva, 2016; Sousa, 2015, 2021) that deal with the relationships between the SSI and moral aspects (da Silva, 2016) or the SSI and argumentation (Mendes, 2012; Penha, 2012). A compilation of studies on socioscientific issues also reveal advances in some of its dimensions, such as decision-making and its development in the classroom, as shown in the books *Formação de professores e questões sociocientíficas no ensino de ciências* [Teacher education and socioscientific issues in science teaching] (Carvalho & Carvalho, 2012), *Questões sociocientíficas: Fundamentos, propostas de ensino e perspectivas para ações sociopolíticas* [Socioscientific issues: Foundations, teaching proposals and perspectives for sociopolitical actions] (Conrado & Nunes-Neto, 2018) and *Formação de professores e questões sociocientíficas: Experiências e desafios na interface universidade-escola* [Teacher education and socioscientific issues: Experiences and challenges in the university-school interface] (Martínez-Pérez et al., 2016).

However, besides the need for advances in the theoretical consistency of the proposal, the socioscientific issues approach must be "metabolized" considering its origin in the northern hemisphere countries and the Brazilian material and historical conditions (Sousa, 2021). Additionally, Brazil was implementing an extensive process of basic education reform, with changes to the National Education Guidelines and Bases Law (Lei de Diretrizes e Bases da Educação Nacional 9.394/1996) and establishing a Common Core Curriculum (Base Nacional Comum Curricular - LDB) (Brasil, 2018). Evidently, research on the SSI could not avoid the discussions involved in this process. Therefore, some studies have also investigated the relationship between the approach to SSI and the official documents that guide the Brazilian educational system.

According to Penha (2012), the socioscientific issues approach has been widely recommended by official documents, such as Project 2061, developed in the United States; the Budapest Declaration, drawn up during the "World Conference on Science for the Twenty-first Century: A New Commitment"; and the National Curriculum Parameters for Brazilian Secondary Education (Parâmetros Curriculares Nacionais para o Ensino Médio Brasileiro - PCN) (Brasil, 2000), a document previously used as a reference in the country. Regarding the PCN, the author underscores, for example, its concern with the development of an education that leads students to understand science and technology as human constructions and their impacts on personal and social life. It also intends that students use their knowledge to diagnose and address social and environmental issues.

Santos et al. (2018), when discussing the results of the analysis of official documents for the teaching of natural sciences in Brazilian high schools, indicate the presence of some harmony with the purposes of STSE education, and that presupposes knowledge, especially about the SSI, that is inherent to science and technology. The authors also present an analysis of the preliminary version of the BNCC, the current normative document for primary and secondary education, signaling that the document explains the importance of addressing SSI in the classroom and that these are highlighted as tools for understanding the inter-STS relationships, indicating an STS-SSI relationship (Santos et al., 2018).

With the approval of the final version of the BNCC in 2018 and the changes made to that document, new analyses are necessary. dos Santos et al. (2022), who analyzed the BNCC version for the final years of elementary school, indicate possible openings for treating SSI from elementary school onwards, despite the dimension of controversy and contradiction not being explicitly allocated as formative perspective. A close look at the guidelines of the document for secondary education also indicates that there are appropriations of expressions and terms from critical lines, but in a vague, superficial way, without explicit commitment to transforming the concrete and historical conditions of the country's deep problems via scientific education. This abstract and non-radical appropriation of critical concepts results in a dispersed and disempowering "no-holds-barred" situation.

To contribute to the understanding of the ongoing reform, below we present some aspects of this broad movement that has been investing in the production of curricular references, often associated with assessment actions, which aim to control teaching production within schools in the Brazilian primary and secondary education and its impacts on scientific and technological education and, consequently, on the approach to socioscientific issues.

7.2.2 Socioscientific Issues and the Brazilian Educational Reform

In the preface to his book *A escola não é uma empresa* [The school is not a company], Christian Laval (2019) analyzes the influences of the neoliberal movement on French education in the 90 s and, at the same time, states that he is surprised to learn how advanced this process is in Brazil. Under the subtopic "O Brasil na vanguarda da escola neoliberal" [Brazil at the forefront of the neoliberal school], the author states that.

> In certain aspects, the Brazilian educational system is already much more "neoliberalized" than the French and many other European educational systems. We can even say that Brazil reached the "school and university capitalism" stage earlier than other countries, characterized by the direct and massive intervention of capital in education. (Laval, 2019, p. 13)

In this sense, despite being underway in Europe for a few decades, this new capital configuration is already deeply rooted in the Brazilian context. Also, it takes on specific contours based on its history, context, and culture. A dialectic necessary for understanding contemporary educational phenomena is underscored, reflecting on how the critical diagnosis regarding this *new world order* can dialogue with resistance and movements against local hegemonics.

Specifically in the Brazilian context, it is vital to list some aspects that resulted in the 1990's educational reform. This decade came after the democratization period in Brazil, which peaked in the mid-1980s. After almost 20 years of an authoritarian regime that suppressed individual freedoms and strongly controlled cultural life, Brazilian citizens won back the right to elect a president through direct ballot. In 1988, the Constitution of the Republic was promulgated, ensuring education as a right for all, with its provision being a duty of the State. The constitutional text provided the definition, among others, of minimum content that should be ensured at all stages of education, stressing what later became known as basic education (dos Santos, 2022).

Besides the promulgation of the 1988 Constitution, there were other ongoing events, such as the economic and administrative reforms that only came to be effectively implemented as of the mid-1990s. At the economic level, in the 1980s, the country experienced high rates of inflation associated with unemployment, economic slowdown, and recession. It is essential to point out that the reforms mentioned were guided by principles that defend the reduction of State action, whether by privatizing public services or promoting managerial administration, focusing on results, and moving away from valuing process development, control, assessment, and improvement. From the perspective of management, defining objectives and achieving results that could be controlled and demanded by citizens became important (Brasil, 1995). Such definitions permeated all areas with services the State offers, including education.

It is, therefore, based on concepts that align with the perspective of managerial administration that the Brazilian curriculum reform was conceived during the first government of Fernando Henrique Cardoso (1994–1997). The notions of competencies and skills disseminated by the PCN for primary education and, later, for secondary education began to circulate in the educational scenario, inserting such concepts in debates and raising them, although not precisely in these terms, to the condition of objectives to be achieved throughout the educational process (dos Santos & Terreri, 2014; dos Santos, 2017, 2022).

In this sense, it is interesting to highlight how such concepts were incorporated into the assessment initiatives that were beginning to take shape in Brazil at that time, as is the case of the Brazilian Basic Education Assessment System (Sistema Brasileiro de Avaliação da Educação Básica—Saeb). In the assessment underlying documents, the link between competencies and skills and the idea of objectives and, consequently, their translation into verifiable results is evident; after all, the results "depict what the student knows, understands, and is capable of doing in the curricular areas examined" (Brasil, 1998, p. 7). This link involves assuming an idea about school knowledge that sees it associated with a pragmatic dimension, with an idea of know-how. In this sense, that knowledge could be evaluated through the demonstration of performances of cognitive operations in questions presented in large-scale assessments. The adherence of the Saeb to this reference of competencies and skills ended up strengthening, reaching its maximum expression at the end of the 1990s (Bonamino, 2002). This period coincides with the creation of the National Secondary Education Examination (Exame Nacional do Ensino Médio - Enem).

We can relate this analysis by Laval with the most recent public educational policies implemented in Brazil: the BNCC and the New High School (Novo Ensino Médio). Both were created in a specific historical and political context: the parliamentary coup suffered by President Dilma Rousseff and the entry of her then-vice-president, Michel Temer. The change in presidency took place on August 31, 2016, bringing an immediate change in democratic procedure. Instead of respecting the participation of organized civil society in elaborating educational policies, President Temer signed an arbitrary decree establishing the New High School, which brought profound changes in the structure and direction of Brazilian teenagers' education. Concomitantly, the BNCC, an institution provided for in the Federal Constitution of Brazil since the country's redemocratization and discussed and built since 2014, took very different directions from then on.

The high investment and financing of digital influencers in their dissemination already revealed the contradictory nature of these public policies, which used historical-educational problems in Brazil to manipulate diagnoses and resolutions in the practical field, placing public opinion at the mercy of superficial speeches and empty and alienating labels. In 2017, research denounced the misinformation contained in advertisements published on television and the internet in the sense that,

> [...] under the guise of democracy, [the advertisements] translate the ideological intention of accepting a measure that was processed without the need for popular support and which, due to structural unfeasibility or lack of planning for the implementation of specific measures, will not fully reach the population, especially the portion that has already been historically marginalized. (Kaspari & de Freitas, 2017, p. 309)

We note the prevalence of the monopoly of discourse and the reforming dynamics of liberal ideology arising from the globalized hegemonic model. There was a somewhat naturalized way of seeing things through business and productivity logic –considered "technical" and "pragmatic"– but which did not reveal its foundations or even the contradictions of that perspective regarding society's desires. Obvious inconsistencies that are not recognized as such.

Among the criticisms of the modus operandi of the neoliberal perspective, the deinstitutionali-zation of educational commitments, the devaluation of teachers as important agents in this field, and the disintegration/pulverization of the formative process stand out. These three movements intend to justify the extinction of institutional processes and place all responsibility on the subjects as "free" individuals. Thus, the school was no longer the centralizing body of education but took on more open contours in that new definition of education –which, in essence, was a conception of civilization. In this sense, "The market thus becomes, in place of the State, the mediating instance seen as responsible for establishing the professional values of individuals" (Laval, 2019, p. 57), resulting in a transfer of commitment and responsibility, an inevitable confusion between what is institutional and what is personal, the loss of the collective character and the construction of education as a social and cultural bond.

This appeasement of contradictions usually brings with it a “terminology transfer.” The notion of “competence” has been established as the center of education and brings with it a great deal of semantic obscurity: the curriculum emptied of content is said to be “flexible,” the subjects explicitly oriented towards technical formation are said to be “interdisciplinary,” the increasingly rigid curricula with little space for teaching autonomy are “innovative,” the correct idea that the human being is a being in constant formation comes to be interpreted as “lifelong learning.” We can interpret these speeches as the “socially necessary illusion” for the functioning of capitalism (Nobre, 2004), which promises only to maintain the system working even when it does not allow it to achieve the self-proposed goal.

Behind the dissemination strategies and advertisements, the reform of secondary education ended up institutionalizing the absence of the State as an institution that ensures the education of Brazilian citizens while aligning the strong presence of the State with the “[...] need of a new worker, conformed to a new pattern of exploitation, a 21st-century pattern” (Ortega & Hollerbach, 2022, p. 19).

As education determinations are not restricted to the classroom, it is not just about different ways of conceiving education but rather a civilizing issue. When business logic prevails in public policies, there is no space left to discuss what society we are and what society we want or can be: everything comes down to replicating the dominant logic of reproduction of capital, which relies on the maintenance of a massive reserve of labor force with minimal education.

Thus, the market believes in its self-created illusion, and our challenge as critical educators experiencing this moment of changes and implementations is to understand and deal with the contradictions that we identify in the educational system, recognizing the deep deformative roots that constitute it but never losing the horizon of transformation.

Among the impacts on scientific and technological education we have, on the one hand, the appropriation of science as an unassisted instrument of political or critical function, inserting it into an educational logic in which scientific knowledge –not deepened and not operated because of denial and criticism of reality– incapacitates the critical powers of teaching. On the other hand, the growing indifference of the working class to science, debate, and scientific work is evident, which reinforces hegemony in the sphere of decisions, productions, and appropriations of science and technology.

With this, we intend to highlight that thinking about the approach to socioscientific issues in Brazil requires rescuing and analyzing the historicity of our educational context. In the large-scale implementation phase, the reform, with the elaboration of normative documents by the country’s federative units and the reconstruction of schools’ political-pedagogical projects, presents a new scenario and new challenges for Brazilian educators and researchers. Beyond changes in the workload and content of subjects in natural sciences, we are committed to reflecting on their role in students’ education. Regarding socioscientific issues, we have the political-epistemological need to deepen the contextualization of this perspective in Brazil, as a country with a colonial history and located on the periphery of the world system (Dussel, 2000), whose modern productive model was built on the superexploitation on the periphery of capitalism (Marini, 2005).

7.3 Theoretical-Political Perspectives and Implications for the Treatment of SSI

In this topic, we initially present some discussions on the historical formation of the Brazilian State, which is essential for thinking about the role of education in the country. Next, we discuss the country's productive matrix and relationship with scientific-technological development in Brazil, highlighting the implications for science education. Having built the contextual basis on which we act, we argue about the need to reflect on the role of morals and ethics in approaching socioscientific issues.

7.3.1 Colonialism and Dependent Capitalism: Bases for the Debate

Debates about coloniality and education have gained considerable visibility and have been increased for decades by researchers and teachers from all over the world, especially those who come from countries that experienced colonial violence in their years of formation. The range of possibilities for treating the colonial problem is also broad and polysemic, involving perspectives with a modern or post-modern, structuralist or post-structuralist, radical or neoliberal epistemological and methodological basis, with a profusion of treatments of the problem and projects for overcoming it. We do not intend to address all the nuances of the debate but to recognize its importance and secure some contributions to outline our object of discussion.

Brazil is a colonized country. First by Portugal, then subjected to the processes of exploitation and domination of the imperialist powers and holders of international capital. The long colonial process, which began in the middle of the sixteenth century, left deep marks on the material structure and culture. Despite major historical changes, it remains latent in the structures, such as the signs of submission, exploitation, alienation, and subjugation. Fanon (2008) shows that to address the issue of colonization, we must consider objective and subjective determinants, claiming that "reality demands a total understanding" (p. 29). For the author, every colonized people experienced the burial of their values and their cultural originality, and, as a consequence, an inferiority complex germinated within their social and psychic conformation resounds. Césaire (2020), in his task of highlighting once and for all that civilization and colonization are antagonistic elements, stresses that the colonial equation is *colonialism = reification*. Historical experience leaves no doubt that the colonial agenda is effective based on the denial of the humanity of the colonized, the subtraction of full social participation, and the delegitimization of their knowledge and practices for understanding and transforming nature.

Both authors, significant Caribbean critics of colonization, establish meaningful relationships between the colonization process, the voracity of capitalist domination and exploitation, and racism as the guarantor of these processes. Moldiz (2013), reaffirming the intrinsic relationship between capitalism and colonialism, argues that the colonial endeavor was fundamental in financing and guaranteeing the development of classical European capitalism. For the author, emancipatory processes must be linked to overcoming conditions of exploitation of work and materiality.

European modernity and the triumph of imperialist nations took their toll, and colonialism was the project that guaranteed payment through enslavement, the super-exploitation of labor forces, and the financing of the Western intellectual heyday with the spoils of the riches of the New World.

Colonial processes have been successful, and as long as objective conditions are favorable, new forms of colonization should nestle within our societies. One of the ways of transmitting knowledge and worldviews and reproducing current hegemonic forces is through formal education. Césaire (2020) also denounces that, in the name of ascending and inconsequential progress, disseminated even within scientific discourses, the agenda of colonialism justified massacre and dehumanization.

> I hear the storm. It tells me of progress, "accomplishments," cured illnesses, and levels of life lifted beyond themselves. But I speak of societies emptied of themselves, trampled cultures, undermined institutions, confiscated lands, murdered religions, artistic magnificences destroyed, extraordinary possibilities suppressed. [...] I'm talking about millions of men in whom fear, an inferiority complex, trembling, kneeling, despair, and servility have been intelligently instilled. (Césaire, 2020, pp. 24–25)

However, we must assert that, from a dialectical perspective, education is also a path to cultural formation and conditions for the production of society, overcoming determinism and announcing possibilities of resistance and emancipatory processes. Thus, colonial and capitalist criticism supports us in evaluating, criticizing, and proposing educational formative projects. After all, education is a human activity, intentional and organized based on institutions, governments, political, and cultural agendas and projects for society and the future; therefore, there is no neutrality; on the contrary, we must highlight that any proposal, program, or education action is the result of disputes, tensions, distribution of privileges and power. As a baseline, scientific education cannot deny the historical and material constitution of the people for whom it is intended. Whether as a form of resistance, denial, or affirmation, the truth of a people must be revealed through contradiction, controversy, and questioning of the status of normality. We will not begin to legitimize and replace epistemological, scientific, and technological systems with ancestral and widespread knowledge by denying the Eurocentric science that gave us some longevity and health security. Nor is it by questioning the effectiveness of the traditional methods of original peoples that we will turn a blind eye to the atrocities committed in the name of science or the vested interests of technoscience.

Therefore, the questions are: Who and what is today's scientific education aimed at? Who produces it? How has it been able (or not) to equip young people and children to analyze concrete reality and to deny and overcome appearances? What do we expect from the future of a nation whose colonial past results in a submissive present, and the controlled memory erases the historical lessons of resistance? How can science education support both the analyses of the past and present and the future projects?

Colonial criticism and overcoming projects require historicity and presuppose profound transformation because they deal with fundamental, structural, and power problems. It is essential to highlight that we are not arguing that classical science should be purged from books and manuals, and, in its place, the knowledge of leaders and masters from traditional communities teaching their healing techniques and practices should triumph. Nor are we recognizing that the scientific contents of Europe from the seventeenth to the nineteenth centuries are sufficient to explain and represent the current world, the people of a nation, its technologies, and its contradictions. We are advocating that, from basic schooling, a people must know deeply about themselves and commit to overcoming contradictions, exploitation, and violence, whether by dominating the oppressor's resources or by authentically inserting themselves into their reality, so that no alternative is left but an unrestricted commitment to the continuity of the dreams of those who have already passed away, whose screams lie in the sepulchral silence of the history of the victors.

In the 1960s and 1970s, Brazilian intellectuals took on the task of thinking about the national reality and appropriating Brazil as a problem that should be faced based on theoretical elaborations whose starting points were deeply linked to the country's concrete, historical, and contradictory problems. In this context, we highlight the dependency theory and some names, such as Theotonio dos Santos, Ruy Mauro Marini, and Vania Bambirra, who sought to analyze, based on Marxist theoretical categories, the dynamics of the reproduction of capitalism in Latin America and the categories to understand underdevelopment on the outskirts of the capital, considering the different historical processes of its formation and maintenance.

Marini (2005) states that Latin American dependence is rooted in super-exploitation. This central concept explains the conditions of participation of the periphery of capitalism in the accumulation of capital in central countries, characterized by the extortion of surplus product from the worker to feed the external market through labor intensification, remuneration below the value of work, strong concentration of income in the upper classes, and as a mechanism to compensate for the low rate of profits in dependent countries. The concept is crucial in understanding the industrialization of peripheral countries and the relationship with central countries regarding the production and consumption conditions of technology and machinery that a world worn down by wars demands.

During the period discussed, several studies in sociology, anthropology, politics, philosophy, and education produced from critical perspectives generated an inexhaustible source of concepts, theories, issues, and authentic ways of reading Brazil and its problems. National and Latin American science was also a crucial problem, especially for Álvaro Vieira Pinto, who, in turn, greatly influenced Paulo Freire's thinking.

Today's Brazil is a relatively new-born democratic republic, as the democratic reopening occurred in 1985, after 20 years of a business-military dictatorship. As of the 1988 Constitution, the new democracy built its projects in the most diverse fields of the nation, including the educational sphere, through the formalization of the Education Guidelines and Bases Law (Lei de Diretrizes e Bases da Educação, 1996) in the decade at the top of neoliberal action in Latin America (expansion of the actions and programs of the OECD and the World Bank, actions and long-term objectives of UNESCO, formalization of the Free Trade Area of the Americas (Área de Livre Comércio das Américas - ALCA) and many other world conferences whose deliberations reinforced the project of expanding neoliberal policy, especially in underdeveloped countries).

Freitas (2014) states that the neoliberal perspective in Brazilian education has operated from the control of the educational process, shaping its notions of meritocracy, ranking, and accountability policies, maintaining the status quo of the poor population by guaranteeing a restrictive (education for all) and non-emancipatory base that helped maintain the class structure. Furthermore, teaching work is restrained when educational goals aim only to fulfill large-scale assessments based on international standards, controlling the pedagogical assessment processes, methodology, objectives, and contents.

Neoliberal, therefore capitalist, reason disposes of the market and morality with disciplinary and governing power over individuals, which are the elements from which adequate resources for human needs would come. Traditional morality here, alluded to so vehemently by neoliberals, whose campaigns "classify as assaults on freedom and morality all social policies that challenge the social reproduction of hierarchies of gender, race, and sex, or policies that promote timid corrections of extreme differences between classes" (Brown, 2019, p. 23) is also required to limit and coerce the actions of the State on the market and individual freedoms. According to Brown (2019), neoliberal morality evokes a traditional conservative morality and a cult of the market wrapped in patriotism, nativism, and Christianity. What is moral, therefore, is completely distant from a popular, emancipatory logic in favor of the working class, as it does not mobilize collective, community meanings in favor of the dispossessed and their complexity of existence, dreams, and concrete needs.

For neoliberal policies for education, the idea of basics is not necessarily the essential or the fundamental, defined from a popular and emancipatory perspective. The basics, in this logic, is the minimum that acts as a force to convince the benefits of capital as a policy of vertical accountability (the offer is here; if you want to progress, you will; or we are offering the basics, you must seek other things by yourself, and the outcome will come from your effort and merit) and reinforces the maintenance of the status quo. Scientific education is affected by the "basics to the base" policy insofar as the content choices, deeply controlled by business reformers (Freitas, 2014), do not allow understanding the nature of science, nor do they equip young people for qualified action and the transformation of reality, summed up in a synthesis of abstract, propaedeutic, and bureaucratized contents, even if behind good intentions.

Thus, the ethics of confrontation are called for[4] based on an unrestricted commitment to the popular classes and emancipation, considering the historical determinants that have resulted in poverty, environmental crises, exploitation, poverty, and marginalization. It is essential to recognize the role of science and technology in this historical project of society, as well as be able to elaborate other possible realities, including them in a place of importance, undertaken from their critical and contesting power and oriented to those projects.

7.3.2 *Status of Science and Technology in the Brazilian Historical and Material Context*

Brazilian productive sectors, which have historically been centered on the plantation, mining, input extraction, and production, are traditionally divided into the primary sector (extractivism, agriculture, livestock, and fishing), secondary sector (processing industry, energy production, and civil construction) and tertiary sector (services such as education, health, tourism, finance, commerce, transport). Even though the third sector is the most representative of the Brazilian economy, the country is one of the largest producers of agricultural commodities in the world, reaffirming Brazil's vocation for the first sector and the difficulty of undertaking a non-vertical industrialization and modernization agenda that does not crush the working class. According to data from the Comex Stat system[5] from the Ministry of Development, Industry, Commerce, and Services, the main products Brazil exports are from the primary and secondary sectors, with emphasis on iron ore, soybeans, crude petroleum oils, sugars and molasses, beef and poultry meat, and cellulose.

Historically, the production matrix was first guaranteed through slave labor on plantations, mining, and extraction—rubber, for example. With the formal abolition of slavery, the productive model continued to require a system of exploitation of the working class, now salaried, to guarantee large monocultures, the strengthening of oligarchies, large estates, and agribusiness, which are the current productive and political focus, safeguarding the differences arising from historical accumulations. Science, a participant in productive work according to the social division of labor, supported productive processes through an elite that was cultivated and educated

[4] Inspired by Mariátegui's (1988) defense that "the ethics of socialism springs from classconfrontation." Mariátegui, José Carlos. Defensa del marxismo [The Defense of Marxism], Lima, Perú, Editora Amauta, 1988, pp. 50–60. (14th ed.).

[5] Access to the system via the website: http://comexstat.mdic.gov.br/pt/home

mainly abroad. Institutionalization of science came very late compared to other countries in Latin America and the Caribbean.[6]

The concept of technoscience is fundamental to allude to the current productive model of science and technology, the historical heir of the models of material and knowledge production established under the aegis of capital. Castelfranchi and Fernandes (2015) argue that technoscience should not be considered a mere juxtaposition of science and technology. Instead, they propose it as a crucial concept for understanding the dynamics of knowledge production, technological changes, and the multiple interplay between science and technology within capitalism. The technoscientific perspective encompasses the assumptions widely accepted in the scientific community of neutrality and technological determinism (Dagnino, 2014) and, as it operates on the basis of capitalism and neoliberal policies, it concerns not only the production or appropriation of the products of science and technology but also the market of technical codes and scientific knowledge (Lacey, 2008).

When alluding to technoscientific citizen formation, Castelfranchi and Fernandes (2015) highlight the capacities of agency and power, recognizing that in the dynamics of capital and technocratic decisions, the development of these perspectives of action on reality requires more concrete conceptions of science and technology, as well as the creation of insistent conditions of resistance and action within the black box, hacking from the field of action itself. In the same sense, Neder (2013), based on the contributions of Latin American Thinking in Science, Technology, and Society (LAT-STS) and the critical theory of technology, also highlights the importance of expanded participation in the production of knowledge, taking into account the contextual specificities, developing educational conditions for the acquisition and sociotechnical adequacy (STA) of science and technology, despite the challenges of doing so under a scenario of overdetermination.

It is worth highlighting that the LAT-STS constituted an autonomous Latin American movement that opposed the uncritical and decontextualized transfer of ideas, conceptual structures, and institutional formats from central to peripheral countries (Vaccarezza, 2011). For example, the discussion emphasized the importance of scientific-technological policies that consider specific material conditions and their historical formation in a local context, instead of applying ready-made, externally devised solutions. This is not about excluding the import of knowledge but adapting it to local demands, understanding science and technology as social processes dependent on the context in which they are introduced or developed (Linsingen, 2007). The concept of sociotechnical adequacy mentioned above helps us understand this process insofar as it considers technoscience as humanly controllable and a bearer of values in such a way that it can be redesigned according to alternative criteria and having its objectives subverted to other models of society (Dagnino, 2010a, b).

[6] Due to Portuguese rule and the colonized culture of Brazilian elites, the first university in Brazil was only founded in the twentieth century, incorporating some existing colleges. In comparative terms, other countries in Latin America and the Caribbean had consolidated universities since the colonial period.

The debate has been the object of study and in-depth study by some Brazilian authors, who have highlighted the contributions of the LAT-STS to scientific and technological education, as well as the need to update their assumptions (Auler, 2018; Roso & Auler, 2016; Strieder, 2012). The non-neutrality of scientific-technological products is disputed, and consequently, the educational import of themes/questions, objectives, conceptualizations, and/or visions of science, technology, and society and their interrelations is problematized (Auler, 2018; Roso & Auler, 2016).

For example, Sousa (2021) affirms that we must "metabolize" the socioscientific issues approach in the Brazilian context. According to the author, the SSI have a space-time dimension that indicates that their approach must have their local manifestations and relationships with universality as ground zero, problematizing the supposed hegemony of a globalized science-technology. According to the author, as discussed by the LAT-STS regarding technology transfer, the approach to SSI in the Brazilian context involves the development of a "sociotechnical adequacy" given its genesis in the socio-political-cultural context of the northern hemisphere. This reinterpretation considers the space-time specificities of a continental and unequal country like Brazil.

In this context, the field of socioscientific issues, especially the most critical appropriations in the Brazilian scenario, benefits from social and cultural studies of science and technology, especially the LAT-STS, critical theories, and other aspects of scientific education, such as critical environmental and emancipatory education, highlighting the situation, the starting points, the educational projects in dispute, the project of society, and the arrival points. It is important to remark that theoretical appropriations are somewhat committed to concrete reality and its objective transformation, depending on their capacity to grasp historical determinants, the depth of the analysis of the situation, the sharpness of criticism, and the enthusiasm they provoke in crafting alternative realities.

In other words, although the approach to the socioscientific issues is developed in Brazil under different approaches, there is a movement towards highlighting the importance of considering the country's historical conditioning when discussing theoretical-methodological propositions from countries from the North. In this essay, we also understand that there is a need to consider political commitment to including SSI in the educational process. After all, one of the main characteristics of this perspective is the emphasis on students' decision-making and moral and ethical development.

Works such as that of Lima and Martins (2013) and Sousa (2021) reveal the attribution of different meanings to "decision making" and few discussions about its structure and role in the educational process in studies in the field of SSI and STS education in Brazil. According to Sousa (2021), there are different understandings:

> About the subject who makes the decisions (individual or collective?), the object involved in the discussion (complex real-life issues or everyday situations that do not require broader reflections?), and the reasoning students are expected to develop (critical positioning or mere choice between given options?). (p. 177)

Regarding what is at stake in decision-making processes, Lopes and Carvalho (2009) highlight how important it is to allow students to do more than "simply choose when faced with products on the supermarket shelves" (p. 5). The authors highlight the need for a "critical positioning, which does not exclude political and emancipated action in debates on controversial issues, such as the implementation of thermonuclear plants, investments in transgenic foods, or stem cell research" (p. 5). Auler (2018), in turn, criticizes the emphasis some works give on discussions only from the point of view of the post-production of science-technology, restricting people's level of participation to using scientific-technological products.

Santos and Mortimer (2001) assert the presence of a strong ideological component in "decision-making" processes related to science and technology; they say it is essential to understand the political and economic context in which they are produced. The authors emphasize that many issues within the social context involve not only the choice *between* but the construction of dialectical syntheses and the overcoming *of* dichotomous alternatives. This process, we must remark, may be guided by different models of society. In this regard, Sousa (2021) problematizes, for example, the foundations of discussions about students' ethical and moral development in the approach to SSI. Furthermore, she asks about the teacher's ethical and moral commitment when selecting and addressing socioscientific issues, challenging an ethical conception coherent with the approach to SSI from an autochthonous perspective. Anchored in Latin American references, the author proposes to resignify the socioscientific issues guided by ethical-critical foundations to contribute to developing a scientific and technological education that enables subjects to have a systemic understanding of their material living conditions with a view to transforming their bordering conditions.

Taking as a reference the discussions presented so far and the defense of the politicization of the approach to socioscientific issues, in the following topic, we will continue the discussions about morals and ethics, presenting a theoretical synopsis of what we are proposing.

7.3.3 SSI and Moral Issues: Which Morality Are We Talking About?

Just as in the capitalist production process, the commodity gains autonomy within the very productive process; it becomes fetishized, alien to its origins, and purified of its history and qualitative characteristics. So, in rationalizing the world into ideas and concepts, they stand as representations of partial moments of reality that can interpret the immediate world but cannot capture the becoming of things. In this case, alienated thinking ends up presenting itself as enlightened, a pseudoactivity of autonomous awareness, a situation that leads to the invariable presence of semi-formation as a process of colonization of awareness (Freire, 2014) due to the instrumental rigging of practically all spheres of sociability.

In scientific education, maintaining the transmission of school knowledge that repeats the image of an unfractured past concurs with the repetition of the barbarity suffered (Mitrovitch, 2011). Within these constraints and challenges, it is imperative to question the purpose of science content –its composition and its pivotal role in shaping subjects. We must ensure that theoretical acquisitions contribute to building the becoming of a different world, accommodating opportunities so we understand the contradictions that constitute it in its current form.

In the Brazilian context, the first moral commitment of socioscientific issues is to always behave in an anti-hegemonic way, rebellious to the normalization of social and environmental processes, interrogating the relationships of power and interest that materialize among several social actors. Only then will the escape from alienated thinking give rise to another educational myth with an empty and propagandistic label. This theoretical commitment can help sustain the moral responsibility of all educational praxis devoted to popular emancipation on the path of decolonization of thinking and on facing the significant civilizational setbacks resulting from the abstract notion of "progress and development."

One of the specificities of the socioscientific issues approach concerns the development of students' ethical and moral reasoning (Sadler, 2004; Zeidler et al., 2005; Zeidler & Nichols, 2009). According to da Silva (2016):

> [...] Moral development is not characterized only by the internalization of society's norms and rules but as a process in which students become responsible for transforming their understanding of morality through reflective processes. This invitation to reflection should direct students to raise critical questions about how these judgments were made and to be more autonomous concerning moral thinking. (p. 110)

That is, if we understand morality as an intersubjective, cultural, and historical concept of values, feelings, intentions, decisions, and actions related to "good" and "evil" and ethics or moral philosophy as more than just having views on what is right and wrong, involving reflection on their reason for being, the problematization and interpretation of moral values (Chauí, 2000), we must ask: What ethical foundations guide the approach to socioscientific issues in the classroom? What is the educator's ethical and moral commitment when selecting and approaching SSI? What assumptions are coherent with an education concerned with the emancipation of subjects, with overcoming social contradictions experienced from the periphery of the world-system, of capitalism?

Such assumptions affect different dimensions of the educational process. When referring to the socioscientific issues approach in particular, they determine, for example, what is meant by SSI, its selection criteria, and the degree of in-depth discussions in class. Therefore, topics such as racism, sexuality, and ethnicity, just to give an example, can be silenced. After all, although they can be understood as controversial topics only from a social point of view, they also involve socioscientific elements in their discussion since, not infrequently, situations of injustice and prejudice seek, among other things, (pseudo) scientific arguments for their historical legitimation, as explained by Dias et al. (2018) when discussing a SSI on racial quotas.

Likewise, the ethical, political, and pedagogical commitment of the educator influences the emphasis of the debate on SSI and its degree of depth. Regarding the first dimension, the generation, distribution, and consumption of electrical energy, frequently discussed in the field of SSI in Brazil, is illustrative. Sousa's work (2021) points to several focuses, more or less broad depending on the foundations of the studies and the material conditions of their development, such as discussions involving the hypothetical construction of a nuclear power plant, a real situation of the construction of a hydroelectric plant, domestic energy savings, rationing situations in the country, etc. Among them, the author stresses the SSI "Energia e Desenvolvimento Humano" [Energy and Human Development] (Lopes, 2010; Lopes & Carvalho, 2009, 2012), in which the authors, concerned with the development of a critical positioning of students –which does not exclude political action emancipated in the debates– chose an issue/situation that reflected their intention to contemplate other dimensions of the production, distribution, and consumption of electrical energy, associating this process with human development. According to Sousa (2021), the authors problematize the very notion of development in the proposal and:

> [...] spotlight the need to rethink the current paradigm of progress, point up that if this is understood as a process of expanding individuals' choices and access to goods and services for a more democratic and egalitarian society, this permeates the discussion of energy production and consumption, which are directly linked to these goods and services. However, they state that this issue is linked to the search for solutions to new energy demands, which involves public policies, the interest of energy companies, and social and environmental issues related to the transposition of rivers and flooding of regions. (Sousa, 2021, p. 179)

Not naively, Lopes and Carvalho (2009) also point to the contradictory relationship between the need to expand production, distribution, and access to electrical energy and the socio-environmental impacts of this process, highlighting the complexity of the issue.

Regarding the degree of depth, the educator's ethical-critical commitment can enable the organization of activities more or less concerned with socio-political action, which, in turn, can present different degrees of engagement. Here, we understand that this dimension must be considered at the level of praxis1. From this perspective, some Brazilian studies have defended the approach of socioscientific issues linked to problem situations in the students' reality. As a result of her mapping, Sousa (2021), as an example, calls attention to the following SSI: "Problemas e injustiças ambientais relacionadas à Represa de São Pedro" [Environmental problems and injustices related to the São Pedro Dam] (de Araújo et al., 2016), addressed as it constitutes a problematic situation experienced by the community; "Produção de soja no norte do Mato Grosso" [Soybean production in the north of Mato Grosso] (Ribeiro & Darsie, 2013), discussed because it is a complex and controversial topic present in the relationships of students in the region; "Dengue Zika e febre Chikungunya" [Dengue, Zika, and Chikungunya fever] (Martins et al., 2016), selected because it is a hot topic and vital for public health in Brazil.

Lopes et al. (2013) also underscore the importance of addressing issues close to students to value not only the relevance attributed by the teacher to a given issue but also to give voice to students and how they perceive it. K. Silva (2016) also points out that discussing socioscientific issues mobilizes cultural, evaluative, political, economic, philosophical, psychological, historical, and ontological aspects that demand a contextualized approach.

Therefore, although we can list several socioscientific issues with global dimensions, stressing the importance of their approach in class and their formative potential, they need to be given new meaning and contextualized. In line with the discussions presented here, Sousa (2021) defends the approach to socioscientific issues from an ethical-critical, autochthonous perspective, which presupposes an approach that has as its starting point of the educational process the understanding of local manifestations of the SSI and that favor a relational, systemic view of them. The author states that considering the specific appropriation of SSI by the place does not mean neglecting its broader dimensions, but it indicates the need to understand them as content-forms situated in human-world relations.

Based on the discussions presented, both on the structure of research on SSI and Brazil's colonial roots and their influences on national scientific-technological development, we intend to call attention to the importance of critical contextualization of this field for the Brazilian educational reality. We seek to spotlight the importance of context not only for engagement purposes, whether from a cognitive or activism point of view, but as a baseline for scientific and technological education dedicated to unveiling and overcoming problems linked to social contradictions experienced by subjects who engender different material conditions of life.

In summary, if the implementation of socioscientific issues in the country's science classrooms is challenging for several reasons ranging from educational tradition, school culture, the historical determinants of education, and scientific education, its political deepening is still more complex. The political-pedagogical demands deepen as the linked political project radicalizes (Fig. 7.1). Remaining on the superficiality of teaching content and actions is as much a political project as undertaking processes in favor of emancipation, liberation from decolonization, and radical transformation of objective conditions and levels of awareness about reality. We recognize that emancipatory projects require not only goodwill or specific initiatives that demand herculean counter-hegemonic efforts and culminate in restricted results but are intrinsically linked with objective social and political transformation. However, we will not advance our agendas without adequately locating this objective in teacher and student formative projects.

7.4 Conclusion and Moral and Political Statement

To conclude this chapter, following the example of what we did at the beginning, we were inspired by Paulo Freire, who is part of one of the Brazilian critical traditions on which we rely for the elaboration of the political pedagogical agenda expressed here:

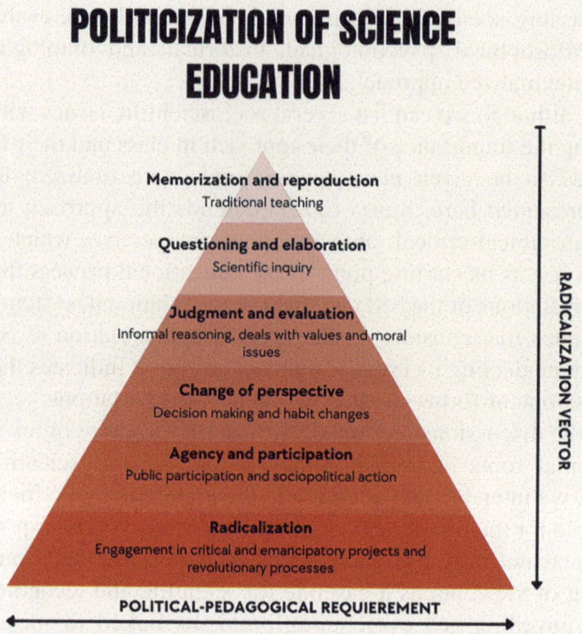

Fig. 7.1 Levels of deepening of political conceptions of scientific education and relevant pedagogical perspectives (Source: The authors)

> [...] A same understanding of educational practice, a same working methodology do not necessarily operate identically in different contexts. The intervention is historical, cultural, and political. This is why I insist so much that experiences cannot be transplanted; rather, they are reinvented. In other words, based on my knowledge of reality as rigorous as possible, I must discover how to apply the same valid principle differently from the point of view of my political option. (Freire, 2007, p. 25)

Given the historical conditions that constituted today's Brazil, the social, political, economic, scientific, and educational panorama and its relationship with global dynamics, the treatment of SSI in Brazilian classrooms must result from a deep understanding of the national reality, as well as clarity on which political agenda scientific education will engage in, whether exclusionary or inclusive, whether in favor of maintaining oppression or overcoming it, whether hegemonic or emancipatory. This requires us to delve deeper into the quality and dimensions of teacher education, scientific knowledge, relevant methodological strategies, curriculum dimension, and public policies to consolidate pedagogical perspectives to overcome the country's material and immaterial needs.

Fanon, in his work Les Damnés de la Terre [The Wretched of the Earth], offers us an important perspective on the politicization of the masses. For the author, politicizing "is opening the spirit, awakening the spirit, giving birth to the spirit" (Fanon, 1968, p. 148), unlike making a political speech or speaking out to the masses about their problems and what they should do. Overcoming a tendency towards infantilization that builds a bourgeois relationship of tutelage, politicization must be authentically popular, attributing commitment and responsibility for the country's project and its consequences to the masses. Politicization, finally, is about strengthening the base, circulating decisions, and overcoming the infantilizing stance that politicizing must be done under relationships of guardianship.

When mobilizing the moral issue within the scope of SSI, it is essential to stress that our defense is in the sense that we are morally constrained by conditions of oppression and exploitation, impelled to objective transformation and oriented to the strategic and emancipatory use of scientific knowledge to favor overcoming these conditions. Thus, we understand that morality must resist purely economic ends and become a critical power to contain the strictly financial argument for complex socioscientific problems, committing itself to raising the level of importance of the other dimensions of the human being's relationship with material and cultural reality, especially to the popular and working classes that compose the great mass of Brazilian public schools.

When we talk about the necessary politicization of the SSI, we want to stress that, no matter how progressive or emancipatory-inspired the perspective may be, there is always a risk of co-option by the trend of neoliberal capitalism. With this, we do not want to offer a formula for harsh criticism of the SSI but to call attention to critical and emancipatory perspectives created in the Brazilian context in dialogue with various research and global experiences. Politicization as a primary educational task requires from us, critical and/or radical Brazilian researchers and teachers, a permanent state of vigilance, critical qualification, establishment of analysis and elaboration networks aligned with popular social movements, and an unrestricted commitment to another possible reality. We must never lose the ability to have hope and, above all, encourage dreams, political anger, and organization for action, as Paulo Freire and so many other educators with a critical affiliation taught us.

References

Araújo, J. C. de, Silva, M. A. da, Santos, F. F. dos, Pereira, C. C. R., & Rodrigues, A. C. (2016). Formação para a justiça ambiental a partir de questões controversas. *Revista de Ensino de Biologia Da Associação Brasileira de Ensino de Biologia—VI Enebio e VIII Eregio Regional, 9*, 5946–5956.

Auler, D. (2018). *CUIDADO! Um cavalo viciado tende a voltar para o mesmo lugar* (1st ed.). Appris.

Bernardo, J. R. (2012). A educação em Ciências e a abordagem baseada em Questões Sociocientíficas: as ligações elétricas irregulares como tema. *Anais do XIV Encontro de Pesquisa Em Ensino de Física*.

Bezerra, B. H. S., & Amaral, E. M. R. (2015). Abordagem de temas sociocientíficos em periódicos nacionais de Ensino de Ciências publicados no período de 2005–2014. *Anais do X Encontro Nacional de Pesquisa em Educação em Ciências*.

Bonamino, A. C. (2002). Tempos de avaliação educacional: o SAEB, seus agentes, referências e tendência. *Quartet*.

Brasil, INEP. (1998). *Resultados do SAEB/95: escalas de proficiência*. INEP.

Brasil, Ministério da Administração e da Reforma do Estado. (1995). *Plano Diretor da Reforma do Aparelho de Estado (PDRAE)*. Câmara da Reforma do Estado.

Brasil, Ministério da Educação. (2000). *Parâmetros Curriculares Nacionais (Ensino Médio)*. MEC.

Brown, W. (2019). *Nas ruínas do neoliberalismo: a ascensão da política antidemocrática no ocidente*. Editora Filosófica Politeia.

Carvalho, L. M. O., & Carvalho, W. L. P. (Eds.). (2012). *Formação de Professores e Questões Sociocientíficas no Ensino de Ciências*. Escrituras Editora.

Castelfranchi, Y., & Fernandes, V. (2015). Teoria crítica da tecnologia e cidadania tecnocientífica: Resistência, "insistência" e hacking. *Revista de Filosofia: Aurora, 27*(40), 167–196. https://doi.org/10.7213/aurora.27.040.DS07

Césaire, A. (2020). *Discurso sobre o colonialismo*. Veneta.

Chauí, M. (2000). *Convite à Filosofia* (13th ed.). Editora Ática.

Conrado, D. M., & Nunes-Neto, N. (Eds.). (2018). *Questões Sociocientíficas: fundamentos, propostas de ensino e perspectivas para ações sociopolíticas*. EDUFBA.

Conrado, D. M., Nunes-Neto, N. F., & El-Hani, C. N. (2016). Análise de argumentos em uma questão sociocientíficas no ensino de Biologia. *Revista de Ensino de Biologia Da Associação Brasileira de Ensino de Biologia—VI Enebio e VIII Eregio Regional, 9*, 5522–5534.

Dagnino, R. (2010a). Um dilema latino-americano: ciência e tecnologia para a sociedade ou adequação sócio-técnica com o povo? In R. Dagnino (Ed.), *Estudos sociais da ciência e tecnologia e política de ciência e tecnologia: abordagens alternativas para uma nova América Latina* (pp. 253–279). EDUEPB.

Dagnino, R. (2010b). Uma estória sobre Ciência e Tecnologia, ou Começando pela extensão universitária. In R. Dagnino (Ed.), *Estudos sociais da ciência e tecnologia e política de ciência e tecnologia: abordagens alternativas para uma nova América Latina* (pp. 282–311). EDUEPB.

Dagnino, R. (2014). A anomalia da política de ciência e tecnologia. *Revista Brasileira de Ciências Sociais, 29*(86), 45–55. https://doi.org/10.1590/s0102-69092014000300004

Delizoicov, D., Angotti, J. A. P., & Pernambuco, M. M. (2011). *Ensino de Ciências: fundamentos e métodos*. Cortez Editora.

Dias, T. L., Fernandes, K. M., Arteaga, J. S., & Sepúlveda, C. (2018). Cotas raciais, genes e política: uma questão sociocientífica para o Ensino de Ciências. In D. M. Conrado & N. F. Nunes-Neto (Eds.), *Questões Sociocientíficas: fundamentos, propostas de ensino e perspectivas para ações sociopolíticas* (pp. 303–324). EDUFBA.

Duso, L., & Hoffmann, M. B. (2013). A discussão das controvérsias sociocientífias na pesquisa em Educação em Ciências: uma revisão narrativa a partir de periódicos no Brasil. *Ensino de Ciências e Tecnologia em Revista, 3*(2), 66–85.

Dussel, E. (2000). *Ética da Libertação: na idade da globalização e da exclusão*. Editora Vozes.

Fanon, F. (1968). *Os condenados da terra*. Civilização brasileira.

Fanon, F. (2008). *Pele negra, máscaras brancas*. EDUFBA.

Ferreira, S., Santana, E., de Almeida, A. C., & Santos, M. (2016). A discussão de temas controversos e abordagem CTSA na formação professores de Ciências e Biologia. *Revista de Ensino de Biologia Da Associação Brasileira de Ensino de Biologia—VI Enebio e VIII Eregio Regional*, 4619–4630.

Freire, P. (2007). *Política e educação* (8th ed.). Villa das Letras.

Freire, P. (2014). *Pedagogia do Oprimido* (58th ed.). Paz e Terra.

Freire, P. (2021). *Pedagogia da Autonomia: Saberes Necessários à Prática Educativa* (67th ed.). Paz e Terra.

Freitas, L. C. (2014). Os Reformadores Empresariais da educação e a disputa pelo controle do processo pedagógico na escola. *Educação & Sociedade, 35*(129), 1085–1114.

Hodson, D. (1994). Seeking directions for change: The personalisation and politicisation of science education. *Curriculum Studies, 2*(1), 71–98. https://doi.org/10.1080/0965975940020104

Hodson, D. (2011). *Looking to the future: Building a curriculum for social activism*. Sense Publishers.
Hodson, D. (2018). Realçando o papel da ética e da política na educação científica: algumas considerações teóricas e práticas sobre Questões Sociocientíficas. In D. M. Conrado & N. Nunes-Neto (Eds.), *Questões sociocientíficas: fundamentos, propostas de ensino e perspectivas para ações sociopolíticas* (pp. 27–58). EDUFBA.
Kaspari, T., & de Freitas, E. C. (2017). Quem conhece aprova? Análise discursiva da propaganda governamental do Novo Ensino Médio. *Cadernos De Linguagem e Sociedade, 18*(3), 294–312.
Lacey, H. (2008). Ciência, respeito à natureza e bem-estar humano. *Scientiae Studia, 6*(3), 297–327.
Laval, C. (2019). A escola não é uma empresa: o neoliberalismo em ataque ao ensino público. *Boitempo*.
Lima, A., & Martins, I. (2013). As interfaces entre a abordagem CTS e as questões sociocientíficas nas pesquisas em educação em ciências. In *IX Encontro Nacional de Pesquisa Em Educação Em Ciências*.
Linsingen, I. (2007). Perspectiva educacional CTS: aspectos de um campo em consolidação na América Latina. *Ciência & Ensino, 1*(especial).
Lopes, N. C. (2010). *Aspectos formativos da experiência com Questões Sociocientíficas no Ensino de Ciências sob uma perspectiva crítica*. Dissertação de Mestrado em Educação para a Ciência. Faculdade de Ciências/UNESP, Bauru-SP.
Lopes, N. C. (2013). *A constituição de associações livres e o trabalho com as Questões Sociocientíficas na formação de professores*. Tese de Doutorado em Educação para Ciência. Faculdade de Ciências/UNESP, Bauru-SP.
Lopes, N. C., & Carvalho, W. L. P. (2009). Energia e desenvolvimento humano: uma abordagem sociocientífica no Ensino de Ciências. *Anais do VII Encontro Nacional de Pesquisa Em Educação Em Ciências*.
Lopes, N. C., & Carvalho, W. L. P. (2012). Aspectos formativos da experiência com questões sociocientíficas no ensino de ciências sob uma perspectiva crítica. In L. M. O. Carvalho & W. L. P. Carvalho (Eds.), *Formação de professores e questões sociocientíficas no ensino de ciências* (pp. 271–296). Escrituras.
Lopes, N. C., Carvalho, W. L. P., & Faria, P. (2013). Agrotóxicos—toxidade versus custos: uma experiência de formação de professores com as questões sociocientíficas no ensino de Ciências. *Anais do IX Encontro Nacional de Pesquisa Em Educação Em Ciências*.
Marini, R. M. (2005). *Dialética da dependência*. Expressão Popular.
Martínez-Pérez, L. F., Lozano, D., & Barragán, I. (Eds.). (2016). *Formação de Professores e Questões Sociocientíficas: Experiências e Desafios na Interface Universidade-Escola*. Editora Unijuí.
Martins, L., Dionor, G. A., Conrado, D. M., & Nunes-Neto, N. F. (2016). Dengue, Zika e febre Chikungunya: a abordagem socioecológica de saúde a partir de uma questão sociocientífica. *Revista de Ensino de Biologia Da Associação Brasileira de Ensino de Biologia—VI Enebio e VIII Eregio Regional*, 9.
Mendes, M. R. M. (2012). *A Argumentação em Discussões Sociocientíficas: o contexto e o discurso*. Tese de Doutorado em Educação. Universidade de Brasília, Brasília.
Mitrovitch, C. (2011). *Experiência e formação em Walter Benjamin*. Editora Unesp.
Moldiz, H. (2013). *América Latina y la tercera ola emancipadora*. Ocean Sur.
Neder, R. (2013). A produção do conhecimento na universidade e os estudos sociais da ciência e tecnologia: contribuições para o ensino e a pesquisa. In R. Neder (Ed.), *CTS ciência tecnologia sociedade—e a produção de conhecimento na universidade* (Observatório do Movimento pela Tecnologia Social na América Latina) (pp. 7–32). UnB/Capes-Escola de Altos Estudos.
Nobre, M. (2004). *A teoria crítica*. Zahar.
Ortega, A. R., & Hollerbach, J. D. G. (2022). Propaganda, media, and education: the official and advertising discourse on the 2017 high school reform. *Educação em Revista, 38*. https://doi.org/10.1590/0102-469837849

Pedretti, E., & Hodson, D. (1995). From rhetoric to action: Implementing sts education through action research. *Journal of Research in Science Teaching, 32*(5), 463–485. https://doi.org/10.1002/tea.3660320505

Penha, S. P. (2012). *Atividades Sociocientíficas em sala de aula de Física: as argumentações dos estudantes*. Tese de doutorado. Universidade de São Paulo. São Paulo-SP.

Ratcliffe, M., & Grace, M. (2003). *Science Education for citizenship: Teaching socioscientific issues*. Open University Press.

Ribeiro, K. (2016). A aprendizagem para ação cidadã na formação de professores de Ciências Naturais por meio de análise de questão sociocientífica. *Anais do XVIII Encontro Nacional de Ensino de Química.*

Ribeiro, K. D. F., & Darsie, M. M. P. (2013). Abordagem sobre alfabetização científica, formação cidadã e questão sociocientífica: um ensaio com alunos de Licenciatura em Ciências Naturais e Matemática. *Anais do IX Encontro Nacional de Pesquisa em Educação em Ciências.*

Roso, C. C., & Auler, D. (2016). A participação na construção do currículo: práticas educativas vinculadas ao movimento CTS. *Ciência & Educação, 22*(2), 371–389.

Sadler, T. D. (2004). Moral and ethical dimensions of socioscientific decision-making as integral components of scientific literacy. *Science Educator, 13*(1), 39–48.

Santos, W. (2002). *Aspectos sociocientíficos em aulas de química*. Tese de Doutorado. Faculdade de Educação da Universidade Federal de Minas Gerais.

Santos, P. G. F. dos, & Carvalho, W. L. P. de. (2012). A constituição de uma questão sociocientífica em um grupo de professores: um processo polifônico de desvelamento da realidade concreta. *AMAZÔNIA—Revista de Educação Em Ciências e Matemáticas, 9*(17), 1–26.

Santos, P. G. F. dos, Carnio, M. P., & Santos, A. V. F. dos. (2022). Brazilian national curriculum and socio-scientific issues: Openings and (im)possibilities for science education in the elementary school. *Revista de Ensino de Ciências e Matemática, 13*(6), 1–24. https://doi.org/10.26843/RENCIMA.V13N6A23

Santos, A. V. F. dos. (2017). *Regularidades discursivas sobre mudança curricular e a produção de subjetividades no Exame Nacional do Ensino Médio (ENEM)*. Tese de Doutorado. Faculdade de Educação da Universidade Federal do Rio de Janeiro.

Santos, A. V. F. dos. (2022). Políticas contemporâneas para o ensino médio brasileiro: entre tradições disciplinares e o novo gerencialismo. *Historia de la Educación, 40*, 303–320.

Santos, A. V. F. dos, & Terreri, L. (2014). Políticas públicas em educação: a avaliação como um problema curricular contemporâneo. *Revista Espaço do Currículo, 7*(1), 24–40.

Santos, W., & Mortimer, E. F. (2001). Tomada de decisão para ação social responsável no Ensino de Ciências. *Ciência & Educação, 7*(1), 95–111.

Santos, W., Mortimer, E. F., & Scott, P. H. (2001). A argumentação em discussões sócio-científicas: reflexões a partir de um estudo de caso. *Anais do III Encontro Nacional de Pesquisa Em Educação Em Ciências.*

Santos, W., Silva, K., & Silva, S. (2018). Perspectivas e desafios de estudos de QSC na educação científica brasileira. In D. M. Conrado & N. Nunes-Neto (Eds.), *Questões Sociocientíficas: fundamentos, propostas de ensino e perspectivas para ações sociopolíticas*. (570 p. EDUFBA.

Silva, K. (2016). *Questões Sociocientíficas e o Pensamento Complexo: tecituras para o Ensino de Ciências*. Tese de Doutorado em Educação. Universidade de Brasília.

Silva, S. M. B. da. (2016). *A moral em Questões Sociocientíficas no Ensino de Biologia*. Tese de Doutorado em Educação em Ciências e Matemática. Universidade de Brasília, Brasília.

Sousa, P. S. de. (2015). *Argumentação centrada em Questões Sociocientíficas e Educação Problematizadora: possibilidades para o Ensino de Ciências*. Dissertação de mestrado em Educação em Ciências. Programa de Pós-Graduação em Educação em Ciências, Universidade Estadual de Santa Cruz, Ilhéus-BA.

Sousa, P. S. de. (2021). *Questões sociocientíficas: sua inserção ético-crítica na educação*. Tese de Doutorado em Educação Científica e Tecnológica. Universidade Federal de Santa Catarina.

Sousa, P. S. de, & Gehlen, S. T. (2017). Questões Sociocientíficas no Ensino de Ciências: algumas características das pesquisas brasileiras. *Ensaio—Pesquisa Em Educação Em Ciências, 19*(e2569).

http://www.scielo.br/scielo.php?script=sci_arttext&pid=S1983-21172017000100202&lng=en&nrm=iso

Souza, A. M., & Medeiros, A. S. (2010). Despertando responsabilidade social através de temas controversos. *Anais do XII Encontro de Pesquisa em Ensino de Física.*

Strieder, R. B. (2012). *Abordagens CTS na Educação Científica no Brasil: sentidos e perspectivas.* Tese de Doutorado em Ciências. Universidade de São Paulo, São Paulo-SP.

Vaccarezza, L. S. (2011). Ciencia, Tecnología y Sociedad: el estado de la cuestión en América Latina. *Ciência & Tecnologia Social, 1*(1), 42–64.

Zeidler, D. L., & Nichols, B. H. (2009). Socioscientific issues: Theory and practice. *Journal of Elementary Science Education, 21*(2), 49–58.

Zeidler, D. L., Sadler, T. D., Simmons, M. L., & Howes, E. V. (2005). Beyond STS: A research based framework for socioscientific issues education. *Science Education, 89*(3), 357–377.

Chapter 8
Encouraging and Enabling Student Socio-political Actions Based on Longitudinal Dispositif Analyses

Lawrence Bencze and Chantal Pouliot

8.1 Introduction

Many more citizens around the world need to rise up, critically examine their circumstances and those of others and, where they perceive harms, design and implement well-informed *sociopolitical* actions to help overcome them. Much of the world is in crisis and, to a great extent, leaders are 'treading water' and, consequently, societies and environments need *revolutionary* actions by 'the people.' Because it is apparent that many personal, social and environmental harms are linked to—although most certainly *not mainly* caused by—fields of science, technology, engineering and mathematics ('STEM'), a major source of such revolutionary actions and possible changes could/should be their educational representatives. In this chapter, we suggest that—based on our study of a long-term case of civic actions to overcome local environmental contamination—revolutionary educational programmes should employ Foucault's (1980) concept of the *dispositif*, which we suggest refers to a material-semiotic network of living, nonliving and symbolic entities ('actants') that are largely aligned to form a kind of 'machine' that can be used to achieve desired outcomes. In making this recommendation, we note several inconsistencies, uncertainties and struggles with its uses. Nevertheless, we make our recommendations with great conviction and hope.

L. Bencze (✉)
University of Toronto, Toronto, ON, Canada
e-mail: larry.bencze@utoronto.ca

C. Pouliot
Université Laval, Québec, QC, Canada
e-mail: chantal.pouliot@fse.ulaval.ca

D. L. Zeidler (ed.), *A Moral Inquiry into Epistemic Insights in Science Education*, Contemporary Trends and Issues in Science Education 61,
https://doi.org/10.1007/978-3-031-63382-9_8

8.2 A World in Crisis

In an aptly named book, Ord (2020), advised that humanity is on a *precipice*, risking catastrophic global destruction. Wryly, he notes that humanity has made such dramatic technological developments that it could easily destroy itself through, for example, nuclear annihilation; yet, unfortunately, development of collective human *wisdom* appears not to have kept pace. Given *existential* risks that he discusses, he suggests that minimizing them (if not *eliminating* them) should be a major global priority. Regarding threats of nuclear annihilation, for example, the Science and Security Board of the Bulletin of the Atomic Scientists (BAS, 2023) moved the hands of its 'Doomsday Clock' forward to "90 seconds to midnight—the closest to global catastrophe it has ever been" (p. 2), adding that, "[i]n this time of unprecedented global danger, concerted action is required, and every second counts" (p. 3). Such dire warnings—unfortunately—have a long history. In 1992, for example, the Union of Concerned Scientists (UCS, 1992) produced a document called, *World Scientists' Warning to Humanity*, in which they gave this major advice:

> Human beings and the natural world are on a collision course. Human activities inflict harsh and often irreversible damage on the environment and on critical resources. If not checked, many of our current practices put at serious risk the future that we wish for human society and the plant and animal kingdoms, and may so alter the living world that it will be unable to sustain life in the manner that we know. Fundamental changes are urgent if we are to avoid the collision our present course will bring about (p. 1).

This warning about threats to atmosphere, water resources, oceans, soil, forests, living species (including much of humanity) was later signed by more than 1700 scientists, representing 71 countries, including all of the 19 largest economic powers. More recently, there have been similar strong warnings to humanity, such as about the climate emergency (scientistswarning.forestry.oregonstate.edu). In their 2021 report (Ripple et al., 2021), for example, scientists noted that 18 of the 31 variables they tracked (e.g., per capita meat production, global tree cover loss, CO_2 emissions, fossil fuel subsidies) reached "all-time record lows or highs" (p. 1). The 2022 warning to humanity (Ripple et al., 2022) seemed even more urgent, stressing 'tipping points' of many variables like those noted above—while suggesting "urgent need to revise economic activities, remove inequities, and shift away from endless economic growth at any cost toward long-term sustainability and justice as described by ecological economics (Daly and Farley, 2011)" (p. 43). Such urgent calls for action are echoed by the IPCC (2023), warning that windows of opportunity are rapidly closing for tipping points, advising that current trends suggest that "it is likely that warming will exceed 1.5°C during the 21st century and make it harder to limit warming below 2°C" (p. 10). Meanwhile, Monbiot (2022) claimed that our current industrialized food production system, both through grazing animals on land and in factories, is as threatening to humanity as the climate crisis—and, like recommendations for abrupt cessation of fossil fuel extraction and uses, urges humanity to replace current industrialized food production systems with those using new technologies like food production from cultured plant and animal cells.

The crises outlined above are, in a sense, the 'tip of the iceberg.' There are numerous others that many analysts suggest are plaguing humanity. In this vein, Tooze (2022) suggests we are in a state of *polycrisis*—in which multiple, some unknown, crises interact in unpredictable ways to together generate results greater than the sum of the 'individual' crises. Lawrence (2022) suggested, moreover, that a polycrisis is a kind of 'wicked' problem—which, largely due to differences in values, principles and ideologies of different interest groups, are largely *intractable*. Although there are, of course, many different—and, sometimes, very conflicting—value systems in our many societies, perhaps of greatest significance are those prioritizing continuous consumer consumption, economic growth and private wealth accumulation. A large cadre of prominent scientists are promoting *de-growth* in order to "minimize the depletion of natural resources, the production of waste and ultimately environmental degradation and social conflicts (Marín-Beltrán et al., 2022, p. 11). In support of this recommendation, they cite dramatic increases in human waste production over the last century (and, likely, in the foreseeable future):

> While human population grew by a factor of 5 between 1900 and 2015, world GDP (Gross Domestic Product) at constant prices increased over 21 times corresponding to an average annual rate of growth of 3% (estimated from Bradford De Long, 1998) ... During the same period, water withdrawal increased by a factor of 6; food production by a factor of 5; the extraction of fossil fuels and energy supply by a factor of 15 and 14, respectively; metal extraction by a factor of 33, and the extraction of other, nonmetallic, minerals by a factor of 50 (p. 2).

In labelling humanity—broadly speaking—'The Society of Waste,' this group of prominent scientists stress that their recommendation for de-growth is likely to be met with strong opposition from "the most powerful actors of the *current consumption-driven economic model*, who benefit from it" (Marín-Beltrán et al., 2022, p. 11; emphasis added). Nevertheless, another group of prominent scientists (Bradshaw et al., 2021) suggest that

> [t]he gravity of the situation requires fundamental changes to *global capitalism*, education, and equality, which include inter alia [among other things] … abolition of perpetual economic growth, properly pricing externalities, a rapid exit from fossil-fuel use, strict regulation of markets and property acquisition, reigning in corporate lobbying, and the empowerment of women (emphasis added).

8.3 Towards Revolutionary Change

If, as discussed above, fundamental changes are required in our socioeconomic systems, we must—of course—have relatively clear conceptions of the nature of them. Broadly speaking, it is apparent that capitalism is *the* dominant global system. Having said that, there are contested conceptions and varying forms of it. To overcome such complexity, diversity and uncertainty, it may be helpful to think of it very broadly. Mueller (2012), for instance, has said that it can generally be thought of as *control of capital exchanges by few individuals and groups*. Such a general conception encompasses capitalist forms varying, for example, from extremely *laissez faire* market-based decision making that was attributed to Hayek (1944), through to *neoliberal* forms featuring *state facilitation* of market-based decisions (Cahill et al., 2018) and on to highly *authoritarian* state domination—as in China—of capitalist enterprises (Liebman & Milhaupt, 2015). With such variety in forms of capitalism, it seems difficult to find a common denominator. However, it we focus on the word "control" in the broad definition above, it seems that capitalist control, while varying, may be inherent to Michel Foucault's definition of a *dispositif* (Rossi, 2013) (*apparatus*, in English). Although Foucault described it in different ways at different times, a version commonly referenced by scholars is as follows:

> [A dispositif is] a thoroughly heterogenous ensemble consisting of discourses, institutions, architectural forms, regulatory decisions, laws, administrative measures, scientific statements, philosophical, moral and philanthropic propositions ... The [dispositif] itself is the system of relations [among heterogeneous elements]. … [A] particular discourse can figure at one time as the programme of an institution, and at another it can function as a means of justifying or masking a practice which itself remains silent … [A dispositif also can be considered [a] formation which has as its major function at a given historical moment that of responding to an urgent need … [or] strategic imperative (Foucault, 1980, pp. 194–195).

This definition appears useful in understanding neoliberal capitalism that has dominated major putative democracies around the world for about the last half century. Although financial and other elite have long had significant power over large fractions of societies, such influences have, apparently, dramatically escalated since about the 1970s when, largely in response to decreases in their shares of wealth, in part due to costs of infrastructure re-building and social programming after the second world war, economic elite, government officials and others worked to establish *neoliberal* socio-economic systems that, broadly, would assimilate governments, corporations, financiers, free-market think tanks (e.g., Atlas Network), transnational trade groups (e.g., World Trade Organization), banks, transportation networks, trade agreements, investor-state dispute settlement mechanisms, special economic zones, militia (Provost & Kennard, 2023), STEM fields, educational programmes, perspectives and practices of large fractions of societies and many more entities into a kind of 'super-dispositif' promoting normalized policies and practices like competitiveness, individualism, cost externalizations, privatization, etc. that, ultimately, would funnel more wealth and wellbeing towards socio-economic elite (Baker, 2006; Harvey, 2005; McMurtry, 2013; Springer et al., 2016). This neoliberal dispositif appears to have, indeed, worked well, dramatically increasing differences between super-rich and most other people on the planet (Oxfam, 2023)—wealth disparities that, moreover, are predicted to dramatically increase, despite such interruptions as the 2008 global financial crisis (Piketty, 2020).

Given 'collusion' by governments and many other entities in neoliberal wealth concentration, it is apparent that challenges to such problematic systems must arise from individuals and groups of citizens (Freire, 1972; Marín-Beltrán et al., 2022). Assuming neoliberal power derives from its distribution across networks of living, nonliving and symbolic actants, it seems clear that opposition to it must emulate such network power; that is, form one or more dispositifs that may differ ontologically, but largely favour values like increased ecojustice (Martusewicz et al., 2021); that is, social justice and environmental vitality—values, more or less, antithetical to those of capitalist (Hardt & Negri, 2019; Lawrence, 2022; McBay, 2019a, b). To mobilize/spread such values across populations, changes in education can help (Bradshaw et al., 2021). Given that fields of science/STEM are relatively 'co-productive' with (reciprocally related to) societal zeitgeists (e.g., normalized values and principles) (Jasanoff, 2004, 2015), an important agent of such education could/should be school science/STEM education (Hodson, 2011; Roth & Désautels, 2002).

8.4 Educating Students About Issues, Choices and Actions

Traditionally, science has been taught in isolation from other subjects, a priority partly fed by claims that its focus on abstract, decontextualized knowledge makes it fundamental to progress in related fields like technology and engineering (Ziman, 1984). Lewis (1995), meanwhile, suggested it was part of our Platonic Legacy, conceiving ideal societies as governed by elite who prioritize abstract knowledge. Since about the early 1970s, however, educational researchers, book publishers, school systems and others claim that it should be taught in relationships with many other disciplines. In this period, for example, there have been numerous efforts around the world to promote different *science-in-context* (SinC) approaches like SAQ (*socially-acute questions*), SSI (*socioscientific issues*) and STSE (*Science, Technology, Society & Environment*) education (Bencze et al., 2020). Although much research and publication has been developed that defends and exemplifies SinC education in different educational situations, mobilization of such education across many school science contexts has been difficult (Hodson, 2011). Among reasons for difficulties in this regard, it appears that a major factor pertains to tendencies of school science systems to prioritize teaching/learning of 'products'—such as laws, theories and inventions—of science and technology that may be perceived not to involve influences from extra-STEM disciplines. When it *has* been implemented, however, a strong tendency appears to be to emphasize *controversies* (Pedretti & Nazir, 2011; Zeidler, 2014). In doing so, students often are invited to debate conflicting positions held by various 'stakeholders'—including politicians, company executives, citizens, activists, scientists, and others—regarding possible harmful effects of products (e.g., biotechnologies) of science and technology on wellbeing of individuals, societies and/or environments (Sadler, 2011). This often involves interrogation of data and claims—frequently in social situations—and then development of highly logical personal positions on issues (Levinson, 2010). Zeidler (2014), for example, who has had significant influences on the nature of socioscientific issues education, states that:

> the SSI framework is aligned with a progressive paradigm of instruction. Here, the justification of associated knowledge is derived through evidence-based reasoning, and challenging the assumptions of dominant knowledge claims is commonplace. SSI curriculum and

> associated pedagogy encourage students to prioritize methods of inquiry while interpreting issues, making decisions including moral judgments, solving problems, and engaging in various forms of discourse including argumentation and negotiation that are, in the long term, aimed at the development of character (p. 698).

On the one hand, there appear to be clear benefits of such argumentation-based approaches, including: development of students' socioscientific *reasoning* skills (Sadler et al., 2007); learning of *products* of science (e.g., laws & theories) (Venville & Dawson, 2010); and, learning about the *nature of science* (Khishfe & Lederman, 2006). Nevertheless, there may be problems when socioscientific issues education practice is limited to *individual*, albeit logical and data-based, decisions. Students often may be, for instance, placed in the role of *receivers* of knowledge from experts and/or people with power—perhaps creating certain dependencies. In analyses of socioscientific issues education approaches, Levinson (2010), for example, concluded that most prioritized 'Deficit' (citizens needing to be informed) or 'Deliberative' (citizens engaged in discussions with fellow citizens) models of citizenship—both of which he claims place most citizens in deference to experts and/or people with power. Such learning outcomes seem reasonable in *representative* democracies, in which citizens only periodically influence governing—as they, for example, participate in elections of leaders every few years (Wood, 1998). However, this can be problematic when leaders and/or experts help withhold or distort information—as they often do when supported, often financially, by private sector interests (Krimsky, 2019; McMurtry, 2013; Michaels, 2020; Oreskes & Conway, 2023; Ziman, 2000). In this light, there have been numerous calls for more *participatory* democratic forms (Chomsky, 2017; Monbiot, 2017). In science/STEM education contexts involving SinC approaches, this implies greater promotion of citizen engagements in critiques of contexts in which they and others live (around the globe) and development and implementation of well-informed and socially negotiated citizen actions to overcome harms of their concern and/or interest (Groleau et al., 2021; Hodson, 2020; Sjöström et al., 2017).

8.5 Stepping Towards Increased Mobilization of Sociopolitical Actions

Since 2006, the first author here has been facilitating action research with school science teachers, after-school science educators and others as they implement (or adapt) the STEPWISE curriculum and pedagogical framework illustrated in Fig. 8.1 (Bencze, 2017). This schema is used to develop teacher lessons and student activities to educate students in ways that they can eventually *independently* design and conduct research-informed and negotiated action (RiNA) projects to overcome harms of their concern in relationships among STEM fields and other societal members and living and nonliving environments (STEM-SE relationships). An essential aspect of the 3-phase pedagogy illustrated in Fig. 8.1 is that teachers are urged to

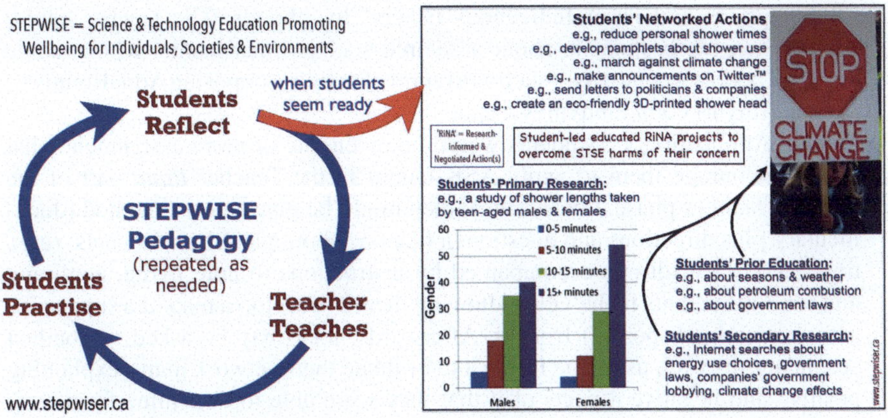

Fig. 8.1 STEPWISE curriculum & pedagogical schema

use *application-based teaching* (in the *Teacher Teaches* phase)—instead of *inquiry-based learning*—to ensure that all students can learn attitudes, skills and knowledge (ASK) that often are unavailable to many of them, at least due 'deficiencies' in:

1. Individual Students Capabilities: Due, for example, to diversity—often, for example, as a bell curve—of students' home access to cultural and social capital (Bourdieu, 1986), many of them may struggle to discover through their inquiries *abstract* ASK, such as reactions in photosynthesis (Abrahams & Millar, 2008; Bencze & Alsop, 2009);
2. STEM Curricula: Many or most students may not receive teacher instruction about problematic influences of for-profit entities on STEM fields, like lack of publication of negative results of tests of new commodities like drugs (Krimsky, 2019), because such ASK often are missing or sanitized in many STEM curricula (Hoeg & Bencze, 2017a, b); and,
3. Public Conceptions: Many or most students may not receive teacher instruction about or discover through secondary research problematic influences of for-profit entities on STEM fields (Krimsky, 2019) because for-profit entities often facilitate omissions or sanitizations of ASK available to the public in order to, for example, cast doubt on them (Michaels, 2020; Oreskes & Conway, 2023).

A possible application-based teaching lesson (in the *Teacher Teaches* phase of the STEPWISE pedagogy, Fig. 8.1) in a biology course could involve:

1. Teacher Input: The teacher might use relatively direct teaching methods to teach students about major aspects of genetically-engineered (GE) salmon (Pierce, 2013), such as semiotics of their size implying abundance of food supplies that can mask network relations (a *dispositif*) involving GE salmon, bioengineers, salt-water growing pens, sea lice (salmon parasites), genetic information (e.g., knowledge or recombinant DNA), AquaBounty™ Corp., Government Regulation Policies (FDA [Food & Drug Administration]) that may, for example, increase

salmon prices and harm Indigenous fishers' livelihoods. The teacher could/should then teach students about a sample sociopolitical action to overcome harms of people's concerns—such as a report of activists opposing AquaBounty™ salmon (tinyurl.com/4ubhvmvy); and,

2. Student Application(s): Students would be given one or more assignments that would encourage them to apply ASK taught in the Teacher *Input* part of the *Teacher Teaches* phase. For instance, they might be given a case method (documentary plus directions and questions), like one about plant-based 'meats' (e.g., tinyurl.com/5ewxduyw) accompanied by instructions to note living, nonliving and semiotic actants in the video allowing them to develop a map of a supportive actor-network (dispositif) from it. Afterwards, they may be asked to conduct secondary research using the Internet to evaluate their network maps, explaining positive and negative aspects of it that they were able to determine from their research. They also may be asked to evaluate the video as a sociopolitical action to overcome harms of interest to the video producers.

After such applied instruction, students may be asked to design and implement a small-scale RiNA project (*Students Practise*) to address problems of their concern related to other biotechnological products and services, possibly supplied by the teacher. At this stage, the teacher may help students who are struggling and/or request assistance. Afterwards, when the teacher feels students have sufficient expertise, confidence and motivation, they could be asked to conduct *Student-led* RiNA Projects to overcome harms of their concern relating to biology.

Over the years, students experiencing STEPWISE-based pedagogies seemed to have had many successes in developing and implementing personal and sociopolitical actions to address harms of their concern. With reference to Fig. 8.2, students first tended to design and carry out 'propositional' (Sign ➔ World) actions, such as posters, slideshows, social media posts, etc. about problems regarding shampoos. Later, especially after new lessons, many conducted more 'practical' actions, like ecojust engineering products (e.g., a newly-designed ecojust shampoo) (Bencze, 2017).

In light of prominence of engineering design in many STEM education initiatives (Pleasant & Olson, 2018), actor-network theory (Latour, 2005) and Foucault's (1980) dispositif concept, Larry and colleagues have—for 4 years (8 semesters)—been working with a technology-capable teacher ('Dan') of high school science and students in his science or biology classes to encourage and enable them to 'mobilize' (distribute across networks) values, principles, etc. that they consider integral to engineering products/services of their design. Although we encouraged such mobilization by first teaching students (with applications) about concepts from *Science and Technology Studies* like engineering design compromises, actor-network theory, dispositifs, governmentality (via the panopticon concept), phenomena-representation dialectics, sociotechnical imaginaries and regulatory capture, many students in Dan's classes, who were ethnically-diverse, socioeconomically-advantaged and in 'academic' (university-qualifying) courses, struggled to develop and apply concepts that would seem to enhance mobilization (Bencze et al., 2023). We believe, accordingly, much work in this regard is still needed.

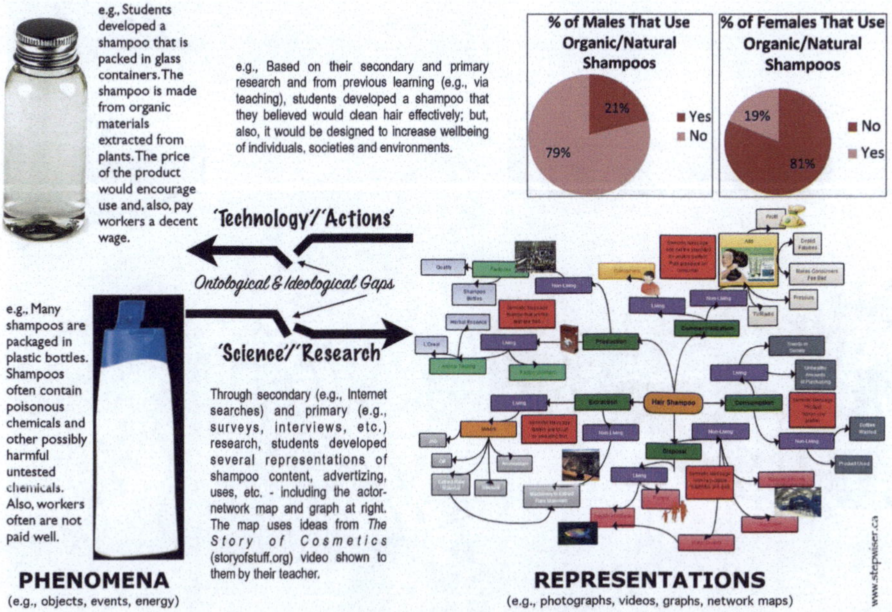

Fig. 8.2 Student RiNA project (e.g., Ecojust Shampoo)

8.6 Context and Methods: Learning Potential of a Longitudinal Study of Civic Actions

Given struggles that 'Dan' (mainly working with Larry) has had in encouraging and enabling students to envisage dispositifs supporting values important to them, the two of us here conducted analyses of long-term changes—over about a decade—in 'a' dispositif that we previously studied (Bencze & Pouliot, 2017). This case concerns events following reddish dust deposits that citizens noticed, in 2012, accumulating in their neighbourhood of Québec City, Canada. Figure 8.3 provides some photographs of scenes around the city that we suggest depict the nature of dust dispersal and accumulation. Soon after they noticed this dust and after city officials told them not to be alarmed and that the dust likely was 'rust,' two citizens of the Limoilou borough of Québec City, Véronique Lalande, and her partner, Louis Duchesne, decided to further investigate this situation. After carefully collecting samples of the dust, they commissioned a local laboratory to test its contents. Results suggested at least two important findings: i) the dust contained several heavy—potentially carcinogenic—metals, including relatively high amounts of: iron, arsenic, chromium, cobalt, copper, manganese, nickel, lead, vanadium and zinc; and, ii) the dust 'signature' (nickel-cobalt ratio) matched that of open piles of nickel ore on the city's marine port. These results spurred them to initiate numerous actions that they hoped would, ultimately, dramatically decrease or eliminate what they perceived to be toxic industrial pollution. Actions included: research about

Fig. 8.3 Reddish Dust Accumulating in Québec City and Its Ocean Port

previous reports of city dust pollution; class action law suits launched against port authorities and a transshipment company (revised once); media reports; an interactive website; and, a protest march. Reacting to early media reports of the dust dispersal and citizen concerns, Chantal began documenting many aspects of the citizens' actions. In 2015, she self-published a short—'citizen-friendly'— documentary in book form (Pouliot, 2015) about the case (available at: tinyurl.com/2phtppty). In this documentary, she highlighted numerous citizen actions and resistance they experienced to them—including from the port authority and shipping company. Soon after learning about her documentary, the two of us began collaborative analyses of the ongoing controversy and, after several presentations at refereed conferences (e.g., Bencze & Pouliot, 2015), we published a book chapter about it (Bencze & Pouliot, 2017). Related to these studies, Chantal has accumulated large amounts of 'data' about the dust controversy since 2012, including:

- News media reports: Many articles published in prominent Québec City and Montréal newspapers (*Le Soleil* and *Le Devoir*, among others) were read and analyzed. Numerous extracts from televised news presentations about the dust situation—and possibly-related port expansion plans—in Québec City were analyzed. Reports often dealt, for example, with acceptable airborne nickel levels, a port expansion proposal, court cases, university support and much more.

- Citizens' website contents: Text and graphics (videos and stills) appearing on the Initiative Citoyenne de Vigilance du Port de Québec website (www.vigilance-portdequebec.com), with material dating from July 2009 to present, were collected.
- Citizen and Government Reports: Reports produced by the public health department and the Ministry of the Environment in connection with the dust were consulted.
- Public Meetings: Numerous public meetings occurred, hosted, for example, by citizens, city officials (e.g., Public Health department members), with anecdotal records written by Chantal.
- Court Case Proceedings: Chantal attended numerous public sessions relating to two Class Action Suits (Citizens vs. Shipping Company and Citizens vs. Port Authority & Shipping Company), spending about 60 days, again, taking notes.
- Dust Collection and Exposure: Chantal, particularly as a resident of Limoilou from 2000–2017, was exposed to much dust and, from 2012, periodically collected samples of it.
- Photographs: In the context of numerous public meetings, Chantal took many photographs.

To analyze these data, we used constant comparison methods based on constructivist grounded theory (Charmaz, 2014), a relatively abductive process that likely involved considerable deduction from our respective background knowledge. Claims reported below resulted from data triangulation; that is, requiring three or more data sources for agreement (Flick, 2022).

8.7 Findings of Our Study of Long-Term Civic Actions

Our analyses, mainly drawing on data as described above collected by Chantal and from her decade-long studies of the dust controversy in Québec City (Bencze & Pouliot, 2017), generated the depictions of 'citizen' dispositifs in three *hypothetical* 'moments in time'; that is, 2012–2013 (Fig. 8.4) and 2015–2018 (Fig. 8.5) & 2021–2022 (Fig. 8.6). We consider these depictions to be, of course, *hypothetical* in light, for example, of our relatively *Naturalist-Antirealist* (Loving, 1991) epistemological views—which posit that efforts to know tend to be influenced by numerous psycho-social factors and that conclusions about phenomena can only be considered best—often socially determined—approximations. As well, actor-networks are considered *dynamic* and, so, we cannot be absolutely certain of precise ontological components, relationships among them (thus, no connecting lines) and more. Nevertheless, given Chantal's *lengthy* immersion in relevant contexts (e.g., attending many civic meetings and court hearings) and our long-term collaborative study (since about 2015) of the ongoing case, we hope readers find our depictions—along with qualitative descriptions of them below by us—relatively reliable, valid and useful.

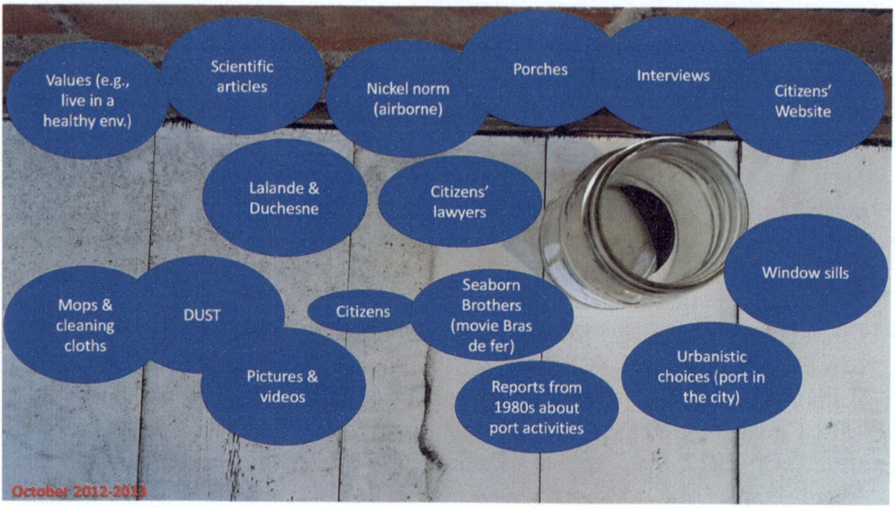

Fig. 8.4 Citizen dispositif (Chantal's French summary), 2012–2013

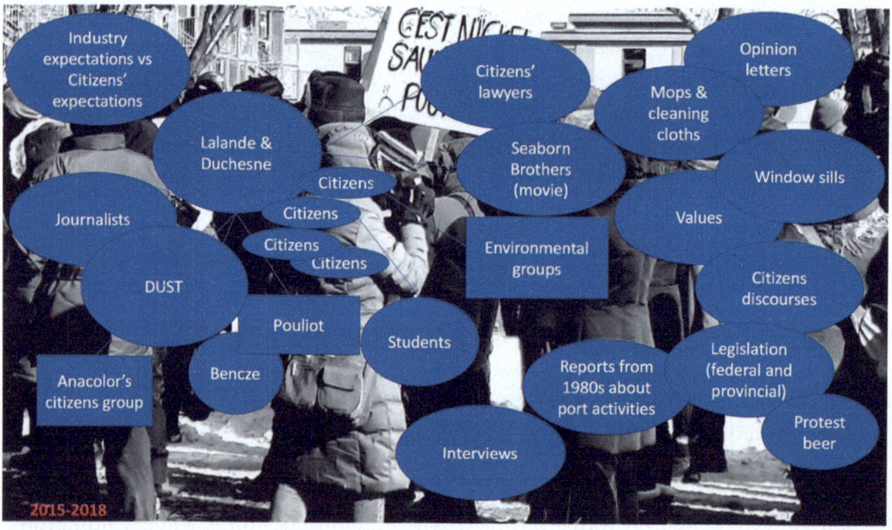

Fig. 8.5 'Snapshots' of citizen dispositif (and Chantal's summary in French) in the 2015–17 period

The image in Fig. 8.4 is meant, in a sense, to serve as a 'baseline' for our study. It depicts key actants that we believe assembled around the two 'instigating' citizens, Véronique Lalande and Louis Duchesne, about a year after they began (in October, 2012) investigating and developing actions to address what they perceived to be toxic dust emanating from their city's port. True to actor-network theory (Latour, 2005) and Foucault's (1980) conceptions of networked power, the growing dispositif supporting healthy air included inanimate objects, like the dust (seen in

Fig. 8.6 ‘Snapshots’ of citizen dispositif (and Chantal’s summary in French) in the 2021–22 period

Fig. 8.4 accumulated on the left side of a porch [cleaned on the right side]), material objects (*inscriptions*) generated by humans, like photographs, 1980s reports about port activities (initially unknown), the movie about the controversy (produced at that time and screened in March 2018), the website, dust data analyses, etc., and values, like rights to live in a healthy environment. Again, however, the image in Fig. 8.4 must *not* be considered to be an exact representation of the 2012–2013 period; but, rather, a possibly helpful depiction of a *dynamic* dispositif—in this case, about ‘citizen’ wishes for healthier air in their community. In that vein, it is important to note that some actants identified in Fig. 8.4 (certainly the dust, but also—for example—the Seaborn brothers’ movie) continued to be engaged in ‘the’ ongoing dispositif, while others seemed to become less active (e.g., some citizens [not shown]) and others (e.g., The Anacolor™ paint company, which was forced by citizen activism to move its plant] citizens group) appeared to be added into ‘it.’ Such dynamism over time is depicted (by us) in Figs. 8.5 and 8.6.

After an initial ‘flurry’ of activities associated with ‘the’ early citizen dispositif depicted in Fig. 8.4, there was somewhat of a lull period until, in 2015, the Québec (City) Port Authority applied for permissions (e.g., environmentally) to expand its surface area and related port transshipment activities. Imagining much more future toxic dust dispersal, the ‘simmering’ citizen dispositif was, relatively suddenly, ‘reignited.’ In Foucauldian (1980) terms, the dispositif responded to a new purpose; that is, threats of escalating pollution and associated health problems. Intense and widespread media reports and opinion pieces increased public interests and concerns surrounding the expansion project to ‘the’ point that it became what Legardez and Simonneaux (2006) have described as a *Socially-Acute Question* (SAQ); that is, a controversy that is popularly considered ‘acute’ in: society, fields of research, and in educational contexts. This acute risk stimulated expansion of ‘the’ citizen dispositif—as illustrated in Fig. 8.5 (2015–2018)—with, for example, citizens writing oppositional opinion letters and briefs for public hearings, journalists adding numerous media publications, Laval university student associations speaking out against the university's support for port expansion, Chantal writing an open-access book about the case for citizens (Pouliot, 2015), and the two of us publishing a book chapter discussing educational implications of the case (Bencze & Pouliot, 2017). Although the port authority persisted with their recommended expansion, even changing the project's name from "Beauport 2020" to "Laurentia" (as, perhaps, a positive semiotic message), its proposal was ultimately rejected—in light of environmental concerns—in 2021. This result seemed like a victory for the citizen dispositif.

The rejection of the port expansion proposal—itself supported by a growing dispositif—did, indeed, seem like a victory for the citizen dispositif; but, unfortunately for 'it,' joy was short-lived. Primarily for *economic* reasons, most prominently—and, perhaps, ironically—due to rapid expansion of needs for nickel for batteries used in apparently more environmentally-sound electric vehicle production, the Québec provincial government *mandated*, in 2021, an *increase* in allowable airborne nickel. This development, perhaps as expected, again spurred ignition of the apparently simmering SAQ that, in turn, appeared to lead to new rapid expansion (and diversification) of 'the' citizen dispositif (Fig. 8.6, 2021–2022). Many more citizens became involved in the dispute—many of them, for example, becoming aware of the concept of 'sacrifice zones'; that is, geopolitical regions with pro-capitalist legal arrangements, such as relaxed environmental and labour laws (Klein, 2007; Provost & Kennard, 2023). But, as well, public health departments in the province of Québec pointed to health risks of increased nickel particles in the air, concerned parents in the Limoilou district wrote an opinion letter to and met with the Québec (provincial) Minister of the Environment to question this increase. Representatives of a democratic socialist and sovereigntist political party (QS [Québec solidaire]) also spoke out against the government's decision. So did an association of doctors and a group of chemists. Even the mayor of Quebec City fiercely opposed the government's decision. Many people also developed creative artistic actions—as illustrated in Fig. 8.7—to challenge those threatening city air quality. Finally, air quality in Limoilou was such an important concern that citizens initiated a research project to document air quality.

More or less in parallel with dynamic changes in the citizen dispositif were, as we published earlier (Bencze & Pouliot, 2017), developments in 'a' 'pro-development' dispositif. In Fig. 8.8, because we have not studied it as extensively and deeply as 'the' citizen dispositif, we provide a depiction of some actants we believe have played major roles in a dispositif that has *intensely opposed* the citizen dispositif and, instead, persistently defended socioeconomic developments relating to Québec City port infrastructure and activities. The representation of the pro-development dispositif can take on different appearances, depending on the moment and the subject. In 2020, the city mayor (at the time) supported the idea of expanding the port. So did the business community. Port officials spoke of the growing need for space and needs to upgrade the port's infrastructure. The 7 million

Fig. 8.7 Citizen Artistic Actions Opposing Polluted Air (and Chantal's summary in French)

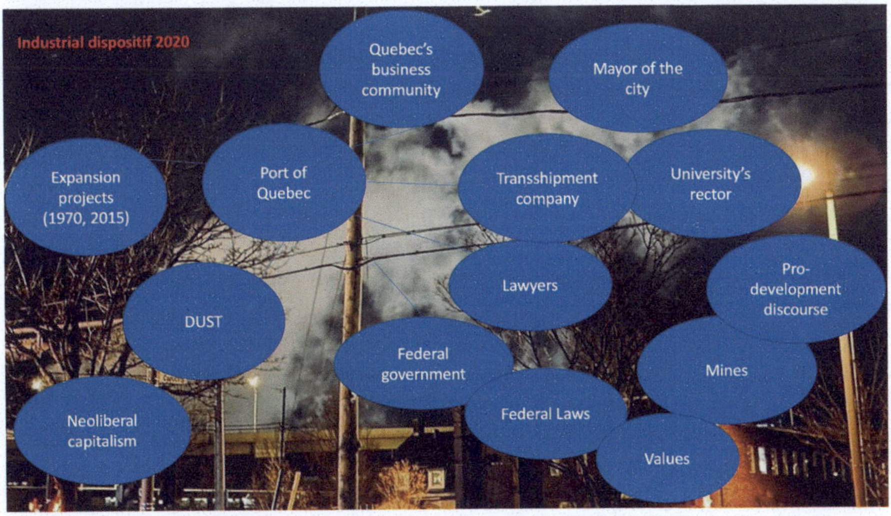

Fig. 8.8 Pro-development Dispositif (and Chantal's summary in French)

containers that would eventually be unloaded each year were expected to travel by truck or train through the Limoilou district. The economic arguments ran counter to those put forward by citizens.

8.8 Coda

Although, admittedly, we began research outlined above with presuppositions about merits of using Foucault's (1980) dispositif concept for analying *existing* situations (and controversies) and visions of ways in which apparently problematic situations may be rectified, we suggest that this concept *also* is quite useful for analyzing *changes* in such situations over relatively long time periods. Accordingly, based on our discussions above, we suggest that students in democracies (and otherwise) can learn much about social and environmental changes through our longitudinal case descriptions above.

Given apparent merits of pedagogical approaches like STEPWISE (Bencze, 2017) promoting civic actions, which has received supports from many educators (e.g., Groleau et al., 2021; Pedretti & Nazir, 2011; Sjöström et al., 2017), we recommend that educators begin their courses—and, possibly, many or most topics—by asking students to reflect on and express their pre-instructional attitudes, skills and knowledge (ASK) regarding topics about which the teacher intends to teach. Such exploratory activities should, we suggest, be largely *student-directed and open-ended* (Lock, 1990) to help ensure that students express *their* ASK, rather than those they assume that teachers would prefer. A common activity for getting students to express their pre-conceived notions is to ask them relatively divergent questions about phenomena related to a topic (e.g., basic chemistry) about which they are to study. Students could be shown examples of common commodities (e.g., dish detergent, stain remover and oil spills) and asked questions like: 'What do you (dis-)like about this and why?,' 'What people and groups support and oppose) this product and why?' and 'For problems of your concern about this product, what could/should be done about them and why might they work?; Explain' (Krstovic, 2017). In theory, students could discuss numerous actants to which the product is attached, including corporations and governments (and their laws and regulations), etc. This varies, of course, from class to class and from product to product, but students often do not mention cooperating networks, per se, or dispositifs and many major actants within them. Rarely, if ever, have any students mentioned—at this stage—transnational actants, like the World Bank or World Trade Organization.

Given most students are unlikely to discover important facts and concepts relating to networks and dispositifs, for reasons like those discussed above, the STEPWISE pedagogy (Fig. 8.1) suggests that—after the *Students Reflect* stage—teachers teach students (in the *Teacher Teaches* phase) about important ASK to which they may otherwise not have access. It would be at this point, that the teacher would teach students—using direct approaches like lectures with slideshows and, perhaps, Socratic questioning to increase student engagement—about concepts and principles like those that might be learned from our depictions of changes in dispositifs over a decade in the context of the Québec City dust pollution controversy. A good place to start, to perhaps give students a sense of enhanced 'power' of dispositifs as increasing numbers of cooperating actants are assembled, is to teach students about Latour's (1990) example of how a hotel manager 'convinces' patrons to leave their room keys at the hotel desk before leaving the hotel; that is, by progressively adding items like: a verbal request, a sign and, then, a heavy weight attached to each key. Teachers may want to adapt such an example to make it relevant to learners' ages and living contexts, such as by using examples like: leaving a toilet seat down after use, returning a washroom key at a gas/petroleum station desk or placing clothing on a hanger after trying them on in a shop. When the teacher feels it is appropriate, students may be exposed to the Québec City dust case—perhaps starting with relevant photographs, proceeding through descriptions and then showing students actant arrays ('dispositifs') like those shown above. In doing so, it would be important for the teacher to discuss likely relationships *within* the arrays/dispositifs—such as between Véronique Lalande and Louis Duchesne, dust samples and dust sample element analyses. To supplement teaching about the dust case (as an 'isolated' phenomenon), students may be asked to develop tentative dispositif maps from their readings of our *Ban the Dust* graphic novel (tinyurl.com/yxa9ptq6). After students appear to have well understood the dust controversy and use of the dispositif concept, the teacher may then share with students different general categories about the nature of *dynamics* of dispositif change. To assist with this, we analyzed (using constant comparative methods based on constructivist grounded theory) our data and associated dispositif depictions above and generated the list of 15 general categories provided in Table 8.1. Teachers may, of course, revise this list—drawing their own conclusions from our descriptions and sketches above.

Once the teacher believes students have developed 'reasonably' strong understandings about characteristics of changes in dispositifs regarding a controversy, students can be engaged in activities (still in the *Teacher Teaches* phase, Fig. 8.1) that ask them to *apply* ideas, concepts, etc. that they have just been taught—and others students generate—in analyses of documentaries of other STEM-SE controversies. Encouraging such application after instruction can deepen students' understand of and commitments to concepts, skills, etc. involved—in light of *knowledge duality theory*, which suggests that such gains may occur as students increasingly gain control over conclusions about phenomena and actions to change them (Wenger, 1998). They may, for example, be asked to review different Multi-Actant Documentaries (MADSs at: tinyurl.com/9acay7ca) and then respond to some teacher requests and questions; e.g., 'Develop a dispositif network supporting this commodity,' 'Develop a dispositif opposing this commodity and how that may change over time,' 'Name [x; e.g., 5] actants that when joined with others may make a stronger dispositif.' Combining teaching of aspects of dispositif change like those we derived from our study of the Québec City dust controversy with students' more student-led relevant activities like the MADs mentioned above can be considered an example of an *Inductive-Deductive Dialectic Immersion* (IDDI) *activity* (Bencze & Elshof, 2004). In such activities, students are first taught something (e.g., about the

Table 8.1 Dispositif dynamism categories

1. 'Acuteness' levels i.e., public consciousness of problem(s) fluctuates ('ebbs & flows') over time **2 .'Reignition' drivers** i.e., different objects, events, etc. may stimulate increases in public consciousness **3. Ongoing 'embers'** i.e., critical consciousness may 'simmer' while problem(s) still exist(s) **4. Interest groups** e.g., different forums (e.g., town meetings, website contributions) for analyses &actions develop **5. Controversy lengths** e.g., resolution of issues or problems can take several months, if not years	**6. Meta-studies** i.e., more researchers study controversy and more education teach about it **7. Dispositif composition** i.e., different living, nonliving & symbolic entities (actants) join the movement **8. Data influences** e.g., respect for citizen data may vary over time, depending on interest groups **9. Relationships** e.g., collaborations change among different 'stakeholders' (e.g., citizens & scientists) **10. Public awareness** i.e., despite fluctuations in 'acuteness' of controversy, general public consciousness increases	**11. Funding** e.g., support from banks, private citizens and other activist groups helps activist causes **11. Symbolic battles** e.g., powerful stakeholders may develop symbolic 'disinformation' and intimidation tactics **12. Leadership & hierarchies** i.e., although group members have different roles, there often are different leaders **13. Personal expertise** e.g., group members develop certain kinds of expertise (e.g., poster making, economic skills, etc.) **14. Legal supports** i.e., support from lawyers, often voluntarily, tends to be crucial to activist causes

dust case) that they can then test/evaluate in a related context, in which they can *also* induce concepts (e.g., discover aspects of dispositifs as they work with them). Such IDDI activities appear to help students to deepen and more personalize their learning.

Lessons and activities like those described above can help students to increase attitudes, skills and knowledge in each of Creswell's (2007) categories of research foci; that is, (i) *Ontologically* (e.g., by gaining understanding of actants in dispositifs); (ii) *Methodologically* (e.g., how to analyze a new controversial case using the dispositif concept); (iii) *Epistemologically* (e.g., learning numerous aspects of dispositif change, as illustrated in Table 8.1); and, (iv) *Axiologically* (e.g., awareness of different value systems influencing assemblage of actants into competing dispositifs).

Once the teacher believes that students have sufficient expertise, confidence and motivation to use the dispositif concept (along with other relevant ASK that would be taught in the STEPWISE *Teacher Teaches* phase, Fig. 8.1), they may be asked by the teacher to design and implement student-directed and open-ended (Lock, 1990) RiNA projects to overcome harms of their concern in STEM-SE relationships. In such projects in democratic societies, students should have freedom to use (or not use) the dispositif concept. Having said that, they may be more likely—as argued above—to use it if they have been taught using application-based teaching. Also as argued above, if students do have expertise, confidence and motivation to use the dispositif concept in framing and possibly changing the world, there may be increased hope in addressing myriad—often worrisomely intertwining—crises like those reviewed above (and more).

Although teachers' uses of the STEPWISE pedagogy appears to have been successful in encouraging and enabling students to develop and implement a great variety of apparently 'extensive' RiNA projects (e.g., Bencze, 2017; Bencze & Alsop, 2014), it also is apparent that such 'successes' seem limited to relatively few contexts in which supportive *educational* dispositifs exist (Bencze & Krstovic, 2017); that is, when actants like the following cooperate: official curricula (e.g., MoE, 2008) aligned with STEPWISE-like values and principles; a teacher supporting *Naturalist-Antirealist* (Loving, 1991) views about science; available relevant teaching and learning materials (e.g., via action research facilitated by a researcher); school administrators and colleagues supportive of teacher reflective practice (praxis); and, students with relatively well developed cultural and social capital. Given relative *rarity* of such educational 'mini-dispositifs,' it seems clear that much work is needed to *expand and deepen* them if values inherent to STEPWISE and similar pedagogies—such as concepts, principles, etc. derived from our analyses of 'the' evolving Québec City dust controversy dispositifs here (Table 8.1) and ecojustice education (e.g., Martusewicz et al., 2021)—are to become more normalized across societies.

References

Abrahams, I., & Millar, R. (2008). Does practical work really work? A study of the effectiveness of practical work as a teaching and learning method in school science. *International Journal of Science Education, 30*(14), 1945–1969. https://doi.org/10.1080/09500690701749305

Baker, D. (2006). *The conservative nanny state: How the wealthy use the government to stay rich and get richer*. Center for Economic and Policy Research.

Bencze, J. L. (Ed.). (2017). *Science and technology education promoting wellbeing for individuals, societies and environments*. Springer.

Bencze, J. L., & Alsop, S. (2009). A critical and creative inquiry into school science inquiry. In W.-M. Roth & K. Tobin (Eds.), *World of science education: North America* (pp. 27–47). Sense.

Bencze, J. L., & Alsop, S. (Eds.). (2014). *Activist science & technology education*. Springer.

Bencze, L., Del Gobbo, D., Zouda, M., El Halwany, S., Ibrahim, S., Guerrero Hernandez, G., Peixoto, M. (2023). *Students' ecojust engineering design and normalization in/through secondary school science*. A presentation at the annual conference of the Society for Social Studies of Science, Nov. 8–12, 2023, Honolulu, Hawai'i.

Bencze, L., & Elshof, L. (2004). Science teachers as metascientists: An inductive-deductive dialectic immersion in northern alpine field ecology. *International Journal of Science Education, 26*(12), 1507–1526.

Bencze, L., & Krstovic, M. (2017). Resisting the Borg: Science teaching for common wellbeing. In J. L. Bencze (Ed.), *Science and technology education promoting wellbeing for individuals, societies and environments* (pp. 227–276). Springer.

Bencze, L., & Pouliot, C. (2015). Activist dispositifs for/though STEM education. A presentation at the annual conference of the *Society for Social Studies of Science*, Denver, CO, Nov. 11–15, 2015.

Bencze, L., & Pouliot, C. (2017). Battle of the bands: Toxic dust, active citizenship and science education. In J. L. Bencze (Ed.), *Science and technology education promoting wellbeing for individuals, societies and environments* (pp. 381–404). Springer.

Bencze, L., Pouliot, C., Pedretti, E., Simonneaux, L., Simonneaux, J., & Zeidler, D. (2020). SAQ, SSI and STSE education: Defending and extending 'Science-in-Context'. *Cultural Studies of Science Education, 15*(3), 825–851.

Bourdieu, P. (1986). The forms of capital. In J. Richardson (Ed.), *Handbook of theory and research for the sociology of education* (pp. 241–258). Greenwood.

Bradshaw, C.J.A., Ehrlich, P.R., Beattie, A., et al. (2021). Underestimating the challenges of avoiding a ghastly future. *Frontiers in Conservation Science, 1* (Jan. 13, 2021). Retrieved from: https://www.frontiersin.org/article/10.3389/fcosc.2020.615419

Bulletin of the Atomic Scientists [BAS] (Science and Security Board). (2023). *A time of unprecedented danger: It is 90 seconds to midnight* (2023 Doomsday Clock Statement, J. Mecklin, Ed.). Bulletin of Atomic Scientists. Retrieved from: https://thebulletin.org/doomsday-clock/current-time/

Cahill, D., Cooper, M., Konings, M., & Primrose, D. (2018). *The SAGE handbook of neoliberalism*. SAGE.

Charmaz, K. (2014). *Constructing grounded theory* (2nd ed.). Sage.

Chomsky, N. (2017). *Optimism over despair: On capitalism, empire, and social change*. Haymarket.

Creswell, J. W. (2007). *Qualitative inquiry & research design: Choosing among five approaches* (2nd ed.). Sage.

Flick, U. (2022). *The Sage handbook of qualitative research design* (1st ed.). Sage.

Foucault, M. (1980). In G. Colin (Ed.), *Power/knowledge: Selected interviews and other writings, 1972–1977*. Pantheon Books.

Freire. (1972). *Pedagogy of the oppressed*. Penguin Books Ltd.

Groleau, A., Pouliot, C., & Arseneau, I. (2021). Educating for the Anthropocene. In F. J. Carrillo & G. Koch (Eds.), *Knowledge for the anthropocene: A multidisciplinary approach* (pp. 98–106). Edward Elgar Publishing.

Hardt, M., & Negri, A. (2019). Empire, twenty years on. *New Left Review, 120*, 67–92.

Harvey, D. (2005). *A brief history of neoliberalism*. Oxford University Press.

Hayek, F. A. (1944). *The road to serfdom*. University of Chicago Press.

Hodson, D. (2011). *Looking to the future: Building a curriculum for social activism*. Sense.

Hodson, D. (2020). Going beyond STS education: Building a curriculum for sociopolitical activism. *Canadian Journal of Science, Mathematics and Technology Education, 20*(4), 592–622. https://doi.org/10.1007/s42330-020-00114-6

Hoeg, D., & Bencze, L. (2017a). Values underpinning STEM education in the USA: An analysis of the Next Generation Science Standards. *Science Education, 101*(2), 278–301.

Hoeg, D., & Bencze, L. (2017b). Rising against a gathering storm: A biopolitical analysis of citizenship in STEM policy. *Cultural Studies of Science Education, 12*(4), 843–861.

Intergovernmental Panel on Climate Change [IPCC]. (2023). *Climate change 2023, Synthesis report: Summary for policy makers*. Intergovernmental Panel on Climate Change. https://doi.org/10.59327/IPCC/AR6-9789291691647.001

Jasanoff, S. (Ed.). (2004). *States of knowledge: The co-production of science and social order*. Routledge.

Jasanoff, S. (2015). Future imperfect: Science, technology, and the imaginations of modernity. In S. Jasanoff & S.-H. Kim (Eds.), *Dreamscapes of modernity: Sociotechnical imaginaries and the fabrication of power* (pp. 1–33). University of Chicago Press.

Khishfe, R., & Lederman, N. G. (2006). Teaching nature of science within a controversial topic: Integrated versus nonintegrated. *Journal of Research in Science Teaching, 43*(4), 395–418.

Klein, N. (2007). *The shock doctrine: The rise of disaster capitalism*. Henry Holt.

Krimsky, S. (2019). *Conflicts of interest in science: How corporate-funded academic research can threaten public health*. Simon & Schuster.

Krstovic, M. (2017). Learning about youth engagement in research-informed and negotiated actions on socio-scientific issues. In L. Bencze (Ed.), *Science and technology education promoting wellbeing for individuals, societies and environments* (pp. 93–114). Springer.
Latour, B. (1990). Technology is society made durable. *The Sociological Review, 38*(S1), 103–131.
Latour, B. (2005). *Reassembling the social: An introduction to actor-network-theory*. Oxford University Press.
Lawrence, M. (2022). *Polycrisis: Why we must turn this meme into a big idea*. Cascade Institute, Dec. 12, 2022. Retrieved from: https://cascadeinstitute.org/polycrisis-why-we-must-turn-this-meme-into-a-big-idea/
Legardez, A., & Simonneaux, L. (Eds.). (2006). *How to teach socially acute questions*. Économie Sociale Familiale (ESF).
Levinson, R. (2010). Science education and democratic participation: An uneasy congruence? *Studies in Science Education, 46*(1), 69–119.
Lewis, T. (1995). From manual training to technology education: The continuing struggle to establish a school subject in the USA. *Journal of Curriculum Studies, 27*(6), 621–645.
Liebman, B. L., & Milhaupt, C. J. (2015). Introduction: The institutional implications of China's economic development. In B. L. Liebman & C. J. Milhaupt (Eds.), *Regulating v visible hand? The institutional implications of Chinese state capitalism* (pp. xiii–xxvi). Oxford University Press.
Lock, R. (1990). Open-ended, problem-solving investigations—What do we mean and how can we use them? *School Science Review, 71*(256), 63–72.
Loving, C. C. (1991). The Scientific Theory Profile: A philosophy of science model for science teachers. *Journal of Research in Science Teaching, 28*(9), 823–838.
Marín-Beltrán, I., Demaria, F., Ofelio, C., et al. (2022). Scientists' warning against the society of waste. *Science of the Total Environment, 811*. https://doi.org/10.1016/j.scitotenv.2021.151359
Martusewicz, R., Edmundson, J., & Lupinacci, J. (Eds.). (2021). *Ecojustice education: Towards diverse, democratic, and sustainable communities* (3rd ed.). Routledge.
McBay, A. (2019a). *Full spectrum resistance (vol. 1): Building movements and fighting to win*. Seven Stories Press.
McBay, A. (2019b). *Full spectrum resistance (vol. 2): Actions and strategies for change*. Seven Stories Press.
McMurtry, J. (2013). *The cancer stage of capitalism: From crisis to cure*. Pluto.
Michaels, D. (2020). *The triumph of doubt: Dark money and the science of deception*. Oxford University Press.
Ministry of Education [MoE]. (2008). *The Ontario curriculum, grades 9 and 10: Science*. Queen's Printer for Ontario.
Monbiot, G. (2017). *Out of the wreckage: A new politics for an age of crisis*. Verso.
Monbiot, G. (2022). *Regenesis: Feeding the world without devouring the planet*. Allen Lane.
Mueller, D. C. (Ed.). (2012). *The Oxford handbook of capitalism*. Oxford University Press.
Ord, T. (2020). *The precipice: Existential risk and the future of humanity*. Bloomsbury.
Oreskes, N., & Conway, E. (2023). *The big myth: How American business taught us to loathe government and love the free market*. Bloomsbury Press.
Oxfam. (2023). *Survival of the richest: How we must tax the super-rich now to fight inequality*. Oxford International.
Pedretti, E., & Nazir, J. (2011). Currents in STSE education: mapping a complex field, 40 years on. *Science Education, 95*(4), 601–626. https://doi.org/10.1002/sce.20435
Pierce, C. (2013). *Education in the age of biocapitalism: Optimizing educational life for a flat world*. Palgrave Macmillan.
Piketty, T. (2020). *Capital and ideology* (trans. A. Goldhammer). Harvard University Press.
Pleasant, J., & Olson, J. K. (2018). What is engineering? Elaborating the nature of engineering for K-12 education. *Science Education, 103*(1), 145–166.
Pouliot, C. (2015). *Quand les citoyen.ne.s soulèvent la poussière*. Carte blanche.

Provost, C., & Kennard, M. (2023). *Silent coup: How corporations overthrew democracy*. Bloomsbury Academic.
Ripple, W. J., Moomaw, W. R., Wolf, C., et al. (2022). Six steps to integrate climate mitigation with adaptation for social justice. *Environmental Science & Policy, 128*(Complete), 41–44.
Ripple, W. J., Wolf, C., Newsome, T. M., et al. (2021). World scientists' warning of a climate emergency 2021. *Bioscience, 71*(9), 894–898.
Rossi, U. (2013). On the varying ontologies of capitalism: Embeddedness, dispossession, subsumption. *Progress in Human Geography, 37*(3), 348–365. https://doi.org/10.1177/0309132512463300
Roth, W.-M., & Désautels, J. (Eds.). (2002). *Science education as/for sociopolitical action*. Peter Lang.
Sadler, T. D. (Ed.). (2011). *Socio-scientific issues in science classroom: Teaching, learning and research*. Springer.
Sadler, T. D., Barab, S. A., & Scott, B. (2007). What do students gain by engaging in socioscientific inquiry? *Research in Science Education, 37*(4), 371–391.
Sjöström, J., Frerichs, N., Zin, V. G., & Eilks, I. (2017). Use of the concept of Bildung in the international science education literature, its potential, and implications for teaching and learning. *Studies in Science Education, 53*(2), 165–192.
Springer, S., Birch, K., & MacLeavy, J. (Eds.). (2016). *The handbook of neoliberalism*. Routledge.
Tooze, A. (2022). Welcome to the world of the polycrisis. *Financial Times*, London, October 29, 2022, 11. Retrieved from: https://www.ft.com/content/498398e7-11b1-494b-9cd3-6d669dc3de33
Union of Concerned Scientists [UCS]. (1992). *World scientists' warning to humanity*. Union of Concerned Scientists.
Venville, G. J., & Dawson, V. M. (2010). The impact of an argumentation intervention on Grade 10 students' conceptual understanding of genetics. *Journal of Research in Science Teaching, 48*(8), 952–977.
Wenger, E. (1998). *Communities of practice*. Cambridge University Press.
Wood, G. H. (1998). Democracy and the curriculum. In L. E. Beyer & M. W. Apple (Eds.), *The curriculum: Problems, politics and possibilities* (pp. 177–198). SUNY Press.
Zeidler, D. L. (2014). Socioscientific issues as a curriculum emphasis: Theory, research and practice. In N. G. Lederman & S. K. Abell (Eds.), *Handbook of research on science education, volume II* (pp. 697–726). Routledge. https://doi.org/10.4324/9780203097267.ch34
Ziman, J. (1984). *An introduction to science studies: The philosophical and social aspects of science and technology*. Cambridge University Press.
Ziman, J. (2000). *Real science: What it is, and what it means*. Cambridge University Press.

Chapter 9
Ocean Literacy as a Socioscientific Issue for Hope in the Anthropocene

Blanca Puig, Isabel María Cruz-Lorite, and Maria Evagorou

9.1 Introduction

The influence of humans is markedly changing Earth systems, including that of the Ocean (IPCC, 2021). Ocean literacy (OL) is an increasingly critical socioscientific issue, drawing attention to the ethical dimension of our relationship with the world's oceans (Leitão et al., 2022). Beyond scientific facts and ecological considerations, OL addresses the moral imperative of responsible citizenship. As human activities such as overfishing, plastic pollution, and climate change continue to degrade marine ecosystems, it becomes clear that our ethical responsibilities extend beyond merely understanding the ocean's role in the Earth's system (Chen & Tsai, 2016; Winks et al., 2020). This is related to the fact that we live in the Anthropocene, the concept that the Earth has moved in a new era characterized by human domination (Malhi, 2017). According to the notion of Anthropocene, "human activity is having a dominating presence on multiple aspects of the natural world and the functioning of the Earth system, and this has consequences on how we view and interact with the natural world—and perceive our place in in" (Malhi, 2017, p. 78). Key features of the Anthropocene include the understanding of two-way interactions between the natural world and the environment, and a realization of the short and long term changes happening to the planet as a whole (White et al., 2023). A need to educate young people to develop an ethic of care and justice is imperative (i.e. Merrett, 2004, and also OECD, 2018). Agency is also important, and the ability to make

B. Puig (✉)
Faculty of Education, Universidade de Santiago de Compostela, Santiago, Spain
e-mail: blanca.puig@usc.es

I. M. Cruz-Lorite · M. Evagorou
Department of Education, University of Nicosia, Nicosia, Cyprus
e-mail: lorite.imc@unic.ac.cy; evagorou.m@unic.ac.cy

D. L. Zeidler (ed.), *A Moral Inquiry into Epistemic Insights in Science Education*, Contemporary Trends and Issues in Science Education 61,
https://doi.org/10.1007/978-3-031-63382-9_9

decisions and act responsibly for themselves and others is a measure of agency (OECD, 2018).

Living in the Anthropocene epoch makes it imperative to emphasize a pedagogy that provides opportunities for learning that are connected to living and engaging with the world and acknowledges that we live in a more than a human world (Gough, 2021). Related to OL in the anthropocene, it calls for acknowledging the value of the ocean and recognizing the interdependence between human well-being and the health of marine environments. Consequently, fostering OL demands a shift in societal values towards recognizing our moral and ethical duty to protect and preserve these vital ecosystems for future generations (Brennan et al., 2019; Schoedinger et al. 2006). Moreover, OL should challenge us to reflect on the ethical consequences of our choices and actions. It underscores the moral obligation to make sustainable choices in consumption, to advocate for policies that protect marine life and habitats, and to support initiatives that mitigate the adverse effects of human activities on the ocean. Finally, OL is also linked to hope, which according to OECD (2018) is "the sense that there is a way towards a possible future that is worth achieving" (p. 6).

The UN Decade of Ocean Science for Sustainable Development 2021–2030 emphasizes the need for a 'rigorous OL program of activities' to increase public awareness and knowledge that can guide sustainable behaviors and inform decision-making (UNESCO, 2018). Focusing on OL as a socioscientific issue supports individuals and societies to address questions of fairness, justice, and the equitable distribution of resources, particularly in the context of communities heavily impacted by ocean-related challenges. As three researchers working in communities linked to the ocean, we have personal interest in the topic and have recently started exploring university students' OL development through the presentation of an SSI related issue. As part of this chapter we provide a framework for OL as an SSI for Hope in the Anthropocene, and a brief overview on the current state of SSI and OL in our respective countries (Spain and Cyprus). We present an example of a learning sequence designed based on the framework. Finally, we reflect on the moral importance of OL and how through this we can essentially contribute to human flourishing, addressing at the same time implications on research, practice and policy.

9.2 Ocean Literacy

The importance of the ocean to humans cannot be overstated (Bennett, 2019); the ocean plays a crucial role to protect the health of the planet and therefore, people and all forms of life on earth. The notion of OL is not a new one; it appeared in 2004 in response to a lack of attention to ocean sciences in National Science Education Standards, as well as to the need of improving public understanding of the ocean (McKinley et al. 2023). This term was coined by ocean scientists and education professionals who recognised a lack of ocean-related subjects in formal education and developed a comprehensive framework to enhance the inclusion of ocean sciences into national and state standard education (UNESCO, 2018). Thanks to this initial effort, OL expanded the borders of the USA and it was accepted worldwide (EMSEA, 2021); the role of NGOs and diverse associations across the world was important to enhance OL across continents. For instance, the United Nations (UN) recognized OL as one of the crucial steps towards the sustainable future of the ocean and in 2021 it commenced a 10-year ambition to foster ocean science for the benefit of humanity, namely the UN Decade of Ocean Science for Sustainable Development (2021–2030). The aim is to achieve transformative science solutions for sustainability, to improve the knowledge on ocean issues and the public engagement in actions to protect the ocean. Thus, the final goal of OL is to change the human relationship with the ocean. The term OL serves then, to connect people and the ocean, a relationship that, unfortunately, is frequently missed in our society and in our educational institutions (Stock, 2010).

OL was defined in its beginnings as "an understanding of the ocean's influence on you and your influence on the ocean" (Santoro et al., 2017; UNESCO, 2018). Therefore, OL cannot be limited to what people know about the ocean, but also to their values, attitudes, behaviors, and the ability to communicate about ocean issues. Based on this definition, an ocean-literate person can be characterized as someone who understands the fundamental concepts about the functioning of the ocean; can communicate about the ocean in a meaningful way; and can make informed and responsible decisions regarding the ocean and its resources (Cava et al., 2005). Ocean issues are included in SDG14 under 2030 Agenda for Sustainable Development, which aims to "conserve and sustainably use the oceans, seas and marine resources for sustainable development". The SDG14 targets within this goal include sustainable management of fisheries, aquaculture and tourism, and an increase in scientific knowledge of the oceans (Brennan et al., 2019). However, despite most studies concerning the relevance of the ocean in the 2030 Agenda tend to focus only on SDG 14, showing little interaction with other SDGs (being the closest interaction with SDG13 about climate change), we concur with Santos (2023), that ocean issues cannot limit to SDG14, and it covers a transversal spectrum in the 2030 Agenda. As Santos (2023) indicates, most SDG (1-3; 6-16) are linked to the ocean, as it is evidenced in their goals and targets: "eradicate extreme poverty for all people everywhere" (SDG1, target 1.1), "end hunger" (SDG2, Target 2.1); "achieve universal and equitable access to safe and affordable drinking water for all" (SDG6, target 6.1);" increase substantially the share of renewable energy in the global energy mix" (SDG7, target 7.2), etc.

Linking the ocean exclusively to SDG 14, might drive a biased view that limits the ocean to its biological dimension, rather than showing its multidisciplinary nature in coherence with the current OL framework. This transversality is also reflected by the role of the ocean in the regulation of the climate, that affects whatever happens in our environment on Earth. A very recent publication in Nature Communications (Ditlevsen & Ditlevsen, 2023) about The Atlantic Meridional Overturning Circulation (AMOC), one of the most important tipping elements in Earth's climate system, predicts that AMOC can potentially collapse around the middle of the century due to global warming, which can have dramatic consequences on Earth as a system. This evidence shows the real impact of the ocean in our lives, well-being and future on Earth. At the same time it highlights the impact humans have on the ocean and the bidirectional relationship between our actions and the environment. The extent, scale and overall collective effects of human activities on the ocean have changed overtime and there is still a gap in understanding how our actions as humans are affecting the ocean over time (Halpern et al., 2019). Even though various studies (i.e. Halpern et al., 2019) show that most of the ocean (59%) is experiencing notable rises in cumulative impact, our understanding of long term consequences is still limited.

For many years experts in ocean sciences and members of education communities pointed to the lack of ocean content and topics in academic standards and to its consequences, and called for ocean inclusion in K–12 curricula (Schoedinger et al. 2006). This led to the development of OL programs of activities to increase public awareness and knowledge that can guide sustainable behaviors and inform decision-making by the UN Decade of Ocean Science for Sustainable Development emphasizes (UNESCO, 2018). However, it needs to be noted that the application of the initial concept of OL in educational settings created an OL framework in which the interactions between the ocean and human beings were understood as linear (Stoll-Kleemann, 2019), an idea that goes against a systemic view to increase OL (Ageitos et al., 2023). The many different ways in which humans and the ocean interact result in complex dynamic systems, composed of multiple interlinked chains of cause and effect; thus, OL requires systems understanding: that is knowledge of our own and others' place and role in the web of interactions (Evagorou et al., 2009; Brennan et al., 2019). OL is not limited to what we "know", but also targets how we feel (our attitudes) about ocean-related issues, our behavior, and the extent to which we communicate about these issues with others. Being an ocean literate person means "being able to make informed and responsible decisions regarding the ocean and its resources" (Cava et al., 2005, p. 5).

9.3 Ocean Literacy Principles and Dimensions

Empowering young people with ocean science in schools is fundamental for making them conscious of ocean problems and equipping them with tools to carry out actions that help to protect the ocean. However this is challenging due to diverse aspects, and also "ocean disconnectedness" (Kelly et al., 2022), since the ocean is something which people consider as distant from their lives. The disconnection from the ocean has been shown to negatively influence people's values, agency, attitudes and their willingness to protect the ocean, and it is linked to limited awareness of marine ecosystems (Kelly et al., 2022). Given the need for introducing the ocean as a topic in the science curricula (Freitas et al., 2022) and for providing students' with basic knowledge on marine issues to protect the ocean, science educators need to make an effort to prepare teachers for introducing OL. OL according to NMEA (2010) encompasses seven Ocean Literacy Principles, as Table 9.1 shows,

Table 9.1 Ocean literacy principles

Ocean Literacy Principles (NMEA, 2010)
P1. The earth has one big ocean with many features
P2. The ocean and life in the ocean shape the features of Earth
P3. The ocean is a major influence on weather and climate
P4. The ocean made Earth habitable
P5. The ocean supports a great diversity of life and ecosystems
P6. The ocean and humans are inextricably interconnected
P7. The ocean is largely unexplored

that can be used as goals or recommendations for introducing ocean in SSI instruction.

These seven principles appear separately in this framework, but they interact with each other in multiple dimensions that have evolved in accordance with the evolution of OL conceptualization. These dimensions for example consider the need to move towards the development of useful tools and approaches to transform ocean knowledge into actions for ocean sustainability (McKinley & Burdon, 2020). Specifically, Brennan et al. (2019) proposed the inclusion of *Knowledge*, *awareness*, *attitude*, *behavior*, *communication* and *activism*. McKinley et al. (2023) suggested including as part of OL: *emotional connections*, *access and experience*, *adaptive capacity*, *trust and transparency*. The aforementioned dimensions can be used for designing OL activities and interventions, ensuring that knowledge development is not the sole focus of initiatives (McKinley et al., 2023), but an emphasis on competences, attitudes, values and agency is also necessary.

9.4 A Framework for Ocean Literacy as an SSI for Hope in the Anthropocene

Given the limitations of the current OL frameworks, our proposal for a more holistic OL framework focuses on presenting ocean issues as SSIs. Specifically, our proposal of OL includes three components, namely: (a) knowledge and skills, (b) attitudes and values, including hope, and, (c) agency. In our proposal we provide a framework of OL as an SSI for "hope" in the anthropocene (Fig. 9.1) that is coherent with the PISA 2025 Framework (OECD, 2018). Knowledge and skills in this framework are considered as intertwined and a component of OL that involves the application of ocean-related knowledge and the articulation of critical thinking and systems thinking skills. Based on the PISA 2025 Framework ocean-related

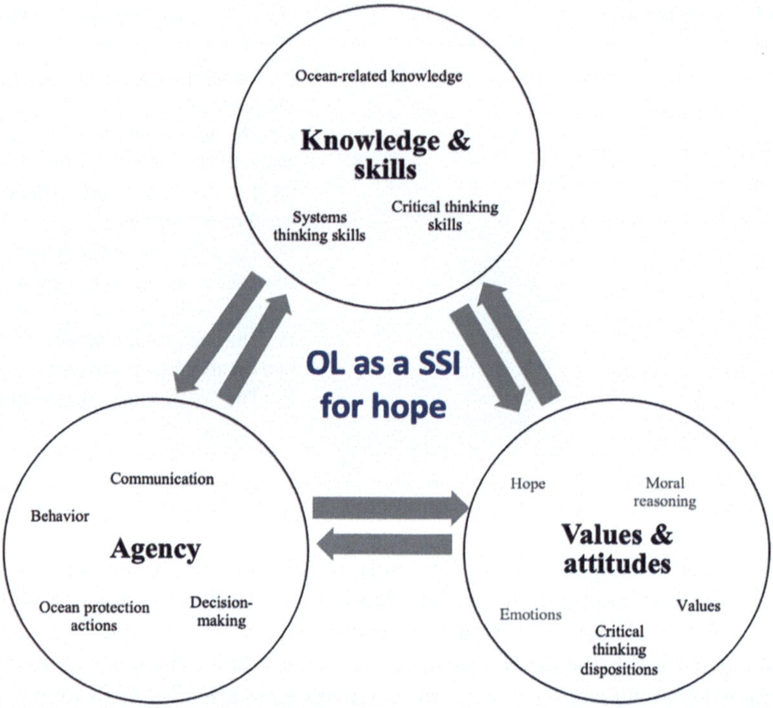

Fig. 9.1 Ocean literacy as an SSI for hope

knowledge cannot only include theoretical knowledge about the ocean and the marine ecosystems, but also epistemic knowledge. This means understanding of science and the ways in which it is generated. This is particularly important in the twenty-first century when humanity faces an uncertain future as it navigates the Anthropocene era and involves, among other aspects, students' understanding of some of the social norms and scientific practices of science to ensure the knowledge it offers is trustworthy (OECD, 2018). Critical thinking skills and systems thinking skills are necessary for OL development, and both are very much related to each other. Critical thinking is a type of thinking involved in reasoning processes that involve cognitive skills as argumentation and critical actions (Jiménez-Aleixandre & Puig, 2022). It requires determining how good a decision you have made has been (Halpern & Danna, 2023), being aware of how you processed the information or evidence available at the moment in question, as well as on your own limitations. This view of critical thinking is aligned with the notion of "intellectual humility" (Osborne & Pimentel, 2023), and supports attitudes for being a critical thinker. In the context of ocean topics, critical thinking involves all the aspects above, as well as the capacity to manage uncertain phenomena in which the ocean plays a crucial role, such as climate change. Systems thinking allows us to tackle complex phenomena as the ones that take place in the ocean in a holistic way (Ben-Zvi-Assaraf

& Orion, 2010), considering, identifying the factors involved and the interdependent relationships among them (Evagorou et al., 2009). Understanding the first principle of OL ("The earth has one big ocean with many features") requires the application of systems thinking skills and to connect the ocean systems to the earth systems. Furthermore, these skills are also necessary in order to exercise agency, as well as attitudes.

Attitudes "reflect a disposition to react to something or someone positively or negatively and can vary according to specific contexts and situations" (OECD, 2018, p. 4). In our OL model, attitudes and values are related to positive dispositions towards the ocean that can be created through the interaction with ocean related issues. They are underpinned by values and beliefs and have an influence on behavior, which is considered a dimension of agency (see Fig. 9.1). As Ruiz-González et al. (2022) point out, only by learning to look at the ocean with the respect, admiration and dedication, people could protect it and stop degrading it. One of the aspects of attitudes and values is linked to hope. One of the conceptualizations of hope which we use in our work, places an emphasis on the emotional character of hope (Ojala, 2023). Hope is perceived as a feeling stimulated by the evaluation that there exists a chance for a problem to improve. Given that this "chance" does not guarantee a definite outcome, hope is related to uncertainty, and therefore directly related to socioscientific issues, making it different from optimism (Chadwick, 2015). According to Ojala (2023), most of the intervention studies focusing on promoting hope target the idea of the possibility of change, but in some cases they trigger the idea that everything will be fine with no further action.

Finally, the third component of the proposed framework is agency which according to the new PISA framework (OECD, 2018) includes decision making and taking actions in the local community and beyond. Agency refers to the manner in which individuals engage with the world and recognize the interconnectedness with ecosystems, rather than viewing themselves as separate entities. It involves acknowledging and valuing all species in the environment. Students with agency according to OECD (2018) collaborate with others, or work individually to comprehend and tackle complex challenges in their communities, but also to communicate their findings and make them visible.

9.5 Connecting Ocean Literacy to Socioscientific Issues

Previous studies suggest that improved OL can help resolve some of the complex socioscientific issues (i.e. Leitão et al., 2022) we are facing as a society globally. However, despite its importance, a review of studies on OL (i.e., Chen & Tsai, 2016; Domegan et al., 2019; Winks et al., 2020) highlights that students and citizens lack the necessary skills and knowledge to understand the impact of their actions on the ocean, and the impact of the ocean on their lives. Also, they rarely engage in environmentally sustainable actions linked to the ocean or the discussion of ocean related socioscientific issues (Ageitos et al., 2023).

Socioscientific issues are complex and subjective, and include a multitude of values such as aesthetics, ecology, economics, morality, education, culture, religion and recreation (i.e. Chiapetta et al., 1998). Socioscientific issues arise from missing knowledge, which is often linked to the novelty of the issue under study and our limited understanding of the overall consequences of the issue. According to Zeidler and Keefer (2003), moral reasoning and emotive beliefs are part of reasoning about SSI and are associated with scientific functional literacy. Dealing with SSI can enable students to understand the relevance of science to everyday life, and gain insight into how people use science (i.e. Evagorou & Dillon, 2020). Despite the increase in studies focusing on SSI during the last decade (Nielsen, 2020), the inclusion of issues related to the ocean is still limited, with more issues focusing on climate change and nuclear power (Kumar et al., 2024).

In our countries, Spain and Cyprus, ocean related issues and socioscientific issues are not often included in the national curricula. For example, in Cyprus socioscientific issues are not included in the national curricula and some issues that have social aspects, as for example environmental issues are presented as part of environmental education (i.e. Zacharia & Georgiou, 2012). Furthermore, the ocean is not listed in the topics that are suggested in the Cypriot curriculum starting from early childhood education until pre-service teacher training. Specifically, the curriculum in Cyprus focuses on issues relevant to various SDGs (i.e. waste management, biodiversity, forest, energy, water, culture and the environment, tourism, poverty), but ocean is not referred in any of the curriculum (i.e. Zacharia & Georgiou, 2012). Students in Cyprus engage in socioscientific issues as part of their formal education only when participating in research projects (i.e. Nicolaou et al., 2015). Teacher preparation programs in Cyprus do not follow a national curricula, but some universities in Cyprus engage pre-service teachers in the discussion of socioscientific issues as part of their Science Methods courses because of the special interest that faculty members have (i.e. Bayram-Jacobs et al., 2019; Evagorou & Puig, 2017; Evagorou et al., 2014; Puig & Evagorou, 2020). In the case of Spain, despite the presence of SDGs in the current science curricula at primary and secondary education (Spanish Ministry of Education and Vocational Training, 2022a, 2022b), socioscientific issues are not present in early childhood, primary and secondary curricula, and ocean related-issues are practically absent (with the sole exception of a mention of plankton, because of its importance for essential human activities such as food production in the secondary curricula). In fact, they continue to focus on earth ecosystems and the application of ecology and biology notions in relation to these ecosystems. Therefore, the situation in Spain is similar to that of Cyprus, in the sense that engagement in socioscientific issues in early childhood, primary and secondary formal education is the result of teachers' interest and/or participation in research projects (i.e. López-Fernández et al., 2022; Tena & Couso, 2023). Although socioscientific issues are included in the science teaching modules of many teacher training programmes in Spanish universities, there are no specific guidelines on their inclusion in the state regulations on teacher education. However, there is an increasing tendency of using socioscientific issues as appropriate contexts in teacher's training programmes in Primary and Secondary Education to engage students in learning science through argumentation processes that involve decision making and taking responsible actions (i.e. Garrido-Espeja & Couso-Lagarón, 2015; Martín-Gámez & Erduran, 2018; Cruz-Lorite et al., 2023). This responds not only to the social needs and environmental challenges we face, but also to the need to prepare future teachers to become scientific literate and provide them with tools to deal with societal and moral problems that are part of our everyday lives.

Despite the lack of emphasis on socioscientific issues and ocean related topics in our national curricula, we appreciate the importance of understanding ocean aspects as socioscientific issues, and working towards introducing them first in the teacher preparation programs, and hopefully later in the local curricula. One example of how OL is directly related to socioscientific issues is provided by Greely (2008) who sought to identify mediating factors that contribute to OL and reasoning. Greely (2008) examined the relationships between the components of OL and the socioscientific elements of functional scientific literacy, proposing the possible relationships shown in Table 9.2.

Based on the relationships proposed by Greely (2008) and our work in socioscientific issues, we make the connection of OL and socioscientific issues by introducing a case-base issue in the classroom linked to the ocean and exploring it in terms of content, environmental morality, informal reasoning and taking action. Two examples of socioscientific issues linked to the ocean that are specific to our countries are described below.

Cyprus is a small island in the Mediterranean sea which despite its location and geography does not have an important culture and economy linked to the ocean (i.e. no history of fishing or culture related to the ocean), other than using the beach to promote tourism. Nevertheless, the island is highly impacted by changes in the ocean indirectly. A very recent example is that of the lionfish (Pterois miles), native to the Indian Ocean that has been spreading in the Mediterranean Sea because of seawater becoming warmer. The invasion of the lionfish affects the marine ecosystem and the biodiversity of the area, and the venomous characteristics of the lionfish can also affect the health of the local population (i.e. Kleitou et al., 2021). This topic can be presented as a local socioscientific issue with an emphasis on promoting the different aspects we have identified in our OL Framework (Fig. 9.1), namely: knowledge and skills, values and attitudes and agency. An example from Spain is

Table 9.2 Possible relationships between components of ocean literacy and the socioscientific elements of functional scientific literacy (Greely, 2008)

OL definition	An ocean literate person	Functional scientific literacy element	Mediating factors
Understanding how the ocean affects you and how you affect the ocean	Understands the science of the ocean	Content knowledge & nature of science	Content & experience (nature context)
	Can communicate about the oceans	Classroom discourse issues & Cultural issues	Environmental morality, content & informal reasoning
	Can make informed decisions about ocean policy	Case-based issues, classroom discourse & cultural issues	Content, morality & behavioral commitment (action)

that of an ecological crisis in the coast of Galicia with millions of plastic pellets polluting the coast of Galicia (Romera-Castillo, 2024). The plastic pellets can potentially end up in the food chain and affect the environment (i.e. water, fish), animals and human beings in multiple ways. This issue can be presented as socio-scientific because of the complexities and uncertainties it entails, and also requires knowledge of the ocean as a system to understand the short term and long term impact on the ocean.

9.6 An Example of Introducing OL as SSI to Pre-service Teachers

In our work we support that OL should be linked to socioscientific issues with an emphasis on promoting hope in the anthropocene. In the proposed framework (Fig. 9.1), we highlight three components as important: (a) knowledge and skills (including critical thinking and systems thinking skills), (b) values and attitudes (including moral reasoning, values, hope), and (c) Agency (including decision making, taking action and communication). In this section we describe a learning sequence that was developed using the specific framework in order to engage pre-service teachers with the topic of white pellet pollution in Galicia. The topic was chosen because it has been in the news recently (Jones, 2024) and is of relevance to pre-service teachers in Galicia, and also because the topic can easily be explored in other countries as well because of the connection to the issue of plastics and microplastics. Table 9.3 outlines the activities in the different sessions and highlights how these are linked to the three components of the framework. The two OL principles addressed are P1 and P6 (see table 9.1). Each session has a duration of 3 h.

The proposed learning sequence is currently under implementation with pre-service teachers studying to become secondary school science teachers. It will be redesigned based on the feedback from the initial implementation. Despite the effort to emphasize all three components during the design, we are aware the component "Agency" poses difficulties, mainly because of constraints related to time and access to the community. We foresee that a new version of the learning sequence will open up the discussion to the wider community and include marine scientists to support the decision making process, and the local community to support in implementing some of the decisions.

This sequence could be easily adaptable to other contexts. Marine litter is a global problem on all coasts around the world, thus, the sequence could be developed without the need to start from a relevant milestone (an accident, as in this case), but could be contextualized by starting from general news about beach pollution and microplastics, adapting it to be relevant to the students.

Table 9.3 Outline of learning sequence "The white pellet tide"

Session/Task	Description of session	Components
Session 1: Introducing the problem	Pre-service teachers are introduced to a news piece which presents the problem of plastic pellets spillage in the shores of Galicia (https://www.euronews.com/green/2024/01/09/spain-investigates-spilage-of-millions-of-plastic-pellets). Through the discussion of the problem pre-service teachers are challenged to think of possible impacts on the ocean and beyond Pre-service teachers engage in a debate around the issue of plastic pellet spillage and provide their personal opinion about the issue. They are asked to reflect on the impact of the issue on the ocean, the environment in general in short and long term and reflect on their own values related to the ocean	**Knowledge & skills** (i.e. systems thinking to understand the overall impact). Additionally, with the debate, moral and ethical dimensions of the issue are under discussion **Values & Attitudes** Hope plays a role at this initial point in dealing with the problem (possible evolution, consequences, expectations, optimism/pessimism, etc.). Emotions are not explicitly asked, but they are also involved
Session 2: Understanding the problem	Information search on the key points of the problem (What are pellets? Are they dangerous for the environment? All its components or only some of them? Where did they come from? What is being done at the moment?) After collecting information about the problem and understanding several aspects of the issue, pre-service teachers are asked to reflect on their emotions about the issue and how these might have changed after learning more about the problem Pre-service teachers are challenged to contrast their ideas here with those expressed in the first session, before going deeper into the problem	**Knowledge & skills** about the composition, origin and impact of plastic pellets. Critical thinking and systems thinking are necessary to analyze the problem **Values & attitudes** (emotions caused by the problem are expressed)

(continued)

Table 9.3 (continued)

Session/Task	Description of session	Components
Session 3: Designing our research	Apart from the information gathered on the current state of the problem, students are asked to make their own diagnosis of the state of an area (beach). Pre-service teachers are asked to consider in their groups the following question: How can we study the impact of the white pellets on a small scale? Pre-service teachers are guided to design their own investigation (i.e. beach sampling) that they will implement during the subsequent session. During this session they are guided to consider their research questions (i.e. What is the quantity of plastic pellets at the beach? How is the ocean being affected by this issue? What changes can be observed in the nearby environment. Based on the questions they also have to decide where to collect data from (i.e. distance from the beach), how to collect data (i.e. tools necessary, quadrads to identify data collection space), what kind of data to collect (i.e. sand, register organisms in quadrads)	**Knowledge & skills** are important: the articulation of epistemic knowledge and procedural knowledge are applied to design the research, and make decisions for data collection Critical thinking skills are exercised as well as dispositions, thus v**alues & attitudes** also emerge
Session 4: Collection of data on site	Pre-service teachers visit the beach and engage in data collection based on their design from session 2. Data can be analyzed either on spot or in the lab	
Session 5: Presenting findings and connecting to the ocean system	Pre-service teachers present their findings and conclusions from the investigation during a whole classroom discussion. During the discussion they are challenged to consider the following questions: What is the effect of their findings on the ocean? What are different parts of the ocean that can be affected? What is the effect of their findings on the environment in general and on humans? If the white tide has no adverse effects for humans, should we be concerned? What might be the short term and long term impact? Can the white tide in Galicia affect other coastal areas in Spain and in other countries? Is the white pellet tide a local or a global phenomenon? How does it affect the lives of people in our country? A role play game is presented in which the pre-service teachers assume different roles (ocean expert, fish farmer, owner of factory making plastic bottles, consumers/citizens, environmentalist) and they are asked to discuss the following question: What kind of damage (understood in general, not only in the environment) can these plastic pellets produce?	**Knowledge & skills** concerning the environmental impact of pellets, with an emphasis on systems thinking and critical thinking. **Values & attitudes** Hope plays a role in terms of identifying short term and long term effects of pellets locally and globally Moral reasoning is involved in the processes of reflection on the white tide's effects (deduced from its research) on both humans and other living organisms, and also in the role play game

Session 6: Searching for solutions	Once the pre-service teachers have analyzed the problem and different points of view have been presented through the role-play, the research carried out can be captured in a controversy map (Cabello-Garrido et al., 2023). In this methodology, a map is configured in which the most relevant actors involved in the problem addressed are included and the existing relationships between them are established. The map produced is a very useful tool when it comes to proposing actions in response to a problem, as it summarizes the results of the research carried out in a visual form. It is possible to do one map in small groups or with the whole class (the latter option would be preferable within the sequence outlined here)	**Knowledge & skills** are required for the identification of actors involved in the problem and systems thinking is needed for the elaboration of the controversy map **Values & attitudes** play a role in the identification of actors, their interactions and effect
Session 7: Proposing actions	After analyzing the possible solutions, which may be of a very diverse nature and not all of which can be addressed by ordinary citizens, students are encouraged to think of actions that may be within their reach (dissemination of the results of their research, awareness-raising campaigns, public demonstrations, beach clean-ups, etc.) Pre-service teachers are asked to consider actions that can take place either in their local communities or online communities that are related to proposing solutions, campaigning to prevent similar situations in the future or informing people of the problem and the potential threats (short term and long term)	**Values & attitudes** are mobilized to propose actions Hope plays an important role for thinking about actions that can be carried out by themselves and that contribute to improving the environmental problems detected **Agency** is present for taking actions in their local communities is also emphasized

9.7 Discussion: Moral Dimensions of Ocean Literacy as SSI

The capacity to make decisions and take responsible actions, what we have called agency, in relation to environmental problems as the one presented in this chapter, depends to a large extent on people's environmental awareness. In this case, awareness of OL. 'Environmental awareness', as OL, is a concept under constant construction, and its study faces a series of limitations, the most relevant being the difficulty in defining the elements that comprise it and the relationship between them. This makes it difficult to operationalise this notion in education.

Some studies have shown that scientific knowledge about environmental problems could influence the adoption of pro-environmental behaviors (i.e. Hines et al., 1987; Muñoz, 2012; Hamid et al., 2017). In the case of the marine environment, Gelcicha et al. (2014) found a strong relationship between the level of concern about marine impacts and the level of information of 10,106 participants in a survey conducted in European countries. However, in contrast to these findings, other studies have shown that there is an inconsistency between the existence of environmental knowledge and the adoption of pro-environmental behaviors by individuals, the so-called 'environmental gap' (Kollmuss & Agyeman, 2002). Thus, knowledge and concern about environmental issues are not variables with predictive power over behavior (Murga, 2008). These different results in the literature have made it difficult for theoretical models relating environmental knowledge to pro-environmental behavior to explain the empirical data.

Given these difficulties, models of pro-environmental behavior have emerged that have incorporated other aspects that could explain environmental behaviors better than knowledge, such as social (Farrow et al., 2017) or moral norms (Schwartz, 1970). In fact, several studies have empirically found a positive and significant relationship between moral norms and problem awareness and attributed responsibility (Han et al., 2015). In the 1970s, Schwartz developed the norm activation model (NAM), which was originally developed in the prosocial domain and has subsequently been widely used in the domain of pro-environmental behavior (Han et al., 2017), proposing a mechanism for the transfer of a moral norm into environmental actions based on personal norms, awareness of consequences and attribution of responsibility (He & Zhan, 2018). This model has subsequently been extended by other researchers to include elements such as external costs, both economic and behavioral (Hunecke et al., 2001), the degree of perceived efficacy of one's own behavior (He & Zhan, 2018) or moral emotions (Rees et al., 2015; Xie et al., 2015); the latter also highlighted in the framework proposed in this chapter. But even having extended the models to other aspects such as moral aspects, there is another difficulty facing the study of environmental awareness and, specifically, in relation to the ocean. Traditionally, people have been considered to have a generic environmental attitude, although people seem rather to have an environmental awareness dependent on the relevance they attach to specific environmental issues they are particularly sensitive to (Corraliza et al., 2002), therefore, measuring instruments have been developed to analyze a global environmental awareness, without much effort having been devoted to the analysis of specific environmental problems (as may be the case of their limitation to the marine environment). This, together with the fact that the ocean is still little studied, contributes to the fact that models for analyzing environmental awareness have been applied to a lesser extent to specifically address ocean environmental problems. Environmental awareness of the ocean depends, among other aspects, on the specific moral standards that people have on the marine environment, which would affect their potential actions to protect the ocean. In this sense, our proposal aims to involve students in moral reasoning processes as a way of raising their awareness of our role in the health of the ocean.

According to Begum et al. (2022), the development of an environmentally conscious society requires educating society about the importance of the environment and getting people to act ethically, which requires a radical change in education to enable the individual to develop a moral conscience that stands firm. But acting 'ethically' often does not have a universal meaning, which leads to conflict. With regard to the marine environment, Miller and Kirk (1992) noted that the resolution of ethical conflicts is problematic, as it depends on the qualitative difference between multiple-use conflicts, where ground rules support consensus and compromise in conflict resolution, and multiple-value conflicts, where negotiated resolution is precluded because consensus is philosophically intolerable. An example of the first case would be conflicts associated with the management of uses and activities in coastal areas, involving various sectors (fishing, tourism, transport, recreational activities, etc.) under the commonly accepted premise of the need and the right to develop such activities, while an example of the second case (multiple-value conflicts) would be the use of marine animals in shows. This often confronts incompatible positions (*a priori*) between those who subordinate animal welfare to the satisfaction of the human desire for recreation and those whose value system orders these aspects in the opposite direction. These two types of conflicts, which shape different contexts, may lead a person to develop different patterns of reasoning when dealing with one or the other, as reflected, for example, in a study with university students on environmental dilemmas (Kortenkamp & Moore, 2001). The researchers found that when the dilemma involved a social conflict, in which needs and effects on humans were highlighted (more in line with a multiple-use conflict), students tended to reason anthropocentrically. However, when the impact of an activity on the environment was emphasized (more aligned with a multiple-value conflict), students tended to reason more biocentrically. Moral reasoning would therefore be influenced by context (Zeidler & Schafer, 1984; Greely, 2008). If we take into account that environmental problems are often crossed by both types of ethical conflicts (multiple uses and multiple values), the position that people take on these problems is often more complex and may not be characterized by a single type of moral reasoning. Our decision to include in the learning sequence activities that allow pre-service to reflect on their ideas and how these have changed after exploring the issue is related to moral reasoning and how this is influenced by the context.

The inclusion of moral reasoning in our framework highlights its importance since it reflects the individual's interpretation of norms and principles in situations of conflict, and the reasons they argue for acting morally when choosing between competing alternatives in such situations reflect an overall mental structure, which emerges through interactions between the individual and their environment (Zeidler & Keefer, 2003). Berkowitz (2002) appealed to the "Moral Anatomy", which would be a set of psychological characteristics that would include "the capacity to think about right and wrong, experience moral emotions (guilt, empathy, compassion), engage in moral behaviors (sharing, donating to charity, telling the truth), believe in moral goods, demonstrate an enduring tendency to act with honesty, altruism, responsibility, and other characteristics that support moral functions" (Berkowitz, 2002, p. 48). In this regard, Zeidler and Sadler (2007) emphasized that pedagogical goals cannot be directed at isolated parts of the student's moral anatomy and that science classrooms that deny the emotive sites of discourse in the discussion of social science issues stifle students' personal development. Scientific decisions cannot be made in the absence of moral reasoning and a concern for human values (Aikenhead, 2006) and this idea needs to be transferred to science classrooms. The SSI holds great promise as a framework for fostering the emotional, behavioral and cognitive features of development to enable students to develop character and offers the possibility of paying attention to the holistic nature of moral growth (Zeidler & Sadler, 2007). This framework contemplates explicit consideration of ethics and moral reasoning by students, using controversial and multifaceted topics to promote science in context, cultivate critical thinking and attempt to improve students' understanding and/or decision-making on socioscientific issues (Bencze et al., 2020). Furthermore, environmental philosophers and ecojustice educators point out that moral responsibility and ethical perspectives should extend beyond ourselves to also consider animals, plants, entire ecosystems and the biosphere as a whole, and a crucial component of developing these ecoethical perspectives is participation in environmental SSI through the use of emotive reasoning (Herman et al., 2020).

However, there are important emerging issues related to environmental issues that current frameworks, including SSI, have barely begun to address at present, the so-called "eco-anxiety" or "climate anxiety" (Pihkala, 2020) which, according to the most commonly cited definitions in the literature, would be the "generalized sense that the ecological foundations of existence are in the process of collapse" (Albrecht, 2012, p. 250), or the "chronic fear of environmental doom" (Clayton et al., 2017, p. 68). In a study of 10,000 young people from 10 different countries, Hickman et al. (2021) found that a large proportion of participants reported feeling emotions such as sadness, anxiety, anger, powerlessness, helplessness and guilty about climate change, and that these feelings negatively affected their daily lives. In the same study, 75% of participants reported that they thought the future was frightening. This context of eco-anxiety poses a challenge to society in general and, in particular, to scientists and researchers in science education, as they play a crucial role in the growing loss of hope among young people related to climate change (Levrini et al., 2019). Futures thinking is one of the most recent current trends that aims precisely to understand these emotions of hope, fear or anger related to the threats of the Anthropocene, integrating the dimension of the future in education with the objectives of "looking ahead to the future, identifying trends of change and their uncertainties, anticipating risks and consequences of decisions and actions,

[…], and imagining solutions or means of action" (Hervé & Panissal, 2022, p. 3), among others. As Hervé and Panissal (2022) point out, educating in the context of the Anthropocene necessarily involves questioning the socio-technical trajectories of modern societies, addressing environmental issues and assessing possible and preferable actions with regard to their future consequences, which has now brought together the pedagogical currents of SSI, futures education and certain approaches to environmental education, thus becoming part of a critical and political tradition of dealing with environmental issues in schools. Thinking about future scenarios related to the ocean is not new in other fields (Kelly et al., 2022), but it seems to be new within the framework of SSI in the field of science education. In our opinion, the OL framework proposed in this work can serve, in addition to articulating elements traditionally less attended to when addressing SSI with students (such as those related to moral issues, values, attitudes, etc.), to explore the role of hope, and the emotions associated with it, when making decisions and acting responsibly, i.e., the role of hope in agency.

Acknowledgments Work supported by the Spanish Ministry of Science, Education and Universities, partly funded by the European Regional Development Fund (ERDF). Contract grant: PID2022-138166NB-C21.

References

Aikenhead, G. S. (2006). *Science education for everyday life: Evidence-based practice*. Teachers College Press.

Ageitos, N., Puig, B., López, A., Ojeda-Romano, G., & Pintado, J. (2023). Desde las orcas hasta la vida marina en suspensión. Promover la cultura oceánica. In B. Puig, B. Crujeiras-Pérez, & P. Blanco-Anaya (Eds.), *Pensar científicamente. Problemas sistémicos y Acción crítica* (pp. 79–93). Graó.

Albrecht, G. (2012). Psychoterratic conditions in a scientific and technological world. In P. H. Kahn & P. H. Hasbach (Eds.), *Ecopsychology: Science, totems, and the technological species* (pp. 241–264). MIT Press.

Bayram-Jacobs, D., Henze, I., Evagorou, M., Shwartz, Y., Aschim, E. L., Alcaraz-Dominguez, S., et al. (2019). Science teachers' pedagogical content knowledge development during enactment of socioscientific curriculum materials. *Journal of Research in Science Teaching, 56*(9), 1207–1233. https://doi.org/10.1002/tea.21550

Begum, A., Liu, J., Qayum, H., & Mamdouh, A. (2022). Environmental and moral education for effective environmentalism: An ideological and philosophical approach. *International Journal of Environmental Research and Public Health, 19*(23), 15549. https://doi.org/10.3390/ijerph192315549

Ben-Zvi-Assaraf, O., & Orion, N. (2010). System thinking skills at the elementary school level. *Journal of Research in Science Teaching, 47*(5), 540–566. https://doi.org/10.1002/tea.20351

Bencze, L., Pouliot, C., Pedretti, E., Simonneaux, L., Simonneaux, J., & Zeidler, D. (2020). SAQ, SSI and STSE education: defending and extending "science-in-context". *Cultural Studies of Science Education, 15*, 825–851. https://doi.org/10.1007/s11422-019-09962-7

Bennett, N. (2019). Marine social science for the peopled seas. *Coastal Management*. https://doi.org/10.1080/08920753.2019.1564958

Berkowitz, M. W. (2002). The science of character. In W. Damon (Ed.), *Bringing in a new era in character education* (pp. 43–63). Hoover Institution Press.

Brennan, C., Ashley, M., & Molloy, O. (2019). A system dynamics approach to increasing ocean literacy. *Frontiers in Marine Science, 6*, 1–20. https://doi.org/10.3389/fmars.2019.00360

Cabello-Garrido, A., Cebrián-Robles, D., España-Ramos, E., González-García, F. J., Cruz-Lorite, I. M., España-Naveira, P., & Blanco-López, Á. (2023). The dominant model of meat production and consumption as a socially acute question for activist education. *Cultural Studies of Science Education*. https://doi.org/10.1007/s11422-023-10188-x

Cava, F., Schoedinger, S., Strang, C., & Tuddenham, P. (2005). Science content and standards for ocean literacy: A report on ocean literacy. *OCEANS, 2005. Proceedings of MTS/IEEE Workshop*. https://www.coexploration.org/oceanliteracy/documents/OLit2004-05_Final_Report.pdf

Chadwick, A. E. (2015). Toward a theory of persuasive hope: Effects of cognitive appraisals, hope appeals, and hope in the context of climate change. *Health Communication, 30*(6), 598–611. https://doi.org/10.1080/10410236.2014.916777

Chen, C., & Tsai, C. (2016). Marine environmental awareness among university students in Taiwan: A potential signal for sustainability of the oceans. *Environmental Education Research, 22*(7), 958–977. https://doi.org/10.1080/13504622.2015.1054266

Chiapetta, E. L., Koball, T. R., & Colette, A. T. (1998). *Science instruction in the middle and secondary school* (4th ed.). Merill.

Clayton, S., Manning, C. M., Krygsman, K., & Speiser, M. (2017). *Mental health and our changing climate: Impacts, implications, and guidance*. APA & EcoAmerica.

Corraliza, J. A., Martín, R., Moreno, M., & Berenguer, J. (2002). El estudio de la conciencia ambiental [The study of environmental awareness]. *Medio Ambiente, 40*, 14–18. https://bitly.ws/u25e

Cruz-Lorite, I. M., Cebrián-Robles, D., Acebal-Expósito, M. C., & Evagorou, M. (2023). Analysis of the informal reasoning modes of preservice primary teachers when arguing about a socio-scientific issue on nuclear power during a role play. *Sustainability, 15*(5), 1–17. https://doi.org/10.3390/su15054291

Ditlevsen, P., & Ditlevsen, S. (2023). Warning of a forthcoming collapse of the Atlantic meridional overturning circulation. *Nature Communications, 14*, 1–12. https://doi.org/10.1038/s41467-023-39810-w

Domegan, C., McHugh, P., McCauley, P., & Davison, K. (2019). Co-creating a sea change social marketing campaign for ocean literacy in Europe: A digital interactive tool for environmental behavior change. In D. Z. Basil, G. Diaz-Meneses, & M. D. Basil (Eds.), *Social marketing in action* (pp. 393–409). Springer.

European Marine Science Educators Association [EMSEA]. (2021). *Ocean literacy history*. https://www.emsea.eu/ocean-literacy-history

Evagorou, M., & Dillon, J. (2020). Introduction: Socio-scientific issues as promoting responsible citizenship and the relevance of science. In M. Evagorou, J. A. Nielsen, & J. Dillon (Eds.), *Science teacher education for responsible citizenship: Towards a pedagogy for relevance through socioscientific issues* (pp. 1–11). Springer. https://doi.org/10.1007/978-3-030-40229-7_1

Evagorou, M., Guven, D., & Mugaloglu, E. (2014). Preparing elementary and secondary pre-service teachers for everyday science. *Science Education International, 25*(1), 68–78.

Evagorou, M., Korfiatis, K., Nicolaou, C., & Constantinou, C. (2009). An investigation of the potential of interactive simulations for developing system thinking skills in elementary school: A case study with fifth-graders and sixth-graders. *International Journal of Science Education, 31*(5), 655–674. https://doi.org/10.1080/09500690701749313

Evagorou, M., & Puig, B. (2017). Engaging elementary school pre-service teachers in modeling a socioscientific issue as a way to help them appreciate the social aspects of science. *International Journal of Education in Mathematics, Science and Technology, 5*(2), 113–123.

Farrow, K., Grolleau, G., & Ibanez, L. (2017). Social norms and pro-environmental behavior: A review of the evidence. *Ecological Economics, 140*, 1–13. https://doi.org/10.1016/j.ecolecon.2017.04.017

Freitas, C., Bellgrove, A., Venzo, P., & Francis, P. (2022). Towards a 2025 national ocean literacy strategy: Current status and future needs in primary education. *Frontiers in Marine Science, 9*, 1–11. https://doi.org/10.3389/fmars.2022.883524

Garrido-Espeja, A., & Couso-Lagarón, D. (2015). Socio-scientific issues (SSI) in initial training of primary school teachers: Pre-service teachers' conceptualization of SSI and appreciation of the value of teaching SSI. *Procedia—Social and Behavioral Sciences, 196*, 80–88. https://doi.org/10.1016/j.sbspro.2015.07.015

Gelcicha, S., Buckley, P., Pinnegar, J. K., Chilvers, J., Lorenzoni, I., Terry, G., Guerrero, M., Castilla, J. C., Valdebenito, A., & Duarte, C. M. (2014). Public awareness, concerns, and priorities about anthropogenic impacts on marine environments. *Proceedings of the National Academy of Sciences, 111*(42), 15042–15047. https://doi.org/10.1073/pnas.1417344111

Gough, A. (2021). Education in the Anthropocene. In C. Mayo (Ed.), *Oxford encyclopedia of gender and sexuality in education*. Oxford University Press. https://doi.org/10.1093/acrefore/9780190264093.013.1391

Greely, T. (2008). *Ocean literacy and reasoning about ocean issues: The influence of content, experience and morality*. USF Tampa Graduate Theses and Dissertations. https://digitalcommons.usf.edu/etd/271

Halpern, D. F., & Danna, D. S. (2023). *Though and knowledge. An introduction to critical thinking* (Vol. 2, 6th ed.). Routledge.

Halpern, B. S., Frazier, M., Afflerbach, J., Lowndes, J. S., Micheli, F., O'Hara, C., Scarborough, C., & Selkoe, K. A. (2019). Recent pace of change in human impact on the world's ocean. *Scientific Reports, 9*(1), 1–8. https://doi.org/10.1038/s41598-019-47201-9

Hamid, S., Ijab, M. T., Sulaiman, H., Md. Anwar, R., & Norman, A. A. (2017). Social media for environmental sustainability awareness in higher education. *International Journal of Sustainability in Higher Education, 18*(4), 474–491. https://doi.org/10.1108/IJSHE-01-2015-0010

Han, H., Hwang, J., Kim, J., & Jung, H. (2015). Guests' pro-environmental decision-making process: Broadening the norm activation framework in a lodging context. *International Journal of Hospitality Management, 47*, 96–107. https://doi.org/10.1016/j.ijhm.2015.03.013

Han, H., Hwang, J., & Lee, M. J. (2017). The value-belief-emotion-norm model: Investigating customers' eco-friendly behavior. *Journal of Travel & Tourism Marketing, 34*(5), 590–607. https://doi.org/10.1080/10548408.2016.1208790

He, X., & Zhan, W. (2018). How to activate moral norm to adopt electric vehicles in China? An empirical study based on extended norm activation theory. *Journal of Cleaner Production, 172*, 3546–3556. https://doi.org/10.1016/j.jclepro.2017.05.088

Herman, B. C., Zeidler, D. L., & Newton, M. (2020). Students' emotive reasoning through place-based environmental socioscientific issues. *Research in Science Education, 50*, 2081–2109. https://doi.org/10.1007/s11165-018-9764-1

Hervé, N., & Panissal, N. (2022). Writing fictional short stories about the anthropocene: Effects on students' futures thinking. *Frontiers in Education, 7*, 1–11. https://doi.org/10.3389/feduc.2022.842252

Hickman, C., Marks, E., Pihkala, P., Clayton, S., Lewandowski, R. E., Mayall, E. E., Wray, B., Mellor, C., & van Susteren, L. (2021). Climate anxiety in children and young people and their beliefs about government responses to climate change: A global survey. *The Lancet Planetary Health, 5*(12), 863–873. https://doi.org/10.1016/S2542-5196(21)00278-3

Hines, J. M., Hungerford, H. R., & Tomera, A. N. (1987). Analysis and synthesis of research on responsible environmental behavior: A meta-analysis. *The Journal of Environmental Education, 18*(2), 1–8. https://doi.org/10.1080/00958964.1987.9943482

Hunecke, M., Blöbaum, A., Matthies, E., & Höger, R. (2001). Responsibility and environment: Ecological norm orientation and external factors in the domain of travel mode choice behavior. *Environment and Behavior, 33*(6), 830–852. https://doi.org/10.1177/00139160121973269

Intergovernmental Panel on Climate Change [IPCC]. (2021). *Climate Change 2021: The physical science basis. Contribution of working group I to the sixth assessment report of the Intergovernmental Panel on Climate Change*. Cambridge University Press. https://doi.org/10.1017/9781009157896

Jiménez-Aleixandre, M. P., & Puig, B. (2022). Educating critical citizens to face post-truth: The time is now. In B. Puig & M. P. Jiménez-Aleixandre (Eds.), *Critical thinking in biology and environmental education: Facing challenges in a post-truth world* (pp. 3–19). Springer. https://doi.org/10.1007/978-3-030-92006-7_1

Jones, S. (2024, January 9). Northern Spain on alert as plastic pellets from cargo spill wash up on beaches. *The Guardian*. https://www.theguardian.com/world/2024/jan/09/northern-spain-plastic-pellets-cargo-spill-beaches

Kelly, R., Foley, P., Stephenson, R. L., Hobday, A. J., Pecl, G. T., Boschetti, F., Cvitanovic, C., Fleming, A., Fulton, E. A., Nash, K. L., Neis, B., Singh, G. G., & van Putten, E. I. (2022). Foresighting future oceans: Considerations and opportunities. *Marine Policy, 140*, 1–12. https://doi.org/10.1016/j.marpol.2022.105021

Kleitou, P., Hall-Spencer, J. M., Savva, I., Kletou, D., Hadjistylli, M., Azzurro, E., et al. (2021). The case of lionfish (Pterois miles) in the Mediterranean Sea demonstrates limitations in EU legislation to address marine biological invasions. *Journal of Marine Science and Engineering, 9*(3), 1–16. https://doi.org/10.3390/jmse9030325

Kollmuss, A., & Agyeman, J. (2002). Mind the gap: Why do people act environmentally and what are the barriers to pro-environmental behavior? *Environmental Education Research, 8*(3), 239–260. https://doi.org/10.1080/13504620220145401

Kortenkamp, K., & Moore, C. (2001). Ecocentrism and anthropocentrism: Moral reasoning about ecological commons dilemmas. *Journal of Environmental Psychology, 21*, 261–272. https://doi.org/10.1006/jevp.2001.0205

Kumar, V., Choudhary, S. K., & Singh, R. (2024). Environmental socio-scientific issues as contexts in developing scientific literacy in science education: A systematic literature review. *Social Sciences & Humanities Open, 9*, 1–10. https://doi.org/10.1016/j.ssaho.2023.100765

López-Fernández, M. M., González-García, F., & Franco-Mariscal, A. J. (2022). *Journal of Chemical Education, 99*(10), 3435–3442. https://doi.org/10.1021/acs.jchemed.2c00223

Malhi, Y. (2017). The concept of the Anthropocene. *Annual Review of Environment and Resources, 42*, 77–104. https://doi.org/10.1146/annurev-environ-102016-060854

Martín-Gámez, C., & Erduran, S. (2018). Understanding argumentation about socio-scientific issues on energy: A quantitative study with primary pre-service teachers in Spain. *Research in Science & Technological Education, 36*(4), 463–483. https://doi.org/10.1080/02635143.2018.1427568

McKinley, E., & Burdon, D. (2020). *Understanding ocean literacy and ocean climate-related behaviour change in the UK- work package 1: Evidence synthesis*. Final report produced for the Ocean Conservation Trust and Defra. https://www.oceanconservationtrust.org/app/uploads/Review-of-Ocean-Literacy-1.pdf

McKinley, E., Burdon, D., & Shellock, R. J. (2023). The evolution of ocean literacy: A new framework for the United Nations Ocean Decade and beyond. *Marine Pollution Bulletin, 186*, 114467. https://doi.org/10.1016/j.marpolbul.2022.114467

Merrett, C. (2004). Social justice: What is it? Why teach it? *Journal of Geography, 103*(3), 93–101. https://doi.org/10.1080/00221340408978584

Miller, M. L., & Kirk, J. (1992). Marine environmental ethics. *Ocean & Coastal Management, 17*(3–4), 237–251. https://doi.org/10.1016/0964-5691(92)90012-A

Ministerio de Educación y Formación Profesional [Ministry of Education and Vocational Training]. (2022a). Real Decreto 157/2022, de 1 de marzo, por el que se establecen la ordenación y las enseñanzas mínimas de la Educación Primaria [Royal Decree 157/2022, of 1 March, establishing the organisation and minimum teaching of Primary Education]. *Boletín Oficial del Estado*, n. 52, 3 March 2022, 24386 to 24504. https://www.boe.es/eli/es/rd/2022/03/01/157

Ministerio de Educación y Formación Profesional [Ministry of Education and Vocational Training]. (2022b). Real Decreto 217/2022, de 29 de marzo, por el que se establece la ordenación y las enseñanzas mínimas de la Educación Secundaria Obligatoria [Royal Decree 217/2022, of 29 March, establishing the organisation and minimum teaching of Compulsory Secondary Education]. *Boletín Oficial del Estado*, n. 76, 30 March 2022, 41571 to 41789. https://www.boe.es/eli/es/rd/2022/03/29/217

Muñoz. (2012). *Concepto, expresión y dimensiones de la conciencia ambiental. El papel de la cultura científica* [Concept, expression and dimensions of environmental awareness. The role of scientific culture]. Editorial Académica Española.

Murga, M. A. (2008). Percepciones, valores y actitudes ante el desarrollo sostenible. Detección de necesidades educativas en estudiantes universitarios [Perceptions, values and attitudes towards sustainable development. Detection of educational needs in university students]. *Revista Española de Pedagogía, 240*, 327–344.

National Marine Educators Association [NMEA]. (2010). *Ocean literacy scope and sequence for grades K-12*. Special Report #3 on the ocean literacy campaign featuring the ocean literacy scope & sequence for grades K-12. http://oceanliteracy.wp2.coexploration.org/ocean-literacy-framework/principles-and-concepts/?page_id=1073

Nicolaou, C. H. R., Evagorou, M., & Lymbouridou, C. (2015). Elementary school students' emotions when exploring an authentic socio-scientific issue through the use of models. *Science Education International, 26*(2), 240–259.

Nielsen, J. A. (2020). Teachers and socioscientific issues—An overview of recent empirical research. In M. Evagorou, J. A. Nielsen, & J. Dillon (Eds.), *Science teacher education for responsible citizenship: Towards a pedagogy for relevance through socioscientific issues* (pp. 13–20). Springer. https://doi.org/10.1007/978-3-030-40229-7_2

Leitão, R., Maguire, M., Turner, S., Arenas, F., & Guimarães, L. (2022). Ocean literacy gamified: A systematic evaluation of the effect of game elements on students' learning experience. *Environmental Education Research, 28*(2), 276–294. https://doi.org/10.1080/13504622.2021.1986469

Levrini, O., Tasquier, G., Branchetti, L., & Barelli, E. (2019). Developing future-scaffolding skills through science education. *International Journal of Science Education, 41*(18), 2647–2674. https://doi.org/10.1080/09500693.2019.1693080

Ojala, M. (2023). Hope and climate-change engagement from a psychological perspective. *Current Opinion in Psychology, 49*, 1–6. https://doi.org/10.1016/j.copsyc.2022.101514

Organisation for Economic Co-operation and Development [OECD]. (2018). *OECD future of education and skills 2030*. Conceptual learning framework. Attitudes and values for 2030. https://www.oecd.org/education/2030-project/teaching-and-learning/learning/attitudes-and-values/

Osborne, J., & Pimentel, D. (2023). Science education in an age of misinformation. *Science Education, 107*(3), 553–571. https://doi.org/10.1002/sce.21790

Pihkala, P. (2020). Anxiety and the ecological crisis: An analysis of eco-anxiety and climate anxiety. *Sustainability, 12*(19), 1–20. https://doi.org/10.3390/su12197836

Puig, B., & Evagorou, M. (2020). Design of a socioscientific issue unit with the use of modeling: The case of bees. *International Journal of Designs for Learning, 11*(1), 98–107. https://doi.org/10.14434/ijdl.v11i1.24142

Rees, J. H., Klug, S., & Bamberg, S. (2015). Guilty conscience: Motivating pro-environmental behavior by inducing negative moral emotions. *Climatic Change, 130*, 439–452. https://doi.org/10.1007/s10584-014-1278-x

Romera-Castillo, C. (2024). Los daños de los pélets de Galicia son la punta del iceberg. *The Conversation*. https://theconversation.com/los-danos-de-los-pelets-de-galicia-son-la-punta-del-iceberg-220947

Ruiz-González, C., Isla, E., & Navarro, J. (2022). Healthy and resilient ocean. In J. L. Pelegrí, J.-M. Gili, & M. V. Martínez de Albeniz (Eds.), *The ocean we want: Inclusive and transformative ocean science* (pp. 39–40). Institut de Ciències del Mar.

Santoro, F., Selvaggia, S., Scowcroft, G., Fauville, G., & Tuddenham, P. (2017). *Ocean literacy for all: A toolkit* (Vol. 80). UNESCO Publishing.

Santos, T. (2023). Blue economy and sustainable development beyond boxes. In M. D. G. García & A. Cortès (Eds.), *Blue planet Las. The ecology of our economic and technological world* (pp. 199–211). Springer.

Schoedinger, S., Cava, F., & Jewell, B. (2006). The need for ocean literacy in the classroom. *Science Teacher, 72*(6), 44–47. http://cosee-ne.cosee.net/documents/NeedforOceanLiteracy.pdf

Schwartz, S. H. (1970). Elicitation of moral obligation and self-sacrificing behavior: An experimental study of volunteering to be a bone marrow donor. *Journal of Personality and Social Psychology, 15*(4), 283–293. https://doi.org/10.1037/h0029614

Stock, J. (2010). *Identifying obstacles to incorporating ocean content into California secondary classrooms*. Master thesis, Dominican University of California. https://doi.org/10.33015/dominican.edu/2010.edu.01

Stoll-Kleemann, S. (2019). Feasible options for behavior change toward more effective ocean literacy: A systematic review. *Frontiers in Marine Science, 6*, 1–14. https://doi.org/10.3389/fmars.2019.00273

Tena, E., & Couso, D. (2023). El diseño de preguntas investigables en el ciclo superior de primaria [Designing research questions with 10-12-year-old students]. *Enseñanza de las Ciencias, 41*(1), 101–123. https://doi.org/10.5565/rev/ensciencias.5573

UNESCO. (2018). *Ocean literacy portal*. https://oceanliteracy.unesco.org/

White, P., et al. (2023). *Agency in the Anthropocene: Supporting document to the PISA 2025 science framework* (OECD Education Working Papers, 297). OECD Publishing. https://doi.org/10.1787/8d3b6cfa-en

Winks, L., Ward, M., Zilch, J., & Woodley, E. (2020). Residential marine field-course impacts on ocean literacy. *Environmental Education Research, 26*(7), 969–988. https://doi.org/10.1080/13504622.2020.1758631

Xie, C., Bagozzi, R. P., & Grønhaug, K. (2015). The role of moral emotions and individual differences in consumer responses to corporate green and non-green actions. *Journal of the Academy of Marketing Science, 43*, 333–356. https://doi.org/10.1007/s11747-014-0394-5

Zacharia, A. & Georgiou, D. (2012). *Guide for the implementation of Programs of Study: Environmental Education and Education for sustainable development*. Ministry of Education and Culture Cyprus. (in Greek). https://archeia.moec.gov.cy/sd/557/odigos_efarmogis_programmatos_spoudon_perivallontikis_ekpaidefsis.pdf

Zeidler, D. L., & Keefer, M. (2003). The role of moral reasoning and the status of socioscientific issues in science education. Philosophical, psychological and pedagogical considerations. In D. L. Zeidler (Ed.), *The role of moral reasoning on socioscientific issues and discourse in science education* (pp. 7–38). Springer. https://doi.org/10.1007/1-4020-4996-X_2

Zeidler, D. L., & Sadler, T. D. (2007). The role of moral reasoning in argumentation: Conscience, character, and care. In S. Erduran & M. P. Jiménez-Aleixandre (Eds.), *Argumentation in science education. Science & technology education library* (pp. 201–216). Springer. https://doi.org/10.1007/978-1-4020-6670-2_10

Zeidler, D. L., & Schafer, L. E. (1984). Identifying mediating factors of moral reasoning in science education. *Journal of Research in Science Education, 21*, 1–15.

Chapter 10
Widening Students' World Views via the Implementation of Socioscientific Issues in Educational Practice

Martha Georgiou

10.1 Introduction—The Value of SSI in Educational Practice

It is well known that students are more easily interested and involved in scientific issues that are relevant to their everyday lives (Layton et al., 1986). Global issues such as the recent (and ongoing) COVID-19 pandemic have once again highlighted the value of education and underlined the active role it should play for the direct application of scientific knowledge to issues of concern to humanity in terms of both empathy and rationality. The reactions of the anti-vaccination movement are, after all, very recent and intense, and the correlation with education is certainly quite evident, mainly in the context of socioscientific issues (SSIs). However, in addition to global incidents of a socioscientific nature, national incidents are also frequently recorded. For example, in Greece, the year 2023 was an environmentally devastating one: Floods in Thessaly wiped out 25% of the country's agricultural land (over 450,000 acres), and the disastrous fire in the national forest of Dadia resulted in 825,000 burnt acres, the two largest such catastrophes in the whole of Europe.

It is clear that the damage caused by these natural disasters is huge on multiple levels, since apart from forest areas, residential and agricultural areas were also affected, thereby disrupting the lives of citizens throughout the country. Proposed solutions to these problems have raised various questions, such as, Should we exploit areas of land to install photovoltaic panels? Should we start reforestation projects, or should we wait for the natural recovery of ecosystems? Should we use the land to grow more edible crops instead of the commercial cotton crops that were produced prior to the disasters? Would it make sense to build tourist-hotel

M. Georgiou (✉)
Department of Biology, National and Kapodistrian University of Athens, Athens, Greece
e-mail: martgeor@biol.uoa.gr

D. L. Zeidler (ed.), *A Moral Inquiry into Epistemic Insights in Science Education*, Contemporary Trends and Issues in Science Education 61,
https://doi.org/10.1007/978-3-031-63382-9_10

complexes in some of these areas to boost national revenues if food exports are to be drastically reduced? These are just some of the questions that arise at first thought, and which clearly have a scientific basis with a definite social impact. However, to achieve a program of successful, sustainable restoration, ordinary citizens will need to collaborate with politicians and state institutions. In other words, citizens will be at the heart of planning and decision making relating to these socioscientific issues, and will need to well informed and knowledgeable to accomplish this.

But are these the only everyday issues that have a strong social impact, and which a person participating in public life has to deal with during their lifetime, either when expressing their opinion or making decisions? Obviously not: Such issues as gene therapy, the use of genetically modified organisms, the consumption of genetically modified foods, and similar topics are part of everyday life, and are both social and individual challenges. This underscores once again the importance of a person's being properly 'equipped' to face such everyday challenges. What should their 'equipment' consist of, and how can they obtain it? It is evident that education aimed at scientific literacy—and particularly when offered from a very young age—can contribute to the above and help students address everyday issues. However, it is important to clarify that by scientific literacy, we mean more than simply the acquisition of knowledge about scientific concepts, phenomena and processes (VISION I) (Roberts, 2007); rather, the emphasis has shifted towards the need to adopt a pedagogical framework that supports scientific literacy for all based on the application of knowledge to everyday life (VISION II) (Roberts, 2007). According to Zeidler and Sadler (2010), such an environment is also where the SSIs are rooted, and indeed this framework extends its boundaries even further by including and emphasizing critical citizenship and scientific "knowing-in-action" (VISION III) (Aikenhead, 2007). It is evident, therefore, that in order for a person and a society to respond to contemporary demands, it is necessary to ensure scientific literacy and ultimately a foundation that "equips" them with knowledge, empathy, critical thinking, and moral/ethical approaches.

To achieve this end, it has become more urgent now than ever before to clarify the role of education, not only in the light of the cognitive completeness of students, but also in the perspective of their active and critical citizenship, i.e., both their ability and will to participate in the public sphere, and to make informed decisions where arguments are based on scientific as well as moral and ethical elements and principles. In other words, it is absolutely essential to strengthen education in the direction of scientific literacy, which seems feasible with the assistance of the SSI framework.

In this chapter, I will present research that has been conducted in Greece, but I will also proceed to comment on the Greek science curricula regarding the use (or not) of the framework of SSIs. Furthermore, I will attempt to make a brief comparison with the corresponding curricula in Cyprus, considered to be the sister country to Greece. Finally, I will refer to some proposals for the integration of SSIs in the classroom, with the ultimate goal of achieving scientific literacy for a prosperous society, one in which humans flourish.

10.2 Greek Contributions to the Study of SSIs

According to Zeidler and Nichols (2009), SSIs have multiple benefits in science education for both students and teachers. As regards learners, SSIs can contribute to the development of reasoning skills, strengthening decision making and moral reasoning abilities, enriching subject matter knowledge, cultivating informal reasoning, and both using and developing previously encountered life experiences. On the other hand, with regard to teachers, SSIs can facilitate the identification of appropriate scientific and social contexts for presenting scientific content, serve to utilize pedagogical content knowledge, knowledge gleaned from multiple real life experiences, and subject matter knowledge (Zeidler & Nichols, 2009). As in other countries around the world, Greek researchers involved and active in science education seem to have recognized the value of SSIs and have been more or less involved in studying them, and research has been conducted in Greece

(a) on the contribution and the role that SSI can play in the teaching of cognitive subjects to Greek students of different ages, and
(b) on the professional development of teachers at multiple levels, e.g. the integration of cutting-edge issues in their teaching, the understanding and use of the nature of scientific knowledge, and others.

10.2.1 The Contribution of SSI to Greek Students

Starting from the research on the implementation of SSI in science teaching to Greek students, it is important to clarify the different aspects that have been studied. In general, the contribution and the role that the SSI framework can have has been explored, but under these terms more specific concepts, such as subject knowledge, strengthening argumentation skills, combination with other pedagogical frameworks, and ability to identify aspects of complex phenomena are housed.

Thus, according to the study by Koulougliotis et al. (2021), 107 students 15–17 years old were involved in SSIs in two thematic areas: climate change and materials recycling. The purpose of the study was to investigate whether SSIs in these two subject areas could enhance students' knowledge of Green Chemistry principles, and at the same time to assess whether students incorporate these principles when forming their opinions on the aforementioned SSIs. The two questions posed to the students under study—'Chemical reactions: How can we mitigate climate change?' (project 1), and 'Chemical bonding: How can we recycle old tires?' (project 2)—aimed at contextualizing chemistry content learning, namely chemical reactions and chemical bonding, by using SSIs combined with inquiry-based learning. The results were positive, as it was found that the students used Green Chemistry principles in their responses (albeit to varying degrees).

A similar approach, i.e. a combination of SSI and scientific inquiry, was used in the research of Smyrnaiou et al. (2015), but this time not with the aim of enhancing content knowledge, but rather aiming at the cultivation of argumentation skills. In this case, the SSIs which the students (16–18 years old) had to negotiate were (1) Future mobility—New approaches in the city, (2) Demography in the city, (3) City and climate change, (4) Energy efficient houses and flats, and (5) Smart city: Life in an urban network. The research was carried out in the framework of the European Students Parliament Programme (EUSP—http://www.student-parliaments.eu/), and revealed that through participatory processes in which the students were involved in dealing with the issues mentioned above, they were able to improve their argumentation skills. In fact, in this particular case, since the students worked in groups for 6 months, they had discussions with expert mentors, and received their teachers' support throughout the whole project. In the process, collaborative communities were formed, and the evaluation questionnaires showed that the benefits to students were not limited simply to the development of argumentation skills, but also went beyond: In particular, a teacher reported that "... students collaborated creatively, overcoming their disputes, managed to put aside their competing for grades and became friends with classmates whom they did not know well..." (Smyrnaiou et al., 2015, p. 1624).

In another study (Kokolaki et al., 2017), students were asked to build an exhibit focusing on three topics: Applications of nanotechnology, Microplastics in the oceans, and Breast milk for an infant's healthy development. Five hundred and fifty primary and secondary school students participated, and 38 exhibits were produced. Prior to working on the exhibits, students were instructed on the specific topics, and they were able to communicate on the content and form of the exhibit with mentors and experts in order to, among other things, make reference to Responsible Research and Innovation (RRI). In other words, the whole project aimed to promote RRI through SSI and exhibit making. It seemed that the students eventually managed to more or less incorporate the principles of RRI in their exhibits, and of interest was the way they perceived its ethical dimension as a set of rules and principles that set limits to research for human benefit. The same project was conducted with Greek pre-service teachers and is reported below.

A benchmark in the literature on students' skills development was the research of Patronis et al. (1999). The researchers used an SSI as a framework for developing students' argumentation and decision making skills, and separated scientific argumentation from social science argumentation, highlighting its value in modern science education. The topic came from the students' immediate everyday life, and evolved alongside the research. It concerned the construction of a road in their neighborhood, and was intended to involve students in the multifaceted consideration of such a project (e.g., whether the road would pass through a residential or non-residential area, the choice of construction materials, labor supply to employees, burden on the ecosystem, and other factors). It is even mentioned that in this research "the aim was to 'socialize' scientific knowledge not by exemplifying directly its use in solving a particular problem, but rather by investigating a social theme familiar to the students and focusing on a particular question. This question is then expected to engage the groups of students in further development of their own conceptions." (Patronis et al., 1999, p. 747). The researchers concluded that students were ultimately able to articulate arguments when they were actively involved in issues that directly affected them, that pedagogical approaches needed to be developed in order to prepare students to be critical citizens, and that SSI context seems to be one such means for achieving this purpose. A similar topic, in the sense of actuality, was used in the case of Roufiktos' (2016) research as a framework for studying the argumentation process of 16–17 year old students. The SSI involved expressing an opinion on the definition and establishment of a waterfront zone in coastal areas according to a government bill.

The improvement of argumentation skills was also studied in the research of Georgiou et al. (2020b). Students aged 15–16 were exposed to the subject of biotechnology through SSI with two different approaches in terms of time of engagement: One group of students (LRP) was taught Biotechnology through SSI for 2 h per week for a total of 20 weeks (two semesters), while the other group (SRP) followed the same content, but for 3 h per week for a total of 13 weeks (one semester). At the end, the level of arguments generated by students in both groups was tested through questions on respective SSIs, and it was revealed that engagement duration shows a positive correlation with the development of strong arguments. Consequently, the SSI framework was also used in this study in order to place students in an environment of appropriate opportunities to develop argumentative discourse (Georgiou & Mavrikaki, 2017; Georgiou et al., 2020b). At the same time, the informal reasoning that students employed in order to express themselves by giving answers to the questions posed to them was also examined (Georgiou et al., 2020b). Students who had been exposed to SSI instruction for a longer period of time (LRP) used more rational informal reasoning and had reduced rates of intuitive informal reasoning compared to their peers in the other group assessed using Sadler and Zeidler's 'rationalistic, intuitive and emotive' model (2005). This means that those students relied more on scientific data when reasoning. However, no differences between the two groups were observed in the emotive type of informal reasoning (i.e. feeling of compassion, concern, etc.), which according to the authors is obviously not an issue that is only influenced by the duration of exposure to SSI, but may require combined approaches (Georgiou et al., 2020b).

In the same research area, Georgiou and Mavrikaki (2015a) examined whether seniors in the upper level of Greek secondary schools have argumentation skills when involved in SSIs relating to everyday life such as the following: "In your area, the mayor has approved the operation of a plant that produces genetically modified microorganisms that help clean up oil spills. Some residents have risen up against its operation, and others support it. What would your position be, and why?" (Georgiou & Mavrikaki, 2015a, p. 734). The 304 responses gathered revealed that the students' arguments were particularly weak. In fact, this did not seem to change even in the case of some students who—unlike the others—had chosen to study Biology (as a subject for the university entrance exams), and therefore clearly had more subject knowledge. In other words, content knowledge alone did not help students in negotiating the SSIs posed, although those students did use more rational informal reasoning. It is evident, therefore, that other factors besides subject matter are necessary to contribute to the development of students' competencies considered necessary for twenty-first century citizenship if it is not placed in an appropriate pedagogical context.

A different proposal for the use of SSI, in the topic area of climate change, was studied by Manoura et al. (2023). Through the question, "How should a digital game be designed to negotiate a CSE, such as climate change, in order to foster the recognition of complexity as a CSE reasoning skill for students?", researchers proceeded to design an educational game and a supporting e-book to engage the 18 K5 students with SSI on climate change. Social and environmental components seemed to be mostly addressed by the students during the game, while little focus was placed on economic and scientific components. Moreover, the degree of complexity of the solutions and decisions proposed by students for the concerns they encountered, considering the number of criteria per content category mentioned, was shown to remain low for the majority. According to the authors, the digital game presents potential in terms of cultivating the skill of SSI reasoning.

Another way of exploiting SSIs was proposed through the research of Maniatakou et al. (2020). The creation of a role-play activity involving an SSI formed the basis of the research, which aimed to eventually study the degree to which students construct strong arguments when participating in simulated public dialogues on everyday life issues. After being given a working hypothesis on an SSI focusing on vaccination, they were divided into pairs and asked to design a role-play involving every 'stakeholder' in a local community. Two students were assigned to be observers, who would be asked to vote for the most convincing opinion at the end. The "actors" were given supplementary written information, and after creating the roles, they were asked to perform the role-play by expressing their character's views on vaccination. An audio recording of the role-play was made for later evaluation, and all students completed a written questionnaire about the activity at the end. In the written responses given after the role-play by the students answering similar SSIs to those of the role-play, there appeared to be a trend of improvement in their reasoning (i.e., pre-test compared to post-test). The results from examination of the audio excerpts were found to be much better than the written responses in that the students performed with stronger arguments. Therefore, according to the authors, the proposed "creative role-playing game" with an SSI axis can yield positive results in terms of students' argumentation skills.

Finally, a combination of an argumentation study through teaching with the help of a digital scenario and an inquiry-based approach against the background of the SSI of karst caves was attempted by Ziogka et al. (2019). Forty 13-year-old students participated in the study: The findings of the evaluation of the digital scenario showed students had difficulties in formulating an informed opinion, although the inquiry-based learning scenario helped them learn concepts around caves.

10.2.2 The Contribution of SSI to the Professional Development of Greek Teachers

As regards research focusing on teachers in Greece, cases involving both in-service and pre-service teachers have been conducted. The studies are mainly focused on the extent to which Greek teachers' professional development can be enhanced to allow them to both manage and utilize the pedagogical framework of SSI according to the objective of their particular teaching. In addition, and again with a focus on broadening Greek teachers' abilities for the benefit of learners, cases have been studied where SSIs can contribute towards mastering other necessary components of science education.

For example, Koumara and Plakitsi (2020), seeking to propose a pedagogical framework for enhancing the Nature of Scientific Knowledge (NOSK) of experienced in-service science teachers, designed a three-phase programme in which 49 science teachers with 10–30 years of experience were engaged in turn with issues related to (a) the history of science (HOS), (b) scientific inquiry (SI), and (c) SSIs. The topics chosen for this purpose were climate change and the use (or not) of nuclear energy versus coal for electricity. What emerged according to the participants' opinion was that HOS is useful for the training of teachers on NOSK, SI aids students in performing tasks, and SSI is preferred for teaching scientifically literate citizens. Participant teachers also noted that indeed students desire a variety of teaching methods, and that SSI seemed to them the most challenging approach to implement within the classroom.

This is certainly not the only case where teachers have been recorded as 'struggling' or hesitating to implement an SSI framework with their students. The potential inclusion of wild animals in a circus was the subject of a study involving pre-service teachers in three countries: Portugal, Spain and Greece (Almeida et al., 2020). All participants argued that animals should not be treated instrumentally by humans; however, only the Portuguese teachers acknowledged the controversial nature of this topic, and this is probably because there are similar topics in the curricula of science teaching/science didactics taught at the university there.

On the other hand, Greek pre-service teachers who were engaged in the construction of exhibits around SSI (i.e., Applications of Nanotechnology, Microplastics in the Oceans, and Breast Milk for Healthy Development) in the framework of RRI, consider that they should not attempt topics with a value-laden nature that may 'complicate' the classroom. According to them, only facts and phenomena that are generally accepted can be negotiated with students, since science teaching is limited to the management of 'settled knowledge' (Kokolaki et al., 2017).

It is therefore evident that teachers need training which not only informs and encourages them about the importance of implementing SSIs to a broad spectrum of everyday, real life questions affecting their students and society, but also provides systematic support to them and their efforts. In this direction, the way in which pre-service teachers tend to use SSIs in their teaching has been studied (Kokolaki & Stavrou, 2020, 2023): When 12 Greek pre-service teachers were asked to construct a teaching scenario for negotiating an SSI raised by the use of plastics-microplastics, and then six more pre-service teachers were assigned the same task but on the topic of nanotechnology, they mainly used SSIs as the beginning or at the end of the scenario to summarize and recapitulate what was taught earlier, i.e. SSIs were used instrumentally, rather than in their actual dimension. More specifically, it was found that the designs of the participants' scenarios displayed three patterns: (a) Multidimensional-driven SSI instructional materials (guided by the negotiation of the diverse factors and perspectives potentially involved in an SSI), (b) Content-driven SSI instructional materials (guided by the negotiation of the scientific dimensions of the issue under study), and (c) SSI instructional materials (utilized SSIs as a framework for engaging students with concepts and principles of scientific content and/or contemporary technological applications) (Kokolaki, 2022). However, although the analysis of the teaching scenarios indicates a balance between the number of activities that focus on scientific content and those that deal with social dimensions, the objectives set and the intended learning outcomes remain focused on the learning of scientific concepts, while the activities focusing on the Nature of Science are much more limited (Kokolaki, 2022; Kokolaki & Stavrou, 2020). These findings indicate the need for appropriate teacher training in the implementation of SSI.

One suggested means of providing teacher training support was the formation of Communities of Learning (CoL). When five mentors worked with 32 in-service teachers, the result was the strengthening of teachers' capacities in engaging and managing pioneering issues within the classroom (Michailidi & Stavrou, 2018). The five CoLs that were formed and lasted for 8–9 months had a positive impact on teachers' adaptation to modules on cutting-edge research topics including SSIs when implementing them with their students. The research, conducted within the IRRESISTIBLE project (www.irresistible-project.eu), showed that on the one hand, the mentors aided the teachers in various ways (i.e. with pedagogical knowledge, feedback, modelling, and administrative issues) on topics such as Nanotechnology, Microplastics and Breast Milk, while teachers on their part seemed to need the support (especially on cutting-edge topics such as Nanotechnology), evidenced by their adopting many of the mentors' suggestions regarding the adaptation of modules (Michailidi & Stavrou, 2018).

Setting SSIs around climate change as the scope of work, the views of 17 Greek pre-service teachers regarding SSIs related to proposed solutions to climate change were studied (Kokolaki et al., 2023). The participants prioritized economic and societal factors relating to climate change over scientific aspects of it, after exploring multiple stakeholders' (i.e. EU, NGOs, etc) perspectives. According to the researchers, the findings, could be used to structure appropriate frameworks for training future educators both in terms of scientific subject matter and SSI argumentation strategies. Another indication that such frameworks should be developed comes from the findings of Anastasopoulou (2022) in a study of a total of 52 Greek pre- and in-service teachers, who demonstrated difficulty in formulating scientific and socioscientific arguments.

As has been shown above, the training of teachers on SSI utilization in teaching is necessary and almost demanded by the teachers themselves. In fact, the combination of teaching strategies (e.g. scientific inquiry) with the SSI framework has been studied among students and seems beneficial to both. Thus, in addition to the aforementioned, workshops have been held during national scientific conferences, and clear supporting guidelines have been presented to facilitate teachers' familiarising themselves with similar topics through edited volumes. Hence, Kathareios et al. (2023) using the SSI of space colonization, had the workshop activities aim for the participants to understand and be able to apply the methodology of approaching SSIs and developing argumentation in their teaching. Similarly, the experiential workshop of Konstantakatos et al. (2020) was designed, this time aiming at solving an SSI (i.e., construction of a hotel unit in a specific island location) by developing an appropriate argumentation through the possibilities provided by satellite images in combination with Geographic Information Systems (i.e., extraction of spatial information, classification, and use). Finally, in the chapter "Scientific literacy in biology, critical thinking and SSI", Korfiatis and Georgiou (2022) present a comprehensive example of SSI implementation in the classroom with step-by-step instructions and commentary, drawing inspiration from the research of Maniatakou et al. (2020) and the case of vaccination.

10.2.3 SSI and Non-formal Education

It is a fact that international and national research has studied various cases of the SSI pedagogical framework in formal education, whether it concerns students or teachers. However, the cases of studying the implementation of SSI in non-formal education have been much less frequent in Greece. For this reason, I will refer to the very few corresponding cases in order to highlight the need to broaden the research in this direction as well. According to Georgiou et al. (2022), there is room for SSIs in non-formal educational settings through museums. In fact, the chapter included in the edited volume "Learning evolution through SSI" (Sá Pinto et al., 2022) lists the proposals for the use of museums (Natural Sciences, Natural History and Zoology) as suitable sites for the development of SSIs around the axis of the concept of biodiversity, since SSIs have already been implemented in museums. Specifically, an educational program around biodiversity for students aged 12–15 years is described for the Zoological Museum of the University of Athens, in which an original combination of SSI and comics is applied in order to make a final evaluation on the achievement of the program's objectives. The results were encouraging. and led to the development of similar projects (Gata et al., 2023) using SSI, this time exploiting biological evolution in general (https://zoolmuseum.biol.uoa.gr/education-m/). Although I refer only to the Greek contribution from that chapter, for reasons of uniformity-consistency with the previous sections where Greek research is presented, I would like to refer at least briefly to the French and Portuguese contributions that were very important and completed the proposal of applying SSI in the context of non-formal education, which we rarely encounter. Thus, overall applied proposals from three Museums (Porto, Paris and Athens) give ideas on how SSI can go beyond the boundaries of formal education.

10.2.4 A Holistic View of Greek Research on SSIs

Summarising the research described above (Table 10.1), it is worth looking at the thematic areas of the SSIs used. Thus, the following thematic areas/topics were noted: nanotechnology, climate change, microplastics, biotechnology, vaccination, alternative energy sources, genetically modified organisms, space colonisation, hotel construction, urban transport and demography, energy efficient houses and flats, wild animals in a circus, recycling of materials, karst caves, breast milk, biodiversity. It is worth noting that the frequency with which climate change is encountered is clearly greater than that of the other thematic areas, and indeed there are topics from the aforementioned that could be included in the broader area of climate change and the environment in general (Fig. 10.1). This leads to the question, Is the thematic focus on the environment coincidental, or could we attempt to provide an interpretation?

Perhaps an interpretation based on the particular interest of the researchers is inferred. Perhaps it has to do with local problems of Greek society that are tangible to students. After all, it was already mentioned at the beginning of the chapter that in the year 2023, Greece was severely affected by a series of natural disasters. This, to a lesser extent, had also happened in previous years and is certainly not an exclusively national incident. Similar phenomena and disasters have occurred in most countries as a result of climate change. Therefore, the interpretation of the predominant choice of such SSIs may be related to the fact that from a moral and ethical point of view, it is easier in multicultural environments such as those of Greek schools to raise issues of a universal nature, like these, and not of another type that might make it difficult for many students to respond, and perhaps even cause an awkwardness and dysfunction in the classroom. It is well known that Greece is a country that gathers a large number of refugees and immigrants, and therefore multicultural classes are formed, which require appropriate management. For example, it is rare to find research on the death penalty, transgenic animals, gene therapy (Georgiou, 2016; Georgiou et al., 2020b). In such less common cases, the mapping of one's personal ethics and the subsequent study of informal reasoning is a requirement if the SSI framework is chosen (Georgiou, 2016; Georgiou & Mavrikaki, 2013). Furthermore, it is very likely that there are strongly differing student voices and opinions, and this may be beyond the ability of untrained teachers to manage in the classroom. It is, therefore, likely that this is a reason for choosing topics around the environment. However, considering that SSIs seek, among other things, the moral/ethical development of those involved, perhaps such subject areas should ultimately not be avoided (if, of course, it is considered that this is what is being

Table 10.1 Greek research on SSI

Author(s), year	Publication Title	Publication type	Participants in the research				Language
			Primary students	Secondary students	Pre-service teachers	In-service teachers	
Patronis et al. (1999)	Students' argumentation in decision-making on a SSI: implications for teaching	Journal			√		English
Koulougliotis et al. (2021)	Probing Greek secondary school students' awareness of green chemistry principles infused in context-based projects related to SSIs	Journal		√			English
Smyrnaiou et al. (2015)	Applying argumentation approach in STEM education: A case study of the European student parliaments project in Greece	Journal		√			English
Maniatakou et al. (2020).	Role-play activities as a framework for developing argumentation skills on biological issues in secondary education	Journal		√			English
Georgiou et al. (2020b)	Investigating the impact of the duration of engagement in SSIs in developing Greek students' argumentation and informal reasoning skills	Journal		√			English
Koumara and Plakitsi (2020)	Using CHAT to Address the Nature of Scientific Knowledge Aspects on a PD-Program for Greek Science Teachers as a Cycle of Expansive Learning	Journal				√	English
Georgiou and Mavrikaki (2017)	Do Greek teenage students argue? The case of socio-scientific issues in biotechnology	Journal		√			Greek
Georgiou et al. (2022)	SSI approach out of schools—How can these approaches be used in science museums and other non formal education contexts?	Edited volume (e-book)	√	√	√	√	English
Korfiatis and Georgiou (2022)	Scientific literacy in biology, critical thinking and SSIs	Edited volume (book)			√	√	Greek

(continued)

Table 10.1 (continued)

			Participants in the research				
Author(s), year	Publication Title	Publication type	Primary students	Secondary students	Pre-service teachers	In-service teachers	Language
Kokolaki (2022)	Teacher training in the negotiation of contemporary SSI in science teaching	Doctoral dissertation			√		Greek
Georgiou (2016)	Students' ability to develop scientific arguments on biological issues with social implications: a focused teaching intervention on biotechnology	Doctoral dissertation		√			Greek
Michailidi & Stavrou (2018)	Teacher training on implementing modules on cutting-edge research topics with SSI aspects	Conference proceedings			√	√	English
Georgiou et al. (2020a)	Is teaching Biology through SSIs enough for the development of argumentation skills?	Conference proceedings		√			English
Ziogka et al. (2019)	"Cave explorers": Using Go-Lab digital scenarios to develop arguments based on geographical features."	Conference proceedings		√			English
Almeida et al. (2020)	How do primary pre-service teachers assess circus with animals and its educational value? A study with Portuguese, Spanish and Greek students	Conference proceedings			√		English
Georgiou and Mavrikaki (2013)	Greek students' ability in argumentation and informal reasoning about SSIs related to biotechnology	Conference proceedings		√			Greek
Kokolaki et al. (2017)	Development of scientific exhibits on SSI by pupils and students	Conference proceedings	√	√	√		Greek
Georgiou and Mavrikaki (2015a)	"Attempting to argue" on SSI: do young Greek citizens succeed?	Conference proceedings		√			Greek
Manoura et al. (2023)	Development of a digital game for the negotiation of SSI of climate change with primary school students	Conference proceedings	√				Greek

Georgiou and Mavrikaki (2015b)	Contributing arguments to the teaching of Biology	Conference proceedings		√			Greek
Kokolaki et al. (2023)	The perspective of PTE students on contemporary SSI of climate change	Conference proceedings			√		Greek
Kokolaki and Stavrou (2020)	SSI: A framework for training future primary school teachers	Conference proceedings			√		Greek
Kokolaki and Stavrou (2023)	The teaching practices of PTE students for the negotiation of SSIs in the light of their epistemological beliefs	Conference proceedings			√		Greek
Kathareios et al. (2023)	Workshop on developing arguments on the SSI of space colonisation	Conference proceedings (workshop)			√	√	Greek
Konstantakatos et al. (2020)	Use of Geographic Information Systems and satellite imagery for the development of argumentation on SSI of geography and environment in secondary education	Conference proceedings (workshop)			√	√	Greek

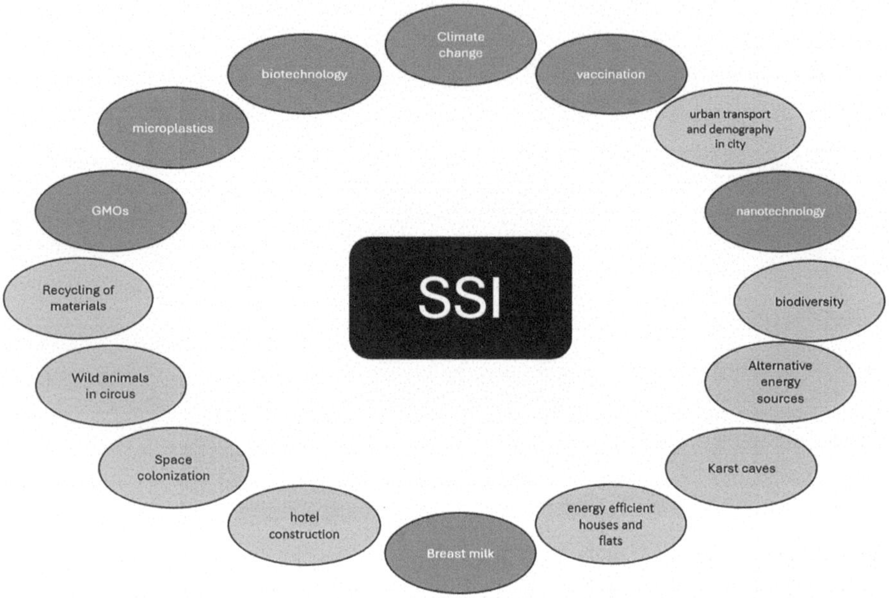

Fig. 10.1 SSI themes found in Greek research (darker colour indicates more frequent reference to the theme)

sought), even if in the light of universal moral education difficulties in approaching and establishing an appropriate framework appear. It is obvious that more research and educational effort is needed, at least in the Greek sphere, in order to find proposals that, apart from strengthening scientific literacy in terms of Vision I, also promote the other two Visions and even more so Vision III.

Perhaps the solution lies in directly approaching the everyday issues of students, however 'difficult' they may be, such as issues whose social dimensions have a religious connotation. For example, in Maniatakou et al.'s (2020) research on attitudes regarding vaccination, students created a role play using characters suggested by the researcher, one of which roles was that of a priest. The students, influenced by the Greek reality associated with a priest's religious identity, attributed an opposition towards vaccination to the character. It is worth noting that the study was conducted and published before the outbreak of the COVID-19 pandemic, while in the end the 'prediction' about the attitude of the character was indeed confirmed during the pandemic, an attitude which clearly affected a large part of Greek society and caused intense controversy. Of course, it could be said that this is possibly a pattern that we might find in any people with similar religious beliefs, but this is not accurate. For example, Greece and Cyprus are two Greek-speaking countries, can be considered as sister countries, since they have similar cultural and social characteristics. Yet according to a study by Georgiou (2016), significant differences were found in the responses of student populations from both countries when asked

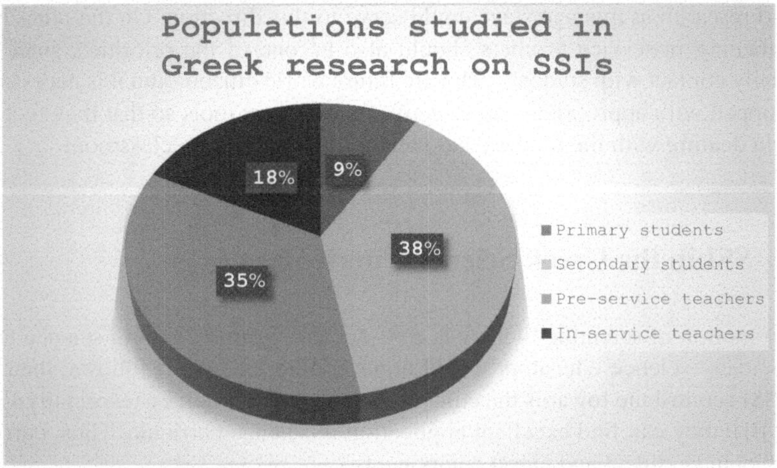

Fig. 10.2 Populations studied in Greek research on SSIs by percentage distribution

about SSIs. Specifically, in a survey conducted through a questionnaire with open-ended questions based on eight different SSIs, one of the questions was the following: 'One of your relatives suffers from thalassemia. For many years, they have been following the treatment recommended by doctors, namely regular blood transfusions. Recently, one of the treating doctors suggested gene therapy as a preferable solution to the problem. Do you think that this relative of yours should proceed with gene therapy or not, and why?' In the responses collected from the Cypriot students, the religious element was particularly strong: Several responses noted comments such as 'we should not interfere with what God has given us', and 'God obviously has his plan, so I don't agree'. However, nothing similar was noted in any of the responses of the Greek students. It seems, therefore, that similar issues are conceptualized differently and have different effects on student populations with very similar but not identical cultures. This perhaps leads to the conclusion that comparative studies involving different cultures may reveal ways in which principles of moral inquiry can be established. It is, therefore, incumbent on educators to help establish a framework for moral inquiry beginning with the SSI topic chosen. This may be a first step in cultivating mutual respect by enhancing the moral dimensions of education. It is certain that the SSI framework can really be a great help in this effort.

Figure 10.2 shows the percentage distribution of populations selected to participate in the research around SSIs. It is evident that the research focused almost equally on pre-service teachers and secondary students (38% and 35%, respectively); perhaps this may be a future suggestion for developing research designs in both primary students and in-service teachers which seem to be selected more rarely as participants of research projects on SSIs. However, it is important to start cultivating scientific literacy in students at a young age, consequently more SSI focused research at this age stage could serve to this direction. On the other hand, strengthening in-service teachers should also be one of the priorities, since they have daily contact with students, who are future world citizens and it is necessary to be equipped with appropriate and effective pedagogical tools so that they feel confident in dealing with the modern and increased demands of a classroom.

10.3 SSI in the Greek Science Curricula

The promotion of scientific literacy is **a** (if not **the**) primary goal of science teaching, as many science educators would argue (Sadler, 2004). It follows, then, that since SSIs contribute towards the cultivation of scientific literacy (especially Vision II and III), they can find excellent application in science curricula. Thus, curricula should include ethical and moral components.

But how is this decided in terms of educational policy, and what influence can different actors have? According to Sadler (2004), in order to include the ethical dimension of students in science curricula, a number of stakeholders and interested parties have to facilitate the process. Teachers, curricula designers, educational policy makers, and other professionals who support them, including researchers and academics, need to play a positive role in order to eventually include SSIs in science classrooms with an emphasis on ethical dimensions.

There are even cases where this has already occurred: An illustrative case is in England in which Lewis and Leach (2006) report that two new subjects, Twenty-First Century Science and Advanced Biology (special chapters), have recently been introduced into the curricula. In these subjects, such issues—SSIs—have been integrated into the courses, ensuring that science teachers have the necessary time, materials and resources to explore them. Therefore, courses of this type are beginning to provide new perspectives in the training of science teachers (Lewis, 2006).

However, Greek science curricula have not made use of SSI, at least until very recently. In fact, it should be clarified that in Greece, Science is taught as a single subject in primary education (K5 and K6), and as separate subjects in secondary education (i.e. Physics, Chemistry, Biology, Geography). At primary level, there is also a lesson in environmental studies for the first grades. For brevity, I will only refer to the single subject in primary and the Biology subject in upper secondary (K10-K12), as a similar pattern is followed in the other subjects. Needless to say, certainly no one should be complacent about the teaching hours of these subjects. At primary level, science as a whole (from K1 to K6) accounts for only 1/9 of the teaching hours! However, this issue is beyond the scope of this chapter; suffice it to say that Greek teachers, researchers, and academics alike recognize the situation and support the position that an increase in science teaching hours is more than necessary (i.e. Kalaitsidaki, 2022). Rather, I will focus on the way the aims and objectives of the curricula are formulated. The last curricula, just before the 2023 renewal, dates back to 2003 (Kalaitsidaki, 2022), that is to say, 20 years ago.[1] The following is an extract from the description of the context in which the 2003 science curricula for K1-K9 (compulsory education) were developed (ΔΕΠΠΣ, 2003):

[...] the teaching of science in compulsory education must contribute:

- *the acquisition of knowledge relevant to theories, laws and principles relating to the individual subjects of the natural sciences, so that students are able to 'interpret' physical, chemical, biological and geological-geographical phenomena, but also situations (e.g. geographical distributions) or processes relating to organisms and their relationships with the environment they live in.*
- *the development of students' personalities by promoting independent thinking, a love of work, the ability to deal with situations in a logical way, and the ability to communicate and cooperate with others.*
- *the acquisition of the ability to recognise the unity and continuity of scientific knowledge in the sciences, and the ability to recognise the relationship between them.*
- *to familiarise the students with scientific thinking, scientific methodology (observation, gathering and exploiting information, formulating hypotheses, testing them experimentally, analysing and interpreting data, drawing conclusions, generalising and constructing models) and the use of information technology, so that they are also capable of research and technological design as future scientists.*
- *the ability to evaluate scientific and technological applications, so that the students, as future citizens, are able to take a critical stance on them and decide on their positive or negative impact on individual and social health, the management of natural resources and the environment.*
- *the acquisition of aesthetic values in relation to the environment.*
- *to recognise the contribution of science to improving the quality of human life.*
- *to gain knowledge of the organisation and processes of the environment (natural and social), and to acquire the ability to participate in efforts to solve social problems by making use of the knowledge and skills acquired.*
- *the acquisition of the ability to communicate, to cooperate with scientific and social bodies, to collect and exchange information, to present the ideas or conclusions of one's studies.*

[1] Although most of the science curricula still in use today are from 2003, it should be noted that there are some minor exceptions (e.g. the K7 physics curriculum dates back to 2011, while the K10-K12 physics curriculum dates back to 1999). There are also cases where some minor changes have been made, such as simply moving modules from one age level to another (e.g. from K12 to K10, and vice versa), which does not constitute an essentially new curriculum in the sense of shaping both in terms of scientific content and pedagogical methods. It is also worth mentioning that there have been attempts to write curricula, before 2023, that were completed but never implemented.

- *the acquisition of basic knowledge, specialised information, methods and techniques that contribute to understanding the structure of geographical space, to understanding and interpreting the interdependencies and interactions between geophysical and social factors, and to justifying the need for harmonious coexistence between man and the environment.*

As one can see at first glance, scientific literacy exists descriptively. Although the objectives of the curricula begin with references to the acquisition of knowledge, the list is then enriched with references to decision-making, participation in the community, harmonious coexistence with the social and natural environment, etc. There is also reference to the development of students' personalities by promoting independent thinking, a love of work, the ability to deal with situations in a rational way and the ability to communicate and cooperate with others. In other words, one could say that scientific literacy is described and pursued (and not only in terms of vision I), but if we move on to consider the teaching objectives according to which the teaching will be developed, then perhaps the conclusion may be different. An example is the case of the digestive system, where the teaching objectives set in the curriculum regarding the teaching of this topic are for K6 students to

- *state the basic functions of the digestive system and relate them to the organs that make up the digestive system.*
- *distinguish the different types of teeth (incisors, canines, premolars and molars) and recognise their role in mastication.*
- *describe the structure of a tooth.*
- *list daily life habits that contribute to the proper functioning of the digestive system.*
- *justify the positive or negative impact of daily life practices on the maintenance of healthy teeth and gums.*

Although some of the objectives mention a connection to everyday life, to decide whether this is enough for a real connection and exploitation of the SSIs, we have to look back at the proposed activities. For example, according to these, to learn about the digestive system, students:

- *describe the pathway of food from the mouth to the large intestine by looking at pictures of the digestive system*
- *are divided into groups and study the role of the individual organs in this process and the factors influencing it, and form a tableau.*
- *construct puzzles with depicting the organs of the digestive system.*
- *observe their teeth in the mirror, and recognise that not all their teeth are the same, that their teeth grow through the gums, that the front teeth are sharper.*
- *bite and chew food and describe the role of different types of teeth in chewing.*
- *identify that teeth, in addition to chewing, help with conversation.*
- *draw an 'x-ray' of a tooth and identify the part of the tooth that is vulnerable to the action of micro-organisms.*
- *discuss habits of daily life and identify those that contribute to good dental health.*

This type of approach stands at a distance from the aforementioned approaches to achieving scientific literacy. At the same time, the pedagogical context is far from that of SSI and what corresponding benefits can be gained by students. Indeed, nutrition issues could be an excellent subject through which to study the digestive system. Health, economic, religious, cultural and, of course, scientific aspects could very effectively be combined when studying an SSI based in this field, and really emphasize elements of ethics and morality especially in multicultural classrooms, as it has already been noted that Greek classrooms are. However, this is not the case with the present primary science curriculum, which was in force until recently and which more or less reflects the prevailing situation in the secondary science curricula as well, with due regard to the age of the children and the concepts taught. 'The Greek science curriculum promotes the idea that citizenship education is strengthened when science education focuses on the acquisition of knowledge concerning the "academic world" of science in order for the students to be able to decide on various SSIs.' (Siatras & Koumaras, 2013, p. 11). As could be expected, the textbooks do not follow an SSI logic either. After all, textbooks stem from curricula and teachers are often relegated to those which they treat as having a guiding role in teaching (Mili & Winch, 2019).

A counter-example to this is Cyprus' educational policy on science. The reference to this country is chosen since, as mentioned in a previous section, Cyprus presents similar characteristics to Greece. Thus, one would expect that a similar evolution in the development of science curricula would prevail in Cyprus, considering that a few years earlier the textbooks used by Cypriot students were the Greek textbooks. Having said that, it would therefore be logical to look for the causes, the subtleties and possible difficulties that obviously affect countries like Greece and Cyprus in following international developments and imperatives regarding education. However, in Cyprus for more than 10 years now (since 2012) things are different. In particular, the curricula have been renewed as well as the textbooks. In their comparative research on Cypriot and Greek curricula Siatras and Koumaras (2013, p. 11) using as an example the citizenship modules (a) Health and human body, and (b) Natural environment, they concluded that the curricula differ in both scientific content and methodology, having found that in Greek curricula, SSIs are mentioned superficially, while in Cypriot curricula, SSIs are the teaching framework. It is noted that

> *the Cypriot science curriculum promotes the notion of citizenship based on the cultivation of knowledge, competencies, and mindset that can contribute to the improvement of children's everyday lives. In this direction, students are strengthened socio-politically to reshape our society towards social justice and equity. We support the latter notion of citizenship and argue that the "scientific literacy for all" movement can radically overthrow the social obstacles that prevent us from moving towards a democratic and human society.*

Giving a description of the new textbooks, Hadjichambis et al. (2013) described this SSI framework adopted through the example of the Reproduction module, which is taught in K7 students in biology classroom. Thus, following their textbook's instructions, students are invited to participate in a SSI in order to make an informed decision and advise a young couple seeking medical advice for childbearing whether to go through *in vitro* fertilization (IVF) or not. The module includes the following six subsections: (1) presentation of the scenario and the mission of the students; (2) male reproductive system and male germ cell; (3) female reproductive system and female germ cell; (4) fertilization—stages of embryo development; (5) *in vitro* fertilization—advantages and disadvantages; (6) completion of mission and decision-making), and attempts to promote implementation and impact of biological knowledge on everyday life and society, incorporating the bioethical dilemmas biological applications may accompany (Hadjichambis et al., 2013). The other modules are taught in a similar way. Based on the above, Georgiou's (2016) comparative research examined the level of reasoning and the type of informal reasoning that Greek and Cypriot adolescent students develop when involved in SSI from the field of Biotechnology with one of the advantages found to be the ability of Cypriot students to know how to manage SSI even when they have an incomplete scientific background on the concepts involved (Georgiou et al., 2020a).

The difference in the philosophy of the two countries' curricula is obvious, and it is this difference that the new 2023 curricula have essentially tried to bridge. Of course, by saying 'bridging', it does not mean that this was sought simply as a response to an informal race for the Greek education system to catch up with advances and align its curricula with others already existing. Rather, it simply means that this revision was made due to the need to provide Greek students with as many skills as science education can effectively offer to enable them to acquire the knowledge and skills necessary for the twenty-first century, a need which has been confirmed by Tsoumanis et al. (2023).

The choice and application of scientific inquiry at all levels for teaching science in Greece has already been mentioned. However, it should be pointed out that while SSI now appears explicitly within the curricula, the new guidelines have either not yet been implemented at all, or only a few have been piloted on a very small part of the student population. Moreover, there are still no accompanying textbooks to support the new curricula—and ultimately the work of teachers; this means that in reality the new curricula exist only on paper. Despite this, it is still worth looking at the way SSIs are integrated into this modernised form: The secondary school biology curricula will be used as an illustrative example (Πρόγραμμα Σπουδών, 2023). According to the curricula and the description of its general objectives, it is stated that

> The reformed biology curricula for all three upper secondary school grades aim, on the one hand, to link the principles of biology as a science with the aims and objectives of the general upper secondary school and, on the other hand, to formulate a framework for teaching and learning, that will allow students to extend their knowledge from the Gymnasium and become scientifically literate in the critical field of knowledge of the life sciences, using this knowledge in their daily individual and social life within the framework of sustainability values of solidarity, social justice and cohesion. More specifically, they seek to equip the students with the ability not only to build their biological knowledge into a coherent whole, but also to formulate attitudes in a set of fields starting from their personal health, extending to their relationship with others and ending in their relationship with society and the environment, such as avoiding the inappropriate use of antibiotics, accepting diversity, rejecting

> racism, protecting biodiversity, etc. More specifically, the reformed curriculum for K10 and K11 seeks to ensure the general education of students, so that they acquire knowledge from the science of biology, cultivate horizontal skills and develop attitudes for solving problems of everyday life and making decisions as critical citizens with individual and public responsibility for issues such as ensuring health, quality of life and ecological balance now and in the future.
>
> Through targeted activities, students practice scientific thinking and an inquiry-based approach to the learning process through microscopic observation, laboratory exercises, experiments, field studies, model building, as well as individual or group research projects in accordance with the general and specific objectives of the Thematic Fields. They also study socio-economic implications of biological processes and phenomena, making use of scientific data, and reflect on technological and scientific achievements in the context of bioethical standards. [...]
>
> *At the same time, it will enable them to reflect on the implications of modern applications and discoveries in biology, which may raise ethical issues and provoke bioethical reflection.*

Although the difference between the 2023 and the 2003 curricula and the shift towards all dimensions of scientific literacy (i.e., Visions I, II, III) is already evident, the 2023 curricula mentioned above is also distinguished by its general objectives and expected outcomes, part of which are stated below. Specifically, the aims of the curricula for students are

- *the connection between scientific knowledge and contemporary everyday life and society,*
  [...]
- *the cultivation and development of cognitive, metacognitive and socio-emotional skills and the capacity for critical thinking and argumentation,*
- *the cultivation of citizenship, where students, as tomorrow's citizens, are called upon to make decisions about the environment, health, safety and security and, in general, about any kind of challenge that will arise from the application of the science of biology in our society.*

As regards the desired outcomes, students are expected to [...]

- *develop an interest in SSIs with a biological background and, through the encouragement of personal expression and initiative, contribute to their further study and development,*
  [...]
- *understand that the application of scientific knowledge can benefit people and the environment and thus form a value system that enables them to take responsibility in the future and to develop capacities for active citizenship,*
- *understand the possible negative consequences of the misuse of scientific advances in biology, to be able to use scientific data and evidence to solve problems in their daily lives, such as health, quality of life, ecological balance (ability to take scientifically based decisions on contemporary issues related to the application of biology in everyday life),*

- *develop a sense of responsibility for the protection of the life of all organisms and to participate actively in actions to mobilise and raise awareness in society to address contemporary environmental issues and health challenges.*

The shift towards a new form of curricula that, apart from scientific inquiry, tries to promote at least some elements of moral inquiry can be easily ascertained both from the general framework set out above and from the individual objectives and expected outcomes that support and promote it. As a final element, I will provide an example (K10 Biology curriculum) of the objectives and proposed activities of some modules so that the difference between old and new curricula at each level is really quite clear and it is also possible to establish support for the general aims of the new as well as the introduction of the SSIs (Tables 10.2 and 10.3).

Table 10.2 2023 Curricula: Module 5, Animal organisms*

Module	Sub-module	Learning objectives	Proposed activities
(5) Socio-economic implications	(5.1) Animals of economic interest	Relate the beginning of animal husbandry to the Agricultural Revolution in the Neolithic period Identify the main categories of animals of economic interest State the role of animal husbandry in the primary production of Greece	Study statistical data on animal husbandry in Greece
	(5.2) SSIs	State basic rules for the protection of animal organisms Investigate social issues arising from the use of laboratory animals to promote scientific research	Create an information spot on the protection of stray animals

*This module was preceded by four other modules: (1) Organisation of life, (2) Nutrition, (3) Responsiveness to the environment, (4) Conservation and continuity of life

Table 10.3 2023 Curricula: Module 5, Plant organisms[a]

Module	Sub-module	Learning objectives	Proposed activities
(5) Socio-economic implications	(5.1) Plants of economic interest	Trace the beginning of the cultivation of plant species to the agricultural revolution Identify the categories of plants cultivated in Greece and relate them to the primary production of the country	Study statistics on agricultural production in Greece and discuss the prospects of growing new plant species
	(5.2) SSIs	Mention the intervention of biotechnology in the production of transgenic plants Discuss the proper use of genetically modified plant organisms	Organize debate to highlight the issues of cultivation of genetically modified plant organism

[a]This module was preceded by four other modules: (1) Organisation of life, (2) Nutrition, (3) Responsiveness to the environment, (4) Conservation and continuity of life

Table 10.4 2023 Curricula: Module 2, Animal organisms[a]

Module	Sub-module	Learning objectives	Proposed activities
(2) Nutrition	(2.5) Nutrition and Man	State the essential components of nutrition for the proper functioning of the human body in its different phases of development. Cultivate attitudes to promote healthy eating.	Select food groups and construct indicative diets to meet the needs of the body. Create an information leaflet on eating disorders.

[a]This module was preceded by one other module, (1) Organisation of life, and followed by three, (3) Responsiveness to the environment, (4) Conservation and continuity of life

Looking at the above examples, we can see that now, and essentially for the first time, SSI has been introduced into the Greek curricula. Of course, the way they are introduced is not identical to that presented in the Cypriot curricula, as previously mentioned. In fact, one might think that since they are the last module of each subject area, the SSIs are used to present everyday applications of scientific concepts. However, this is not the case since similar cases of activities are found in previous modules of the Greek curricula. A final example will clarify this (Table 10.4).

Based on the 2023 revised curricula objectives and activities, therefore, it seems that SSIs are not only implemented in a concluding module but are inherent in other parts of the curricula as well: This, indeed, is a different proposal for utilising SSI in teaching than that of the Cypriot curricula. In other words, while based on the Cypriot curricula and the corresponding teaching manuals each unit is taught through one SSI, in the new Greek curricula it has been chosen that students come in contact with more SSIs that stem from the same subject area and therefore the same scientific concepts.

The 2023 Greek curricula has ensured that students will come in contact with more SSIs stemming from the same subject area, and therefore the same scientific concepts. The aim is to bring students into contact with even more aspects of the application of scientific knowledge, but also with possible social, economic and cultural implications, so that they can develop moral and ethical reflection, enhance respect for the natural and social environment, become aware of their fellow human beings, and practice critical citizenship. The other 2023 Greek science curricula are more or less constructed similarly.

Of course, it should be noted that textbooks play a very important role in all this effort. However, at the time of this writing, the textbooks have not yet been published. This means that the earliest they will be released is September 2024, although this is quite an optimistic scenario. In any case, if the textbooks do not follow the spirit of the new curriculum, it is very likely that its objectives will be thwarted and, of course, the expected results will not be the desired ones. Therefore, it is necessary to harmonise the educational material with the guidelines and standards that have emerged from the curricula and, of course, it is then necessary to follow up with relevant teacher training. It is important to be optimistic, not because we already have the results of an evaluation, but because on the one hand we have a renewal after many years, and on the other hand, new pedagogical frameworks that have been proven to be effective internationally have been included.

10.4 Proposals for Future Directions

In the previous sections, an attempt was made to reflect on the research that has been done in Greece on and around SSI. Studies at the level of students of different ages and teachers (pre-service and in-service) were presented. It has already become clear that Greek research is in line with international research on both the benefits of SSI in science education and the difficulties that teachers face in their implementation, and the resulting hesitancy (Lee & Yang, 2017; Kokolaki & Stavrou, 2020; Sadler et al., 2016). On the other hand, as societies evolve, it becomes increasingly imperative that people should be able to participate in processes that concern the immediate and the distant future. To accomplish this, the role of education is crucial. The 28th United Nations Climate Change Conference (COP28) highlighted the central role that education must play not just to provide solutions to climate change, but also to lead us in a unified effort towards a sustainable future. Specifically, commenting on the critical role of education, His Excellency Dr. Tariq Al Gurg, Chief Executive Officer and Vice-Chairman of Dubai Cares said: 'Education sits at the heart of human development, yet its role has not been prioritized in global agendas addressing global challenges. The COP28 Presidency is setting a new precedent in not only recognizing the role of education as a catalyst for climate action, but also positioning it as a prerequisite for a sustainable future for all. The RewirEd Summit 2023 at COP28 is a moment for education to take its central role as the engine for a prosperous and sustainable future through solutions, actions, investments and commitments that will catalyze and accelerate its transformation (COP28, 2023). It is, therefore, critical for us as educators, to be able to propose ways to use and enhance the SSI pedagogical framework in this direction. Thus, a first step is to introduce SSI into the curricula. This is an integrated and planned approach that can work efficiently and effectively if combined with appropriate teacher training. In this way, teachers will feel more secure in managing such issues in the classroom, and will be more self-confident. In other words, it is not just a matter of professional development, but a more general issue of educational policy. This need has also been documented in the case of curricula for second chance schools in order to offer suggestions on how pedagogical practices associated with SSI inquiry 'are leverage to move students towards thinking about moral principles that transcend particular contexts (Kollas & Halkia, 2016). If more social and political elements familiar to students were embedded in SSIs, it could be a means of strengthening the necessary struggle for the achievement of social equity and justice (Kollas & Halkia, 2016).

On the other hand, if we consider that such a thing is far from the standards of a country and its curriculum, then should we accept that SSIs be excluded from use in learning processes? Greece is one such country, which, at least for the moment, has not entered the process of using SSIs. However, even in these cases, there is room to introduce SSIs. Georgiou (2016) proposed the engagement of students in SSIs in the context of the 'research project' course (a course introduced in 2012 in upper secondary education that enabled students to engage with topics from different fields, including science). Obviously, teachers could choose a different methodology, since, in essence, there was great flexibility in the course. This situation was exactly what created the opportunity to develop SSI within the allotted course time (Georgiou & Mavrikaki, 2015b). Similar courses and zones exist for other levels of education, and could at least in this informal way accommodate SSI, as long as the teachers are familiar with them. It has already been mentioned, for example, how learning communities can work effectively (Michailidi & Stavrou, 2018). Thus, even when SSIs are not explicitly present in the Science curriculum (such as the research project mentioned above), then it is at least important for the teacher to be given the flexibility that can allow room for the integration of SSI.

Moreover, the choice of subject matter to be focused on is a challenge. As already mentioned, environmental issues are not only important, not only topical, but also have a truly international impact and a universal character; clearly, they must be studied in depth. However, it is important to enrich the subject content by applying it in other contexts which, although more 'difficult' to manage collectively in a classroom, may ultimately further promote a moral education and help to formulate its principles. For example, what conclusions would be drawn if instead of *in vitro* fertilization (the example used in Cyprus), the teaching of the reproductive system started with a question to students about surrogacy? And what if the question included the aspect of same-sex couples? These are certainly difficult questions to answer, and certainly more difficult to discuss in a school environment. However, let us consider that although in Greece the above issues have been debated for years, no answers have yet been provided by legislation due to strong disagreements between different stakeholders. Is it not the case, then, that the role that education and the pedagogical frameworks that govern it must play is becoming increasingly clear? It might, therefore, be possible to see the 'moral inquiry potential' of SSIs to enhance students' scientific literacy as they contain economic, religious, social, cultural, health, environmental, moral, and ethical dimensions.

At this point, apart from the reasons stated above and the proven value of the SSI framework when part of an education program that really aims and seeks the scientific literacy of citizens and therefore a prosperous society, I would like to express my purely personal first perception of SSI: In my contact with students of all levels but especially secondary school students, the most frequent question I encountered was "and where/how will this knowledge be useful to me?" In other words, scientific concepts were constantly being questioned by students, who in their own way sought the connection between scientific knowledge and everyday life. In the early years of my teaching career, therefore, I intuitively tried to create an intimacy for the students between science and life. When I came into closer contact with pedagogical frameworks and eventually started to study SSI via my doctoral thesis, I felt that through SSIs, I could then answer my students' question by creating appropriate context and opportunities. In other words, the SSIs themselves were the context that underpins the relationship between science and real life, a relationship that is self-evident for scientists but not for students. I owe this insight in part to Professor Evangelia Mavrikaki, my PhD dissertation supervisor, who urged me to work on this topic. I thank her warmly because due to her encouragement, I began a journey into SSIs, and I have continued to deal with and discover their value to this day.

Through this ongoing involvement, I have had the pleasure and honour of contributing to the new Greek curricula as a member of the writing team, suggesting the explicit entry of SSIs. I would also like to say that most of the authors in this volume have also been an inspiration to me and my "SSI journey", with Professor Zeidler being the primary one, along with Professors Sadler and Dawson, although the list goes on!

10.5 Conclusions

Studies relating to the way in which SSIs are incorporated into the Greek curricula have been reported and reflected on in the previous sections of this chapter in order to focus attention not merely on the Greek context, but as a means of underscoring the multiple ways in which SSI frameworks properly implemented in the classroom can offer to students, regardless of the subject content, the educational level, and the teaching context.

In an age that seeks educators' contribution to guide students to acquire both inner balance and balance with the social and natural environment, strengthening the development of empathy, not just in the sense of compassion and caring, but in the sense of combining a rational and moral-ethical approach to the world is imperative and perhaps ultimately a way through which we can achieve true and universal well-being.

It is the duty and obligation of educators to promote scientific literacy in all its dimensions so that they can readily answer the students' question about how the knowledge they acquire will be useful to them and the society, and that this answer not only looks forward to material applications, but also to ethical ones. Education and educators have a key role to play in the achievement of functional scientific literacy which at the same time promotes responsible citizenship. On the other hand, it is the duty of the governments of every country to listen to educators by trusting and institutionally adopting their proposals, creating a trust for a future of prosperity.

References

Aikenhead, G. S. (2007). Expanding the research agenda for scientific literacy. In C. Linder et al. (Eds.), *Promoting scientific literacy: Science education research in transaction*. Uppsala: Geotryckeriet.

Almeida, A., García Fernández, B., & Papadopoulou, P. (2020). How do primary pre-service teachers assess circus with animals and its educational value? A study with Portuguese, Spanish and Greek students. In O. Levrini & G. Tasquier (Eds.), *Electronic proceedings of the ESERA 2019 Conference. The beauty and pleasure of understanding: Engaging with contemporary challenges through science education* (pp. 50–59). ALMA MATER STUDIORUM–University of Bologna. 978-88-945874-0-1978-88-945874-0-1.

Anastasopoulou, V. (2022). *Developing arguments on contemporary biological and socio-scientific issues by pre-service or in-service biology teachers*. Master thesis. In Greek.

ΔΕΠΠΣ Διαθεματικό Ενιαίο Πλαίσιο Προγραμμάτων Σπουδών των Φυσικών Επιστημών Δημοτικού—Γυμνασίου (2003). ΦΕΚ 304B/13-3-2003 In Greek.

Gata, D., Valakos, E., & Georgiou, M. (2023). Engaging and assessing students via a museum educational program. *Eurasia Journal of Mathematics, Science and Technology Education, 19*(10), em2334. https://doi.org/10.29333/ejmste/13574

Georgiou, M. (2016). *Students' ability to develop scientific arguments on biological issues with social implications—Focused teaching intervention on biotechnology*. Doctoral dissertation. National and Kapodistrian University of Athens. https://freader.ekt.gr/eadd/index.php?doc=39526&lang=el

Georgiou, M., Fonseca, M. J., Fortin, C., Turpin, S., & Roux-Goupille, C. (2022). SSI approach out of schools—How can these approaches be used in science museums and other non-formal education contexts? In X. Xa-Pinto, A. Bennierman, T. H. Børsen, M. Georgiou, A. Jeffries, P. Pessoa, B. Soussa, & D. Zeidler (Eds.), *Learning evolution through socio-scientific issues*. UA Editora. https://doi.org/10.48528/4sjc-kj23

Georgiou, M., & Mavrikaki, E. (2013). Greek students' ability in argumentation and informal reasoning about socio-scientific issues related to biotechnology. In C. P. Constantinou, N. Papadouris, & A. Hadjigeorgiou (Eds.), *Proceedings of the 10th conference of the European science education research association* (pp. 1158–1166).

Georgiou, M., & Mavrikaki, E. (2015a). "Attempting to argue" on socio-scientific issues: do young Greek citizens succeed? In D. Psyllos, A. Molochidis, & M. Kalleri (Eds.), *Proceedings of the 9th Panhellenic conference on science teaching and new technologies in education—Teaching and learning in science and technology: Research, innovations and practices* (pp. 733–737). ENEFET. ISBN 978-960-243-702-5 In Greek.

Georgiou, M., & Mavrikaki, E. (2015b). Contributing arguments to the teaching of Biology. In A. Polyzos, D. Schizas, P. Stasinakis, & G. Vardakostas (Eds.), *Proceedings of the 3rd Panhellenic conference of PEB. Biology in education* (pp. 108–112) ISBN: 978-618-81159-1-0. In Greek.

Georgiou, M., & Mavrikaki, E. (2017). Can Greek adolescent students express arguments? The case of socio-scientific issues of biotechnology. *Themes in Science and Technology Education, 9*(3), 137–149. In Greek.

Georgiou, M., Mavrikaki, E., & Constantinou, C. P. (2020a). Is teaching biology through socio-scientific issues enough for the development of argumentation skills? In B. Puig, P. B. Anaya, M. J. G. Quilez, & M. Grace (Eds.), *Biology education research. Contemporary topics and directions* (pp. 177–186). Servicio de Publicaciones [Publications Service], Universidad de Zaragoza. https://doi.org/10.26754/uz.978-84-16723-97-3

Georgiou, M., Mavrikaki, E., Halkia, K., & Papassideri, I. (2020b). Investigating the impact of the duration of engagement in socioscientific issues in developing Greek students' argumentation and informal reasoning skills. *American Journal of Educational Research, 8*(1), 16–23. https://doi.org/10.12691/education-8-1-3

Hadjichambis A., Georgiou G., Paraskeva-Hadjichambi D., Kyza E., Mappouras D. (2013). Human Reproduction by Inquiry: conceptual understanding and motivation of students of the 7th grade of Cyprus, In E. Mavrikaki, P. Stasinakis, & C. Zahopoulos (Eds.), *Proceedings of the 2nd Conference of Panhellenic Bioscientists Association, Biology in Education* (pp 21–28). Athens, Greece. In Greek.

Kalaitsidaki, M. (2022). Biology in the kindergarten and primary school curricula. In K. Kambourakis, M. Ergazaki, K. Korfiatis, & Stasinakis (Eds.), *Didactics of biology* (pp. 65–96). Crete University Press. In Greek.

Kathareios, V., Galani, L., Babasidis, G. & Mavrikaki, E. (2023) *Workshop on developing arguments on the SSI of space colonization*. In 13th Panhellenic conference on teaching of science and new technologies in education: Proceedings of extended sessions of papers. https://doi.org/10.12681/codiste.5454 In Greek.

Kokolaki, A. (2022) *Teacher training in the negotiation of contemporary socio-scientific issues in science teaching*. Doctoral dissertation. University of Crete. https://freader.ekt.gr/eadd/index.php?doc=52698&lang=el In Greek.

Kokolaki, A., Kapelonis, N. & Stavrou, D. (2023) *The perspective of primary education university students on contemporary socio-scientific issues of climate change*. In 13th Panhellenic conference on science education and new technologies in education: Proceedings of extended sessions of papers. https://doi.org/10.12681/codiste.5561 In Greek.

Kokolaki, A., Michailidi, E., & Stavrou, D. (2017). Development of scientific exhibits on socio-scientific issues by pupils and students. In D. Stavrou, A. Michailidi, & A. Kokolaki (Eds.), *Proceedings of the 10th Panhellenic conference on science education and new technologies in education—Bridging the gap between science, society and educational practice* (pp. 674–681). Rethymno, University of Crete. ISBN 978-960-86978-3-6 In Greek.

Kokolaki, A., & Stavrou, D. (2020). Socio-Scientific Issues: A framework for training future primary school teachers. In A. Spirtou, P. Papadopoulou, A. Zoupidis, G. Malandrakis, & P. Kariotoglou (Eds.), *Electronic proceedings of the 11th Panhellenic conference on science teaching and new technologies in education. Redefining the teaching and learning of science and technology in the 21st century* (pp. 435–442). University of Western Macedonia. ISBN 978-618-83267-7-4 In Greek.

Kokolaki, A., & Stavrou, D. (2023). The interaction between pre-service primary teachers' SSI teaching practices and their epistemological beliefs. In K. Skordoulis, K. Stefanidou, A. Mandrikas, & H. Boikos (Eds.), *The role of education in science in the society of the 21st century. Proceedings of the 12th Panhellenic conference on science education and new technologies in education* (pp. 157–165). National and Kapodistrian University of Athens. ISBN 978-618-82007-4-6 In Greek.

Kollas, S., & Halkia, K. (2016). Second chance schools in Greece: a critical analysis of science teachers' views and practices on designing scientific literacy curricula. In N. Papadouris, A. Hadjigeorgiou, & C. Constantinou (Eds.), *Insights from research in science teaching and learning. Contributions from science education research* (pp. 289–305). Springer.

Konstantakatos, G., Georgiou, M., Galani, L., Skordoulis, K., & Koutromanos, G. (2020). Utilization of Geographic Information Systems and satellite imagery for the development of argumentation on socio-scientific issues of Geography and environment in secondary education. In A. Spirtou, P. Papadopoulou, A. Zoupidis, G. Malandrakis, & P. Kariotoglou (Eds.), *Electronic proceedings of the 11th Panhellenic conference on science teaching and new technologies in education. Redefining the teaching and learning of science and technology in the 21st century, Florina* (pp. 1125–1130). University of Western Macedonia. ISBN: 978-618-83267-7-4 In Greek.

Korfiatis, K., & Georgiou, M. (2022). Scientific literacy in biology, critical thinking and socio-scientific issues. In K. Kambourakis, M. Ergazaki, K. Korfiatis, & Stasinakis (Eds.), *Didactics of biology* (pp. 97–125). Crete University Press. In Greek.

Koulougliotis, D., Antonoglou, L., & Salta, K. (2021). Probing Greek secondary school students' awareness of green chemistry principles infused in context-based projects related to socio-scientific issues. *International Journal of Science Education, 43*(2), 298–313.

Koumara, A., & Plakitsi, K. (2020). Using CHAT to address the nature of scientific knowledge aspects on a PD-program for Greek science teachers as a cycle of expansive learning. *Cultural-Historical Psychology, 16*(2).

Layton, D., Davey, A., & Jenkins, E. (1986). Science for specific social purposes (SSSP): Perspectives on adult scientific literacy. *Studies in Science Education, 13*, 27–52.

Lee, H., & Yang, J. E. (2017). Science teachers taking their first steps toward teaching socioscientific issues through collaborative action research. *Research in Science Education*, 1–21.

Lewis, J. (2006). Bringing the real world into the biology curriculum. *Journal of Biological Education, 40*(3), 101–106.

Lewis, J., & Leach, J. (2006). Discussion of socio-scientific issues: The role of science knowledge. *International Journal of Science Education, 28*(11), 1267–1287.

Maniatakou, A., Papassideri, I., & Georgiou, M. (2020). Role-play activities as a framework for developing argumentation skills on biological issues in secondary education. *American Journal of Educational Research, 8*(1), 7–15. https://doi.org/10.12691/education-8-1-2

Manoura, I., Kokolaki, A., Stavrou, D. & Papadopoulou P. (2023) *Development of a digital game for the negotiation of social-scientific issues of climate change with primary school students.* In 13th Panhellenic conference on teaching of science and new technologies in education: Proceedings of extended sessions of papers. https://doi.org/10.12681/codiste.5428. In Greek.

Michailidi, E., & Stavrou, D. (2018). Teacher training on implementing modules on cutting-edge research topics with socio-scientific aspects. In O. E. Finlayson, E. McLoughlin, S. Erduran, & P. Childs (Eds.), *Electronic proceedings of the ESERA 2017 conference.Research, practice and collaboration in science education* (pp. 2285–2293). Dublin City University. ISBN978-1-873769-84-3.

Mili, & Winch, C. (2019). Teaching through textbooks: Teachers as practitioners of a discipline? *Theory and Research in Education, 17*(2), 181–201.

Patronis, T., Potari, D., & Spiliotopoulou, V. (1999). Students' argumentation in decision-making on a socio-scientific issue: implications for teaching. *International Journal of Science Education, 21*(7), 745–754.

Πρόγραμμα Σπουδών Βιολογίας Α, Β και Γ Λυκείου (2023). ΦΕΚ 138Β/18-1-2023 In Greek.

Roberts, D. A. (2007). Scientific literacy/science literacy. In S. K. Abell & N. G. Lederman (Eds.), *Handbook of research on science education* (pp. 729–780). Lawrence Erlbaum.

Roufiktos, P. (2016). *Study of argumentation of high school students in socio-scientific issues on the occasion of the example of privatization and exploitation of coastal areas.* Master thesis. In Greek.

Sá-Pinto, X., Bennierman, A., Børsen, T. H., Georgiou, M., Jeffries, A., Pessoa, P., Soussa, B., & Zeidler, D. (2022). *Learning evolution through socioscientific issues.* UA Editora. https://doi.org/10.48528/4sjc-kj23

Sadler, T. D. (2004). Informal reasoning regarding socioscientific issues: A critical review of research. *Journal of Research in Science Teaching: The Official Journal of the National Association for Research in Science Teaching, 41*(5), 513–536.

Sadler, T. D., Romine, W. L., & Topçu, M. S. (2016). Learning science content through socio-scientific issues-based instruction: A multi-level assessment study. *International Journal of Science Education, 38*(10), 1622–1635.

Sadler, T. D., & Zeidler, D. L. (2005). Patterns of informal reasoning in the context of socioscientific decision making. *Journal of Research in Science Teaching, 42*(1), 112–138.

Siatras, A. & Koumaras, P. (2013). *Science education as public and social wealth: The notion of citizenship from a European perspective.* Paper presented at the American Educational Research Association conference, San Francisco, CA.

Smyrnaiou, Z., Petropoulou, E., & Sotiriou, M. (2015). Applying argumentation approach in STEM education: A case study of the European student parliaments project in Greece. *American Journal of Educational Research, 3*(12), 1618–1628. https://doi.org/10.12691/education-3-12-20

Tsoumanis, K., Stylos, G., & Kotsis, K. (2023). A comparative study between Greek pre-service teachers and primary school students' scientific literacy levels. *Science Education International, 34*(2), 121–131.

Zeidler, D. L., & Nichols, B. H. (2009). Socioscientific issues: Theory and practice. *Journal of elementary science education, 21*(2), 49–58.

Zeidler, D. L., & Sadler, T. D. (2010). An inclusive view of scientific literacy: Core issues and future directions. In *Exploring the landscape of scientific literacy* (pp. 186–202). Routledge.

Ziogka, K., Georgiou, M., Galani, L. & Mavrikaki, E. (2019) *"Cave explorers": Using Go-Lab digital scenarios to develop arguments based on geographical features.* Eurogeo 2019 Conference, Paris, France, 14–15/3/2019.

Chapter 11
"Content-Teaching First, Socioscientific Issue Next" as Mutant SSI Instruction: Making Sense of Hong Kong Preservice Science Teachers' Pedagogical Decisions

Kennedy Kam Ho Chan and Maurice Man Wai Cheng

11.1 Background of the Research

Socioscientific issues (SSIs) are complex scientific problems that cannot be adequately resolved without considering their societal implications, such as ethical, moral, political, and economic dimensions (Zeidler & Sadler, 2023). SSI-based instruction involves intentionally incorporating personally relevant, controversial, and open-ended scientific issues into teaching, fostering student engagement through dialogue, discussion, debate, and argumentation (Zeidler & Sadler, 2023). SSI-based instruction 'offers learning contexts that include both conceptual connections to science and ill-structured and compelling problems that engage learners in discussion, critical thinking, and decision-making' (Klosterman & Sadler, 2010, p. 1018). There is a growing recognition of the benefits of integrating SSI-based pedagogical framework into science teaching for enhancing students' understanding of science, as well as fostering critical competencies for functional scientific literacy, such as argumentation skills, media literacy as well as democratic citizenry and civic virtue (Sadler & Dawson, 2012; Zeidler et al., 2019).

Various instructional models have been proposed to guide science teachers in integrating SSIs into their teaching practices (e.g., Ke et al., 2021; Owens & Sadler, 2023; Presley et al., 2013). One example is the socio-scientific teaching and learning (SSI-TL) model (Sadler et al., 2017), which highlights the sequence of student learning experiences as well as the desired learning outcomes for SSI-based

K. K. H. Chan (✉)
The University of Hong Kong, Hong Kong, Hong Kong
e-mail: kennedyckh@hku.hk

M. M. W. Cheng
University of Auckland, Auckland, New Zealand
e-mail: maurice.cheng@auckland.ac.nz

D. L. Zeidler (ed.), *A Moral Inquiry into Epistemic Insights in Science Education*, Contemporary Trends and Issues in Science Education 61,
https://doi.org/10.1007/978-3-031-63382-9_11

instruction. This model consists of three key phases: (1) Encountering the focal SSI issue, (2) Engaging with the SSI issue, and (3) Synthesis of ideas and practices. In the SSI-TL model, students first encounter an SSI and develop an awareness of its connections to science and societal dimensions. They then engage in learning experiences that integrate scientific practices, core scientific ideas, and socio-scientific reasoning. These experiences involve analyzing the complex SSI from multiple perspectives, identifying issues for ongoing inquiry, critically evaluating information, and exploring the role of science and its limitations while learning important science content and scientific practices. In the final phase, students synthesize the ideas and practices they have encountered and reflect on their own perspectives regarding the SSI. Similarly, Levinson (2018) proposed the 3-step socio-scientific inquiry-based (SSIBL) model, which combines inquiry-based learning, SSIs, and citizenship education. The model follows an *Ask-Find out-Act* sequence, beginning with contextualizing learning by engaging students in authentic questions about the SSI. Students then explore open-ended questions by integrating social and scientific inquiry. Finally, students formulate solutions and take action to enact changes.

While nuances exist among different instructional models, several common features can be identified in terms of learning goals, instructional sequence, and the advocated teaching pedagogies. First, all the models emphasize learning goals that go beyond the acquisition of canonical science content, focusing on development of competencies crucial for functional scientific literacy such as informed decision-making skills, argumentation skills, media literacy, character and value for global citizens, and etc. These competencies are cultivated through active engagement with socially and politically controversial issues. This is in line with the philosophy that "learning science concepts alone has never been the primary focus of the SSI movement" (Zeidler et al., 2019, p. 2). Second, these models incorporate the foal SSI early in the instructional sequence, not only to activate students' prior knowledge but also to provide a genuine context for students to engage with rigorous science content and cultivate valued competencies. The focal issue is revisited near the end of the end of the unit, ultimately culminating in a resolution of the issue (Sadler, 2011). In other words, learning of SSI goes along *with* the application and practice of content (please refer to Fig. 11.1 for a simplified version of ideas associated with SSI). Finally, pedagogies associated with the models include the use of student-centred pedagogies, such as inquiry-based instruction, dialogic argumentation, debates, which engage students in high-order thinking, reasoning, and deliberation of the moral and ethical dilemmas involved in the SSIs.

Although existing instructional models offer valuable guidance to teachers, integrating SSIs into day-to-day teaching practices can impose significant challenges to teachers. Chen and Xiao (2021) conducted a recent review of 25 empirical studies to identify the challenges at the teacher, student, and policy levels that in-service science teachers face when teaching SSIs. Challenges identified include insufficient teacher knowledge, skills, and interests, as well as concerns about students' abilities and learning habits. Additionally, time constraints, pressures from standardized testing, and a lack of high-quality curriculum materials further compound the

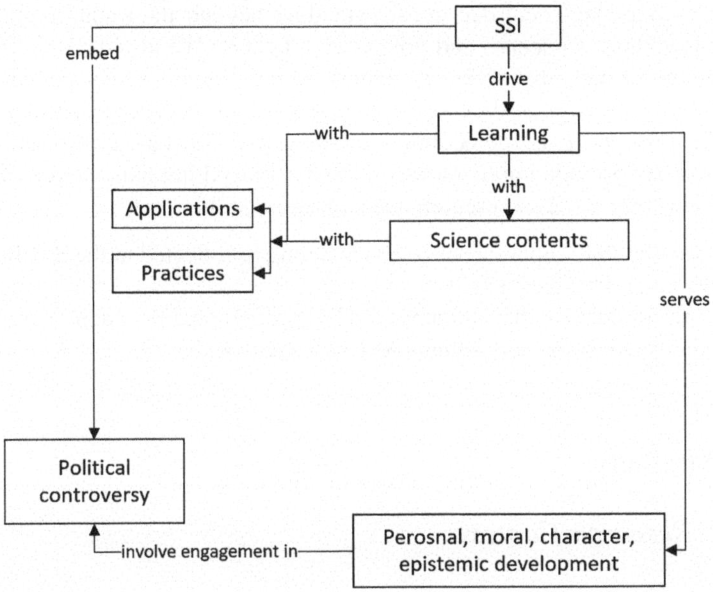

Fig. 11.1 Roles of SSI in commonly accepted SSI instructional models

challenges. Similar findings have been identified in studies focusing on pre-service science teachers (PSTs) (e.g., Amos et al., 2020; Kara, 2012; Kinskey & Zeidler, 2021, 2024). These and other similar studies have provided valuable insights into how to better support PSTs in their learning to integrate SSIs into their future science teaching practices.

This chapter contributes to this line of literature by examining the pedagogical decisions underlying PSTs' development of SSI-lessons in sociocultural contexts not necessarily conducive to SSI-based instruction. We argue that it is particularly valuable to examine the variations in approaches to SSI teaching and identify the relevant factors that influence PSTs' pedagogical decisions situated in unique contexts as teachers in diverse sociopolitical and cultural contexts likely encounter considerations that impede their interest and commitment to engage in full-fledged SSI teaching. For instance, in many parts of the world, standardized curriculum and high-stakes external assessments are seen as crucial for academic success and cultural achievement. Open debates on government policies may not be encouraged in certain contexts. Globally, only an estimated 6.4% of the population lives in full democracies (which represents the majority of contributing authors in this volume), while 37.1% live under authoritarian regimes, and the remaining >50% live in systems that fall between these two extremes (Economist Intelligence Unit, 2022). These statistics raise questions about how teachers in different contexts and systems would approach the selection, planning, and teaching of SSI-based lessons. We have chosen to adopt an asset-based perspective in our analysis of PSTs' pedagogical decisions, aiming to understand their considerations within sociocultural contexts. Our goal is to expand the understanding of these various contextual factors that influence their pedagogical decisions, extending beyond the traditional considerations of teachers-, students-, and policy-level factors. We also seek to illuminate the distinct challenges that PSTs encounter and the important resources they bring to the learning process as they strive to adopt a new pedagogical framework in potentially unsupportive sociocultural contexts for SSI-based instruction. This information can provide implications for the design and implementation of teacher preparation programs. Two research questions guided this study:

1. What are the PSTs' approaches to SSI teaching as reflected in the SSI-based lessons developed by the PSTs?
2. What are the factors that influence PSTs' pedagogical decisions based on the PSTs' considerations in designing SSI-based lessons?

11.2 Methods

11.2.1 Research Context

The study is situated in a five-year Bachelor of Science and Bachelor of Education program that certifies preservice teachers to teach science in secondary school in Hong Kong. The participants were final year undergraduate students (age: 21–25 years old) who were participating in a teacher preparation course aimed at enhancing PSTs' understanding of SSIs and the nature of science (NOS). These PSTs had just completed a group project (4 students in a team) in which they had to demonstrate their ability to integrate SSIs into science teaching. Specifically, they were asked to work in a design team to develop a 90-minute lesson plan, instructional materials (PowerPoint slides (PPTs)/toolkit, etc.), and assessment items and rubrics. They also gave a group presentation in which they introduced the activity, including (1) the rationale for the design and (2) how their design integrates SSIs with other curriculum components (e.g., content knowledge, NOS).

It is worth pointing out that SSIs are not mentioned in any of the Hong Kong science curriculum and assessment guide. The junior and senior secondary science curricula, however, advocate Science, Technology, Society and Environment (STSE) as the curriculum emphasis (Curriculum Development Council, 2017). STSE, which has little emphasis on students' development of democratic citizenry and civic virtue, is advocated as a curriculum trend in Hong Kong.

11.2.2 Data Collection

All the artifacts related to the group project were collected from the PSTs. These include lesson plan, completed Content Representations (CoRes) (Loughran et al., 2004), teaching and learning materials as well as group presentation PPTs. Four students from three different groups were interviewed. The interview comprised three parts. In the first part, the PSTs were asked about their habits/hobbies related to science and science teaching. In the second part, the PSTs were invited to explain their pedagogical decisions in relation to their selection and development of the chosen SSI issue. In the final part, the PSTs were given the learning ecology diagram (based on Cowie et al., 2023) and to explain how each source may have influenced their selection and development of the SSI issues and curriculum materials. The interview lasted from 51 min 57 s to 65 min 55 s.

11.2.3 Data Analysis

The interviews were transcribed verbatim, and multiple data sources, including lesson plans, PPTs, and completed CoRes, were thoroughly reviewed.

To address RQ#1, we first identified the focal SSI and the intended instructional goals of the SSI based lessons. We then summarized the key activities, both before, during, and after the double lessons. This summary was based on a careful examination of the lesson plans, triangulated with other data sources including the interview transcripts, the completed CoRes, final presentation PPT. These supplementary data sources provided a window into the PSTs' pedagogical decisions behind their selection of the focal SSIs and their development of curriculum materials associated with the SSI-based lessons. Distinct differences from existing SSI instructional models (e.g., Sadler et al., 2017) were identified.

For RQ#2, initial codes were developed to capture the PSTs' reasons for selecting or discarding specific SSIs, as well as their challenges when designing and developing the associated curriculum materials for the SSI lessons. These initial codes were coalized to form codes of higher level and then re-organized into themes to gain a comprehensive understanding of the underlying factors influencing the PSTs' pedagogical decisions. To capture the interconnectedness of the concepts associated with each theme and across the themes, we chose to construct a concept map that represents our findings (refer to Fig. 11.2).

To ensure the trustworthiness of the data, we employed data triangulation, combining multiple data sources, and investigator triangulation (Denzin, 2017), where the two authors independently analyzed the data and then discussed their findings in research meetings. A member check was performed, involving sharing our interpretation with the PSTs to ensure that their ideas were accurately represented.

11.3 Findings

In this section we first provide an overview of the design of the SSI-based lessons, including the focal issue selected to reveal PSTs' approaches to SSI teaching (RQ#1). We then present our analysis of PSTs' considerations to illuminate the factors that influenced their approaches to SSI teaching (RQ#2).

11.3.1 PSTs' Approaches to SSI Teaching

Table 11.1 presents the SSI-based lessons that were designed by the three design teams. The focal issues addressed in these lessons were all related to the content covered in the Hong Kong senior biology curriculum (Curriculum Development Council, 2007). In terms of the learning goals, Design Teams 1 and 3 aimed to develop students' science content knowledge while design team 2 focused on *application* of previously taught science content. Their lesson plans also include other learning goals commonly associated with SSI-based instruction, such as understanding the NOS (Design Teams 1–3), argumentation skills (Design Team 3), information evaluation skills (Design Teams 1, 2), and value development (Design Team 3).

In terms of the instructional sequence, it is noteworthy that none of the design teams introduced the focal SSI at the start of the lessons to contextualize or legitimize the learning of scientific ideas and practices, which diverges from widely accepted models of SSI instruction (e.g., Sadler et al., 2017). Instead, the learning of science content knowledge appeared to be disconnected from the SSI itself. For example, Design Team 1 focused on teaching content related to stem cell applications and the NOS, with argumentation regarding the SSI "whether stem cell applications/technologies should be developed" taking place *outside* of class time through online forums. Similarly, Design Team 3 provided students with the necessary content knowledge *before* engaging them in discussion and negotiation of the focal SSI. Only Design Team 2 elicited students' initial views on the SSI in a pre-lesson task and facilitated students in revisiting and reflecting on their changed views about the SSI. Interestingly, Design Team 2 started the lessons by revising the previously taught content knowledge but not the SSI. Overall, the SSI discussion appeared to be treated as an "add-on" rather than being integrated into the learning of *new* content knowledge.

The lesson plans proposed various student-centred pedagogies commonly used in SSI-based instruction. These included role-playing (Design Team 2), debates (Design Team 3), online discussion forums (Design Team 1), and jigsaw discussions (Design Teams 2 and 3).

In summary, the learning goals and pedagogies proposed in the SSI-based lessons were largely aligned with those advocated in SSI-based models. However, there were differences observed in the placement of the SSIs and their role in contextualizing student learning of science content compared to existing SSI-based instructional models.

11.3.2 Factors Underlying PST' Pedagogical Decisions in Designing SSI-Based Lessons

In this section, we present four interrelated themes related to the factors underlying PSTs' pedagogical decisions based on our analysis on the PSTs' considerations when selecting SSIs and developing curriculum materials for the SSI-based lessons.

Table 11.1 SSI-based lessons developed by the design teams

		Design team 1	Design team 2	Design team 3
Focal SSI		Should stem cell applications/ technologies be developed?	Should detoxification diets be used?	Should contraceptive methods be used?
Instructional goals		Content knowledge (Applications of STEM cells) NOS ideas Information search and evaluation skills Decision making skills Appreciation of the relationship between science and society	Application of content knowledge (Evaluation of the effectiveness of detoxification diets) NOS ideas Information evaluation skills Decision making skills	Content knowledge (Methods and biological basis for different contraceptive methods) NOS ideas Argumentative and communication skills Value related to sex education
Key activities	*Before the lessons (Pre-lesson tasks)*	Pre-lesson worksheet (Concepts about stem cells)	Pre-lesson questionnaire (Views about detoxification diets)	
	During the lessons	Checking pre-lesson worksheet Viewing a video of a news report to identify NOS ideas related to application of stem cells in eye disease treatment Reading articles about STEM cell applications, on-line information research and presentation (NOS Bingo activity) Summarizing concepts related to application of STEM cells and related NOS ideas	Reviewing prior knowledge on balanced diet and food substances Reading materials on different stakeholders' views and discussion (Role playing) Sharing of the perspectives of different stakeholders in a mixed group (Jigsaw discussion) Making own judgement on the SSI Group discussion to make group judgement on the SSI Summarizing the different stakeholders' views on the SSI and the NOS ideas related to the SSI	Jigsaw reading activity on different types and the biological basis of different contraceptive methods Constructing a concept map to summarize the different methods of contraception and their biological basis Reading different stakeholders' views on whether contraceptive methods should be used (InfoCards) and sharing of different stakeholders' views Whole class debate on whether contraceptive methods should be used Reflective discussion on the NOS ideas related to the SSI

(continued)

Table 11.1 (continued)

		Design team 1	Design team 2	Design team 3
	After the lessons (Post-lesson tasks)	Writing summary on the 4 different applications of stem cells from the group presentations Watching a video related to possible drawbacks of STEM cells in treatment of eye disease Writing on on-line forums to express views on whether stem cell applications/ technologies should be developed	Completing post-lesson reflection (Reflection on SSI and the application of content knowledge)	Designing posters on one or multiple perspectives on use of contraceptive methods

These findings go beyond the general challenges of SSI teaching identified in prior studies (Chen & Xiao, 2021; Kara, 2012; Kinskey & Zeidler, 2021). We also illustrate how to make sense of PSTs' pedagogical decisions within the unique sociocultural contexts of Hong Kong to identify important resources PSTs bring to the learning of a new pedagogical framework. It is worth noting that we recognize that the linear textual presentation of themes may not fully capture their interconnectedness. To address this limitation, we use a concept map that visually represents the connections across the four themes and their associated concepts. As we discuss each theme, we encourage readers to refer to relevant parts of the concept map Fig. 11.2.

1. PSTs value SSI but prioritize the delivery of content knowledge prescribed in the curriculum

All the design teams unanimously concurred that SSI deserves a prominent position within the science curriculum. However, the PSTs' primary concern when designing the SSI-based unit was about delivering the content prescribed in the curriculum, as embedded in their sayings below:

> Basically, the first assumption that we had when we were developing this lesson package was that it needs to [be] related to the content knowledge prescribed in the DSE [Diploma of Secondary Education, i.e., the school-leaving qualification that is based on the public examination] textbooks. (Design Team 3)
>
> If we spend a whole lesson just talking about one application [of stem cells], and its connection to our curriculum is not very clear, then it might not seem worth spending the entire class time on that particular topic. We would rather focus more on teaching students about biology content at the same time [in the SSI-based lessons]. (Design Team 1)

According to Design Team 1, the key factor in selecting an SSI is its relevance ("connection") to the curriculum content. Due to limited teaching time for the

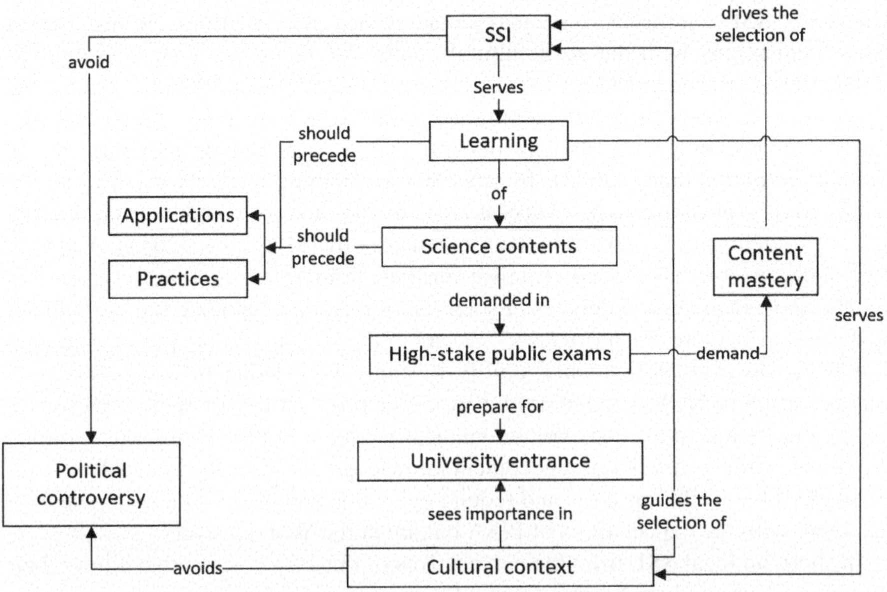

Fig. 11.2 A model of socioculturally embedded selection and development of SSI

biology curriculum, they need to be cautious and prioritize content delivery. Their guiding principle suggests that they are unlikely to choose an SSI with science content that falls outside the DSE curriculum. However, SSI can be complex and require an understanding of multiple scientific concepts and practices, posing a significant challenge for these teachers in finding an SSI that aligns with the curricular contents. Given the high demand for matching SSI with curriculum content, it becomes apparent that PSTs encountered difficulties in integrating SSI with the teaching of content knowledge in the curriculum. An example illustrating this challenge is provided below:

> It [the challenge] is about determining how much extra content we should include, as not all SSIs are covered in their textbooks. Since we are taking some of the students' time to learn exam-related knowledge to discuss these extra topics, another difficulty lies in deciding how much additional information we should provide. (Design Team 1)

Phrases and words such as "extra", "additional", "taking some of the students' time to learn exam-related knowledge" reflect the PSTs' priority to cover the content within the boundary of the mandated curriculum. It seems the PSTs perceive SSIs as a means to achieve the end of effectively delivering content for the high-stakes examination. Design team 2 chose to avoid *new* content knowledge and used an SSI issue that students already had the prior knowledge:

> We were struggling to strike a balance between delivering the concept and the issue itself. We thought that initially, it's better not to tackle the difficulty head-on, but rather to focus on the basic concepts while developing a richer discussion around the SSI. (Design Team 2)

The use of SSI as a means to support a content-driven curriculum and high-stakes assessment aligns with the sociocultural context of Hong Kong, where approximately 90% of the population is of Chinese ethnicity. Examinations hold great importance in Hong Kong, as gaining admission to a university is regarded as a cultural achievement, arguably an integral part of personal development in the Chinese culture (Huang, 2023). In other words, helping students excel in public exams could contribute to their success as members of society and their families by helping students move upward in the social hierarchy (Marginson, 2011). Although high-stakes exams have faced criticism for their limitations, such as not assessing collaborative inquiry, oral debates or socio-emotional development, they should not be dismissed as mere "memory tests". We are of the view that some public examination items do evaluate students' ability to apply knowledge or assess students' understanding of NOS in meaningful ways. As a result, it is important to recognize that emphasizing exams does not automatically equate to promoting rote learning. While the exams primarily assess content knowledge, some of the items also evaluate higher-order thinking skills and require epistemic and procedural understanding that align with the expectations of PISA but are at the Year-12 level.

In short, an ideal SSI, from the perspectives of our PSTs, is one that allows their students to learn and apply more curricular content while minimizing the inclusion of content that goes beyond the official curriculum. This approach allows students to focus on mastering the prescribed contents in the curriculum, which is crucial as achieving high examination scores and securing a university placement are considered culturally significant. (Please refer to the central area of the Fig. 11.2 (from "SSI" to "Cultural context"; and "Content mastery" in the right) as a way to make sense of the PST's pedagogical decisions and selection of their SSIs).

2. PSTs' pedagogical considerations are in line with traditional Chinese notion of learning

The PSTs maintained that some content knowledge needs to be delivered *before* engaging the students in the focal issue. This is exemplified by the following:

> The entire SSI activity was designed to be placed *after* the conceptual inputs. The recap is intended to activate students' prior knowledge on diet-related scientific concepts. In the subsequent lessons, the students need to apply these scientific concepts in the lesson activities. (Design Team 2, our emphasis in italics)
>
> When should be SSI should be introduced? … We are considering if the students don't know those concepts [different contraceptive methods and their biological basis] *before* talking about the SSI, then they might not be able to come up with those ideas [perspectives]. (Design Team 3, our emphasis in italics)

PSTs' statements above reveal a tendency to view SSI as an add-on rather than an integrated part of content learning. This inclination is apparent in their descriptions of how to position SSI in relation to a stand-alone activity for content input. Their ways of thinking about SSI may reflect some of the cultural norms about learning. The traditional classic text *Analects* (ca. 300 BCE) by Confucius emphasizes the importance of practice in learning (Sterckx, 2019). Nevertheless, the concept of "learning" in the Chinese language aligns well with the idea of prioritizing the acquisition of basic content before practice or understanding its relevance. Concepts about learning 學習 (learning and then practising), 學以致用 (to learn something that will be of use in the future), 溫故知新 (reviewing the old to understand the new) and a famous saying in *Analects*「學而時習之，不亦說乎?」("To learn and then repeatedly apply what you have learned, is that not a source of delight" (English translated in Sterckx, 2019, p. 175) emphasize the importance of mastering fundamental knowledge *prior to* its application and practice in real-life situations. (Please refer to the connections between "Learning", "Application" and "Practice" in Fig. 11.2).

3. Political considerations impacted PSTs' selection and development of SSI-based instruction.

Another factor rather unique to Hong Kong that influenced the PSTs' pedagogical decision-making is the political considerations, particularly its recent political developments.

> When choosing a topic, we will try to avoid anything related to politics, so that our classes can focus on scientific discussions, discussions about science, and the relationship between science and society. Ethical discussions are also acceptable, but we will try to avoid political controversies as much as possible. (Design Team 1)
>
> I think it's related to political or politically sensitive issues. Usually, in this aspect, we try to express the controversial nature itself without inclining to either side. ... We hope to have a more objective stance when looking at the SSI issue, and we do not try to incline to any side. (Design Team 2)

Political considerations not only influenced the choice of focal SSIs, as evident from design team 1's decision to "avoid anything related to politics", but also impacted how PSTs orchestrate discussions around the SSIs, as they refrain from taking a stance on political matters. Additionally, design team 3 mentioned how they "censored" the content in their package, resulting in limited exposure to stakeholder views:

> I think as a teacher we do somehow censor our content in the lessons … [We chose not to mention] [t]he political and also typical citizens who avoid to give birth to where we selectively avoid some parts as they don't like the current political environment in Hong Kong. So they pretend not to give birth because they don't [want] their offspring to live in Hong Kong. (Design Team 3)

Despite the significance of political considerations in shaping PSTs' pedagogical decisions (see "Political controversy" and its connections with "Cultural context" and "SSI" in Fig. 11.2), such considerations are often overlooked in literature from countries such as the United States, United Kingdom, Australia, Canada, and New Zealand, where only 8% of the world population lives in full democracy. For broader development of SSI-based instruction, we argue that researchers must acknowledge these crucial considerations that teachers share, even if not explicitly expressed.

4. SSI-based lessons more in line with the key features of SSI instruction were based on the PST's funds of knowledge and identity

The observations mentioned above do not imply that the SSI-based instruction is entirely incompatible with the Hong Kong context. We observed that the plans and curriculum materials developed by the students did incorporate some essential elements of SSI units.

> Our intention is to engage students in this controversial issue [related to pseudo-science] and encourage them to experience internal struggles while attempting to evaluate the issue. … We believe that through SSI, we can nurture students in decision-making and critical thinking processes that may not be directly addressed through our usual formal instruction on scientific concepts. (Design Team 2)
>
> If we're having many different stakeholders' information with students, the students in the discussion can be pretending like tiny societies that is quite similar to the real society, because a real society also have many stakeholders that will argue within its attitude. … We can also simulate this, society view into the discussion so that they can understand how the societies are really working in reality. (Design Team 3)

Information evaluation skills, decision making skills, discussion on the social and moral dimensions of science are otherwise rarely incorporated into science teaching in the classrooms in Hong Kong. When the PSTs attempted to inject these elements into their lessons, we found a variety of depth, quality, and commitment to the essential aspects of the SSI-based instruction (e.g., the role of moral reasoning in discussing an SSI) across the units designed by the three design teams. The best-aligned unit was developed based on the PST's funds of knowledge and identity (Design Team 3) (please refer to the connection between "Cultural context" and "SSI" in Fig. 11.2). Furthermore, the moral and ethical components of the SSI reflect Hong Kong's cultural environment, as illustrated below:

> [In Hong Kong], people are always thinking whether they should have children or not, especially for Hong Kong residents. I believe that this question raises various factors and values that students should be aware of. (Design Team 3)
>
> There is a common perception that all Christians are against the use of contraception. However, my groupmates questioned this notion, as in Hong Kong, there are no limitations or censorship regarding the use of contraceptive methods. During our research, they (my groupmates) discovered that there are also Orthodox Christian beliefs that consider contraception to be a sin. (Design Team 3)

Thus, while we value students' funds of knowledge and identity, it is also crucial to recognize the importance of the PST's funds of knowledge and identity and help PSTs' leverage those resources for successful planning of SSI-based instruction.

The detox diet unit (Design Team 2) also aligns with key features of SSI teaching. Students were engaged in perspective-taking in which they analysed the issue from the perspective of stakeholders to make decisions based on different resolutions and reasoning. We believe that to further develop its potential, we could explore the potential of Chinese values related to diet (such as the concepts of "qi" 「氣」, "cold" 「寒」 and "heat" 「熱」 in food).

11.4 Discussion

We investigated the approaches of Hong Kong PSTs to SSI teaching and explored the factors that informed their pedagogical decisions as they learned to adopt a new pedagogical framework within a sociocultural context that is not inherently supportive of SSI-based instruction. Our analysis identified additional dimensions of the factors that shape the pedagogical decisions of PSTs and illuminated valuable resources PSTs bring to their learning. In the following sections, we provide a summary of our major findings and discuss how our asset-oriented approach to analysis of PSTs' pedagogical decisions can provide implications on future research on SSI and teacher preparation.

Our findings indicate that the proposed learning goals and pedagogies of the SSI-based lessons proposed by the PSTs were in line with those advocated in the SSI literature. However, the instructional sequence and concepts of learning generally did not align with existing SSI instructional models. Overall, we observe a disparity in how our PSTs perceive the value of SSI (Fig. 11.2) compared to what is embedded in commonly accepted instructional models (Fig. 11.1).

While some may perceive the SSI lessons developed by our PSTs as lethal mutations (Brown & Campione, 1996), we view these mutations as beneficial rather than detrimental to the widespread adoption of SSI-based instruction in diverse cultural contexts worldwide. We believe that these variations serve as valuable raw materials for natural selection. Variations that retain essential elements of SSI instruction, while being better adapted to the sociocultural context, are more likely to be embraced by other teachers and adopted on a larger scale. It is worth noting that we consider the variations introduced by our PSTs as significant advancements to decontextualized, siloed or value-free science education, especially in a context like Hong Kong and in many contexts over the world, where teacher-led content-focused instruction is the dominant instructional approach and the norm.

In their recent review of literature on preservice science teacher knowledge and learning, Gray et al. (2022) highlighted the dominance of studies adopting a cognitive deficit perspective over the past 11 years. They "urge[d] teacher educator colleagues to examine PST learning in relation to the opportunities that are provided, or denied, in the preparation programmes we design and enact" (p. 77). We deliberately chose to shift away from a deficit perspective that views PST learning through a deficit lens of what they fail to do and perceives their "misconceptions" about SSI as obstacles to be overcome. Instead, our focus was on understanding the underlying pedagogical decisions that inform PSTs' approaches to SSI teaching with due consideration of their specific sociocultural context. This approach allowed us to highlight dimensions of factors, particularly those related to sociocultural influences, that extend beyond the general challenges of SSI teaching identified in previous studies (e.g., Kara, 2012; Kinskey & Zeidler, 2021, 2024). For example, we provided evidence to demonstrate that sociocultural factors, such as Chinese conceptions of learning and the cultural significance of examinations, impacted PSTs' pedagogical decisions. Furthermore, we showed how political factors can shape the pedagogical decisions of PSTs, an aspect that has frequently been neglected in the existing literature on SSI. As education and learning are culturally embedded (National Academies of Sciences & Medicine, 2018), it becomes crucial for educators to acknowledge the cultural environment in which SSI-based instruction is implemented. Moreover, it is crucial to recognize and consider the political considerations that teachers face when planning SSI curricula. The key features of SSI and moral reasoning are intricately connected within the sociocultural and political sphere. Although our data is specific to Hong Kong, we believe that this claim holds relevance in the majority of political jurisdictions. More broadly, we also aim to encourage future research on teachers' pedagogical decisions and teacher learning to prioritize the significance of sociocultural contexts.

Another salient finding is that the lessons most consistent with the key features of SSI-based instruction were those developed based on the PSTs' funds of knowledge and identity. Empowering teachers and PSTs to value and leverage their own funds of knowledge and identity can be a valuable starting point for selecting focal SSIs and developing SSI-based curriculum materials. This finding resonates with a recent study on the selection of SSI among experienced teachers. Hancock et al. (2019) examined how a design team of experienced teachers co-designed a focal SSI and discovered that teachers translated their passions into a valuable resource when identifying the focal SSIs. When considered alongside our findings, it suggests that leveraging teachers' existing passions should be emphasized in the process of selecting and developing SSI-based instruction.

11.5 Implications for Teacher Preparation

We concur with Zeidler et al. (2011) that adoption of the SSI paradigm "necessitates transformative mind-shifts in how we think about science education" (p. 279). This can be particularly challenging for PSTs as they endeavor to adopt a pedagogical framework with an underlying philosophy that may not be entirely compatible with the sociocultural values of the context in which they are situated. Therefore, it is crucial to recognize that PSTs' journey towards learning SSI-based instruction should be acknowledged as a slow and gradual process that necessitates ongoing support.

To initiate such development, instead of presenting SSI as yet another new pedagogical framework that PSTs must learn, we propose leveraging PSTs' funds of knowledge and identity. A crucial initial step is to facilitate PSTs in identifying science topics and issues that they possess intrinsic motivations, passion, and drive to learn more about and bring into their classrooms. It is equally crucial to recognize that PSTs bear the cultural responsibility of facilitating their students' acquisition of content knowledge mandated in the curriculum to succeed in high-stakes examinations, which often fail to assess the valued goals associated with SSI-based instruction, such as moral reasoning, the formation of virtue, and character development. Therefore, it becomes imperative to offer support through the provision of tools and scaffolds that aid PSTs in integrating content knowledge with the SSI they feel passionate about. Teacher educators should recognize that the learning process for PSTs extends beyond coursework and should actively engage in preparing them to navigate values and perspectives about science learning that may be incompatible with the SSI paradigm. It is important to acknowledge that these conflicting views may be prevalent in the schools where PSTs undergo their student teaching or in their future schools. By supporting PSTs to reflect on what it means to teach science within these discrepancies, such conflicting views can serve as valuable resources, ultimately enhancing their learning experiences. Overall, we believe that adopting a more asset-based approach to teacher preparation, where the unique sociocultural contexts of PSTs are given prominence, is more likely to sustain their learning and assimilation of the values and philosophy underpinning the SSI paradigm.

In closing, we acknowledge the limitations of our study, particularly the small sample size that our findings are based on. Future studies with larger sample sizes may provide further insights into additional sociocultural factors that influence PSTs' pedagogical decisions when designing SSI-based instruction. We also encourage future studies to examine teachers' pedagogical decisions with due considerations of their unique sociocultural backgrounds. We believe that adopting a more asset-based approach to analyzing teacher learning of the SSI framework holds great promise for advancing SSI research.

References

Amos, R., Knippels, M.-C., & Levinson, R. (2020). Socio-scientific inquiry-based learning: Possibilities and challenges for teacher education. In M. Evagorou, J. A. Nielsen, & J. Dillon (Eds.), *Science teacher education for responsible citizenship: Towards a pedagogy for relevance through socioscientific issues* (pp. 41–61). Springer.

Brown, A. L., & Campione, J. C. (1996). Psychological theory and the design of innovative learning environments: On procedures, principles, and systems. In L. Schauble & R. Glaser (Eds.), *Innovations in learning* (pp. 289–325). Routledge.

Chen, L., & Xiao, S. (2021). Perceptions, challenges and coping strategies of science teachers in teaching socioscientific issues: A systematic review. *Educational Research Review, 32*, 100377.

Cowie, B., Cheng, M. M. W., & Byrant, N. (2023). Unleashing the full potential of teachers: Personal ecologies and funds of knowledge/identity as resources for curriculum making. *New Zealand Annual Review of Education, 28*, 104–115.

Curriculum Development Council. (2007). *Biology curriculum and assessment guide (Secondary 4–6)*. Government Logistics Department.

Curriculum Development Council. (2017). *Science education key learning area curriculum guide (Primary 1–Secondary 6)*. Govt. Logistics Dept.

Denzin, N. K. (2017). *The research act: A theoretical introduction to sociological methods*. Transaction Publishers.

Economist Intelligence Unit. (2022). *Democracy index 2021: Less than half the world lives in a democracy*. https://www.eiu.com/n/democracy-index-2021-less-than-half-the-world-lives-in-a-democracy/

Gray, R., McDonald, S., & Stroupe, D. (2022). What you find depends on how you see: examining asset and deficit perspectives of preservice science teachers' knowledge and learning. *Studies in Science Education, 58*(1), 49–80.

Hancock, T. S., Friedrichsen, P. J., Kinslow, A. T., & Sadler, T. D. (2019). Selecting socio-scientific issues for teaching. *Science & Education, 28*(6), 639–667.

Huang, Y. (2023). *The rise and fall of the EAST: How exams, autocracy, stability, and technology brought China success, and why they might lead to its decline*. Yale University Press.

Kara, Y. (2012). Pre-service biology teachers' perceptions on the instruction of socio-scientific issues in the curriculum. *European Journal of Teacher Education, 35*(1), 111–129.

Ke, L., Zangori, L. A., Sadler, T. D., & Friedrichsen, P. J. (2021). Integrating scientific modeling and socio-scientific reasoning to promote scientific literacy. In W. Powell (Ed.), *Socioscientific issues-based instruction for scientific literacy development* (pp. 31–54). IGI Global.

Kinskey, M., & Zeidler, D. (2021). Elementary preservice teachers' challenges in designing and implementing socioscientific issues-based lessons. *Journal of Science Teacher Education, 32*(3), 350–372.

Kinskey, M., & Zeidler, D. (2024). Elementary preservice teachers' pedagogical decisions about socioscientific issues instruction. *Journal of Research in Science Teaching*.

Klosterman, M. L., & Sadler, T. D. (2010). Multi-level assessment of scientific content knowledge gains associated with socioscientific issues-based instruction. *International Journal of Science Education, 32*(8), 1017–1043.

Levinson, R. (2018). Introducing socio-scientific inquiry-based learning (SSIBL). *School Science Review, 100*(371), 31–35.

Loughran, J., Mulhall, P., & Berry, A. (2004). In search of pedagogical content knowledge in science: Developing ways of articulating and documenting professional practice. *Journal of Research in Science Teaching, 41*(4), 370–391.

Marginson, S. (2011). Higher education in East Asia and Singapore: Rise of the confucian model. *Higher Education, 61*, 587–611.

National Academies of Sciences, E., & Medicine. (2018). *How people learn II: Learners, contexts, and cultures*. National Academies Press.

Owens, D. C., & Sadler, T. D. (2023). Socio-scientific issues instruction for scientific literacy: 5E Framing to enhance teaching practice. *School Science and Mathematics*.

Presley, M. L., Sickel, A. J., Muslu, N., Merle-Johnson, D., Witzig, S. B., Izci, K., & Sadler, T. D. (2013). A framework for socio-scientific issues based education. *Science Educator, 22*, 26–32.

Sadler, T. D. (2011). Socio-scientific issues-based education: What we know about science education in the context of SSI. In T. D. Sadler (Ed.), *Socio-scientific issues in the classroom: teaching, learning and research* (pp. 355–369). Springer.

Sadler, T. D., Foulk, J. A., & Friedrichsen, P. J. (2017). Evolution of a model for socio-scientific issue teaching and learning. *International Journal of Education in Mathematics, Science and Technology, 5*(2), 75–87.

Sadler, T. D., & Dawson, V. (2012). Socio-scientific issues in science education: Contexts for the promotion of key learning outcomes. In B. Fraser, K. Tobin, & C. McRobbie (Eds.), *Second international handbook of science education* (pp. 799–809). Springer.

Sterckx, R. (2019). *Chinese thought: From confucius to cook ding*. Penguin.

Zeidler, D. L., Applebaum, S. M., & Sadler, T. D. (2011). Enacting a socioscientific issues classroom: Transformative transformations. In T. D. Sadler (Ed.), *Socio-scientific issues in the classroom: Teaching, learning and research* (pp. 277–305). Springer.

Zeidler, D. L., Herman, B. C., & Sadler, T. D. (2019). New directions in socioscientific issues research. *Disciplinary and Interdisciplinary Science Education Research, 1*(1), 11.

Zeidler, D. L., & Sadler, T. D. (2023). Exploring and expanding the frontiers of socioscientific issues. In N. G. Lederman, D. L. Zeidler, & J. S. Lederman (Eds.), *Handbook of research on science education* (pp. 899–929). Routledge.

Chapter 12
The State of the Art of Socioscientific Issues in the Arab World

Rola Khishfe

12.1 Introduction

12.1.1 Socioscientific Issues

Socioscientific issues (SSI) are complex, socially relevant, and controversial issues that involve science, technology, and society (STS). SSIs not only encompass the full scope of what STS entails, but also incorporate ethical aspects of science, moral reasoning of students and their emotional development (Sadler et al., 2007; Zeidler et al., 2002). Real-world problems such as genetic testing and global warming are examples of SSI that include ethical components (Sadler et al., 2007). SSIs are often characterized by uncertainty, value conflict, and the need for decision-making. SSI can be found in different domains, such as environmental issues, health issues, and technological development.

12.1.2 The Need for SSI in Arab Countries

The teaching of SSI in Arab countries is important for several reasons. First, it can help students develop critical thinking skills and the ability to evaluate scientific information. Second, it can help students understand the complex relationship between science, technology, and society (STS). Third, it can help students develop the skills they need to participate in decision-making about SSI related to their everyday lives.

R. Khishfe (✉)
American University of Beirut, Beirut, Lebanon
e-mail: rk19@aub.edu.lb

D. L. Zeidler (ed.), *A Moral Inquiry into Epistemic Insights in Science Education*, Contemporary Trends and Issues in Science Education 61,
https://doi.org/10.1007/978-3-031-63382-9_12

Arab countries are undergoing rapid social and technological change. In addition, these countries are diverse in terms of their cultures, religions, and values. Therefore, the teaching of SSI can help Arab countries address these challenges. By learning about SSI, students can develop the skills they need to participate in public decision-making and to make informed choices about their own lives. Furthermore, the teaching of SSI can help to (a) develop students' understanding of the nature of science (NOS) and their argumentation skills, (b) promote students' interest in science and technology, and (c) prepare students for the challenges of the twenty-first century, which will require to think critically and solve complex problems. Overall, the teaching of SSI should be an important component of science education in Arab countries because it can contribute to helping students develop the skills that they need to be informed citizens and to make a positive contribution to society.

12.2 Arguments for the Use of SSI

Arab researchers provided various arguments to justify integrating SSI in the teaching of science in Arab countries. Al-Jrrah et al. (2013) in Jordan as well as Khishfe (2012) in Lebanon used SSI to investigate the argumentation skills among students, as it provides an appropriate context for expressing one's opinion, weighing on ethical questions, and making informed decisions. In Lebanon, SSI has been utilized as a platform for the teaching of two pillars of scientific literacy (BouJaoude, 2002; Khishfe, 2014): nature of science (NOS) (e.g., Khishfe, 2022) and argumentation (e.g., Khishfe, 2023a). Al-Zunaidi and Al-Salim (2020) in the UAE and Alsaaideh (2016) in Jordan emphasized the development of decision-making skills, scientific knowledge, and reflective thinking skills in students using SSI. The studies showed that SSI-based programs can improve these skills.

Alsaaideh (2016) considers the integration of the social aspect into science education as vital, especially when students face issues such as pollution and warfare. The researcher showed that when students develop an understanding of the connections between science and society, they are better able to address relevant SSI that relate to their daily lives.

Several authors (Almarashi & Alshahri, 2021; El Arbid & Tairab, 2020) acknowledge the rapid advancement of scientific knowledge and its effect on social issues. These authors stress the importance of integrating SSI and science, technology, society, and environment (STSE) into school science in the Kingdom of Saudi Arabia (KSA) to address societal concerns that interplay with advancements in technology and science. In addition, the inclusion of such issues would encourage students to connect the relationship between society and science (Almarashi & Alshahri, 2021). Issues such as environmental pollution, genetically modified food, and cloning can challenge students' scientific understanding and ethical reasoning. El Arbid and Tairab (2020) recognize the complexities in integrating SSI into curricula of conservative communities in the United Arab Emirates (UAE), given that cultural, religious, and personal beliefs of educators influence their teaching practices and the development of students' argumentation skills. El Arbid and Tairab (2020) believe that including SSI in the science curriculum will "afford students the opportunity to develop discourse and argumentation skills that emphasize the development of competence in decision making, ethical reason, and critical thinking" (p. 737). Similarly, Almarashi and Alshahri (2021) emphasized the need to include Science, Technology, Society, and Environment (STSE) issues in curricula. They noted the importance of comprehensive coverage and balance in addressing STSE issues in science textbooks.

In Iraq, Frhod and Abdulwahed (2022) argue that modern science education should go beyond simply conveying content knowledge, but rather engage students in discussions about complex scientific issues to enable them to navigate the complexities of a rapidly changing world. Teachers play a crucial role in fostering critical thinking and ethical considerations among students, and using SSI in such contexts can enable lifelong learning in students. Similarly, Al-Zunaidi and Al-Salim (2020) views the teaching of SSI as an opportunity to enhance scientific knowledge, critical thinking, decision-making skills, and the exploration of scientific concepts.

In Egypt, the Vision 2030 for Sustainable Development emphasizes educational improvement and human capital development. As technology advances, our world becomes increasingly knowledge based. STEAM (Science, Technology, Engineering, Arts, and Mathematics) education, which combines science, technology, engineering, arts, and mathematics, is essential for solving complex societal problems and preparing students for the future. Al-Khayal (2022) argues that educators should be trained to integrate STEAM with SSI to improve learning quality and holistic understanding in students. Al-Khayal (2022) views the integration of STEAM with SSI as crucial to improve students' attitudes, motivation, and educational development. In addition, SSI issues such as gene editing can be of high importance for the social, ethical, and economical realms of the country. Indeed, the Egyptian government, through the Egyptian Reference Human Genome project, aims to study the genome of ancient Egyptians and genes related to common diseases in the country. Therefore, Abou-El-Fetouh Sheira (2023) believes it is necessary to enhance teachers' knowledge and capabilities to address SSI issues such as gene editing. This will allow them to make informed decisions.

12.3 Methodology

In this study, a comprehensive methodology was employed to conduct a literature review focused on the Arab region. The primary objective was to gather a diverse range of scholarly articles. To achieve this, a multifaceted approach was utilized. First, we consulted the Shamaa platform, a reputable website dedicated to educational research in the Arab world. Shamaa served as an invaluable source for free access to research articles, books, reports, conferences, theses, and PhD dissertations, specifically within the educational context. Shamaa's database includes articles written in Arabic, English, and French. The inclusion of Shamaa in our methodology ensured that we captured a comprehensive body of knowledge related to the Arab region, which might not be readily available through other conventional databases.

In addition to the Shamaa platform, we expanded our search further by including Google Scholar searches. Both English and Arabic search terms were used to ensure the inclusivity of the review. The dual-language approach was essential, as it acknowledged the cultural and linguistic diversity of the Arab region, where many valuable contributions are presented in Arabic. We utilized a combination of relevant keywords and phrases in both languages to identify potential articles. The inclusion criteria for the studies to be included were:

The search was done in accordance with the following inclusion criteria: (a) articles that were published in journals; (b) articles that report empirical studies; (c) articles that address studies involving SSI; (d) articles that address studies with teachers and/or students; and (e) articles that are published in either English or Arabic languages. This approach allowed access to a wide array of research materials, providing a holistic understanding of the use of SSI in the Arab world. Our comprehensive methodology, spanning both English and Arabic languages and leveraging multiple databases, facilitated a thorough literature review, contributing to the richness and depth of our analysis on the topic under investigation.

12.4 SSI Research in the Arab Countries

12.4.1 SSI and Scientific Literacy: NOS and Argumentation

For research conducted in Jordan, SSI was used as a context to assess levels of argumentation and thought process among students. In their study, Al-Jrrah et al. (2013) conducted semi-structured interviews with thirty 10th grade Jordanian students to explore their levels of argumentation and reasoning. The SSI topics covered in the interviews were cloning, genetic engineering, endogamy, and premarital genetic screening. The results of the study showed that most students used intuitive thinking in their argumentation, a large majority used emotional thinking, and a minority of students used rational thinking when presented with the SSI topics. This shows that most of the arguments the students put forth did not rely on scientific understanding, but rather on sentiments towards the genetic and social issues discussed. The authors found that these findings align with previous research from other countries (e.g., Osborne et al., 2004) and shows that students may have a limited capacity to construct strong arguments due to their tendency to jump to conclusions without providing evidence or sufficient data to back their arguments. Al-Jrrah et al. (2013) postulate that this could be due to teaching strategies that do not promote active student participation in problem-solving and scientific discussions. In addition, given the SSI topics selected for this research, the authors believe that the social context and cultural beliefs may contribute to students' weak argumentation skills.

The authors recommend that teachers emphasize the connection between evidence and claims in the classroom, and guide students in interpreting the support or refutation of claims using evidence.

Global technological advancements have led to a continuous movement towards a knowledge-based society. STEAM education plays a significant role in addressing complex societal issues and preparing students for future challenges. For this reason, Al-Khayal (2022) found the preparation of educators to integrate STEAM with SSI is a means to enhance learning quality and holistic understanding in students. The researcher developed and assessed a STEAM-based program for female early childhood education students. The methodology incorporated descriptive, experimental, and qualitative approaches to achieve these goals, emphasizing both theoretical framing and practical evaluation through pretest-posttest designs and content analysis of lesson plans. The results of the study showed a positive impact on motivation across various dimensions, except for cognitive aspects. The study highlights the following factors that contributed to the program's success: interactive discussions, addressing societal issues, diverse information sources, and guidance. Recommendations include educator training in STEAM lesson planning, student teaching strategies, hands-on learning, and integrating cultural and general education. The research suggests investigating the effectiveness of the program on educators' motivation and comparing it to other approaches. However, Al-Khayal (2022) state that the study's limitations include its focus on a specific sample of female students, certain societal issues, and teaching methodologies influenced by the ongoing Covid-19 pandemic. Overall, the study underscores the importance of STEAM and the program's potential to enhance motivation and engagement.

Another study conducted in Egypt (Abd-El-Latif, 2019) aimed to create a program centered around SSI relevant to the local context and assess its efficacy in enhancing the argumentation skills and students' knowledge and attitudes concerning these issues in secondary schools. To achieve this goal, the most prevalent SSI within Egyptian society were identified, and the program's foundations were established. Issues include electronic cigarettes, nuclear energy, meat consumption, and fish farming. A test measuring the argumentation skills, a knowledge assessment, and an Attitude scale were developed. The study employed a quasi-experimental pre-test/post-test design with both experimental and control groups. The findings revealed that the utilization of the proposed program for teaching local SSI had a positive impact on students' argumentation skills, while also enhancing their knowledge and attitudes toward local SSI.

Once more in Egypt, Abou-El-Fetouh Sheira (2023) highlight the advancement of gene editing in the country, and the plans of the Egyptian government to apply genetic engineering in various fields such as agriculture, medicine, and food production. The current curriculum in Egypt lacks, however, an in-depth coverage of modern genetic applications, especially as it relates to its ethical and societal considerations. For this reason, Abou-El-Fetouh Sheira (2023) finds it necessary for teachers to be able to address gene editing and have an enhanced knowledge on the subject. Abou-El-Fetouh Sheira (2023) develops a program that focuses on gene modification issues and allows participants to actively engage in scientific argumentation by questioning, answering questions, and making decisions. The intervention included 30 students enrolled at the Faculty of Education in a university. The participants were administered pre and posttests, and the scores showed improvement in both scientific argumentation and decision making in the students. Based on the positive findings, the study suggests the development of teacher programs related to genetic modification concepts and applications.

In KSA, Khishfe et al. (2017) examined the understanding of NOS and argumentation skills in the context of SSI among grade 11 students. A questionnaire was administered to 74 participants from six public schools in Riyadh. The instrument contained four scenarios covering the following SSI: global warming, genetically modified food, acid rain, and human cloning. The results showed that participants' stances on various SSI were influenced by factors like religious beliefs, perceptions of the natural world, and environmental concerns. The study found that most participants struggled to develop well-structured arguments and had limited knowledge about NOS aspects, which aligns with previous research indicating students often have naïve views about NOS. More notably, the study participants who generated well-developed arguments tended to have a more informed understanding of NOS aspects. Overall, the study showed that integrating SSI into science education would help advance scientific literacy by helping students understand NOS and engaging them in meaningful argumentation about real-world problems. The researchers believe that this approach prepares students to address scientific and societal challenges effectively.

Once again in KSA, Alasmari et al. (2019) evaluated the effectiveness of a new teaching model for SSI on high school biology students' argumentation skills. The research adopted a design-based research (DBR) approach that blends both quantitative and qualitative methods. The study involved 46 students, divided into two groups: one received instruction using the new model (experimental group), and the other group was taught using traditional methods (control group). To achieve the research objectives, data were gathered through open tests, individual and group interviews, teacher diaries, students' discussion worksheets, as well as audio and video recordings. An analysis of the data, both quantitative and qualitative, revealed significant differences in the post-test for the experimental group. Students in the experimental group were able to support their claims with data, visual aids, asking questions, and effectively communicating their ideas. Alasmari et al. (2019) recommended including the proposed model when teaching SSI.

When compared to scientific contexts, SSI can provide a suitable context for the teaching and learning of NOS and argumentation. Indeed, in Khishfe's (2022) study in Lebanon, both SSI and scientific contexts were compared to examine their impact on students' NOS understandings and argumentation abilities. The study involved undergraduate students who were divided into two groups: one focused on SSI and the other on scientific topics. The instructional process consisted of two stages, where NOS and argumentation were taught through various activities for both groups initially. Subsequently, each group engaged in different scenarios, one group addressed SSI, and the other tackled scientific topics. Data collection included pre- and post-questionnaires and semi-structured interviews. The findings revealed that participants in the SSI group demonstrated more significant enhancements in their argumentation skills when dealing with SSI scenarios compared to the science group. Concerning NOS conceptions, students in the SSI group exhibited greater improvements than students in the science group when responding to SSI scenarios. Generally, the participants improved their argumentation skills more than their NOS comprehension. The author interpreted the results by noting that argumentation can be viewed as practice while NOS as content.

In another study in Lebanon, Khishfe (2012) examined how high school students' understandings of three aspects of NOS relate to their ability to construct arguments concerning two different SSI. Five schools located in different parts of Beirut were involved in the study. A total of 219 grade 11 students responded to an open-ended questionnaire that addressed genetically modified food and water fluoridation. Results showed some correlation between argumentation components and NOS aspects. However, this correlation was found to be more robust when students were presented with the SSI of water fluoridation. The author attributed that to the notion that students were familiar with the topic, had prior knowledge about it, and showed interest in it. Therefore, the findings imply that when students are given a pertinent SSI that they are previously exposed to, this may improve their understanding of NOS and their capacity to engage in argumentation.

In yet another study in Lebanon by Khishfe (2013), participants were 10th-grade students from one high school divided into two distinct groups. The instructional intervention consisted of a six-week unit centered around genetic engineering. One group- termed the "non-NOS group"- received no explicit NOS instruction, while the other group- known as the "NOS group"- was provided with explicit instruction covering three NOS aspects (subjective, empirical, and tentative), integrated within the genetic engineering unit. The main objectives of this research were twofold: (a) to assess the effectiveness of explicit instruction of NOS within a SSI context, and (b) to investigate the potential transfer of these newly acquired NOS understandings, which were explicitly taught in one SSI scenario, to other similar contexts (whether familiar or unfamiliar). To assess the change in participants' NOS understandings and their ability to apply their acquired knowledge, a questionnaire comprising two open-ended scenarios, along with semi-structured interviews, was employed. The first scenario pertained to a familiar context concerning genetically modified food, while the second focused on an unfamiliar context related to water fluoridation. The results indicated that there was no enhancement in the NOS understanding among participants in the non-NOS group concerning both familiar and unfamiliar SSI contexts. Conversely, participants in the NOS group demonstrated overall improvement in their NOS understandings, applicable to both familiar and unfamiliar SSI scenarios. This suggests that NOS understandings may not be context-bound but can be transferred and applied to various SSI scenarios.

Another study in Lebanon by Khishfe (2021) aimed to explore how teaching argumentation, especially within the context of SSI and teaching NOS, when compared to solely teaching argumentation, can enhance the argumentative skills and NOS understandings of 10th-grade students. The study involved 36 students from two randomly assigned groups: Treatment and Comparison. Over four months, both groups received explicit instruction in argumentation related to SSI. However, the Treatment group received additional explicit instruction in NOS. To assess their argumentation skills and NOS understandings, participants underwent pre- and post-testing through open-ended questionnaires and individual semi-structured interviews. The results indicated that both groups showed improvements in their argumentation skills due to explicit argumentation instruction, with greater improvements observed in the Treatment group. Moreover, the Treatment group, which received explicit instruction in both NOS and argumentation, demonstrated advancements in their NOS understandings. These findings were analyzed in the context of three key themes: argumentation instruction, NOS instruction, and the combination of instruction in argumentation and NOS. The study highlights the positive impact of explicit NOS instruction, when combined with argumentation instruction, in improving participants' NOS conceptions and fostering links between NOS conceptions and argumentation components. These findings underscore the potential benefits of integrating NOS instruction into science education, particularly within the framework of argumentation in SSI contexts.

The retention of the students in Khishfe's (2021) study was examined three months after the intervention in another study (Khishfe, 2020). It was found that after the treatment, 39% of participants in the Treatment group continued to demonstrate informed categories for the argument component three months later, which was the same percentage observed immediately after the treatment. On the other hand, only 17% of the comparison group had informed arguments, which was less than the percentage observed after the intervention. As for NOS conceptions, the treatment group largely preserved their NOS conceptions three months post-treatment. The successful results of the treatment group can be explained due to the following premises. First, the participants had prolonged exposure to the content, 16 weeks. Second, the treatment group benefited from multiple reflective experiences that were distributed and spaced out throughout the intervention. Third, the coupling of explicit instruction in both argumentation and NOS allowed for meaningful connections to be built between these concepts and facilitated better retention. Fourth, using NOS as a metacognitive tool allowed the treatment group participants to engage in reflection as well as question scientific knowledge and its development. The combination of these four premises, added to the context of SSI, may have played a crucial role in the retention of NOS conceptions and argumentation skills in the treatment group.

To further explore the relationship between argumentation and NOS in SSI contexts, Khishfe (2023a) administered the *Controversial Socioscientific Issues Questionnaire* (CSI) to 125 grade 10 students which featured two scenarios centered around SSI (genetically modified food and water fluoridation). The study, conducted in Lebanon, specifically investigated whether having well-informed NOS concepts contributes to improved argumentation skills. The data obtained from the participants' responses were subjected to stepwise regression analyses. The three explored NOS aspects (subjective, tentative, and empirical) were treated as independent variables, while the argumentation components (argument, counterargument, and rebuttal) were treated as dependent variables. The results revealed a positive correlation between argumentation skills and NOS understandings. Indeed, for genetically modified food, the results revealed that the tentative aspect of NOS significantly predicted the argument, the empirical aspect predicted the counterargument, and both the tentative and subjective aspects predicted the rebuttal in students' argumentation. In the case of water fluoridation, the subjective aspect of NOS was a significant predictor of the counterargument, while the tentative aspect significantly predicted the rebuttal in students' argumentation. The results from this study and from previous studies provide compelling evidence of the relationships between argumentation and NOS knowledge within the context of SSI and set the context for connecting between the two in instruction.

Towards examining how a connected learning approach can affect students' NOS comprehension and argumentation skills, Khishfe (2023b) combined explicit instruction on NOS and argumentation and compared it to a non-connected approach. For this study conducted in Lebanon, 42 grade 10 students were randomly assigned to two groups, one receiving connected instruction (where NOS and argumentation were taught together) and the other receiving non-connected instruction (where they were taught separately). This was done within two different SSI contexts: one scenario was about global warming, while the second covered acid rain. The study lasted for two months, and assessments were conducted through open-ended questionnaires and individual interviews before and after the study. Results showed that the connected group performed better in generating arguments and counterarguments. Both approaches led to improved NOS understandings. These findings suggest that connecting the teaching of NOS and argumentation can enhance the learning outcomes related to NOS and argumentation. In addition, the study emphasizes the importance of curriculum coherence and an explicit approach in enhancing students' comprehension of NOS and argumentation, particularly in the context of SSI.

At another level, the effect of having different SSI contexts on the NOS views of students was explored by Khishfe (2017, 2019). In the first study, 261 high school students from eight schools located in different parts of Lebanon responded to a questionnaire that contained five topics covering SSI. There was no intervention in the study. Using a qualitative research design, the results showed that the students' NOS views were not consistent across the aspects (Khishfe, 2017). This could be due to individual differences in students and their capacity to recognize similarities between different tasks or contexts. Consequently, the author advised that instructional strategies need to be tailored to address these issues in science education. In the second study (Khishfe, 2019), however, an intervention was conducted with grade 7 students. The participants were taught about NOS explicitly during a unit about volcanoes, plate tectonics, and earthquakes. The total number of participants was 38, divided into two groups: a control and a treatment group. When comparing the two groups, the students who did not receive the NOS instruction did not improve their NOS understandings. As for the group with NOS instruction, participants showed more improvements in applying their understandings when the contexts were similar and familiar to the original context (scientific) and fewer improvements with the contexts (SSI) that were not similar to the original context.

On the other hand, a study by Khishfe (2015) showed that the long-term retention of newly acquired NOS understandings can be retained by some but not all participants. The study involved 24 tenth-grade students enrolled in a private school in Lebanon who received explicit NOS instruction in the SSI context of genetic engineering. Findings showed that students were able to improve their naïve NOS conceptions after the instruction, but after four months, not all students retained their newly acquired NOS understandings. Some participants reverted to a more naïve NOS understanding after the four-month retention period. During the intervention, there were two different scenarios of SSI, water fluoridation and genetically modified food. The results showed no significant differences in the retention of students' NOS understandings between the two SSI scenarios. The study suggested that connections between the newly acquired knowledge and the existing knowledge play a key role in long-term retention. Additionally, prolonged exposure to the domain and distributed opportunities for learning contribute to improved retention.

12.4.2 SSI and Environmental Issues

In Jordan, Alsaaideh (2016) explored the coverage of SSI in eighth-grade science textbooks. Then, the researcher administered a multiple-choice test to 115 eighth-grade students to measure their comprehension of the SSI and sociological aspects covered in the textbook. The study showed that the textbook contained numerous social and sociological issues, primarily centered around environmental problems, with certain topics conspicuously absent from its content. The test results showed that on average, the students' understanding of SSI was low, but female students had a significantly better comprehension than their male counterparts. In addition, results showed that some issues in the textbook were better understood than others. More specifically, the topic of conservation of plants and animals received the highest score of understanding among the students, followed by the environmental issue of global warming and ozone depletion. In addition, these topics were extensively addressed in the textbook. On the other hand, the topics that were briefly mentioned in the textbook, such as the social and environmental impact of technological warfare, nuclear reactor effects, communication technology, and space pollution, were less understood. The author recommends further emphasis on SSI by using contemporary teaching strategies, and when necessary, providing professional development to science teachers. In addition, the author calls for balanced and comprehensive coverage of SSI in textbooks.

12.4.3 SSI and Scientific Reasoning Skills, Critical Thinking Skills, and Decision-Making Skills

In Jordan, Alsaaideh et al. (2017) emphasized the need to address challenges arising from the interplay between science, technology, and society in today's world. Their study underscores the adoption of teaching methods focusing on real-life problems and utilizing the SSI approach to enhance students' critical thinking and decision-making skills. It also highlighted the importance of environmental education. Specifically, the study investigated the impact of the SSI approach on students' understanding of modern environmental issues at Al-Balqa Applied University. The study used a quasi-experimental design, involving an experimental group taught using the SSI approach and a control group using traditional methods. Results showed that the SSI approach significantly improved students' comprehension of contemporary environmental problems, especially among those with higher scientific inclinations. Alsaaideh et al. (2017) recommended adopting the SSI approach, conducting teacher training workshops, nurturing students' scientific interests, and further researching the effectiveness of the SSI approach in teaching. The study's significance lies in its integration of scientific and societal concerns in education, promoting active engagement with real-world challenges.

Again in KSA, Al-Zunaidi and Al-Salim (2020) emphasized the importance of nurturing various critical thinking skills among learners to enrich their problem-solving abilities and their ability to apply these skills in addressing future challenges. Hence, the study used SSI to develop students' decision-making skills, scientific knowledge, and reflective thinking skills by introducing an ethical analysis approach. The ethical analysis approach is a teaching sequence that consists of three main steps: explanation, analysis, and argumentation, and finally making a judgment on the presented scientific issue. Participants in the study were secondary female students enrolled in a chemistry class. Al-Zunaidi and Al-Salim (2020) developed a reflective thinking assessment tool to measure the following skills in students: observation and reflection, detecting fallacies, reaching conclusions, providing explanations, and proposing solutions. Results of the study showed that the ethical analysis approach was moderately effective in developing observation and reflection skills and reaching conclusions, and effective in developing the skill of detecting fallacies among the students. In addition, students' abilities to reach conclusions and justify their decisions were significantly improved after engaging in ethical analysis.

Frhod and Abdulwahed (2022) explored the decision-making skills related to SSI among high school biology teachers in Iraq. The researchers found it essential for teachers to be able to implement SSI in their classroom discussions since these issues foster personal and moral development in students. The study focused on the ethical and societal implications of rapid scientific progress, such as climate change, genetic therapy, and surrogacy. To measure teachers' decision-making abilities, a sample of 250 teachers was selected at random, and a test designed by the authors was administered to the teachers. The results revealed that these teachers exhibit a suitable level of decision-making regarding SSI. Gender and years of service were also analyzed, but no significant differences were found in decision-making based on these factors. Considering these findings, the authors recommended the integration of SSI into classroom discussions to develop students' ethical reasoning and decision-making skills. Additionally, they suggested further research on decision-making among physics and chemistry teachers and an analysis of the inclusion of SSI in biology textbooks.

12.4.4 Representation of SSI in Science Textbooks

Almarashi and Alshahri (2021) highlight the rapid development of science and technology and its negative effect on the environment and society. The authors believe that the inclusion of science, technology, society, and environmental (STSE) issues, which are similar to SSIs, in educational curricula would increase students' opportunities to understand and analyze the relationship between the world, technology, society, and the environment. Almarashi and Alshahri (2021) gathered a list of STSE issues that should be included in the curriculum of the first intermediate grade level in KSA and were mentioned in previous studies. This list included 11 main issues and 44 sub-issues. Then, the authors analyzed the degree of inclusion of the listed STSE issues in the science textbook of the same grade level (first intermediate). The results of the study showed that the issues related to human health and illness had the highest frequency in the textbook, but at a low rate (16%). The topic of energy had the least frequency (4%), while issues related to war technology were not included at all. The authors recommended restructuring the intermediate science courses by increasing the inclusion of STSE issues.

In Palestine, Ayta (2013) examined the inclusion of STSE concerns within the primary elementary level General Science textbooks. The researcher employed the analytical descriptive approach and compiled a roster of STSE topics for the analysis. The study identified a lack of integration and balance in addressing STSE issues in these textbooks. In addition, the inclusion of STSE issues appeared disorganized and scattered throughout the books. Ayta (2013) noted that this hinders effective learning and understanding of the STSE topics. The researcher also noted the following issues as entirely neglected in the textbooks: air pollution, water depletion, soil erosion, and the negative impact of technology on health. Ayta (2013) recommended the inclusion of these topics in the Palestinian curricula to enhance the quality of science education and better prepare students to address modern contemporary environmental challenges.

In KSA, Ali Dalan et al. (2015) analyzed a total of 12 books, which were the physics students' and teachers' guidebooks of grades 10, 11, and 12, for their incorporation of SSI. The research aimed to assess the extent to which physics textbooks and teacher guides cover SSI, by using a content analysis technique. It also investigated how teachers approach SSI in their teaching. To explore this, the researchers conducted interviews with a sample of 12 physics teachers to investigate their understanding and handling of SSI in the curriculum. The findings revealed that SSI were inadequately covered in physics textbooks and materials, with only 12 out of 60 teachers addressing these issues in their teaching. The study identified factors such as time constraints, the length of the curriculum, a lack of training, and teachers' perceptions of the importance of SSI as reasons for their limited inclusion. In summary, the research underscores the importance of placing greater emphasis on SSI in physics education and integrating these issues more effectively into the curriculum.

Also in KSA, Alhomaidy and Alosaimi (2020) aimed to identify SSI that could be incorporated into middle school science textbooks and determine the extent to which these issues were included in the first and second semesters within the textbooks for the first, second, and third intermediate stage grade levels. The study found several key results, including the identification of four main SSI: environment, food and nutrition, genetic engineering, and medical treatment issues, containing a total of 27 scientific issue subsets. It was observed that science textbooks in middle school included all of the main SSI, but the extent of inclusion varied across different grades. Additionally, certain societal scientific issues were not included in these textbooks, such as in vitro fertilization, stem cells, electronics' impact on health, hormonal therapy, and gene therapy. The analysis revealed that environmental issues were the most frequently included in the content of science books for all three grade levels, followed by genetic engineering, food and nutrition issues, and finally issues related to the production of medicine and medical treatment. The study concluded that the second-grade science book included the highest number of SSI, followed by the third grade, and finally the first grade. As a result, the study offers recommendations, emphasizing the need to focus on both the main and subsidiary SSI, particularly those that have received very few mentions in the textbooks, and enhancing attention to issues concerning both science and society.

12.4.5 Views of Teachers About SSI Integration in the Curriculum

In the UAE, there is a strong emphasis on high-quality education, and the country's educational reforms are focused on the integration of SSI into the curriculum, emphasizing "the importance of providing students with problem-solving and critical-thinking skills" (El Arbid & Tairab, 2020, p. 736). However, the various religious and cultural backgrounds of teachers and students present challenges in implementing SSI in the classrooms. For this reason, El Arbid and Tairab (2020) chose to study the factors that may facilitate or impede teachers in the UAE from including SSI in the curriculum. A survey consisting of 24 Likert-type statements was administered to 131 secondary science teachers who were selected at random from the UAE secondary school science teachers' population. The participants consisted of 22% of teachers with less than 5 years' teaching experience, while the rest had more than 5 years of experience. Thirty four percent of the participants were males, the rest were females. In addition, most of the participants had taken classes that covered SSI in their professional development or former education. The instrument assessed the teachers' views on the necessity of implementing SSI in the curriculum, their views on the factors that could impede or facilitate its integration, and information about the teachers' backgrounds. Findings showed that most teachers support the inclusion of SSI in the curriculum and noting that it will improve students' decision-making skills, interest, and engagement in SSI and science topics. Teachers with prior knowledge of SSI were more positive about their inclusion of SSI in the science curriculum. However, teachers were less agreeable to the inclusion of SSI in ethics, religion, and social curricula. Furthermore, teachers reported students' maturity, language skills, and ability to contextualize SSI as factors that could hinder its inclusion. Overall, this study showed that teachers' prior knowledge and professional development positively affect their attitudes towards including SSI in education. However, teachers' personal beliefs may pose challenges in implementing SSI and they need to be addressed.

12.5 Implications for Use of SSI in the Teaching and Learning of Science

SSI are complex, controversial, and value-laden issues that involve science, technology, and society. They can be used as a powerful tool to teach science, as they can help students to develop their critical thinking skills, moral and ethical reasoning, and decision-making skills. In the Arab World, there has been a growing interest in the use of SSI in science education (Abou-El-Fetouh Sheira, 2023; Almarashi & Alshahri, 2021; Alsaaideh, 2016; El Arbid & Tairab, 2020). Several studies have shown that the use of SSI can be effective in improving students' understanding of the relationship between science, technology, and society (e.g., Almarashi & Alshahri, 2021) and other scientific literacy components as NOS (e.g., Khishfe, 2022).

For example, a study by Alsaaideh (2016) found that students who were taught using SSI had a better understanding of environmental problems than students who were taught using traditional methods. Another study by Alsaaideh et al. (2017) found that the use of SSI can improve students' understanding of modern environmental issues. These studies suggest that the use of SSI can be a valuable tool for teaching science in the Arab countries. However, it is important to note that the use of SSI can be challenging, as teachers need to be prepared to facilitate discussions about these issues in a safe and respectful environment.

The following are some of the implications of using SSI in the teaching and learning of science: SSI can help students to develop their critical thinking skills (Al-Zunaidi & Al-Salim, 2020; Khishfe, 2012). When students are presented with an SSI, they are forced to weigh the evidence and make decisions. Second, SSI can help students to develop their moral and ethical reasoning skills. SSI often involves ethical dilemmas, and students need to think about the ethical implications of their decisions. Third, SSI can help students develop their decision-making skills. SSI often require students to make decisions about complex issues, and they need to be able to weigh the pros and cons of different options. Fourth, SSI can help students understand the relationship between science, technology, and society (Almarashi & Alshahri, 2021; Ayta, 2013). Indeed, SSI often involves the intersection of these three areas, and students need to understand how they interact with each other. Fifth, studies conducted in Lebanon, UAE, Egypt, Jordan, and Saudi Arabia, highlight the importance of integrating SSI into the science textbooks (e.g. Alsaaideh, 2016). These studies emphasize the need to include a variety of SSI topics, such as cloning, genetic engineering, environmental issues, and societal impacts, in the school curricula. It is crucial for educators and curriculum developers to ensure a comprehensive coverage of SSI topics, addressing both main and subsidiary issues, while recognizing the cultural and environmental context of the region. Arab countries need to focus on creating an effective balance in the inclusion of SSI issues, paying attention to those that are underrepresented such as electronic cigarettes, nuclear energy, and technological warfare.

Finally, SSI can help students become more engaged in science learning. SSI are often more relevant to students' lives than traditional science topics, and they can help to make science learning more meaningful. In addition, a study by Khishfe (2013) found that using SSI as a context for explicit NOS instruction can be efficient as students are able to apply their acquired NOS comprehension in familiar contexts to other familiar and unfamiliar contexts. Overall, the use of SSI can be a valuable tool for teaching science.

12.6 Recommendations for Future Research

The research on SSI in science education, especially in the Arab World, is still in its early stages, but it has the potential to significantly improve the way science is taught and learned. Future research in this area is essential to fully understand the benefits of SSI-based instruction and to develop effective ways to implement it in the science classrooms in a global context.

One recommendation is to conduct more research on the effectiveness of SSI-based instruction in developing students' scientific literacy-including NOS understandings, argumentation skills, and decision-making skills. Some of the studies discussed above provide some evidence that SSI-based instruction can be effective in these areas, but more research is needed to confirm these findings and to identify the specific factors that contribute to its effectiveness. For example, it would be useful to compare the effectiveness of SSI-based instruction to other approaches in science education, such as traditional textbook-based instruction. It would also be helpful to investigate the effects of SSI-based instruction on students with different cultural backgrounds and with different levels of prior knowledge.

Another recommendation would be to develop and evaluate SSI-based curriculum materials that are appropriate for different grade levels and cultural contexts. The studies reviewed in this chapter suggest that the effectiveness of SSI-based instruction can be influenced by the quality of the curriculum materials used. However, there is currently a lack of high-quality SSI-based curriculum materials available, particularly for use in low-resource settings. Future research should focus on developing and evaluating SSI-based curriculum materials that are appropriate for different grade levels and cultural contexts.

In addition to these two recommendations, there are many other areas where future research on SSI in science education could be conducted. For example, researchers could investigate the effects of SSI-based instruction on students' motivation to learn science, their understanding of NOS, their attitudes towards science, and their civic engagement. Moreover, research can explore several ways that can possibly improve the learning and retention of NOS understandings within SSI contexts by looking at: (a) spacing out various experiences of NOS instruction in the SSI context; (b) linking new knowledge with prior knowledge; and (c) giving many chances for explicit NOS instructions (Khishfe, 2015). Furthermore, programs that focus on enhancing scientific debate skills and incorporating cognitive skills development in the context of local SSI can be researched and developed (Abd-El-Latif, 2019).

At the same time, researchers could also investigate the ways in which SSI-based instruction can be used to promote equity and social justice in science education. Lesnefsky et al. (2023) noted that students care more about science when it is used for equity, social justice and helping others. Along these lines, they claim that socio-scientific modelling is a strategy that privileges equity and access to science practices. Horak (2022) claimed that teaching socioscientific issues is beneficial because it offers an opportunity to integrate equity and social justice into the decision making of a social science-based problem. The integration of equity into the science curriculum is considered as valuable and essential and socioscientific issues are grounded in social justice (Bailey, 2022). By using real-world issues to engage students in learning science for social justice and where students make reasoned decisions about the impact of inequity on their daily lives, science classrooms can become comfortable spaces where social justice becomes part and parcel of science teaching and learning (Powell, 2023). This will lead students to realize that institutions in society can "operate in a manner that promotes and reinforces different forms of inequities" (p. 4). Towards fulfilling that, Miller et al. (2023) argue for an urgent call for science teachers to implement practices that foster equity-centered environments and social justice-orientations that promote system change.

References

Abd-El-Latif, O. J. A. (2019). A program based on local socioscientific issues to develop scientific argumentation skills, knowledge, and attitudes towards those issues among secondary school students. *Studies in Curriculum and Teaching Methods Journal, 243*, 58–107. [ISSN 2535-213X].

Abou-El-Fetouh Sheira, S. M. (2023). A program in the issues related to gene editing applications based on socio-scientific inquiry-based learning (SSIBL) for developing the scientific argumentation and decision-making skills towards these issues of biology section students at the Faculty of Education. *Egyptian Journal of Science Education, 26*(3), 155–204. https://doi.org/10.21608/mktm.2023.309173

Alasmari, I. M., Alshaya, F. S., & Alzaghibi, M. A. (2019). The impact of a proposed model for teaching socio-scientific issues in the third secondary grade biology class on developing students' level of scientific argumentation. *Zarqa Journal for Research and Studies in Humanities, 19*(2), 148–163.

Alhomaidy, T. H., & Alosaimi, H. H. (2020). An analytical study of science books content for inter mediate stage in the light of socio-scientific issues (SSI) in Saudi Arabia: دراسة تحليلية لمحتوى كتب العلوم بالمرحلة المتوسطة في ضوء القضايا العلمية المجتمعية (SSI) 96–120, (8)4, مجلة العلوم التربوية و النفسية .في المملكه العربيه السعوديه.

Ali Dalan, F. S., Alshayeh, L., & Alzoghbi, M. (2015). The extent of inclusion of socioscientific issues (SSI) in high school physics textbooks in Saudi Arabia, and the level of teachers' awareness of these issues. 197–161, (2)*16*, مجلة العلوم التربوية و النفسية.

Al-Jrrah, Z. A., Khataibeh, A. M., & Bni Khalaf, M. H. (2013). Basic tenth grade students' arguments about socio-genetic issues in Biology as related to reasoning patterns. *Jordan Journal of Educational Sciences, 9*(3), 307–318.

Al-Khayal, N. (2022). A program based on STEAM approach to develop concepts of some socioscientific issues, interdisciplinary planning, and the science learning motivation among Kindergarten pre-service teachers. *Al-Majala Al-Tarbaweya L'kuliyat Al-Tarbeya B'Jameaat Sohag, 94*(3), 1314–1244. https://doi.org/10.12816/EDUSOHAG.2021

Almarashi, A. K., & Alshahri, S. Z. G. (2021). Science, technology, society, and environment (STSE) issues included in the first intermediate grade science course in the Kingdom of Saudi Arabia. 94–67, (16) 5, المجلة العربية للتربية النوعية. https://doi.org/10.33850/ejev.2021.138595.

Alsaaideh, M. H. (2016). The socioscientific issues and sociology of science integrated in the science textbook of the eighth grade and students understanding level of it. *Studies in Educational Sciences, 43*(Supplement), 1967–1986.

Alsaaideh, M. H. A., Olimmat, A. B. R., & Alsaaideh, J. A. (2017). The effect of using the socioscientific issues approach on the comprehension of current environmental problems in light of scientific interests among Al-Balqa Applied University Students. 125–193, (2)*41* مجلة كلية التربي ة- جامعة عين شمس.

Al-Zunaidi, T. A. R., & Al-Salim, G. H. I. S. (2020). Effectiveness of a program for teaching socio-scientific issues based on an ethical analysis approach on developing reflective thinking skills among high school students in Riyadh. 24) 4, مجلة العلوم التربوية), (Part 3), 71–122.

Ayta, B. Z. S. (2013). Science, Technology, Society, and Environment issues in the general science curriculum of the primary elementary level in Palestine. 150–113, (1)*21*, مجلة الجامعة الإسلامية للدراسات التربوية والنفسية. ISSN 1726-6807. [Link to the journal: http://www.iugaza.edu.ps/ar/periodical/].

Bailey, S. M. (2022). *The use of social justice socioscientific issues in secondary biology classes: An action research* [Doctoral dissertation, University of South Carolina]. Scholars Common.

BouJaoude, S. (2002). Balance of scientific literacy themes in science curricula: The Case of Lebanon. *International Journal of Science Education, 24*, 139–156.

El Arbid, S. S., & Tairab, H. H. (2020). Science teachers' views about inclusion of socio-scientific issues in UAE science curriculum and teaching. *International Journal of Instruction, 13*(2), 733–748. https://doi.org/10.29333/iji.2020.13250a

Frhod, W., & Abdulwahed, A. A. (2022). Decision making in socioscientific issues of biology teachers for secondary stage. اشراقات تنموية مجلة علمية محكمة *(Illuminations of Development, a Peer-Reviewed Scientific Journal), 34*, 203–228. Al-Qadisiyah University/College of Education.

Horak, R. (2022). Teaching science literacy about socioscientific issues. *American Society for Microbiology.* https://asm.org/Articles/2022/June/Teaching-Science-Literacy-About-Socioscientific-Is

Khishfe, R. (2012). Relationship between nature of science understandings and argumentation skills: A role for counterargument and contextual factors. *Journal of Research in Science Teaching, 49*(4), 489–514.

Khishfe, R. (2013). Transfer of nature of science understandings into similar contexts: Promises and possibilities of an explicit reflective approach. *International Journal of Science Education, 35*(17), 2928–2953.

Khishfe, R. (2014). A reconstructed vision of environmental science literacy: The case of Qatar. *International Journal of Science Education, 36*(18), 3067–3100. https://doi.org/10.1080/09500693.2014.951980

Khishfe, R. (2015). A look into students' retention of acquired nature of science understandings. *International Journal of Science Education, 37*(10), 1639–1667.

Khishfe, R. (2017). Consistency of nature of science views across scientific and socio-scientific contexts. *International Journal of Science Education, 39*(4), 403–432.

Khishfe, R. (2019). The transfer of nature of science understandings: A question of similarity and familiarity of contexts. *International Journal of Science Education, 41*(9), 1159–1180.

Khishfe, R. (2020). Retention of acquired argumentation skills and nature of science conceptions. *International Journal of Science Education, 42*(13), 2181–2204.

Khishfe, R. (2021). Explicit instruction and student learning of argumentation and nature of science. *Journal of Science Teacher Education, 32*(3), 325–349.

Khishfe, R. (2022). Nature of science and argumentation instruction in socioscientific and scientific contexts. *International Journal of Science Education, 44*(4), 647–673.

Khishfe, R. (2023a). Connected learning: An approach for teaching nature of science aspects and argumentation components. *Journal of Research in Science Teaching.* https://doi.org/10.1002/tea.21855

Khishfe, R. (2023b). Relationship between nature of science and argumentation: A follow-up study. *International Journal of Science and Mathematics Education, 21*(4), 1081–1102.

Khishfe, R., Alshaya, F. S., BouJaoude, S., Mansour, N., & Alrudiyan, K. I. (2017). Students' understandings of nature of science and their arguments in the context of four socioscientific issues. *International Journal of Science Education, 39*(3), 299–334. https://doi.org/10.1080/09500693.2017.1280741

Lesnefsky, R. R., Kirk, E. A., Yeldell, J., Sadler, T. D., & Ke, L. (2023). Socioscientific modelling as an approach towards justice-centred science pedagogy. *London Review of Education, 21*(1), 30. https://doi.org/10.14324/LRE.21.1.30

Miller, E., Berland, L., & Campbell, T. (2023). Equity for students requires equity for teachers: The inextricable link between teacher professionalization and equity-centered science classrooms. *Journal of Science Teacher Education.* https://doi.org/10.1080/1046560X.2023.2170793

Osborne, J., Erduran, S., & Simon, S. (2004). Enhancing the quality of argumentation in school science. *Journal of Research in Science Teaching, 41*(10), 994–1020.

Powell, W. (2023). A socioscientific issues approach to ninth-graders' understanding of COVID-19 on health, wealth, and educational attainments. *PLoS One, 18*(3), 1–22. https://doi.org/10.1371/journal.pone.0280509

Sadler, T. D., Barab, S. A., & Scott, B. (2007). What do students gain by engaging in socio-scientific inquiry? *Research in Science Education, 37*, 371–391. https://doi.org/10.1007/s11165-006-9030-9

Zeidler, D. L., Walker, K. A., Ackett, W. A., & Simmons, M. L. (2002). Tangled up in views: Beliefs in the nature of science and responses to socioscientific dilemmas. *Science Education, 86*, 343–367.

Chapter 13
Approaches to Socioscientific Issues in Portugal: A Scoping Literature Review

Patrícia Pessoa, Maria João Fonseca, Bruno Sousa, António Pedro Costa, and Xana Sá-Pinto

13.1 Introduction

Many extant global and local issues are complex, with diverse causes and consequences. These involve and/or affect diverse stakeholders, some of which holding different world visions, values, and motivations, thus resulting in diverse solutions (Funtowicz & Ravetz, 1993; Rittel & Webber, 1973; Räthzel & Uzzell, 2009). Addressing these issues requires the engagement of a diverse range of stakeholders—including those that are directly affected by the problem- in the research about these problems and in the development and evaluation of solutions. This is expected to further enrich the research with meaningful information and better account for risk and uncertainty in the development of solutions (Funtowicz & Ravetz, 1993).

P. Pessoa (✉)
University of Trás-os-Montes e Alto Douro (UTAD), Vila Real, Portugal

Research Centre on Didactics and Technology in the Education of Trainers (CIDTFF), Department of Education and Psychology, University of Aveiro, Aveiro, Portugal
e-mail: ppessoa@ua.pt

M. J. Fonseca
Natural History and Science Museum of the University of Porto (MHNC-UP), Porto, Portugal
e-mail: mjfonseca@mhnc.up.pt

B. Sousa
Albufeira Poente School Group, Albufeira, Portugal
e-mail: executivo.brunosousa@gmail.com

A. P. Costa · X. Sá-Pinto
Research Centre on Didactics and Technology in the Education of Trainers (CIDTFF), Department of Education and Psychology, University of Aveiro, Aveiro, Portugal
e-mail: apcosta@ua.pt; alexandrasapinto@ua.pt

D. L. Zeidler (ed.), *A Moral Inquiry into Epistemic Insights in Science Education*, Contemporary Trends and Issues in Science Education 61,
https://doi.org/10.1007/978-3-031-63382-9_13

Education needs to empower students with attitudes, values, skills, and knowledge that allow them to understand and navigate the complexity and uncertainty of these issues and to research, develop, evaluate, negotiate, implement, and improve solutions to address these and to contribute to sustainable, resilient, and equitable societies (UNESCO, 2018). This requires the development of students' key competencies in sustainability which, according to Wiek (2011; but see UNESCO 2018 for additional competencies), include: (i) systems thinking; (ii) anticipatory competency; (iii) normative competency; (iv) strategic competency; and (v) interpersonal competency. Aligned with this view, the official document guiding mandatory schooling in Portugal states that, when leaving the 12th grade (the mandatory schooling in Portugal), students should be able to "mobilize values and competencies that allow them to intervene in individuals' and societies' lives and histories, make free and informed choices about natural, social and ethical questions, and have the capacity to participate civically, actively, conscientiously and responsibly" (translated by the authors from Martins et al., 2017, p. 10). The same document states that education in Portugal should be guided by eight principles, one of which is sustainability, and should develop students' sustainability consciousness.

Science should contribute to this major goal and empower students to positively and actively act in society to contribute to sustainable development. The major aim of Science Education, the development of students' scientific literacy (National Research Council, 2012), has been acknowledged by several international institutions (OECD, 2019; International Commission on the Futures of Education, 2020) as fundamental to address global complex problems. Initially conceived as a kind of students' scientific culture and more related to the memorization of scientific facts and laws (Vision I), the concept of scientific literacy has evolved to address the ability to relate science with social-cultural contexts (Vision II) and, more recently, to recognise the importance of developing students' engagement, social participation and emancipation (Vision III, see review at Valladares, 2021). In line with the concept of scientific literacy, educational approaches used in science education have also evolved through time. Besides approaches such as problem-based learning (PBL), project-based learning (PrBL), or challenge-based learning (CBL) that are used across several disciplines to foster students' learning and engagement with local and global problems (Barrows & Tamblyn, 1980; Gallagher & Savage, 2023; Kokotsaki et al., 2016; Merritt et al., 2017), two educational approaches, that are more specific to science, are frequently employed to develop students' scientific literacy: the Science Technology Society and Environment (STSE) approach and Socioscientific Issues (SSI) approach (Sarikaya & Topçu, 2022; Zeidler et al., 2005; Zeidler & Sadler, 2023). However, while both approaches promote students' contextualized learning of science about science and sustainability issues, the SSI approach recognises as expected learning goals the development of character, moral and ethical values, and socioscientific argumentation (Sarikaya & Topçu, 2022), with several articles focusing on how this approach allows promoting these learning goals (Clarkeburn, 2002; Fowler et al., 2009; Hogan, 2002; Kolstø, 2006; Lee et al., 2012; Tuncay-Yüksel et al., 2023; Zeidler & Sadler, 2007; Zohar & Nemet, 2002).

A recent literature review supports the potential of the SSI approach to develop all the key competencies in sustainability (Pessoa et al., 2022), although the studies analyzed showed lower potential to support the development of anticipatory and strategic competencies. This problem is expected to be overcome when students are engaged in real contexts acting in the society to address SSI (Zeidler et al., 2019). In these immersive contexts, students are expected to develop more sophisticated ways to conceptualize and address the SSI by having first-hand experiences with diverse stakeholders who are expected to share their visions and perspectives (Zeidler et al., 2019). This may further contribute to open schools to the society, promoting links between schools and local stakeholders that, besides contributing to a better understanding of the complexity of the problem, may support students in the development, evaluation, and refinement of possible solutions they provide, contributing to the development of their strategic competency. This is expected to contribute to educating future citizens for the extended peer communities so needed in the post-normal science era (Hauge & Barwell, 2017).

In Portugal, the official curricula for the disciplines that include science education until the sixth grade present several opportunities to explore diverse SSI ranging from technology, health, environment, and social conflicts and biases based on human diversity issues (Pessoa et al., submitted). It also includes diverse opportunities to explore dimensions of socioscientific reasoning and socioscientific identity (as defined by Sadler et al., 2017; see results in Pessoa et al., submitted). However, although the Portuguese official curricula explicitly mention the STSE approach, no explicit mention of the SSI approach is found (Portuguese Government/Ministry of Education, 2018). Despite this, a recent study found 24 dissertations and a thesis produced in Portugal about the SSI approach (Krupczak et al., 2020). According to this study, SSI addressed in these publications include global warming, ozone layer depletion, acid rains, pollution (including solid waste and plastic in marine ecosystems), pesticides, deforestation, biodiversity loss, erosion, human genetics, vaccination, human health, and territorial management. Most of the Portuguese dissertations and theses produced until 2020 mostly focus on mandatory education (up to the 12th grade; 84%), and the remaining (16%) focus on teacher training. According to Li and Guo (2021) and Dogan (2023), Portugal is recognized as one of the leading countries in science education research, ranking within the top 20–25 in terms of Portuguese contribution for papers on scientific literacy and science education. However, when it comes to research articles specifically on SSI education, Portugal has a significantly lower representation of Portuguese researchers as first authors when compared to other countries worldwide (Ban & Mahmud, 2023). Up to our knowledge, no study has reviewed what has been published about Portugal or by Portuguese researchers about SSI in peer-reviewed international journals and scientific conferences proceedings. With the current paper, we aim to address this lack of knowledge by performing a scoping review to analyze what has been published about SSI by Portuguese authors or in studies developed in Portugal and identify: (i) the social structure of the research community working on this topic in Portugal; (ii) the studied topics; (iii) the target groups which the practices/research have been focusing on; (iv) the SSI addressed; (v) the main results and (vi) the overview of recommendations and policies and/or advises conveyed. Based on this information, we aim to identify knowledge gaps and areas that deserve further research.

13.2 Methodology

To achieve our goals, we conducted a scoping review (Peters et al., 2021). Scoping reviews are invaluable for examining the extensive range of literature on a subject, especially in cases where the area is intricate or has not been thoroughly researched (Peters et al., 2015). Landa et al. (2011) frequently employs them to provide insights for future studies and pinpoint deficiencies in the existing body of knowledge. In addition to the scoping review, we conducted a bibliometric analysis (Aria & Cuccurullo, 2023) to examine the production, identify the most influential authors, and analyze the social structure of the publications. The qualitative analysis of the studies was performed using the content analysis technique (Costa & Amado, 2018).

13.2.1 Scoping Review

Web of Science, Scopus, and B-on were the databases chosen as data sources. The Web of Science platform was selected because it is one of the global leaders in scientific citation search and analytical information, providing access to the oldest publications (Falagas et al., 2008; Li et al., 2018). Scopus is widely recognized for its comprehensive coverage and multidisciplinary nature, containing one of the largest collections of searchable literature, including citations and abstracts (Falagas et al., 2008). B-on is a Portuguese online knowledge library that provides students, teachers, and researchers access to an extensive and varied collection of electronic journals, e-books, and databases (B-on, 2004).

To search for studies published by Portuguese authors about SSI and studies conducted about SSI in Portugal, we combined the keywords 'socioscientific issues' and 'socio-scientific issues', as both are used in the literature. Additionally, we incorporated the keyword 'Portugal' in our search and looked for their presence in all the fields of the papers. Thus, our search expression was: ('socio-scientific issues' OR 'socioscientific issues') AND 'Portugal'. No time restrictions were made when conducting the search. In the first phase, all the research results were examined to ensure that these were either: (i) conducted by Portuguese authors; (ii) applied in Portugal; or (iii) literature reviews that encompassed data from Portugal. Studies that did not meet at least one of these criteria were excluded. We also excluded at this phase all the studies that were not available to this research team (books and book chapters). In the second phase of analysis, studies that did not have a connection to education or that did not use the term SSI to mention an educational approach were excluded from the analysis. The phases of the paper selection are presented in Fig. 13.1.

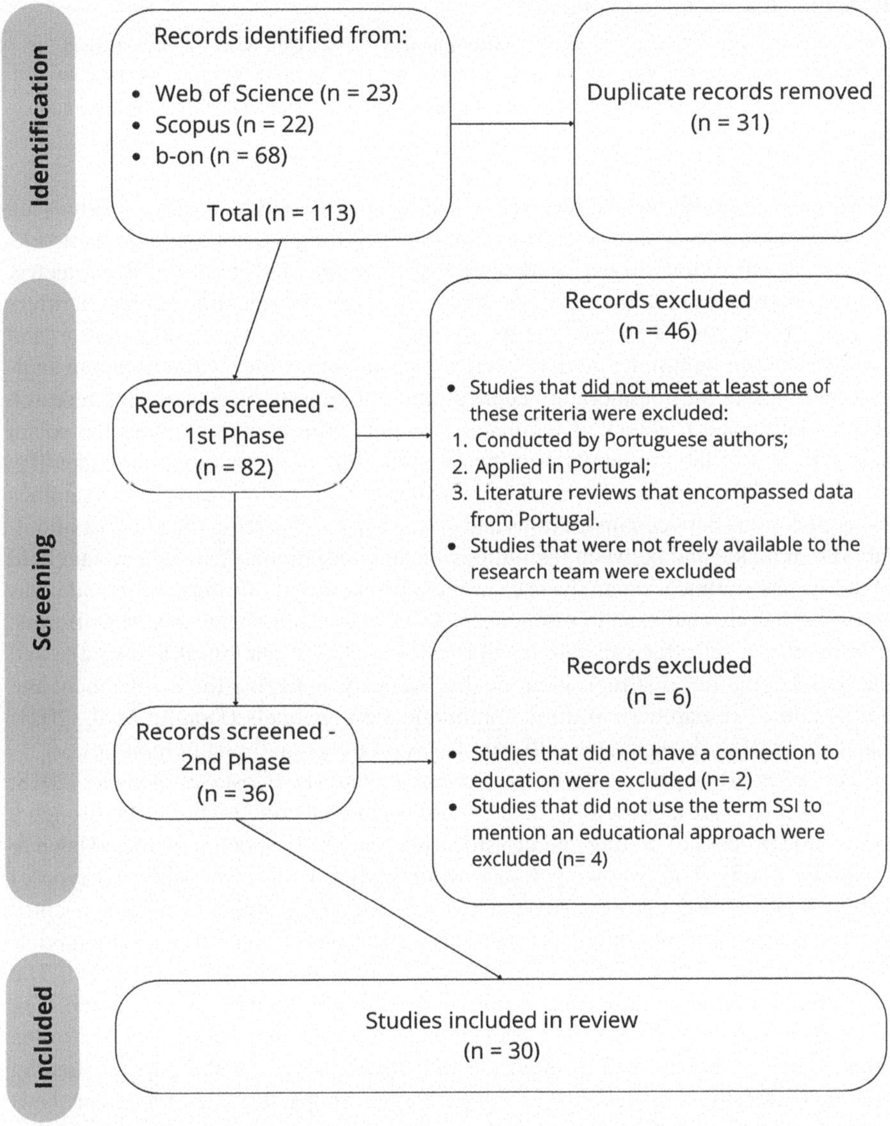

Fig. 13.1 Phases of the scoping review

13.2.2 Analysis of the Papers

The software Bibliometrix was used to analyze the production over time, the most relevant authors, and the social structure of the publications (Aria & Cuccurullo, 2023; see Table 13.1).

Table 13.1 Bibliometric social structure

Structure	Definitions	Bibliometric technique	Unit of analysis	Statistical techniques
Social	How authors interact with each other	Co-authorship	Authors	Collaboration network

When doing bibliometric research, social structure analysis focuses on studying the social elements within scientific domains, including the collaborative networks and social connections among academics, organizations, and countries. Bibliometrix is an invaluable tool for analyzing the social structure of a scientific subject. It offers a range of approaches to examine the dynamics and patterns of collaboration and communication within the field, delivering significant insights. Co-authorship analysis investigates the collaboration connections among writers by analyzing research articles published together. It facilitates researchers in comprehending the social networks and collaboration patterns inside a scientific field, encompassing identifying significant authors and research organizations. Collaboration analysis examines the connections between institutions and countries, emphasizing the patterns of collaboration inside and between institutions on an international scale in a certain field of study. Communication analysis examines the exchange of information and ideas among researchers through conferences, workshops, and social media platforms. Bibliometrics can offer valuable insights into the dissemination of knowledge and the social dynamics that fuel scientific discovery by analyzing the involvement and interaction of researchers in these communication channels (Donthu et al., 2021; Zhao, 2022). In this paper, we will only focus on the co-authorship relations.

The selected papers underwent a content analysis (Costa & Amado, 2018; Krippendorff, 2018), chosen for their suitability in analyzing textual data through a systematic process of coding and classification to identify specific themes (Hsieh & Shannon, 2005). The following topics were analysed for each paper: (i) type of publication; (ii) target groups; (iii) the SSI it focused on (categories defined according to Pessoa et al. (submitted)); (iv) research questions or objectives; (v) methodology; (vi) main results; and (vii) policy and/or educational advice provided. The categories for types of publications followed those proposed by SCOPUS and Web of Science databases. Except for topics (i) and (iii), the categories of analysis for the other topics were obtained through the inductive analysis of the papers' floating reading (Merriam & Tisdell, 2015). For the target groups, the categories of analysis were also informed by the formal division of the 12 years of compulsory education in Portugal, that includes 9 years of basic education with a shared curriculum (split into three cycles: first cycle from the 1st to the 4th grade, second cycle from the 5th to the 6th grade, and third cycle from the 7th to the 9th grade) and 3 years of secondary education (from the 10th to the 12th grade). Kindergarten was also included to scope for studies targeting students in early childhood. Finally, we also included in-service and pre-service teachers, higher education students (who were not pre-service teachers), and the general public.

The researchers individually analyzed each of the papers and categorized them based on the presence or absence of the categories. The examples of evidence found in each paper were recorded. Any doubts that arose during the analysis were identified and discussed with all the researchers until a final decision was reached. Interrater reliability was estimated as the percentage of agreement between raters in the initial independent coding phase (McHugh, 2012).

13.3 Results and Discussion

The search was carried out in the databases Web of Science, Scopus and B-on and recovered 82 publications, of which only 30 were considered for further analysis after applying the exclusion criteria (Fig. 13.1).

13.3.1 Bibliometric Results

The 30 publications considered for analysis were authored by a total of 69 authors. Among these, there is a strong predominance of publications authored by Pedro Reis (n = 12), followed by Cecília Galvão (n = 4) and Mónica Baptista (n = 3), all affiliated to the same institution. Pedro Reis is one of the earliest Portuguese authors to publish on SSI, having started his publications in 2004, and has also been the most prolific author in recent years (2017–2023, Fig. 13.2), being at the center of a large network of Portuguese and international co-authors (Fig. 13.3).

Cecília Galvão is also one of the earliest Portuguese authors to publish on SSI, but no studies authored by her in this field have been identified since 2014. In the last 10 years, more Portuguese authors have published on SSI, although not

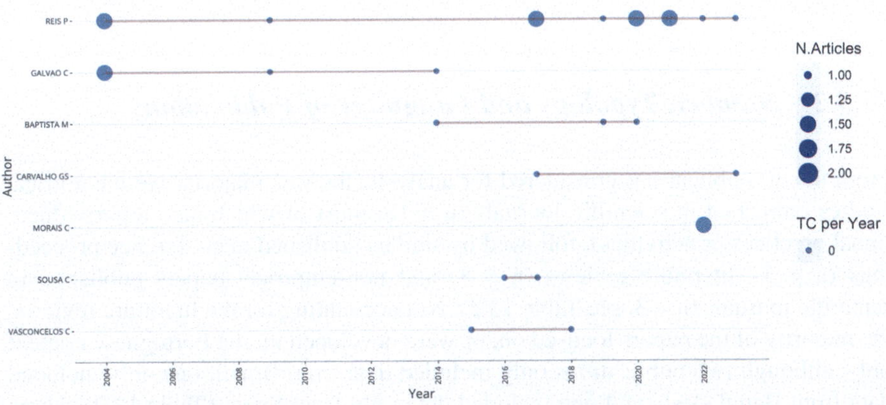

Fig. 13.2 Number of studies published by the most productive authors over time

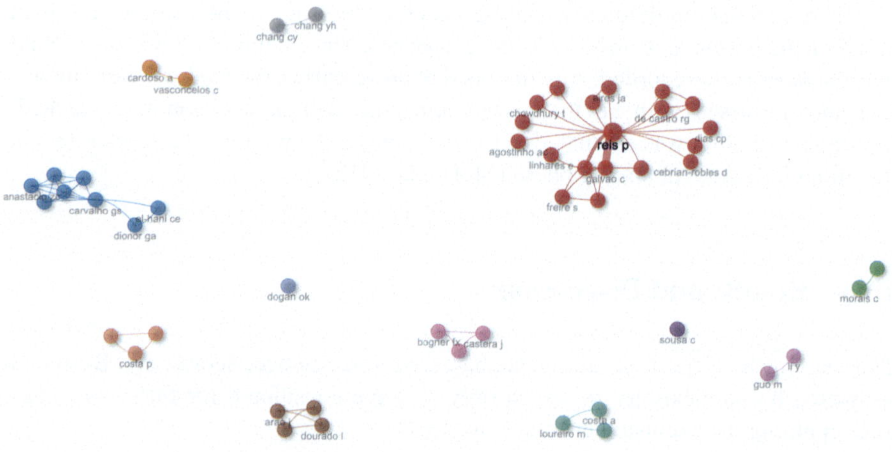

Fig. 13.3 Co-authorship network

constantly. More recently, Graça Carvalho from the University of Minho and Carla Morais, C. Sousa and Clara Vasconcelos, all from the University of Porto, also published on SSI, being authors of two papers each, suggesting that this research topic is consolidating and being adopted by other Portuguese institutions. However, the fact that very few authors in Portugal have published more than one study on SSI (with most of them having published just one), and the fact that these are coming mostly from three institutions, suggest that this research topic is far from being a central research topic in the Portuguese science education research community. Nevertheless, there is notable evidence indicating the emergence of new networks in recent years. Despite the predominant centrality of Pedro Reis and Cecília Galvão within the largest network of co-authorships, the publication activities of other Portuguese researchers (such as Graça Carvalho, Carla Morais, C. Sousa, and Clara Vasconcelos), have catalyzed the formation of distinct networks, as illustrated in Fig. 13.3.

13.3.2 *Number, Typology and Language of Publications*

From the 30 publications considered for analysis, the vast majority were empirical studies published in scientific journals (n = 18, most of which also report educational practices or activities), followed by studies published in conference proceedings (n = 5), literature reviews (n = 4) and non-empirical papers published in scientific journals (n = 3, see Table 13.2). Not accounting for the literature reviews, the majority of the papers focused on, or were developed in, the Portuguese context only, although two publications only included data from Brazil, one only included data from Bangladesh, and one included data only from Spain (Table 13.2). Three papers included data from Portugal and other countries.

Table 13.2 Studies analyzed in this study and its characteristics

Study code	Short reference	Type of study	Target group	Research question	Methodology	SSI addressed
S1	de Freitas et al. (2023)	Empirical paper[b]	Secondary Education	Socioscientific reasoning, Argumentation, Strategic competency	Focus group	Environmental issues
S2	Cebrián-Robles et al. (2021)	Empirical paper[b]	Pre-service teachers	Socioscientific reasoning, Awareness of the issue	Participants' productions	Environmental issues
S3	Reis and Galvão (2004a)	Empirical paper[a,b]	Secondary Education	Conceptions about science and scientists	Interviews or open questionnaires, Participants' productions	Technology issues, Human health issues, Environmental issues, Exobiology issues, Social conflicts and biases based on human diversity issues
S4	Conceição et al. (2019)	Empirical paper[a,b]	3rd Cycle of Basic Education (7th to 9th grade)	Science learning, Strategic competency	Interviews or open questionnaires, Participants' productions, Observations	Environmental issues
S5	Chowdhury et al. (2022)	Empirical paper	In-service teachers	Teachers perceptions	Closed questionnaires	The teaching through SSI
S6	Reis et al. (2020)	Empirical paper	1st, 2nd and 3rd Cycle of Basic Education (1st to 9th grade)	Socioscientific reasoning, Strategic competency, Conceptions about scientific competencies and science classes	Closed questionnaires	Technology issues, Human health issues, Environmental issues

(continued)

Table 13.2 (continued)

Study code	Short reference	Type of study	Target group	Research question	Methodology	SSI addressed
S7	Reis and Galvão (2004b)	Empirical paper[a]	In-service teachers	Teachers perceptions and practices	Interviews or open questionnaires, Participants' productions, Observations	Technology issues, Human health issues, Environmental issues, Social conflicts and biases based on human diversity issues, The teaching through SSI
S8	Fonseca et al. (2012)	Empirical paper[a,b]	Secondary Education	Socioscientific reasoning, Awareness of the issue, Science learning	Interviews or open questionnaires, Participants' productions, Observations	Human health issues
S9	Leite et al. (2023)	Empirical paper	University Education	Opinions and attitudes towards specific SSI	Closed questionnaires	Human health issues
S10	Morais et al. (2022)	Empirical paper[a]	General Public	Opinions and attitudes towards specific SSI	Closed questionnaires	Environmental issues
S11	Faria et al. (2014)	Empirical paper[a]	3rd Cycle of Basic Education (7th to 9th grade)	Science learning	Participants' productions	Human health issues
S12	Anastácio et al. (2023)	Empirical paper[a]	Pre-service teachers	Opinions and attitudes towards specific SSI	Closed questionnaires, Interviews or open questionnaires	Human health issues
S13	Castéra et al. (2018)	Empirical paper	In-service teachers	Opinions and attitudes towards specific SSI	Closed questionnaires	Human health issues, Environmental issues
S14	Agostinho and Reis (2021)	Empirical paper[a,b]	Secondary Education	Science learning, Strategic competency	Closed questionnaires, Interviews or open questionnaires, Participants' productions, Observations	Human health issues

S15	Reis and Galvão (2009)	Empirical paper[a,b]	In-service teachers	Teachers practices	Interviews or open questionnaires, Observations	Technology issues
S16	Ribeiro and Vieira (2014)	Empirical paper[a,b]	1st Cycle of Basic Education (1st to 4th grade)	Socioscientific reasoning, Science learning, Argumentation	Participants' productions	Technology issues, Environmental issues
S17	Linhares and Reis (2017)	Empirical paper[a,b]	Pre-service teachers	Science learning, Awareness of the issue, Strategic competency	Closed questionnaires, Interviews or open questionnaires, Observations	Environmental issues
S18	Dias and Reis (2017)	Empirical paper[a,b]	3rd Cycle of Basic Education (7th to 9th grade)	Socioscientific reasoning, Science learning	Interviews or open questionnaires	Technology issues, Environmental issues
S19	Sousa (2016)	Non-empirical paper[a,b]	3rd Cycle of Basic Education (7th to 9th grade) and Secondary Education	Description of educational practice	Not applicable	Social conflicts and biases based on human diversity issues
S20	Sousa (2017)	Non-empirical paper[a]	3rd Cycle of Basic Education (7th to 9th grade) and Secondary Education	Description of educational practice	Not applicable	Technology issues, Human health issues, Environmental issues, Social conflicts and biases based on human diversity issues
S21	Morais (2022)	Non-empirical paper	Not applicable	Characterization of citizen science	Not applicable	Not applicable
S22	Chang et al. (2010)	Literature review	Not applicable	Provide an overview in the field of science education research	Scientometrics study based on published literature	Not applicable

(continued)

Table 13.2 (continued)

Study code	Short reference	Type of study	Target group	Research question	Methodology	SSI addressed
S23	Li and Guo (2021)	Literature review	Not applicable	Generate a comprehensive view of scientific literacy-related literature	Bibliometric analysis	Not applicable
S24	Krupczak et al. (2020)	Literature review	Not applicable	Analyze how socioscientific issues are addressed in Brazilian and Portuguese theses and dissertations.	Literature review from Portuguese and Brazilian theses and dissertations on SSI	Not applicable
S25	Dogan (2023)	Literature review	Not applicable	Understand the topics that the articles in the Journal of Research in Science Teaching focused on over the last 20 years.	Bibliometric study	Not applicable
S26	Maurício and Teodoro (2011)	Proceedings papers and abstracts[a]	University Education	Training of socio-cultural animators	Not applicable	Not applicable
S27	Costa et al. (2017)	Proceedings papers and abstracts	Not applicable	Scientific literacy	Not applicable	Not applicable
S28	Dionor et al. (2017)	Proceedings papers and abstracts[b]	Not applicable	Description of educational practice	Closed questionnaires	Human health issues
S29	Vasconcelos et al. (2018)	Proceedings papers and abstracts[a,b]	3rd Cycle of Basic Education (7th to 9th grade)	Awareness of the issue, Science learning	Interviews or open questionnaires, Participants' productions	Human health issues, Environmental issues
S30	Vasconcelos et al. (2015)	Proceedings papers and abstracts[a]	In service teachers	Geoethics and sustainability	Interviews or open questionnaires	Human health issues, Environmental issues, The teaching through SSI

Legend: [a]studies with data collected in Portugal or focusing the Portuguese context; [b]studies that describe an educational practice/activity

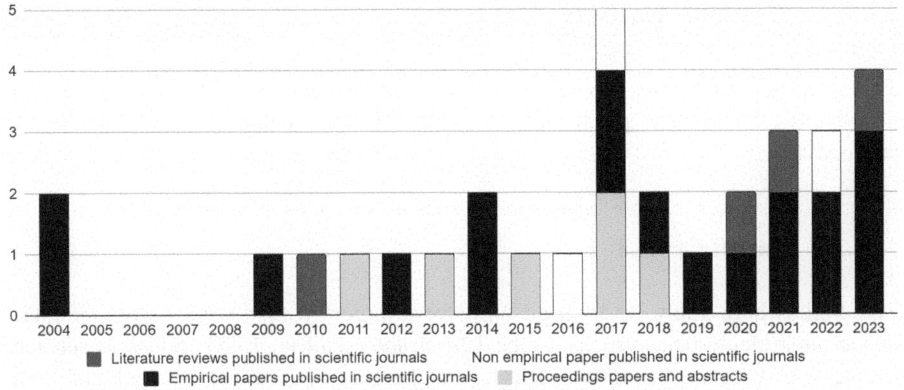

Fig. 13.4 Absolute frequency of publications by year and type of publication

Five publications were written in Portuguese (1 proceedings paper, 2 empirical papers, 1 non-empirical paper, and 1 literature review), and one empirical paper was written in Spanish. All the remaining publications were written in English.

The first publications on SSI developed in Portugal were published in 2004. After that, 4 years elapsed until a new publication on SSI was done in Portugal, but since then there has been at least one publication per year (Fig. 13.4). In 2017, there was an increase in the number of publications, including empirical and non-empirical studies and conference proceedings. Since 2020, there has been a progressive increase in the number of publications, with a higher frequency of papers published in scientific journals, especially empirical studies. This pattern observed in Portugal is aligned with expectations from the international community. The year of the first publications in Portugal on SSI, 2004, was also identified in a study that analyzed research and trends in SSI education through an analysis of journal publications from 2004 to 2022, as the year in which we expect to see a significant increase in research and papers on SSI (Ban & Mahmud, 2023). According to the authors, this increase in research is the result of UNESCO's strong promotion of education for sustainable development and of values-based learning and interdisciplinary approaches that emphasize critical thinking during the period 2005–2014, themes considered to be aligned with the SSI approach (Zeidler, 2014). Regarding studies focused on environmental SSI, Kumar et al. (2024) also report an increase in the number of publications in 2017 and a robust research activity from 2018 to 2021. Ban and Mahmud (2023), with regard to studies on SSI in general, also report a doubling in the number of publications between 2015–2017, and further growth between 2018–2022, suggesting that SSI has become a popular and important field of research for many researchers and has been recognized by many well-known journals.

13.3.3 Content Analysis

This section will report the results of the content analysis of the empirical and non-empirical studies published in scientific publications and conference proceedings.

13.3.3.1 Target Groups

The studies here analyzed focused on diverse target groups, with a stronger incidence on students from the 7th to the 12th grade (see Fig. 13.5) and on in-service teachers. If we consider that pre-service teachers are also university students, then there is also a high percentage of studies focusing on this target group. Kindergartens, students from the 1st to the 6th grade, and general audience are either not covered or very infrequently covered by these studies. This tendency to focus research and resources on older students rather than on students from primary education has been identified by several studies not focused on Portugal (Chen & Xiao, 2021; Kinskey & Zeidler, 2021; Ozden, 2020; Zangori et al., 2018).

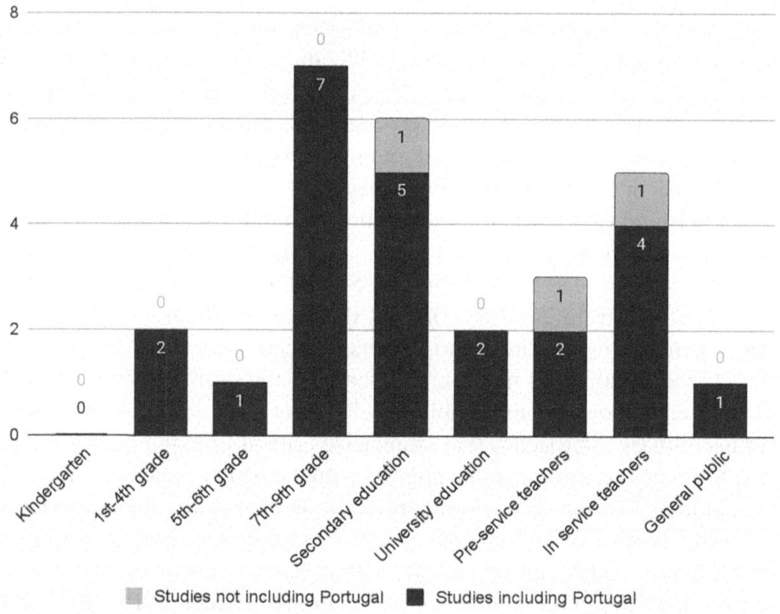

Fig. 13.5 Number of publications (including or not including Portuguese participants) focusing on each category of target groups

13.3.3.2 Research Questions and Methodologies

Most of the empirical studies published in scientific papers included in this study aimed to assess how the use of SSI affected the development of knowledge and skills in students (n = 8): namely, science learning (n = 7), socioscientific reasoning (n = 5), strategic competency (or social activism; n = 5), awareness of the SSI (n = 3), and argumentation (n = 2). Several studies have been done on how people's characteristics, like education, knowledge, and social background, affect their feelings about certain SSI (n = 4). There have also been studies on how teachers use SSI in the classroom and how students feel about it (n = 3). A smaller number of studies look at how exploring SSI changes people's ideas about science, how science lessons work, and their scientific skills (n = 2). Other types of studies include those that ask future teachers to describe the problems and solutions of an SSI (n = 1) and those that look at the pros and cons of a science exhibition in teacher training (n = 1).

Of the three non-empirical papers, two describe curricular proposals to address gender equity and bioethics. The third one aimed to present a characterization of citizen science as a way of involving citizens in authentic scientific research processes developed to address SSI under a STSE approach.

Studies published in conference proceedings focused on the use of SSI to develop science learning and awareness of the SSI (n = 1), on the description of educational practices (n = 1), on how scientific literacy is defined (n = 1), on the links between teaching geoethics and the context of sustainability (n = 1), and on the training of socio-cultural animators to work with SSI in communities (n = 1).

While the studies included in this analysis predominantly focus on the benefits of the SSI approach in education, such as science learning, socioscientific reasoning, argumentation, and SSI awareness, the topics covered do not align completely with the results of international studies. Mirroring what is found at the international level, our results show a predominance of argumentation and reasoning (Ban & Mahmud, 2023; Kumar et al., 2024). However, while studies focusing on the nature of science are frequent at the international level (Ban & Mahmud, 2023; Chang et al., 2010; Dogan, 2023), we did not find any studies about this topic in Portugal. Another difference is the high frequency of studies dedicated to the development of strategic competency (mostly represented by studies on social activism), a theme not identified in international literature reviews (Ban & Mahmud, 2023). On the other hand, the low proportion of studies focused on teacher training and pre-service teachers observed in studies conducted in Portugal and by Portuguese authors mirrors the pattern found at the international level (Ban & Mahmud, 2023; Nielsen, 2020).

The instruments most often used to collect data in the empirical studies published in scientific papers were open questionnaires and/or interviews (n = 9), immediately followed by closed questionnaires (n = 8) and analysis of participants' productions (n = 8). Observation was also frequently used to collect data (n = 6) while focus groups were used in a single study. Eight studies used more than one instrument to collect data and triangulated the results. The studies published in conference proceedings used open questionnaires and/or interviews (n = 2), analysis of participants' productions (n = 1), and closed questionnaires (n = 1) as data collection instruments. The higher predominance of interviews/open questionnaires, analysis of participants' productions, and observations suggest that qualitative methods approaches are frequent in these studies, as is observed in other international literature reviews from the international community (Ban & Mahmud, 2023).

13.3.3.3 SSI Addressed in the Studies

When we consider all the empirical papers and all the proceedings papers with empirical data, the SSI related to human health and the environment were most often addressed, followed by issues related to technology (Fig. 13.6). Very few studies addressed issues related to social conflicts and biases based on human diversity and exobiology, which were mostly mentioned in the empirical papers where the participants were asked to choose SSI either to work with or to teach. For a non-exhaustive list of examples of SSI addressed in the analyzed studies in each category, see Table 13.3.

The tendency towards a wider focus on environmental and health-related issues is also reported by studies conducted on the publication of SSI papers in the global context (Ban & Mahmud, 2023) and is aligned with the opportunities to explore SSI present in the Portuguese national curricula in which these categories of SSI are more frequently addressed (Pessoa et al., submitted). Curiously, there were

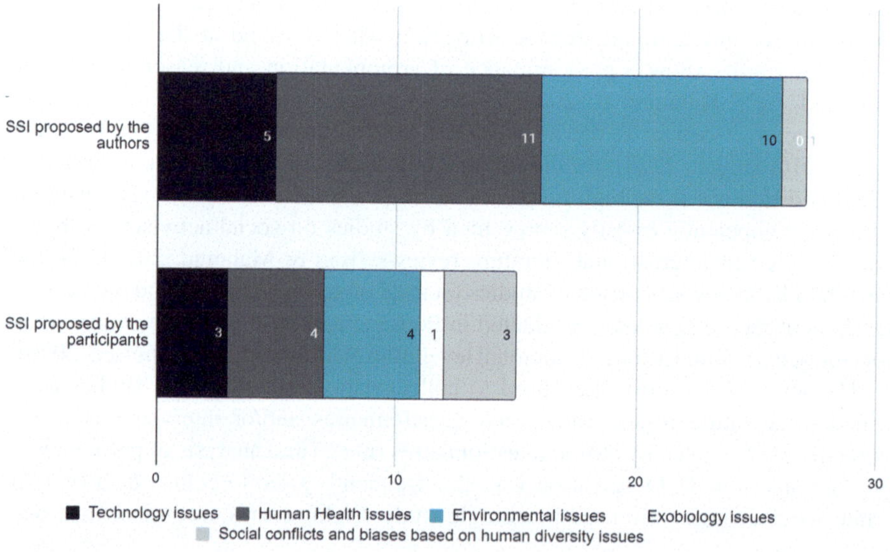

Fig. 13.6 Number of publications addressing different SSI categories accounting for who chooses the problems to be addressed

Table 13.3 Examples of SSI addressed/proposed in the analyzed studies per category and subcategory of analysis. Information between brackets in the co-occurrence column refer to the codes of the studies as identified in Table 13.2

Category/subcategory	Co-occurrences/ articles	Indicators
Technology issues		
Technology issues in general	4 (S6, S7, S15, S20)	Nanotechnology; New technologies affecting living beings
Medicine	4 (S3, S7, S15, S20)	Production of human tissue or organs for transplants; Xenotransplants; In vitro fertilization (IVF).
Agronomy/Agriculture	5 (S3, S7, S13, S16, S20)	Genetically modified organisms (GMO); the use of pesticides;
Environment	2 (S6, S17)	Treating toxic waste; Climate Geoengeneering
Genetics	5 (S3, S7, S13, S15, S20)	Cloning; genetic engineering; GMO
Industry	0	
Animal experiments	3 (S15, S16, S20)	The use of human embryos in research; human cloning; the use of animals in medical research;
Information and communication technology	1 (S20)	DNA databases; consent, privacy and confidentiality;
Human health issues		
Human health issues in general	4 (S3, S7, S29, S30)	Geoethics in individual and community health; the case of the Siamese twins whose survival depended on eliminating one; the production of human tissue or organs for transplants
Diseases	6 (S3, S7, S9, S14, S15, S28)	Research for cure of AIDS or cancer; issues related to COVID19, dengue, zika e chikungunya; genetic diseases testing and treatment
Food	5 (S3, S7, S8, S13, S20)	Carbohydrates in breast milk; Consumption of GMO,
Access to resources	2 (S20, S29)	Water as a right; use of geological resources for skincare
Use of medicines	4 (S8, S12, S18, S20)	Antibiotic use and resistance; vaccination
Human diversity	5 (S3, S7, S11, S15, S20)	Postmenopausal health; IVF; abortion; birth control;
Non-nutrient substances intake	1 (S7)	Drug addiction

(continued)

Table 13.3 (continued)

Category/subcategory	Co-occurrences/ articles	Indicators
Environmental issues		
Environmental issues in general	4 (S6, S15, S17, S18)	environmental problems affecting populations; Climate Change geoengineering; rising ocean;
Energy	0	
Pollution	7 (S2, S3, S4, S6, S7, S16, S20)	Illegal mining; treating toxic waste; environmental pollution; plastic pollution in oceans; climate change; pollution and co-incineration of toxic waste in concrete factories; ozone depletion; taxing cars entering in cities; rivers pollution
Consumerism	5 (S16, S18, S20, S29, S30)	Fishing; The use of sand from islands for skin care; sustainable use of the resources
Biodiversity	3 (S1, S18, S20)	The disappearance of bees; fishing; protecting biodiversity and the biosphere; exploitation versus preservation of natural resources
Territory management	2 (S6, S10)	Extension of the Portuguese continental shelf; deep sea management
Exobiology issues	1 (S3)	Space
Social conflicts and biases based on human diversity issues		
Social conflicts and biases based on human diversity issues in general	1 (S3)	Human evolution
Social conflicts and biases based on human diversity issues due to biological diversity	1 (S19)	Gender equity
Social conflicts and biases based on human diversity issues due to cultural diversity	2 (S7, S15)	Abortion, sexual orientation, birth control,

subcategories of SSI, such as energy, that despite being recently highly discussed in the Portuguese media due to the construction of very controversial dams (Reis & Galvão, 2004b) and offshore wind meal parks, were not addressed in these papers nor mentioned by the participants.

13.3.3.4 Main Results of the Analyzed Papers

Several studies report that, after engaging in SSI educational activities, students developed (or report to have developed in some cases) science learning and/or socioscientific reasoning (n = 8; Agostinho & Reis, 2021; Cebrián-Robles et al., 2021; Conceição et al., 2019; de Freitas et al., 2023; Fonseca et al., 2012;

Linhares & Reis, 2017; Reis et al., 2020; Ribeiro & Vieira, 2014; Vasconcelos et al., 2018) and awareness of the SSI (n = 7; Agostinho & Reis, 2021; Cebrián-Robles et al., 2021; Conceição et al., 2019; de Freitas et al., 2023; Fonseca et al., 2012; Linhares & Reis, 2017; Reis et al., 2020; Vasconcelos et al., 2018). Two studies that engaged students in SSI educational projects/activities also report the development of students' argumentation skills (de Freitas et al., 2023; Ribeiro & Vieira, 2014). Only one study reported strong limitations on students' scientific knowledge and ability to reason about SSI, but in this study, participants were not subjected to any educational intervention (Faria et al., 2014).

Studies that involved students in "activities of social activism" such as the development of public exhibitions (Agostinho & Reis, 2021; Linhares & Reis, 2017; Reis et al., 2020) show that these fostered the development of strategic competency (as defined by Wiek et al., 2011). These students perceived themselves as more competent "in developing exhibitions in science classes as a way of creating awareness on topics relating to science, technology, and society" and "of influencing other citizens' decisions and behaviors about issues related to science, technology, and the environment, through the development of scientific exhibits" (Reis et al., 2020, p. 8). The students participating in the study of Conceição et al. (2019) also prepared and publicly presented a theatre play, but their learning outcomes, in terms of strategic competency, were not assessed. On the other hand, studies that engaged students in an educational approach that did not include a dimension of development and implementation of solutions in the community report a modest development of their strategic competencies, with students struggling to plan initiatives to address the problem, identifying several obstacles to active participation in social transformation (de Freitas et al., 2023). The results of these studies support the potential of SSI to promote strategic competency if it includes students' engagement in the development and implementation of solutions to the problem. This supports Zeidler et al. (2019) claim that it is important to provide students with immersive SSI activities in which they engage in real contexts and address real society problems. Despite this, strategic competency is not a research topic usually addressed in the literature about SSI (Ban & Mahmud, 2023; see also Pessoa et al., 2022 to see how the different skills for sustainability are present in SSI studies that address evolution education).

The studies focusing on teachers' perceptions and practices about the SSI approach in Bangladesh (Chowdhury et al., 2022) and Portugal (Reis & Galvão, 2004b, 2009) reveal that teachers from both countries value this educational approach that they consider to be important. A similar result was obtained in a systematic literature review on the perceptions of science teachers in teaching SSI (Chen & Xiao, 2021). In the studies analysed by Chen and Xiao (2021) the majority of teachers perceived SSI as necessary and vital in school science. Despite this, 136 teachers from Bangladesh stated that the implementation of SSI practices in the classroom was not high (Chowdhury et al., 2022). The few teachers interviewed in Portugal reported implementing SSI educational approaches in their classrooms (Reis & Galvão, 2004b, 2009). However, the few teachers engaged in the Portuguese studies preclude the generalization of these results to the Portuguese context and suggest the importance of quantitative studies that include a larger number and higher diversity of teachers.

In Portugal, the interviewed teachers identify several curriculum topics that can be used to apply SSI approaches (Reis & Galvão, 2004b) and see it aligned with bioethics teaching (Vasconcelos et al., 2015). In fact, a recent study that analyzed the science curricula up to 6th grade in 3 European countries, including Portugal, identified several opportunities to address SSI at these school levels (Pessoa et al., submitted).

The analyzed studies report several factors that may affect teachers' conceptions and classroom practices related to SSI. These include gender, class size, and teacher training programs in Bangladesh (Chowdhury et al., 2022). In Portugal these factors included national curriculum, national exams, teachers' previous experience as scientists, personal educational priorities or aims (Reis & Galvão, 2004b), as well as teacher's convictions about the educational relevance of these activities and the knowledge needed for their design, management and assessment, and teacher training (Reis & Galvão, 2009). These factors that affect teachers when implementing SSI are also identified in the literature. Particularly at the teacher level, Chen and Xiao (2021), in their literature review, also refer to challenges related to insufficient knowledge base, lack of instructional skills, feelings of insecurity when addressing SSI in class, and lack of personal interests or beliefs. Policy-related factors are also documented and are similar to those indicated by Portuguese teachers. Namely, the lack of time to carry out activities related to the discussion of controversial issues, the lack of time for overloaded curricula and the pressure of standardized tests, the fact that SSI are not explicitly addressed in curriculum standards, the lack of adequate teaching materials, resources, and relevant professional training, and the difficulty of assessing SSI learning are mentioned as constraints (Chen & Xiao, 2021). In addition to the challenges identified in our analysis, the literature also mentions teachers' concerns related to students, such as their scientific knowledge, levels of understanding, abilities, learning habits, and personal values which were not mentioned by Portuguese teachers (Chen & Xiao, 2021). Together, these studies suggest the importance of promoting pre-service and in-service teacher training programs that empower educators to effectively and frequently implement SSI educational approaches in their classrooms. This is in fact a recommendation left by several studies here analyzed (Chowdhury et al., 2022; Krupczak et al., 2020; Linhares & Reis, 2017; Reis & Galvão, 2009).

In fact, the studies addressing teacher training show that engaging pre-service teachers in SSI approaches promoted the development of their socioscientific reasoning, awareness of SSI, science knowledge, and strategic competency (Cebrián-Robles et al., 2021; Linhares & Reis, 2017). But no studies here have evaluated the impact of these teacher training programs on the participants' pedagogical knowledge or on their educational practices, which highlights the importance of not only developing but also evaluating the effect of programs to train teachers to implement SSI in Portuguese classrooms. Internationally, some studies indicate that both shorter workshops and SSI training programs for pre-service science teachers have been found to have positive immediate effects, but they often have limited impact on these teachers' subsequent practices (Nielsen, 2020). Given that the main reason behind these unsatisfactory international results is the status of SSI teaching in teacher training programs, which are sometimes used as auxiliary aspects and/or superficially added to an existing teacher training program (Nielsen, 2020), these findings should be used to inform teacher training in Portugal in order to empower teachers to bring SSI into their classrooms.

Studies that targeted participants' conceptions and attitudes towards science or specific SSI show that these may be strongly influenced by how scientific controversies are handled by the media (Reis & Galvão, 2004a) and that it may change from country to country (Castéra et al., 2018; Leite et al., 2023). Anastácio et al. (2023) show that participants' conceptions and attitudes towards science or specific SSI are influenced by the instruction received on related scientific topics (Anastácio et al., 2023), but Leite et al. (2023) report less scientific knowledge about COVID-19 in science undergraduates than education students. The results from Anastácio et al. (2023) are supported by other studies that report students' changes in attitudes towards an SSI when they engage in educational approaches that provide them an active role in the collection and interpretation of the information to make sense of an SSI and in discussing the issues and/or proposals and implementation of solutions to these (Fonseca et al., 2021; Pessoa et al., 2024). This is also aligned with the elaboration likelihood model, which states that the likelihood and strength of attitudes' changes are dependent on the time and cognitive effort spent by a person on the interpretation of information important to evaluate the attitudes' object (Petty et al., 1999). Together with attitudes towards scientific exploration, perceptions of the deep sea, gender, education, and science engagement were also shown to be associated with social representations of deep sea (Morais et al., 2022). Evidence supporting (Castéra et al., 2018) and contradicting (Morais et al., 2022) the 2-major environmental values model was found in the articles addressing teachers' attitudes towards genetically modified organisms and deep sea respectively.

The studies of Dionor et al. (2017) and of Dias and Reis (2017) resulted in the development of educational strategies and proposals validated by teachers and researchers to address SSI, such as the spread of diseases by *Aedes aegypti* (Dionor et al., 2017) or vaccination, sea management, and climate change under inquiry-based science education and through the use of Web2.0 tools (Dias & Reis, 2017).

13.3.3.5 Studies' Recommendations

The analyzed papers provide recommendations for integrating SSI into science education. Emphasizing individual and collective actions, educators are encouraged to address the impacts of these issues in cultural, social, and economic contexts, empowering students to navigate complex socio-political processes with an environmentally fair perspective (de Freitas et al., 2023; Morais et al., 2022; Reis et al., 2020). Additionally, the importance of deepening relationships in school practices is highlighted, urging educators to move beyond the environment-society relationship in discussions, reflecting on coexisting dimensions, and considering power dynamics (de Freitas et al., 2023; Reis et al., 2020). The incorporation of hands-on activities is emphasized for its educational benefits to underscore the significance of practical activities in science education (Fonseca et al., 2012).

The recommendations further advocate for the development of argument-based activities, promoting trans-disciplinary dialogue in health education within science curricula (Dionor et al., 2017; Faria et al., 2014; Leite et al., 2023). Engaging in socio-critical reflection and immersive real experiences is encouraged, positioning socio-scientific activism as a form of citizenship and fostering reasoned student engagement in science-related topics in real contexts (Agostinho & Reis, 2021; Conceição et al., 2019). Educational strategies focusing on current issues are proposed (Morais et al., 2022), enriching higher education programs with socioscientific topics, and promoting a culture that values students' environments as valuable resources for science literacy (Reis & Galvão, 2009). Utilizing SSI is suggested as a tool to stimulate critical-reflexive capacity among students (Reis & Galvão, 2009). Finally, the articles underscore the need for teachers' professional development (Chowdhury et al., 2022), integration of bioethics into science education (Castéra et al., 2018; Sousa, 2017), enhancing scientific literacy with cultural considerations (Li & Guo, 2021), linking science education with socio-cultural animation (Maurício & Teodoro, 2011), and aligning SSI with science curricula (Vasconcelos et al., 2018). These recommendations collectively advocate for a SSI approach to science education, fostering critical thinking, engagement, and a deeper understanding of the complexities within socio-scientific issues.

13.3.4 Literature Reviews

In what concerns the four literature reviews analyzed, only one was co-authored by a Portuguese researcher (Krupczak et al., 2020). This review provides a comparative analysis of the discussion of SSI as a didactic strategy in Brazil and Portugal, based on the analysis of theses and dissertations. It covers outputs available from the CAPES Thesis and Dissertations Catalog (defended until 2017) and from the online repository of the Education Institute of the University of Lisbon (defended until 2019).

When we focus only on the results reported by these authors for Portugal, between 2010 and 2017, the discussion of SSI was mainly used to promote citizenship and activism skills and to foster the development of scientific literacy and other competencies, such as argumentation skills, which is in line with the findings of the present review focused on studies published in conference proceedings and scientific journals. Also aligned with the results that we report here, the activities carried out in the studies reviewed by Krupczak et al. (2020) were mainly focused on secondary education, 3rd cycle of basic education and involved didactic sequences and activities, but a higher frequency of studies targeting the 2nd cycle of basic education are found in Portuguese thesis and dissertations. Krupczak et al. (2020) also identified thesis addressing teacher training actions, involving courses, study or research groups, isolated disciplines or activities in pre- or in-service teacher training, but to a much lesser extent. Nevertheless, a thesis was found covering the link between SSI and the nature of science and discussing the beneficial impacts of capacitation actions based on the discussion of SSI in changing teachers' misconceptions about the nature of science.

The SSI topics that most frequently emerged from the review of the thesis and dissertations were grouped by Krupczak et al. (2020) according to an identical structure to the one we propose in regard to environmental (e.g., pesticide use, deforestation, biodiversity loss, climate change, plastic pollution), health (e.g., human genetics, vaccination, human health maintenance), and social (e.g., anthropic occupation and traffic education) issues. Nevertheless, technological issues were not grouped in their own independent category, and biotechnology issues were grouped together with health-related issues. The categories of SSI most frequently addressed in the thesis and dissertations are similar to those found in our study and included environmental and health related issues.

In what concerns the other three literature reviews considered in this analysis (Chang et al., 2010; Dogan, 2023; Li & Guo, 2021), only one (Li & Guo, 2021) looked explicitly into SSI, and even so, indirectly, through its link with scientific literacy. The other two (Chang et al., 2010; Dogan, 2023) looked for trends and topics in science education by assessing papers in major indexed journals—*International Journal of Science Education*, *Journal of Research in Science Teaching*, *Research in Science Education*, and *Science Education*. In all of them, SSI was frequently associated with the nature of science. This connection was not specifically reported for Portugal, and it is not consistent with the findings of Krupczak et al. (2020) and the results of the present study, which show that the nature of science is not frequently addressed in studies focusing on SSI. And while Li and Guo (2021) and Dogan (2023) identify Portugal as one of the top 20–25 European countries most contributing with papers on scientific literacy and science education, respectively, there is no clear reference to SSI, which again stresses the limited scope of research in this field in our country. This perspective is further sustained by the results of Dogan (2023), which present 2013 as the year with the highest number of references to SSI, at least in what concerns papers published in the *Journal of Research in Science Teaching* from 2000 to 2020; according to our findings, we witness a four-year delay in this trend in Portugal, with 2017 coming across as the year with the highest frequency of papers on the topic.

13.4 Main Conclusions and Recommendations

The results of our analysis indicate that research on SSI in Portugal is still at an early stage and that there are still few researchers and institutions publishing consistently about this educational approach. However, the significant growth in the number of researchers and institutions working in this domain and the increased volume of empirical studies published in recent years suggest that this research field is consolidating, growing and diversifying. The research topics addressed by Portuguese authors mirror the trends from the international community but have also distinctive features, namely the strong focus on immersive SSI contexts within the scope of social activism (aligned with the strategic competency) and the lack of research focusing on the relationship between SSI and the nature of science.

Research studies show that there are plenty and diversified opportunities to explore SSI in alignment with Portuguese official curricula (Pessoa et al., submitted), and the studies here reviewed show that Portuguese teachers recognise these opportunities and the value of the SSI pedagogical approach (Reis & Galvão, 2004b, 2009). However, the studies focusing on teachers' perceptions and practices are of qualitative nature, focusing on a very small number of teachers from the Lisbon region, and there is a clear need for studies that allow us to understand how widespread and how often the SSI educational approach is used in Portuguese classrooms. Furthermore, the studies reviewed suggest the need to develop more research focused on students until the 6th grade, as well as on the use of the SSI approach in science communication to the general public. Additionally, the types of SSI addressed could also be more diversified, exploring not only issues related to the environment and human health, but also social conflicts and biases based on human diversity issues, exobiology issues and technology issues, for example.

Several studies here analyzed and from the international community stress the importance of offering comprehensive and effective teacher training programs to equip educators with the requisite skills and knowledge to effectively introduce SSI in their instructional practices. Gaining an understanding of how teacher training programs contribute to teachers' perceptions, self-efficacy, and instructional strategies is essential to effectively enhance their capacity to teach SSI and address their concerns. However, in our research, very few studies focused on teacher training and none addressed in-service teacher training. Conducting studies on the impact of these training programs would yield valuable insights.

It would also be important to understand how often and how the SSI approach is addressed in pre-service training provided by the different institutions in Portugal and to encourage greater exploration of SSI within research institutions, leading to the provision of teacher training and the implementation of new classroom practices. This collaboration between researchers, institutions, and educators can foster meaningful changes in teaching and learning approaches related to SSI.

We should highlight that our study has important limitations that could affect the representativeness of the results in the Portuguese context. We are aware that our choice of databases excluded from our analysis non-indexed publications, even if these are published in scientific journals such as the publication of Paixão et al. (2021). Additionally, the use of English keywords to perform the search may not capture studies only published in Portuguese. We should also stress that, in Portugal, the STSE approach is widely disseminated both in the researcher and teacher communities, being mentioned and encouraged by the official national science curriculum (Portuguese Government/Ministry of Education, 2018). SSI and STSE share some important features (Bencze et al., 2020; Sarikaya & Topçu, 2022; Zeidler, 2014) and, in fact, several of the studies included in the present analysis mentioned both approaches (see for example Conceição et al., 2019; Morais, 2022; Reis et al., 2020). By not looking for or including studies focusing only on STSE in Portugal, we may have lost information about some dimensions that are also important for research on SSI. However, the SSI approach differs from the STSE approach because it strongly values and supports the development of character, moral and ethical values, and socioscientific argumentation (Bencze et al., 2020; Sarikaya & Topçu, 2022; Zeidler et al., 2005; Zeidler, 2014). This difference strongly argues for the need to more strongly focus on and value SSI as a distinctive educational approach and explicitly mention it in the official curricula, supporting the development of moral, ethical and socioscientific reasoning in Portuguese science classes since the first years of schooling, in alignment with the recommendations of UNESCO (2018) to promote key competencies in sustainability. Despite the limitations mentioned, we feel that our work provides an important contribution to achieve this goal by characterizing the status of the research conducted in Portugal about SSI and providing important information to identify areas that deserve further studies.

Aknowledgements PP is funded by Portuguese national funds through FCT—Fundação para a Ciência e a Tecnologia, I.P., within the scope of the PhD grant 2020.05634.BD. APC is funded by Portuguese national funds through FCT—Fundação para a Ciência e a Tecnologia, I.P., under the Scientific Employment Stimulus—Institutional Call—[CDL-CTTRI-248-SGRH/2022] and the CIDTFF (projects UIDB/00194/2020 and UIDP/00194/2020). XSP is funded by Portuguese funds (OE), through FCT –Fundação para a Ciência e a Tecnologia, I.P., under the Scientific Employment Stimulus—Individual Call—2022.06070.CEECIND and the CIDTFF (projects UIDB/00194/2020 and UIDP/00194/2020).

References

Agostinho, A., & Reis, P. (2021). Contributos Educativos de Exposições Interativas online sobre imunidade Como Prática de Ativismo Coletivo em Contexto de Pandemia. *UNIPLURIVERSIDAD, 21*(2), 1–22. https://doi.org/10.17533/udea.unipluri.346064

Anastácio, Z. C., Da Silva, T. C., Antão, C. M., Braide, A. S., Battisti, I. D., & Carvalho, G. S. (2023). Influence of Human Biology and Health (HBH) teaching–learning process on students' conceptions of the COVID-19 vaccine. *Sustainability, 15*(1), 41. https://doi.org/10.3390/su15010041

Aria, M., & Cuccurullo, C. (2023). A brief introduction to bibliometrix. https://www.bibliometrix.org/vignettes/Introduction_to_bibliometrix.html

Ban, S., & Mahmud, S. N. D. (2023). Research and trends in socio-scientific issues education: A content analysis of journal publications from 2004 to 2022. *Sustainability, 15*(15), 11841.

Barrows, H. S., & Tamblyn, R. M. (1980). *Problem-based learning: An approach to medical education* (Vol. 1). Springer Publishing Company.

Bencze, L., Pouliot, C., Pedretti, E., Simonneaux, L., Simonneaux, J., & Zeidler, D. (2020). SAQ, SSI and STSE education: Defending and extending "science-in-context". *Cultural Studies of Science Education, 15*, 825–851.

B-on. (2004). B-on—Research service. Biblioteca do Conhecimento Online—b-on (Online Knowledge Library). https://www.b-on.pt/

Castéra, J., Clément, P., Munoz, F., & Bogner, F. X. (2018). How teachers' attitudes on GMO relate to their environmental values. *Journal of Environmental Psychology, 57*, 1–9. https://doi.org/10.1016/j.jenvp.2018.04.002

Cebrián-Robles, D., España-Ramos, E., & Reis, P. (2021). Introducing preservice primary teachers to socioscientific activism through the analysis and discussion of videos. *International Journal of Science Education, 43*(15), 2457–2478. https://doi.org/10.1080/09500693.2021.1969060

Chang, Y.-H., Chang, C.-Y., & Tseng, Y.-H. (2010). Trends of science education research: An automatic content analysis. *Journal of Science Education and Technology, 19*(4), 315–331. https://doi.org/10.1007/s10956-009-9202-2

Chen, L., & Xiao, S. (2021). Perceptions, challenges and coping strategies of science teachers in teaching socioscientific issues: A systematic review. *Educational Research Review, 32*, 100377.

Chowdhury, T., Holbrook, J., Reis, P., & Rannikmäe, M. (2022). Bangladeshi science teachers' perceived importance and perceived current practices in promoting science education through a context-based, socio-scientific framework. *Science & Education, 31*(2), 487–523. https://doi.org/10.1007/s11191-021-00236-9

Clarkeburn, H. (2002). A test for ethical sensitivity in science. *Journal of Moral Education, 31*(4), 439–453. https://doi.org/10.1080/0305724022000029662

Conceição, T., Baptista, M., & Reis, P. (2019). La contaminación de los recursos hídricos Como Punto de Partida para el activismo socio-científico. *Revista Eureka Sobre Enseñanza y Divulgación de Las Ciencias, 16*(1), 1–13. https://doi.org/10.25267/rev_eureka_ensen_divulg_cienc.2019.v16.i1.1502

Costa, A. P., & Amado, J. (2018). *Content analysis supported by software* (1st ed.).

Costa, A., Loureiro, M., & Ferreira, M. E. (2017). Scientific literacy at the beginning of the 21st century. In L. G. Chova, A. L. Martinez, & I. C. Torres (Eds.), *9th International conference on education and new learning technologies (Edulearn17). EduLearning proceedings* (pp. 1708–1717).

de Freitas, A. C., do Nascimento, L. A., de Castro, R. G., Motokane, M. T., & Reis, P. (2023). Biodiversity and citizenship in an argumentative socioscientific process. *Sustainability, 15*(4), 2987. https://doi.org/10.3390/su15042987

Dias, C. P., & Reis, P. (2017). O Desenvolvimento de Atividades Investigativas com Recurso à Web 2.0 no Âmbito da Investigação e Inovação Responsáveis. *Sisyphus, Journal of Education, 5*(3).

Dionor, G. A., Martins, L., El-Hani, C. E., & Carvalho, G. S. (2017). Dengue, Zika e Chikungunya: superando uma limitação do livro didático a partir de uma questão sociocientífica. XI Encontro Nacional de Pesquisa em Educação em Ciências—XI ENPEC. Universidade Federal de Santa Catarina, Florianópolis, SC—3 a 6 de julho de 2017.

Dogan, O. K. (2023). Trends and issues in science education in the New Millennium: A bibliometric analysis of the JRST. *Science Insights Education Frontiers, 16*(1), 2375–2407. https://doi.org/10.15354/sief.23.or249

Donthu, N., Kumar, S., Mukherjee, D., Pandey, N., & Lim, W. M. (2021). How to conduct a bibliometric analysis: An overview and guidelines. *Journal of Business Research, 133*, 285–296. https://doi.org/10.1016/j.jbusres.2021.04.070

Falagas, M. E., Pitsouni, E. I., Malietzis, G. A., & Pappas, G. (2008). Comparison of PubMed, Scopus, web of science, and Google scholar: Strengths and weaknesses. *The FASEB Journal, 22*(2), 338–342.

Faria, C., Freire, S., Baptista, M., & Galvão, C. (2014). The construction of a reasoned explanation of a health phenomenon: An analysis of competencies mobilized. *International Journal of Science Education, 36*(9), 1476–1490. https://doi.org/10.1080/09500693.2013.783723

Fonseca, M. J., Santos, C. L., Costa, P., Lencastre, L., & Tavares, F. (2012). Increasing awareness about antibiotic use and resistance: A hands-on project for high school students. *PLoS One, 7*(9). https://doi.org/10.1371/journal.pone.0044699

Fonseca, C. A., Sá-Pinto, X., Dinis, H. A., & Vasconcelos, R. (2021). Shooting skinks for good: Producing a movie improves attitudes towards a threatened species. *Science of the Total Environment, 15*(791), 148356. https://doi.org/10.1016/j.scitotenv.2021.148356

Fowler, S. R., Zeidler, D. L., & Sadler, T. D. (2009). Moral sensitivity in the context of socioscientific issues in high school science students. *International Journal of Science Education, 31*(2), 279–296. https://doi.org/10.1080/09500690701787909

Funtowicz, S. O., & Ravetz, J. R. (1993). Science for the post-normal age. *Futures, 25*(7), 739–755. https://doi.org/10.1016/0016-3287(93)90022-L

Gallagher, S. E., & Savage, T. (2023). Challenge-based learning in higher education: An exploratory literature review. *Teaching in Higher Education, 28*(6), 1135–1157.

Hauge, K. H., & Barwell, R. (2017). Post-normal science and mathematics education in uncertain times: Educating future citizens for extended peer communities. *Futures, 91*, 25–34. https://doi.org/10.1016/j.futures.2016.11.013

Hogan, K. (2002). Small groups' ecological reasoning while making an environmental management decision. *Journal of Research in Science Teaching, 39*(4), 341–368. https://doi.org/10.1002/tea.10025

Hsieh, H. F., & Shannon, S. E. (2005). Three approaches to qualitative content analysis. *Qualitative Health Research, 15*(9), 1277–1288.

International Commission on the Futures of Education. (2020). *Education in a post-COVID world: Nine ideas for public action*. UNESCO.

Kinskey, M., & Zeidler, D. (2021). Elementary preservice teachers' challenges in designing and implementing socioscientific issues-based lessons. *Journal of Science Teacher Education, 32*(3), 350–372. https://doi.org/10.1080/1046560X.2020.1826079

Kokotsaki, D., Menzies, V., & Wiggins, A. (2016). Project-based learning: A review of the literature. *Improving Schools, 19*(3), 267–277.

Kolstø, S. D. (2006). Patterns in students' argumentation confronted with a risk-focused socio-scientific issue. *International Journal of Science Education, 28*(14), 1689–1716. https://doi.org/10.1080/09500690600560878

Krippendorff, K. (2018). *Content analysis: An introduction to its methodology*. Sage publications.

Krupczak, C., Aires, J. A., & Reis, P. G. R. (2020). Controvérsias sociocientíficas: análise comparativa entre Brasil e Portugal. *Amazônia. Revista de Educação em Ciências e Matemáticas, Belém, 16*(37), 89–105. https://periodicos.ufpa.br/index.php/revistaamazonia/article/view/8584/6691

Kumar, V., Choudhary, S. K., & Singh, R. (2024). Environmental socio-scientific issues as contexts in developing scientific literacy in science education: A systematic literature review. *Social Sciences & Humanities Open, 9*, 100765.

Landa, A. H., Szabo, I., Le Brun, L., Owen, I., & Fletcher, G. (2011). An evidence-based approach to scoping reviews. *Electronic Journal of Information Systems Evaluation, 14*(1), 46–52.

Lee, H., Chang, H., Choi, K., Kim, S.-W., & Zeidler, D. L. (2012). Developing character and values for global citizens: Analysis of pre-service science teachers' moral reasoning on socio-scientific issues. *International Journal of Science Education, 34*(6), 925–953. https://doi.org/10.1080/09500693.2011.625505

Leite, L., Dourado, L., Mapatse, M. V., Arão, J., Moraes, J., & Castellar, S. (2023). Higher education students' perceptions of issues related to covid-19: A study on three Portuguese-speaking countries. Cogent. *Social Sciences, 9*(1). https://doi.org/10.1080/23311886.2023.2186578

Li, Y., & Guo, M. (2021). Scientific literacy in communicating science and socio-scientific issues: Prospects and challenges. *Frontiers in Psychology, 12*. https://doi.org/10.3389/fpsyg.2021.758000

Li, K., Rollins, J., & Yan, E. (2018). Web of Science use in published research and review papers 1997–2017: A selective, dynamic, cross-domain, content-based analysis. *Scientometrics, 115*(1), 1–20.

Linhares, E. F., & Reis, P. (2017). Interactive Exhibition on Climate Geoengineering: Empowering Future Teachers for Sociopolitical Action. *SISYPHUS, JOURNAL OF EDUCATION, 5*(3), 85–106.

Martins, G. O., Gomes, C. A. S., Brocardo, J., Pedroso, J. V., Camilo, J. L. A., Silva, L. M. U., Encarnação, M. M. G. A., Horta, M. J. V. C., Calçada, M. T. C. S., Nery, R. F. V., & Rodrigues, S. M. C. V. (2017). *Perfil dos alunos à saída da escolaridade obrigatória*. Ministério da Educação, Direção-Geral da Educação.

Maurício, P., & Teodoro, A. (2011). Qualified intervention/qualifying formation: The socio-cultural animator in a science based society. In I. C. Torres, L. G. Chova, & A. L. Martinez (Eds.), *Proceedings of the 4th international conference of education, research and innovation (Iceri)* (pp 1988–1996).

McHugh, M. L. (2012). Interrater reliability: The kappa statistic. *Biochemia Medica, 22*(3), 276–282. https://doi.org/10.11613/BM.2012.031

Merriam, S. B., & Tisdell, E. J. (2015). *Qualitative research: A guide to design and implementation*. Wiley.

Merritt, J., Lee, M. Y., Rillero, P., & Kinach, B. M. (2017). Problem-based learning in K–8 mathematics and science education: A literature review. *Interdisciplinary Journal of Problem-Based Learning, 11*(2).

Morais, C. (2022). Ciência cidadã e educação CTS/CTSA perspectivando contributos, desafios e oportunidades. *CTS: Revista iberoamericana de ciencia, tecnología y sociedad, 17*(51), 157–178. https://dialnet.unirioja.es/servlet/articulo?codigo=8738128

Morais, C., Moreira, L., Teixeira, A. S., & Aguiar, T. (2022). No waves from surface knowledge: Diving into the social representations of the Deep Sea. *International Journal of Science Education, Part B, 12*(1), 22–41. https://doi.org/10.1080/21548455.2021.2017507

National Research Council. (2012). *A framework for K-12 science education: Practices, crosscutting concepts, and core ideas*. National Academies Press.

Nielsen, J. A. (2020). Teachers and socioscientific issues–An overview of recent empirical research. In *Science teacher education for responsible citizenship: Towards a pedagogy for relevance through socioscientific issues* (pp. 13–20). Springer.

OECD. (2019). PISA 2018 science framework. In *PISA 2018 assessment and analytical framework*. OECD Publishing. https://doi.org/10.1787/f30da688-en

Ozden, M. (2020). Elementary school students' informal reasoning and its' quality regarding socio-scientific issues. *Eurasian Journal of Educational Research, 20*(86), 61–84.

Paixão, V. V. M., Cruz, M. C. P., Mortimer, E., Silva, A., Cassiani, S., & Lopes, J. B. (2021). Desafios Sócio-científicos às práticas educativas e à investigação tendo como (Pre)texto um artigo relatando a construção de um biodigestor numa escola do Brasil. *APEduC Journal, 2*(1), 150–167.

Pessoa, P., Lopes, J. B., Pinto, A., & Sá-Pinto, X. (2022). Evolution education through SSI for sustainable development. In X. Sá-Pinto, A. Beniermann, T. Børsen, M. Georgiou, A. Jeffries, P. Pessoa, B. Sousa, & D. L. Zeidler (Eds.), *Learning evolution through socioscientific issues* (pp. 29–47). UA Editora. https://doi.org/10.48528/4sjc-kj23

Pessoa, P., Aboim, S., Afonso, L., Lopes, J. B., & Sá-Pinto, X. (2024). Tasting to preserve: An educational activity to promote children's positive attitudes towards intraspecific diversity conservation. *PLoS One, 19*(1), e0285649.

Pessoa, P., de Lima, J., Valentina Piacentini V., Realdon, G., Jeffries, A., Ometto, L., Lopes, J. B., Zeidler, D. L., Fonseca, M. J., Sousa B., Pinto, A., Sá-Pinto, X. (submitted). A framework to identify opportunities to address socioscientific issues in the elementary school curricula: A case study from England, Italy, and Portugal. Journal of Research in Science Teaching.

Peters, M. D., Godfrey, C. M., Khalil, H., McInerney, P., Parker, D., & Soares, C. B. (2015). Guidance for conducting systematic scoping reviews. *JBI Evidence Implementation, 13*(3), 141–146.

Peters, M. D., Marnie, C., Tricco, A. C., Pollock, D., Munn, Z., Alexander, L., et al. (2021). Updated methodological guidance for the conduct of scoping reviews. *JBI Evidence Implementation, 19*(1), 3–10.

Petty, R., Wegener, D., Chaiken, S., & Trope, Y. (1999). The elaboration likelihood model: Current status and controversies. In *Dual process theories in social psychology* (pp. 41–72). Guilford Press.

Portuguese Government/Ministry of Education. (2018). Essential learning goals: Basic Education. http://www.dge.mec.pt/aprendizagens-essenciais-ensino-basico

Räthzel, N., & Uzzell, D. (2009). Transformative environmental education: A collective rehearsal for reality. *Environmental Education Research, 15*(3), 263–277. https://doi.org/10.1080/13504620802567015

Reis, P., & Galvão, C. (2004a). Socio-scientific controversies and students' conceptions about scientists. *International Journal of Science Education, 26*(13), 1621–1633. https://doi.org/10.1080/0950069042000205413

Reis, P., & Galvão, C. (2004b). The impact of socio-scientific controversies in Portuguese natural science teachers' conceptions and practices. *Research in Science Education, 34*(2), 153–171. https://doi.org/10.1023/b:rise.0000033760.04656.a1

Reis, P., & Galvão, C. (2009). Teaching controversial socio-scientific issues in biology and geology classes: A case study. *Electronic Journal of Science Education, 13*(1), 162–185.

Reis, P., Tinoca, L., Baptista, M., & Linhares, E. (2020). The impact of student-curated exhibitions about socio-scientific issues on students' perceptions regarding their competences and the science classes. *Sustainability, 12*(7), 2796. https://doi.org/10.3390/su12072796

Ribeiro, F., & Vieira, C. T. (2014). Abordagem de Questões Sócio-Científicas Controversas no 1° Ciclo do Ensino Básico. *Revista Educação: Teoria e Prática, 24*(47), 97–117.

Rittel, H. W. J., & Webber, M. M. (1973). Dilemmas in a general theory of planning. *Policy Sciences, 4*, 155–169. https://doi.org/10.1007/BF01405730

Sadler, T. D., Foulk, J. A., & Friedrichsen, P. J. (2017). Evolution of a model for socio-scientific issue teaching and learning. *International Journal of Education in Mathematics, Science and Technology, 5*(1), 75. https://doi.org/10.18404/ijemst.55999

Sarikaya, E., & Topçu, M. S. (2022). Using socioscientific issue approach to promote students' scientific literacy. In X. Sá-Pinto, A. Beniermann, T. Børsen, M. Georgiou, A. Jeffries, P. Pessoa, B. Sousa, & D. L. Zeidler (Eds.), *Learning evolution through socioscientific issues* (pp. 29–47). UA Editora. https://doi.org/10.48528/4sjc-kj23

Sousa, C. (2016). Inquiry learning for gender equity using history of science in life and Earth Sciences' Learning Environments. *Multidisciplinary Journal for Education, Social and Technological Sciences, 3*(1), 84. https://doi.org/10.4995/muse.2016.3762

Sousa, C. (2017). Integrating bioethics in sciences' curricula using values in science and socio-scientific issues. *Multidisciplinary Journal for Education, Social and Technological Sciences, 4*(1), 122. https://doi.org/10.4995/muse.2017.6481

Tuncay-Yüksel, B., Yılmaz-Tüzün, Ö., & Zeidler, D. L. (2023). Epistemological beliefs and values as predictors of preservice science teachers' environmental moral reasoning. *Journal of Research in Science Teaching*, 1–34. https://doi.org/10.1002/tea.21889

United Nations Educational, Scientific and Cultural Organization [UNESCO]. (2018). *Issues and trends in education for sustainable development.* In UNESCO Publishing. https://www.bic.moe.go.th/images/stories/ESD1.pdf

Valladares, L. (2021). Scientific literacy and social transformation. *Science & Education, 30*, 557–587. https://doi.org/10.1007/s11191-021-00205-2

Vasconcelos, C., Torres, J., Vasconcelos, L., & Moutinho, S. (2015). Geoethics in the context of sustainability and its teaching across the curriculum. In L. G. Chova, A. L. Martinez, I. C. Torres (Eds.), *Iceri2015: 8th International conference of education, research and innovation. ICERI proceedings* (pp. 2555–2560).

Vasconcelos, C., Cardoso, A., & Vasconcelos, M. L. (2018). Socio-scientific issues and scientific literacy. In *Proceedings of the 11th international conference of education, research and innovation (Iceri2018)* (pp. 7500–7505).

Wiek, A., Withycombe, L., & Redman, C. L. (2011). Key competencies in sustainability: A reference framework for academic program development. *Sustainability Science, 6*(2), 203–218.

Zangori, L., Foulk, J., Sadler, T. D., & Peel, A. (2018). Exploring elementary teachers' perceptions and characterizations of model-oriented issue-based teaching. *Journal of Science Teacher Education, 29*(7), 555–577.

Zeidler, D. L. (2014). Socioscientific issues as a curriculum emphasis. Theory, research, and practice. In N. G. Lederman & S. K. Abell (Eds.), *Handbook of research on science education* (Vol. 2, pp. 697–726). Routledge/Taylor & Francis Group.

Zeidler, D. L., & Sadler, T. D. (2007). The role of moral reasoning in argumentation: Conscience, character, and care. In S. Erduran & M. P. Jiménez-Aleixandre (Eds.), *Argumentation in science education* (Vol. 35, pp. 201–216). Springer Netherlands. https://doi.org/10.1007/978-1-4020-6670-2_10

Zeidler, D. L., & Sadler, T. D. (2023). Exploring and expanding the frontiers of socioscientific issues. In S. K. Abell & N. G. Lederman (Eds.), *Handbook of research on science education: Vol. III* (pp. 899–929). Lawrence Erlbaum Associates.

Zeidler, D. L., Sadler, T. D., Simmons, M. L., & Howes, E. V. (2005). Beyond STS: A research-based framework for socioscientific issues education. *Science Education, 89*, 357–377. https://doi.org/10.1002/sce.20048

Zeidler, D. L., Herman, B. C., & Sadler, T. D. (2019). New directions in socioscientific issues research. *Disciplinary and Interdisciplinary Science Education Research, 1*, 11. https://doi.org/10.1186/s43031-019-0008-7

Zhao, X. (2022). Customer orientation: A literature review based on bibliometric analysis. *SAGE Open, 12*(1), 215824402210798. https://doi.org/10.1177/21582440221079804

Zohar, A., & Nemet, F. (2002). Fostering students' knowledge and argumentation skills through dilemmas in human genetics. *Journal of Research in Science Teaching, 39*(1), 35–62. https://doi.org/10.1002/tea.10008

Chapter 14
SSI as An Educational Innovation in Spain: Easier Said Than Done

Silvia Alcaraz-Dominguez

14.1 Introduction

It is a pleasure to present this chapter, consisting of a review about recent uses of Socioscientific Issues (SSI) in formal education in Spain, as well as research about it. This chapter is motivated by a determination to collect and organise current knowledge on this matter in this specific sociocultural context. But why is this necessary? I will argue for a gap between theory and practice. In other words, SSI is easier said than done.

In many countries SSI are applied in educational settings as an innovation, this is to say, they are directly or indirectly linked to changes that ultimately aim to make education better (Alcaraz-Dominguez & Barajas, 2021). These changes may be found in the curriculum, in the teaching methodologies, or in other aspects (Zeidler et al., 2019). It is safe to state that Spain is one of these countries. The potential of SSI to deliver better education for all has been repeatedly made clear in the literature, either theoretically or empirically. From a theoretical perspective, it has been argued that because SSI are societal problems with a link to science, they are an opportunity to translate into classroom practices the current understanding of scientific activity as developed in the context of a society (Romero Ariza et al., 2017). Also, to the extent that they represent real-world problems without a clear solution, with social implications and that call to action, Spanish scholars have conceptually positioned SSI as a tool to teach in a competence-based way (Díaz Moreno & Jiménez-Liso, 2014; Solbes & Torres, 2012). According to this approach, the objective of basic education is not only to equip students with a set of knowledge, skills and attitudes but to help them learn to use these contents to be competent. This

S. Alcaraz-Dominguez (✉)
Universitat de Barcelona, Barcelona, Spain
e-mail: silvia.alcaraz@ub.edu

D. L. Zeidler (ed.), *A Moral Inquiry into Epistemic Insights in Science Education*, Contemporary Trends and Issues in Science Education 61,
https://doi.org/10.1007/978-3-031-63382-9_14

includes solving problems or coming up with creative solutions that respond to nowadays societal challenges, including those related to science. In the specific case of science education, SSI have been said to respond to the need to teach not only scientific knowledge but also how it is created (Puig et al., 2012). Another theoretical justification to include SSI in education in this country is to inspire or promote curricular changes that could make basic education more consistent with the needs of society, such as fighting misinformation or pseudoscience (Solbes Matarredona, 2013a, b).

In addition to those theoretical justifications, Spanish scholars have provided empirical evidence supporting the potential of SSI to tackle challenges of formal education in this country, such as students' lack of motivation or interest towards learning not only science but towards learning in general, as well as to foster their interest towards science (Ariza et al., 2021; Vázquez-Alonso et al., 2016). What is more, the curriculum in Spain is progressively including aspects that could facilitate the incorporation of SSI approaches, such as the Nature of Science (Alcaraz-Dominguez et al., 2018; García-Carmona, 2022).

In other words, there is a discourse and favorable conditions that sufficiently endorse the application of SSI in formal education in Spain. However, a recent study reveals that Science, Technology and Society approaches (including SSI) are far from being present in classroom practice (Reverte et al., 2022).

Many reasons could explain this gap. In the specific case of SSI, there is a lack of discussion or representation of SSI approaches in teacher continuing professional development, or innovation forums. As an example, a quick search in two of the main science education dissemination journals, which are mostly aimed at teachers, lead to less than ten results under the keyword "Socioscientific Issues". Similarly, the repository of teaching resources hosted by the National Institute of Educational Technology and Teacher Education only returns a few entries when searching for "Socioscientific Issues". Last, in its 2022 edition, the international conference of teachers of science and technology held at the Complutense University of Madrid included no reference to SSI or related aspects.

These spaces are very important considering that public administration and national and international organizations are generally not references for teachers in Spain, even when these organizations promote evidence-based policies (Pattier & Olmos Rueda, 2021). In this scenario, disseminating recent research and practice of SSI seems to foster a discussion among teachers or in other domains that have an influence to shape teaching practice. These include initial and continuing teacher education, or teacher associations and professional networks. More specifically, this chapter aims to:

- Offer visibility to recent research and practices related to Socioscientific Issues that show their potential to improve science education in Spain. Making this knowledge avaliable could motivate future research in Spain, since this work includes recent, updated content on this subject. Maybe this knowledge could also help to interpret findings of studies already performed in this sociocultural context. In addition to researchers, practitioners in Spain may find in this work ways to apply SSI that they may not have considered.

- Conceptually organize available knowledge in topics, each of them addressing different aspects of the potential of SSI to improve formal education in Spain. In this way the reader shall be provided with an overview of the most common trends in research and practice of SSI in formal education in Spain. The awareness of these topics may be useful for researchers to associate their investigations in one or more of these, as long as they make sense to them. In fact, a positive result of this review would be to ignite or resume discussions about the very nature of these topics, for example to what extent they accurately represent the application of SSI to formal education at a conceptual level. Continuing reflection about the key characteristics of an educational approach is essential for its sustainability in time.

It is also worth stating that this chapter is grounded on the assumption that SSI are applied in similar ways in other countries. In this regard, international researchers, practitioners or policy makers may also benefit from this compendium, at least as long as they work on sociocultural contexts that share commonalities with Spain in their understanding and application of SSI as a way to make formal education better.

Having considered the motivation and objectives of this chapter and before delving into its main part, it is time to move on to its theoretical constructs and assumptions.

The first construct is, not surprisingly, Socioscientific Issues. As it has been widely acknowledged, this construct has been operationalized in different ways and from different disciplines, both in Spain and internationally—see, for instance, (Alcaraz-Dominguez & Barajas, 2021) and (Tekin et al., 2016). The review of SSI education research and practice made in this chapter is grounded in the understanding of this concept as "complex and contentious social issues with substantial connections to science ideas and principles" (Zeidler, 2014). According to this definition, SSI are, above all, social issues, to the extent that they represent some sort of concern to people within a given sociocultural context. This approach is not only consistent with my social sciences background, but garnering strong consensus among educational researchers and practitioners, as it can be seen by the high number of publications that directly or indirectly build on this understanding. Moreover, this definition is suited for a review that aims to collect as many works as possible, as opposed to more restrictive understandings for example based on the controversial nature of SSI, as other authors have proposed—see, for example, (Levinson, 2006). Last but not least, and even if the title of this chapter is SSI in formal education in general, most of the work reviewed in this chapter is framed in science education, to the extent that it reports on teaching and learning of science or science-related knowledge, skills, attitudes or competences. Therefore, it made sense to adopt a definition of SSI originated in the field of science education.

The second and last theoretical construct to clarify is "formal education". It is true that SSI are being applied in a wide range of educational contexts, including informal (see, for instance, Birmingham & Calabrese Barton, 2014 and Zeidler et al., 2019). However, research and practice of SSI in education in Spain is mostly found on formal education. For the purposes of this work, formal education is defined as any teaching and learning situation that belongs to a structured, recognized program offered by a formal education provider: school, high school, university or similar.

Having exposed the objectives and theoretical groundings of this chapter, it is worth mentioning its methodological aspects. In this regard, this is not an empirical work, i.e. this is not a systematic literature review. Nevertheless, it is guided by a few principles.

In terms of the sources, it includes bibliography from academic and non-academic journals, books, and conference proceedings. Although these publications are not necessarily aimed at the same audience, broadening the type of information sources is useful to increase the chances of obtaining (and providing) the highest possible representation of the voices that exist on this matter at the moment. On the other end, this review excudes educational policy-related sources such as laws or the curriculum because these are publicly available in case the reader wants to access them. Besides, the objective is not to depict the status of formal education in Spain, but to present recent and innovative approaches based on SSI.

In terms of time period, this review does not aim to show the evolution of this subject over time but to do a state of the art, understood as a how things are at this time that can be called recent. But what does "recent" mean? This chapter reviews publications from the last five years for a main reason, wich is to keep the number of sources to review manageable. Nevertheless, these sources make reference to others that made important contributions. These references have also been included in this review, as long as they contributed to moving the discussion of SSI in formal education in Spain forward.

On the basis of these theoretical groundings and following the methodological guidelines stated above, this chapter is the result of reviewing more than 50 sources of information and establishing relations between them. This process led to classifiying the information from these sources in what emerged as semantical units of meaning, which I call topics. These topics are the following:

1. Students dealing with SSI that are relevant at the moment of speaking. This topic can be considered as a line of scholarship that aims to inspire and promote changes in the science curriculum so that it correctly reflects new, science-related issues, as per the educational level with which these works are concerned (mainly primary, secondary, and teacher education);
2. The pedagogical potential of SSI to develop Functional Scientific Literacy or other, equivalent operationalizations of this term in this sociocultural context. The majority of works reported in this topic explore the potential of SSI to help students develop argumentative and critical thinking skills;
3. Maintaining and sustaining the practice of SSI education in the classroom. Most of the literature quoted in this topic focusses on evaluating SSI-based instructional models, either from students' point of view or as a practice that teachers enact. It is mainly concerned with factors that can influence the correct

implementation of SSI education in practice, such as teachers' knowledge, skills and attitudes or beliefs, but also the strength of existing SSI-based instructional models.

By providing this visibility and organization, this chapter may provide ideas to discuss or investigate the gap between SSI theory and practice in this specific context. This discussion could involve policy makers, researchers, practitioners and even students or other stakeholders. Hopefully, such discussion would lead to bridging this gap in Spain or other, similar contexts.

14.2 Topic I: Students Dealing with SSI That Are Relevant at the Moment of Speaking

In the last years formal education in Spain has shifted from delivering content knowledge or even helping students to develop skills towards helping students to be active participants of a society that faces challenges such as rapid technological and scientific developments and a high level of uncertainty. Although each of the publications compiled under this topic may have other, secondary goals, they share a common objective, which is to provide information about students' opinions and intentions to act on science-related issues. In these works, these issues are approached as SSI, either when setting the scene for a learning situation, or in students' lives. As it will be detailed further on, these works focus on one of two goals: one, eliciting students' knowledge or opinions about SSI. The other, determining what can be done to influence their understanding or actions to address these issues.

Most of the studies aiming to elicit Spanish students' knowledge or opinions reveal a generally poor understanding of the SSIs. A study with a relatively large sample (n = 308) was developed to describe compulsory and non-compulsory secondary students' perception of risk to the Covid-19 virus, and their intention to be vaccinated against it, depending on their age and gender (Pozuelo Muñoz, 2022). First, it was found that students had different opinions about the purpose of the Covid-19 vaccine compared to other vaccines that work in the same way. Moreover, their perception of the risks of the vaccine was higher on students from higher grades, i.e. those that have more science knowledge. Characterizing students' perceptions and intentions over an SSI was also the goal of the research by Cebrián-Robles et al. (2021), who concentrated on preservice teachers' views on illegal mining in the Amazon forest. One of the results of this study is that what students identify as problems of this SSI are harm to the environment, whereas the sociocultural aspects are way less represented. These works can be complemented by Pozo-Muñoz et al.'s, who provide evidence of a low degree of responsibility towards water issues on the basis of an analysis of elementary school students' drawings (2023).

Among the studies reporting actions to influence students' understanding of SSI, that by Portillo-Blanco et al. stands out. The authors did a study on more than n = 800 Secondary Education, Upper Secondary Education, BA in Primary Education and Master's Degree in Secondary Teacher Training students' decisions to get the COVID-19 vaccine (Portillo-Blanco et al., 2022). Similarly, in their study on the antivaccine movement with n = 39 preservice teachers, Puig and Ageitos show not only that for all critical thinking skills except for one, students mobilize scientific knowledge, but that such knowledge is correlated with a better performance in critical thinking skills (Puig & Ageitos, 2021). Both these studies constitute evidence supporting the idea that in the specific case of the Covid-19 pandemic, which was characterized by fake news or not scientifically accurate news, scientific knowledge may be a good resource to equip students with. However, it is not clear to what extent such scientific knowledge will lead students to reach a better understanding of the SSI. In this regard, Sergio Calavia et al., inquired about the dimensions that n = 46 11th graders consider to justify their COVID-19 vaccination decision, namely (D1) Scientific; (D2) Political-economic; (D3) Ethical; (D4) Ideological-Religious; (D5) Environmental, (D6) Personal opinion; and (D7) Social-cultural. The authors report on how students combine and integrate these dimensions showing that they combined more than one dimension in a single argument (Calavia et al., 2022). Likewise, Uskola et al. set out to characterize preservice teachers' (n = 72) decision-making about vaccination as an integration of scientific knowledge and argumentation. The authors observe that although those who expressed a more complex immune system model argue better, only 15% referred to this scientific knowledge in their decisions (Uskola Ibarluzea et al., 2021).

The interaction of scientific and other types of knowledge in students' decision-making about SSI is also demonstrated for SSI other than Covid. In their inquiry about preservice teachers' choice of a type of diet, Jiménez-Aleixandre and Brocos' observe argumentative interactions between students in in small groups. In this regard, the authors stress the influence of cultural values in negotiation processes about this SSI (Jimenez-Aleixandre & Brocos, 2017). Similarly, in their study about students' contradictions in their choice of omnivorous with meat reduction, omnivorous, vegetarian or vegan diet, they find out that these contradictions are supported by cultural, pragmatic, and anthropocentric considerations. The authors report that "even if the data about environmental impact and animal suffering caused by diets were explicitly considered in most [student] reports, their consequences in their choices (…) were limited" (Jiménez-Aleixandre & Brocos, 2021)—p. 1807. Similarly, the works by Cruz-Lorite and colleagues analyze the arguments that preservice teachers give during a role-play on nuclear power, under the lens of informal reasoning. Results show that students mostly use environmental arguments, followed by financial, whereas social-oriented arguments appear way less frequently, thus confirming the results of similar studies carried out in other, international sociocultural contexts (Isabel María Cruz-Lorite et al., 2023). What is interesting from this study is that 1) it understands informal reasoning, i.e. everyday logic, as part of making a decision about an SSI. Last, López-Fernández et al. (2022) include considering different views on the issue (chemical, environmental, financial, social, healthcare, etc.) as part of the activities that students must develop on an SSI-based activity in Chemistry education. Other authors have gone a step further to consider students' awareness of their own reasoning in their interpretation of students' discussions about an SSI (Jimenez-Aleixandre & Brocos, 2017).

All this leads to concluding, first of all, that Spanish scholars are very interested in diagnosing students' opinions about SSI. We can see an explicit intention to give visibility to students' views, even in the case of an issue that is defined as against the law such as ilegal mining (Cebrián-Robles et al., 2021). This focus on students has led to valuable data not only about what they think about these issues but in the reasoning process about them. More specifically, these studies shed light about a very specific SSI-like situation, i.e. Covid-19 pandemic, characterized by fake news, pseudoscience, and risk of science negationism. These results can serve as needs analysis providing lines on action for educational policy-makers to support students and teachers in learning about these kinds of SSI that even now in a post-pandemic context are likely to occur. Nevertheless, there are other, pressing SSI that are also characterized by a flow of misinformation and such as the care for the environment. For this reason, it may be worth considering to apply the learnings from Covid to other, similar SSI.

Second, knowing students' opinions about SSI is useful to make recommendations on actions to be taken at a policy level. For example, in the study on primary school students' drawings, the authors suggest to address content related to water care and conservation in the curriculum from a multidimensional and interdisciplinary approach, in order to understand the origin and impact of water problems (Pozo-Muñoz et al., 2023).

Last, the information about how students reason and make decisions about different SSI points to the importance of arguments of different kinds (other than scientific) in these processes. Since considering social, economic and political aspects of science is part of the Spanish curriculum, the results of these studies may be taken as more reasons to implement SSI in practice. New lines of research in this direction would be to further investigate the weight of each of these arguments in students' forming opinions or course of action in order to provide appropriate support.

14.3 Topic II: SSI as a Means to Achieve Scientific Literacy in Science Education

As stated in the introduction of this chapter, the application of SSI to formal science education in Spain is analysed in this review from the point of view of educational innovation, i.e. as a change that can improve learning. In the case of science education, one of the most convincing arguments to introduce SSI is as a way to promote functional scientific literacy (Zeidler, 2014). This is different from other attempts to improve science education through a social approach, such as for example the Science, Technology and Society (STS) approach (Zeidler et al., 2005). More recently, the author has argued that the SSI approach allows students to engage in scientific practices linked to an idea of citizenship that seem to underlie recent, vision III of scientific literacy, which promotes science education for social transformation (Sjöström et al., 2016; Zeidler & Sadler, 2023).

In this regard, SSI are in many cases introduced with arguments that rather ressemble the STS approach. However, a change can also be observed towards a conceptualization of SSI that resembles the functional scientific literacy framework, at least since the definition of scientific literacy in the Spanish curriculum started to resemble Roberts' Vision II (Díaz-Moreno et al., 2019; Díaz Moreno, 2019). In this context, and as it is true in other countries, several works by Spanish scholars have a common objective, namely assessing SSI-based teaching approaches for student development of scientific literacy as defined in Visions II or III.

A large part of these works responds to what is known as empirical, positivist or interpretivist research, in the sense that they aim to observe a reality and create knowledge about it. What follows is an analysis of those works, grouped in what emerged as the most frequent components of scientific literacy that they focus on, namely argumentation, critical thinking, and Nature of Science (NoS).

Most of these publications assess the potential of SSI to teach argumentation, especially among secondary school students. As an example, Esquivel-Martín and others report on a teaching and learning experience in 10th grade, i.e. students who are typically 15 or 16 years old, where students rank cities from the most to the least polluted on the basis of the problems of reproduction found in their ecosystems. The authors analyze learners' performance in using evidence and formulating causal explanations to justify their rankings, as well as in proposing reasoned solutions considering different perspectives. Among the conclusions of the study is that the design of the activity encourages students to establish frequent causal relationships between human, animal, and environmental health data. Moreover, students could engage in evidence-based reasoning without having to master scientific content (Esquivel-Martín et al., 2023). Also dealing with an environmental issue, González-Sánchez and others study the effect of a role-playing and dramatization activity in students' argumentations about a Socioscientific Issue—building a marine eolic energy field near their city. Results are positive in the sense that the level of argumentation is higher after the activity, as operationalized in quantity of claims, arguments that support these claims, and modifiers (González-sánchez et al., 2021). Last, it is worth highlighting the study by Cuevas-Aldunate and others on secondary school students' argumentation on essays about an SSI about genetically modified foods. This work analyses the argumentation of 15-year-old students about genetically modified organisms. The study is made on an intervention that links argumentation on SSI to students' reading profile, on the assumption that by knowing which of these profiles students fit into, better scaffolding could be provided and hence a more effective learning experience of argumentation through SSI (Cuevas-Aldunate et al., 2021).

The second, most common group of works aiming to assess the affordances of SSI in formal education concentrate on its potential to help students develop critical thinking. One of its examples is by López-Fernández et al., who report on a teaching and learning sequence to foster secondary school students' critical thinking in chemistry education throughout a whole academic year. The sequence is based on a systems-thinking approach, where students go from simple to complex in their understanding of the problem of plastics and the pollution they cause. As measured through a questionnaire, the authors provide evidence of students' development of critical thinking as a result of the activity (López-Fernández et al., 2022). A similar use of SSI is made in two other works, both performed in in preservice teacher education by scholars at the University of Santiago de Compostela, in the north-west of Spain. In their study with preservice primary school teachers enrolled in a course about natural sciences, the authors demonstrate that students are able to use critical thinking skills in their assessment of five premises used in an SSI, framed as an anti-vaccine movement, thus suggesting that this is approach is worthwhile using in this specific context (Puig & Ageitos, 2021).

In the study quoted in topic I on students' decisions on a type of diet (omnivorous, omnivorous with meat consumption reduction, vegetarian, or vegan), Jiménez-Aleixandre and Brocos also contribute to understanding the affordances of SSI to teach critical thinking. The authors provide evidence of students' awareness of contradictions between their choice on an SSI related to human alimentation and the claims that support the arguments that justify this choice, which they consider a stance of critical thinking in science education. Therefore, it is a worthy use of SSI.

The last aspect covered in this empirical or positivist approach to demonstrating the affordances of SSI to develop students' key competences is Nature of Science (NoS). Although to a lesser extent than in other countries, recent Spanish scholarship pays attention to the affordances of SSI to teach this aspect. In fact, the interest by Spanish education researchers in the Nature of Science as part of science education goes back decades, and doing so through SSI is no exception. As it is not possible to review this whole tradition here, let us focus on two of the most recent studies. In their experimental (pre and post-test) study with 30 participants, Crujeiras-Pérez and others provide evidence of a positive impact of an SSI-based teaching and learning sequence on preservice teachers' knowledge of certain aspects of the NoS, namely: tentative nature, creativity, sociocultural integration, and observation and inference. According to this study, the teaching and learning sequence, based on the SSI of the possible installation of a nuclear deposit, constitutes an example of effective SSI-based instruction for this specific learning objective in this sociocultural context and educational level. Almost equally interesting is another result of this research—the NoS dimension that increases the least after the activity is sociocultural integration, i.e. the influence of society in scientific developments (Crujeiras-Pérez et al., 2021). Since it is precisely an intervention based on SSI, it would be necessary to dig deeper into the details of the sequence to understand this result, and to find out how to enhance this aspect, which would otherwise be expected to be the one that increases the most. The author of another study also describes as "moderate" the evolution of the understanding of the nature of technology of a group of students who participated in a teaching and learning experience contextualized in the historical controversy between Tesla and Edison. Four out of the six groups of students went one level above in their definitions of technology and its components, whereas two groups stayed in the same level (García-Carmona,

2021). Compared to the study on the nuclear deposit SSI, the students in this research participated in a much shorter teaching and learning activity, which would at least in part explain their reduced progress.

A not so large group of works in this topic follow a rather critical paradigm to research. They have the same goal as those covered above, i.e. provide evidence of the affordances of SSI-based instruction for student development of competences or skills associated with scientific literacy. However, the research paradigm of these studies is different, in the sense that student learning or performance is not measured objectively, but as perceived by the participants.

Two of these studies elicit preservice teachers' opinions of SSI-based teaching and learning experiences that aim to develop their critical thinking. Verdugo Perona and others applied a Likert-scale type questionnaire to a group of n = 62 participants, which was made of 10 categories including their assessment of an SSI-based instructional experience in terms of developing reflective action, autonomy, and critically question their ways to do things. The results show a generally positive perception, especially regarding the potential of the activity to promote critical and reflective thinking (Verdugo Perona et al., 2022). In a slightly different study, Hierrezuelo Osorio and others analyze two groups of preservice teachers' perceptions of their critical thinking skills before and after participating in an SSI-based training program. The authors discover that students from both groups perceive an increase in their skills, with a greater increase found in group 1, corresponding to primary school preservice teachers (Hierrezuelo Osorio et al., 2022). One more study to be highlighted is that by Palma-Jiménez and others, about preservice teachers' assessment of three pedagogical strategies to deal with breastfeeding as an SSI, namely (a) an activity about argumentation, (b) a cartography of controversy, and (c) role playing. The findings point to the role-playing activity as the one that students perceived as most useful not only to learn about the SSI, but in terms of participation, motivation and fun.

To sum up, the publications collected in this topic generally support the effectiveness of SSI-mediated teaching and learning experiences to teach key components of scientific literacy, in this case argumentation, critical thinking, and Nature of Science. In the case of argumentation, they provide tools to scaffold it and frameworks of analysis, which could constitute assessment frameworks for teachers, mostly for secondary school. In the case of critical thinking, the studies offer evidence of this effectiveness both in secondary education and preservice teacher education on a diversity of SSI, including plastics, vaccines, and food related issues. As per the Nature of Science, these studies constitute evidence of the effectivity of SSI for teaching NoS mostly in preservice teacher education. This is significant as teachers' inadequate knowledge about the NoS is one of the obstacles for teaching NoS in Spanish schools (Aguirre, Haggerty & Linder, 2007, quoted in Crujeiras-Pérez et al., 2021).

Last, the works report on learners' generally positive opinions about pedagogical strategies in relation to their learning in SSI-based instruction, especially role play. This is important as it gives information about the preferences of the receivers of SSI-based lessons.

Additionally, some of these studies also make important contributions to the design of SSI-based lessons. In the aforementioned article by López-Fernández and others (2022), two contributions can be found. First, the development of an SSI-based learning sequence to foster critical thinking in chemistry education, based on an analysis of the curriculum and solid pedagogical groundings, that includes SSI aspects in a teaching and learning sequence for science education, such as analyzing the trustworthiness of the source of scientific information. Considering that not all teachers are used to this way of teaching, this sequence can promote SSI-based instruction in this subject and level in Spain. Second, the focus on translating what has been done in class into real life, i.e. the passing of a hypothetical law that would regulate the use of single-use plastics. This is significant as it presents a not so frequent yet valuable application of SSI in formal education. In turn, the work by Cuevas-Aldunate and others explicits the relation between scientific and linguistic competences by having students work with scientific essay as a learning methodology. This is an innovative use of SSI that may be of interest to Spanish teachers in the context of the current tendency towards interdisciplinary teaching, especially in secondary school.

14.4 Topic III: Maintaining and Sustaining the Practice of SSI Education

Like any educational innovation, the effective application of SSI in formal education in Spain depends on many factors, one of them being who promotes it and from which position. In many cases, innovation does not emerge from its very actors, i.e. teachers or even students, but as a result of educational intervention: developing resources, changes in the curriculum, or incorporating this aspect in teacher education programs. In these cases, it meets resistances and faces threats due to its top-down nature.

This seems to be the underlying assumption of several of the publications and reports found in this review, which are compiled under topic III. These works analyze SSI in formal science education in the light of existing practices: strengths, weaknesses, obstacles and threats. They try to explain how this approach is being adopted, if so, and what can be done to ensure its sustainability. By doing so, these studies aim to counterpart possible resistances to the implementation of SSI approaches in science education, mostly with empirical evidence. The works found to belong in this topic are exposed next through two main foci: a) SSI-based instructional models, and b) in-service teachers.

Let's start by the first—what are SSI-based instructional models? As it has been shown by the examples so far, there are different ways to apply SSI to teaching practice. These constitute models of SSI-based instruction. In short, the works with this focus aim to validating or proposing a specific model of SSI-based instruction. This literature seems to pursue the ultimate objective of providing SSI with entity as a stand-alone teaching practice or reinforce it, so that teachers or policy makers can practice and promote.

Some of these works constitute operationalizations of SSI in science education, in the form of learning sequences, mostly for secondary education. The learning sequence has been identified as one of the main components of STEM lessons where SSI have been "inserted" (Alcaraz-Dominguez & Barajas, 2021). One of the ways to organize these sequences is Socio-Scientific Inquiry-Based Learning (SSIBL). This model was proposed to contribute to Responsible Research and Innovation (RRI), a European policy of citizen engagement in science, through formal science education (Amos et al., 2020; Romero Ariza et al., 2017). The method combines Inquiry-Based Learning (IBL), Socio-Scientific Issues and Citizenship Education, and it is based on three principles, namely authenticity, mapping controversy, and action (Romero Ariza et al., 2014). A group of scholars, mainly from the university of Jaén, in the South of the Spanish peninsula, implemented this model in classrooms as a way to establish it, and lately as applied to environmental education (Ariza et al., 2021).

Around the same period, scholars at the University of Barcelona, in collaboration with other European institutions promoted, tested and validated a similar approach to structuring science lessons around SSI. In this model, students start by encountering the issue, then they collect evidence, and lastly, they propose a solution to the issue on the basis of the evidence collected. These phases are called "Engage", "Extend" and "Evaluate" respectively (Alcaraz-Dominguez et al., 2018). Evidence of how teachers use this model to teach scientific literacy was recently published (Alcaraz-Dominguez et al., 2024).

These or other models inform applications of SSI in formal education reported in other studies, mainly for secondary school students. As an example, the work presented by Ruiz-González and colleagues at the 11th International Conference on Research on Science Education reports on the design of a didactic proposal to improve argumentation skills of students in post-compulsory secondary school. In this proposal, students use scientific knowledge to deal with a socio-scientific issue, in this case the production of insulin with biotechnology. The proposal includes four activities. First, the SSI is presented by a piece of news from the Media, and students express their initial knowledge in small groups. Then, students collaboratively read further documents about the issue to engage with the scientific content and classify the texts according to whether they support insulin creation from a bacterium. After that, a discussion takes place, first in small groups and then with the whole class through group representatives. Finally, each student is asked to write an essay expressing their opinion about the issue (Ruiz González et al., 2021).

Some of these SSI-based instructional models have been validated. A few years ago, Romero-Ariza and others assessed n = 60 preservice teachers' SSI-based lesson plans, which they developed in the context of a semester course on Didactics of Science where they played the roles of learner, reflective practitioner, and lesson designer. The researchers analyzed students' productions in two main dimensions, namely higher order thinking skills and Nature of Science (NoS) aspects. It was found that 90% of students' proposals offer opportunities to work on one or more of the desired aspects, mostly communication and different perspectives, closely followed by critical thinking and decision-making (Romero Ariza et al., 2017).

SSI-based instructional models have also been put forward for college-level education. Fernández-Oliveras and others created a didactic proposal to promote the development of critical thinking in preservice science teacher education, framed within a socioscientific context: the management of nuclear waste. The proposal comprises six lessons, which go from eliciting students' previous knowledge to deciding about the installment of a nuclear deposit in their city. Along the process, students engage in different activities, including readings, role-playing and visually representing the controversy around the issue. As a way to add more credibility to the proposals, the authors tested the sequence with a group of students. The results highlight that in all sessions cognitive processes of the highest level and higher order skills of critical thinking are encouraged (Fernández-Oliveras et al., 2022).

Last, worth highlighting as part of the works related to SSI-based learning sequences is that by Dr. Jordi Solbes, who in the past decade has been proposing pedagogical applications of SSI to develop preservice teachers' critical thinking. The author argues that compared to other ways to get to know the world, science is intrinsically critical, as it has its own language and system that is separate from others, for example religion. However, it is only transformative when it threatens political or economic interests. Following the same reasoning, science education that teaches only knowledge and procedural aspects of science will fail to foster critical thinking unless it addresses socially relevant issues. In this sense, the author argues that SSI are suitable to achieve this in science education (Solbes, 2019).

As part of the works proposing SSI-based instructional models, a lot of them operate at a more granular level than a teaching and learning sequence. These works propose the introduction of SSI as pedagogical strategies, understood as ways for students to engage with the SSI that are part of bigger, self-contained units of learning organization such as lessons or projects. Role-play, controversy and dilemmas are, in this order, the most represented strategies in Spanish literature.

Regarding role-play, Crujeiras-Pérez and colleagues characterize different aspects of students' arguments in one of these activities, consisting of a discussion on installing a nuclear deposit. What is interesting from this study are the trends identified, such as that students' argumentation operations do not differ depending on the position they take in the role play, or that they are unable to identify the limitations of their arguments after they have been questioned (Crujeiras-Pérez et al., 2020). The outcomes from this study can be used in the design of role-playing activities for the specific case of SSI. In their work on the same topic, i.e. nuclear energy, and the same student skills, i.e. argumentation, Cruz-Lorite and others also provide valuable insight on this matter. More specifically, they examine the extent to which role play promotes the development of scientific arguments and the use of evidence, justifications and conclusions during an experience with preservice primary teachers. The authors find out a much higher use of evidence in their arguments, compared to other elements such as conclusions and justifications (Isabel Maria Cruz-Lorite et al., 2022).

The work of other authors to validate the use of certain pedagogical strategies for SSI-based instruction has been structured around the construct of "controversy". Díaz-Moreno sets out to help teachers who want to address SSI in class, by proposing criteria to select SSI from the press to the extent that they include controversy or not (Díaz Moreno, 2019). Another approach, i.e. the "cartography of controversy" is similar to role playing, to the extent that it aims to explicit all the points of view over an SSI. However, it goes beyond people as the only ones who play a role over the issue, as it uses the concept of "actant". In this framework, an actant is anything that alters the course of action and the relations that the other actants have at the time of speaking (España Naveira et al., 2023). When applied to science education, the cartography of controversy guides students into an inquiry that aims to understand and analyze a socioscientific controversy and represent its complexity by visualizing the actions of those who participate in it. Although this work explicitly draws from the socially acute questions (SAQ) line of scholarship (Bencze et al., 2020; Simonneaux, 2011), it is true that in essence it shares the same principles as other, more widespread constructs such as socioscientific perspective taking (Newton & Zeidler, 2020).

Related to the concept of controversy are the works that try to promote what is known as socioscientific dilemmas as a pedagogical strategy in science education. As an example, Hierrezuelo-Osorio and others collect preservice teachers' perceptions to design and implement three different formats of socioscientific dilemmas (text, video and opinions) aimed to develop critical thinking through argumentation and decision making in secondary school. Overall, participants perceive this strategy as useful for the classroom and with a medium design difficulty. However, they consider that the opinions format is the one that could be the most difficult to design, and the most complex to implement in the secondary classroom (Hierrezuelo Osorio et al., 2022).

Having covered one of the two main foci of this topic, i.e. studies that attempt to model the practice of SSI-based instruction in order to legitimate its use in science education, we can move on to the other, i.e. science teachers. These works share the assumption that science teachers are gatekeepers in SSI education, as it has been noted in other countries (Borgerding & Mulvey, 2022; Ekborg et al., 2013; Faisal & Martin, 2022; Kinskey & Zeidler, 2021; Tidemand & Nielsen, 2017). In the case of Spain, this line of scholarship concentrates mostly on teacher beliefs about SSI education.

These studies focus on what authors almost indistinctively define as attitudes, perceptions, or even opinions about using SSI-based instructional approaches in the classroom. In this regard, the work from colleagues from the University of Málaga, in the Southern part of Spain, stands out. In their oral presentation at the 11th international conference on research in science education, Hierrezuelo Osorio and Franco Mariscal report on preservice teachers' perceptions about the design and implementation of three different formats of socioscientific dilemmas (text, video and opinions) aimed to develop critical thinking in secondary school. To that goal, they report on an experiment with n = 43 participants who undertook SSI-based instruction in the three formats. In the text format, they are asked to write, two texts based on the news, each of them with opposing ideas on the problem to be addressed, including different arguments for and against the problem that make the student think about it. In the video format, they must search or edit two videos with opposing ideas about the problem with arguments for and against the problem. Last, the opinions format consists of developing a set of opinions on the problem by specialists. Opinions from blogs and forums can also be included as long as they contain contrasted information with arguments for and against and allow reflection on the subject. Overall, participants perceive SSI-based instruction as useful for the classroom and with a medium design difficulty regardless of the format. More specifically, they consider that the opinions format is the one that could be the most difficult to design, and the most complex to implement in the secondary classrooms (Hierrezuelo Osorio & Franco Mariscal, 2021). Another example is the work already quoted in topic II by (Díaz Moreno, 2019). This article shows a clear interest in facilitating the task of teachers in the selection of SSI to use in the classroom, through criteria that emerge from rigorous research.

As a synthesis of this topic, it is safe to state that in Spain, the application of SSI-based instruction in science lessons is sufficiently backed by theoretical models. These models share commonalities with other, well-known approaches such as project-based learning and Inquiry-Based Learning, to the extent that they are learner-centered and based on students collecting data about the SSI and making a decision or a synthesis. The SSI is, in these cases, in the center of the lesson, thus potentially allowing to address a wide range of curricular contents through the SSI approach. This is what other authors have called "full-fledged SSI instruction" (Tidemand & Nielsen, 2017). It is also true that there is a solid body of evidence that teachers are, in general, in favor of SSI-based instruction.

However, looking into detail most of the studies that empirically assess the effectivity or efficiency of SSI-based instruction in real settings focus on just one of the activities that can be carried out as part of an SSI-based instructional model, such as role-play, controversies or dilemmas. In a context where both educational policy makers and teachers are looking for certainties when making the effort and taking the risk to innovate in the classroom, this means that in this specific sociocultural context SSI instruction may mostly be enacted (and hence conceived) as isolated pedagogic activities, as opposed to a model of instruction.

The partial nature of empirical studies on SSI-based instruction is also true for studies on teachers' knowledge, beliefs or attitudes. Compared to other countries, while still being far from a truly participative approach in the desing of SSI-based instructional methods or materials, teachers' opinions about SSI in education are a focus of interest in Spanish educational research. This is shown, for example, by the prevalence on studies on teachers' opinions about SSI-based instructional approaches, as opposed to to studying how they deliver it. However, it is also true that most recent studies report on preservice teachers. Whereas understanding preservice teachers' opinions on SSI instruction can lead to taking measures to change these opinions in case they are negative, it is also true that initial teacher education has become only a part of their training. Instead, the concept of lifelong learning has reached the professional profile of teachers, meaning that their education is considered a continuum that goes from their initial training until the end of their career, including continuing professional development (CPD). CPD is, thus, a valuable scenario to work on teachers' knowledge, beliefs and attitudes towards SSI instruction. Therefore, more research on actual in-service teachers would be very useful to inform CPD priorities, curricula etc. Also, because of their methodological design, the results of studies on preservice teachers found in Spanish literature cannot be applied widely across the country. Although Spain can ge defined as a federal state, at least in the sense that Education competences are transferred to its autonomous regions, in the case of an educational innovation such as SSI-based instruction it would be beneficial to carry out studies leading to results that apply to as many teachers as possible. In addition to whatever scholars in this field consider worth researching, such studies could help to understand to what extent the prevalence of the "easier" ways to introduce SSI in science classroom is due to teachers' lack of knowledge, skills, negative attitude towards the new teaching approach, or other factors.

14.5 To Conclude

Having covered the three topics that emerged as a result of this review on SSI as an educational innovation in Spain, it is time to move on to conclusions. The first part of these conclusions starts by resuming the ambition of this chapter. As stated in the introduction, this chapter aimed to reach two main objectives. One, giving visibility to recent research and practices in Spain on SSI applied to formal education. Another, conceptually organizing the information found in three topics, which could help to gain an overview of the current trends of SSI in formal education in this country and eventually inspire current or future research. Achieving these objectives would ultimately lead to bridging the gap between SSI theory and practice that exists in Spain.

As per the first objective, a reasonable number of studies and educational interventions based on SSI were found. Methdologically speaking, most of these studies follow a qualitative research approach, and use ethnographic research designs. This choice makes sense, as Spain is a federal state with different education laws depending on the region. Therefore, if the goal is to influence practice or policy, educational research must be constricted to the specific context where these decisions can be made. This is especially true for preservice teacher education, where another layer of decision-making is added, since universities (i.e. initial teacher education providers) have autonomy to decide on key aspects of the programmes offered, including the curriculum (López Ramos et al., 2020; Manso et al., 2019). However, the differences between the studies in terms of the phenomenon being studied, as well as the participants makes it very challenging to answer essential questions about SSI-based instruction when applied as an innovation in Spanish classrooms.

This diversity is also true for the SSI-based educational interventions reported or analysed as part of these studies. They are very diverse in terms of what SSI is being addressed, curricular contents, objectives, pedagogical approaches, and duration.

In terms of the SSI being addressed, a trend can be found to frame SSI-based instructional activities in the context of the Covid-19 pandemic. Seeing that this topic may not be so relevant at the present time, it is worth highlighting other issues that are addressed through SSI-based instruction, such as water, food consumption, or energy production, i.e. environment-related.

In terms of content, most of the research reported in this chapter is performed in departments of science education, also known as didactics of science, as opposed to general didactics. Authors of the publications belong to science education departments, and educational research in Spain is still performed within the boundaries of academic disciplines. For this reason, it would be expected that most interventions aim to help students learn scientific knowledge, as this is one of the most intuitive uses of SSI, i.e. teaching science in context. Indeed, rich and valuable studies have been found in this review, to the extent that they tackle very specific aspects of how students learn specific content or engage in specific cognitive processes, such as informal reasoning. This is beneficial, as there is a large volume of international research done from the field of science education, and in this way, knowledge is more likely to be exchanged. However, a surprisingly high amount of the SSI-based educational interventions reported aim to help students develop skills, mainly argumentation and critical thinking. From the interventions aiming to help students develop knowledge, most of them focus on knowledge about the Nature of Science (NoS), which is refers to how science is carried out and developed, i.e. a meta-knowledge about scientific facts and principles. This could be due, again, to the time in which these studies were carried out. During the Covid-19 pandemic, fake news, pseudoscience and misinformation were more common and concerning. Therefore, although these skills traditionally belong to other areas of the curriculum, such as language, it may have felt necessary to devote time in science education to protect students against these threats. Maybe for this reason, in this chapter there are studies where dealing with SSI is intertwined with scientific reading and writing, thus linking it to linguistic competence. In this way, applications of SSI-based instruction to contents that are not traditionally from science subjects are backed not only by intellectual justifications but by empirical research, showing the effectiveness of certain approaches in specific groups of students. This emerging line of research is going in the right direction seeing that educational reforms are promoting interdisciplinarity and that students' linguistic competences are lowering according to recent international assessments.

In terms of pedagogical approaches, most publications only provide an outline of the SSI-based intervention. For this reason, it is difficult to understand what the most used instructional methods are, i.e., how is SSI-based instruction operationalised in Spanish classrooms. If anything, it is true that SSI-based educational interventions in this country are student-centered and cooperative. In terms of duration, the interventions reported last between one lesson and one year.

Within this diversity, it came clear that SSI approaches are applied in a wide range of settings and levels within formal education, including primary, secondary and higher education. Other, still formal educational contexts, such as vocational training or senior education, were not significantly represented.

Leaving aside the methodological aspects of the sources of information reviewed, a significant result of this review is the specific understanding of SSI in education in Spanish research and practice. In this regard, two main conceptions were found. One, anything realted to what students know, think about, reason or do about SSI. The studies and experiences grounded on this conception are valuable to the extent that techniques, resources or policies can be put in place to make students more literate and hence participate in the social resolution of a given SSI. The other application of SSI to formal education is mostly pedagogical, i.e. the SSI is a mediator between students and the curriculum. These studies and experiences include models of SSI-based instruction, as well as studies on its impact and teachers' opinions, among others.

Regarding the second objective of this chapter, the three topics contain works that ultimately wish to inform the application of SSI-based instruction in formal education, by a) providing knowledge about what or how students think, b) assessing the effectiveness, efficiency or suitability of SSI-based instructional approaches, or c) aiming to tackle the challenges of SSI-based instruction. As it turned out, one of these two topics is concerned with what students know or think about certain SSI, whereas two of the three topics focus on the application and sustainability of SSI-based approaches. Although similar studies can be found in other countries, it is worth considering to what extent the very emergence of topic I is due again to the Covid-19 pandemic, as it constituted an easily available SSI to be addressed. This being said, studies should still be done on students' knowledge, opinions or intentions to act about the other SSI found in the literature, or others. Recent policies point to a greater integration between science and society, such as Responsible Research and Innovation in Europe (European Commission, 2012). In this light, most of what are now considered scientific issues will soon become SSI.

Another conclusion in this regard is that topic II, focused on assessing the potential of SSI to help students develop components of scientific literacy, turned out to be the longest and richest in terms of the information to be extracted and organized. In contrast, only a few sources were found in topic III, aiming to provide certainties about SSI-based instruction that can facilitate its widespread application.

Going back to the structure of three topics, it is worth noting that the boundaries between them can be blurry: if we ask a preservice teacher about the usefulness of an SSI-based lesson for developing their own critical thinking skills, are we doing research about the potential of this methodology in this educational context as it is perceived by its receivers (topic II), or are we providing information about future teachers' perceptions of this instructional method, thus identifying challenges, if any (topic III)? In this regard, the objective to organise the information in these topics has only partially been achieved. The concern is not about a possible lack of legitimacy of these topics, as they were not aimed to be systematic. However, and precisely for this reason, realising the weaknesses of this system confirm that this is just one of the many possible ways to organise this information. To be completely sure, the reader should check the objectives of any publication quoted here that is of further interest.

Having considered the consecution of the objectives of the chapter, it is time to critically analyse the works reviewed. Let's start with the definition of the constructs, which is an essential aspect of research to the extent that it allows to build knowledge from different studies and experiences that operationalize a construct in the same way. Although in most of the studies SSI are operationalized as mediators between students and the curriculum, it is also true that some of the reports of SSI-based instructional experiences found in this review aim to help students learn contents or develop skills that can be defined in many different ways depending on the culture. In many of these cases, no explicit definition is given of this construct. In others, not only the construct is not explicitly defined, but the link between it and the SSI is not clear. For instance, some articles argue that that SSI allow to improve social values. Although it is possible to imagine what the authors mean, these statements make the study vulnerable to criticism for not clarifying their specific undestanding of social values. Beyond the concerns risen by the fact that social values can be defined differently even within the same culture depending on the ideology, this example shows that the attention can be distracted by what is important, i.e. the contributions of the article, thus hindering its potential to add on to academic discussion. Another problem in terms of construct definition observed in the sources analysed in this chapter is a minsuderstanding between the SSI approach and other, science-in-society based teaching approaches. Several papers include SSI in the title, abstract or body, but they report on educational interventions that rather fit into a Science, Technology and Society/Environment (STS/E) approach, as they imply that students can explain, solve or make a decision about the SSI solely after having understood its scientific facts and principles (eg. Hierrezuelo Osorio et al., 2022; Uskola Ibarluzea et al., 2021; Puig & Ageitos, 2021). This is not bad in itself, as this approach may prove very valuable in the reality of Spanish classrooms. However, a more homogeneus understanding of the application of SSI into formal education would make it easier, again, to build more reliable knowledge about this phenomenon.

In addition to a clearer definition of key constructs, there should be more research on in-service teachers and SSI in Spain. A large amount of the works reviewed in this chapter analyse preservice teachers. This is not a concern to the conclusions or the three topics that are put forward in this chapter, as the studies on inservice teachers apply SSI-based instruction in courses related to their disciplinary training, as opposed to pedagogical. In this regard, the science content taught at certain levels or schools within secondary education may not be too different from that from preservice teacher education. Therefore, its results may be applied to one another. What is most concerning is the reason why SSI-based instruction is so prevalent in inservice teacher education, as opposed to other, key educational stages such as compulsory education. The question is: to what extent the amount of studies of SSI in higher education responds to a real interest in the potential of SSI to improve teaching and learning in this educational level, or whether these studies are developed with this kind of students because they are an easily accessible group for educational researchers. Indeed, the authors and the context of the studies reported are often the same. Moreover, several of the works aim to present an educational intervention and to evaluate it, which is not a common practice in educational research. In other words, for the sustainability of SSI-based instruction in Spain, more research and practices should focus on in-service teachers, seeing that they are gatekeepers in educational innovation.

Last, as it emerged from the results of topic III, most research aiming to assess or describe instances of SSI-based instruction in real settings are performed on students interacting with SSI in usually isolated, short learning experiences, mostly role-play. In a context of a gap between SSI theory and practice, even in the best case scenario that teachers accessed this evidence and put it into practice, the application of SSI to formal education would be partial and far from full-fledged. This is not bad in itself, as long as it meets the goal to improve learning in this context. However, increasing the number of studies on full-fledged SSI-based instructional experiences is essential in the journey towards bridging the gap between SSI theory and practice. Although some of the studies reviewed provide invaluable insight about specific affordances of SSI in education, for example to practice argumentation, the only way to bridge the gap between SSI theory and practice at this stage where it seems to be quite unknown is to provide arguments and ideas that encourage in-service teachers and alike to implement it. This is of crucial importance in a scenario where teachers, give little credit to what is promoted through policy, as it has been stated in the introduction of this chapter.

Looking at the future, a shift can be observed in educational curricula worldwide, not only in the definition of scientific literacy but in the very purpose of basic education. In the context of the Sustainable Development Goals (United Nations, 2017), the basic education curriculums are shifting from the concept of citizenship to include references to what can be defined as activism, i.e. a responsible way to be in a world that includes justice and prosperity. SSI research and practice should take this into account if it aims to keep making an impact in formal education.

References

Alcaraz-Dominguez, S., & Barajas, M. (2021). Conceiving socioscientific issues in STEM lessons from science education research and practice. *Education Sciences, 11*(5), 238. https://doi.org/10.3390/educsci11050238

Alcaraz-Dominguez, S., García Ruiz, A., & Barajas, M. (2018). Scientific controversies in pre-service teacher education: An Experience on the Crimea-Congo virus. *EDULEARN18 Proceedings, 1*(July), 10271–10277. https://doi.org/10.21125/edulearn.2018.2495

Alcaraz-Dominguez, S., Shwartz, Y., & Barajas, M. (2024). SSI-based instruction by secondary school teachers: What really happens in class? *International Journal of Science Education*, 1–19. https://doi.org/10.1080/09500693.2024.2303779

Amos, R., Knippels, M. C., & Levinson, R. (2020). Socio-scientific inquiry-based learning: Possibilities and challenges for teacher education. *Contemporary Trends and Issues in Science Education, 52*, 41–61. https://doi.org/10.1007/978-3-030-40229-7_4

Ariza, M. R., Christodoulou, A., van Harskamp, M., Knippels, M. C. P. J., Kyza, E. A., Levinson, R., & Agesilaou, A. (2021). Socio-scientific inquiry-based learning as a means toward environmental citizenship. *Sustainability (Switzerland), 13*(20), 1–22. https://doi.org/10.3390/su132011509

Bencze, L., Pouliot, C., Pedretti, E., Simonneaux, L., Simonneaux, J., & Zeidler, D. (2020). SAQ, SSI and STSE education: Defending and extending "science-in-context". *Cultural Studies of Science Education, 15*, 825–851. https://doi.org/10.1007/s11422-019-09962-7

Birmingham, D., & Calabrese Barton, A. (2014). Putting on a green carnival: Youth taking educated action on socioscientific issues. *Journal of Research in Science Teaching, 51*(3), 286–314. https://doi.org/10.1002/tea.21127

Borgerding, L. A., & Mulvey, B. (2022). Elementary teachers' trust in science and scientists throughout a COVID-19 SSI unit. *Journal of Science Teacher Education.* https://doi.org/10.1016/B978-0-323-60984-5.00062-7

Calavia, S., Bravo-Torija, B., & Mazas, B. (2022). Which socio-scientific dimensions do 11th graders refer to when deciding whether to be vaccinated against COVID-19? *Journal of Biological Education, 00*(00), 1–14. https://doi.org/10.1080/00219266.2022.2118354

Cebrián-Robles, D., España-Ramos, E., & Reis, P. (2021). Introducing preservice primary teachers to socioscientific activism through the analysis and discussion of videos. *International Journal of Science Education, 43*(15), 2457–2478. https://doi.org/10.1080/09500693.2021.1969060

Crujeiras-Pérez, B., Martín-Gamez, C., Díaz-Moreno, N., & Fernández-Oliveras, A. (2020). Learning argumentation through role-playing: Should we install a nuclear deposit? *Ensenanza de Las Ciencias, 38*(3), 125–142. https://doi.org/10.5565/REV/ENSCIENCIAS.2888

Crujeiras-Pérez, B., Fernández-Oliveras, A., Martín-Gámez, C., & Díaz-Moreno, N. (2021). Evolución de las concepciones sobre naturaleza de la ciencia de maestros y maestras en formación inicial. In *XI Congreso internacional en investigación en didáctica de las ciencias 2021* (pp. 1383–1386).

Cruz-Lorite, I. M., Evagorou, M., Cebrián Robles, D., & Acebal Exposito, M. del C. (2022). Preservice primary teachers' argumentative skills during their participation in a role play on nuclear energy. In *30 Encuentros Internacionales de Didáctica de las Ciencias Experimentales. La enseñanza de las ciencias en un entorno intercultural* (pp. 179–184).

Cruz-Lorite, I. M., Cebrián-Robles, D., Acebal-Expósito, M. del C., & Evagorou, M. (2023). Analysis of the informal reasoning modes of preservice primary teachers when arguing about a socio-scientific issue on nuclear power during a role play. *Sustainability (Switzerland), 15*(5). https://doi.org/10.3390/su15054291

Cuevas-Aldunate, M. E., Domènech-Casal, J., & Marbà-Tallada, A. (2021). Calidad y riqueza de las argumentaciones en relación al perfil lector del alumnado en una Controversia sobre Transgénicos. In *XI Congreso internacional en investigación en didáctica de las ciencias 2021* (pp. 1855–1858).

Díaz Moreno, N. (2019). Caracterizando controversias sociocientíficas en la prensa escrita. Una herramienta para el desarrollo de la alfabetización científica. *Revista Eureka Sobre Enseñanza y Divulgación de Las. Ciencias, 16*(1), 1102. https://doi.org/10.25267/rev_eureka_ensen_divulg_cienc.2019.v16.i1.1102

Díaz Moreno, N. C., & Jiménez-Liso, M. R. (2014). Las controversias sociocientíficas como contexto en la enseñanza de las ciencias. *XXVI Encuentro de Didáctica de Las Ciencias Experimentales*, 693–701. Retrieved from http://www.uhu.es/26edce/actas/docs/comunicaciones/orales/pdf/084.5-Diaz-Moreno.pdf

Díaz-Moreno, N., Caparrós Martín, E., & Sierra Nieto, J. (2019). Using socioscientific issues as an educational tool to develop scientific literacy. *International Journal of Educational Research and Innovation (IJERI), 12*, 261–281.

Ekborg, M., Ottander, C., Silfver, E., & Simon, S. (2013). Teachers' experience of working with socio-scientific issues: A large scale and in depth study. *Research in Science Education, 43*(2), 599–617. https://doi.org/10.1007/s11165-011-9279-5

España Naveira, P., Cruz Lorite, I. M., Cebrián Robles, D., Cabello Garrido, A., España Ramos, E., González García, F. J., & Blanco López, Á. (2023). Enfoque de cartografía de controversias para abordar cuestiones socialmente vivas desde la enseñanza de la ciencia y la tecnología. *Revista Eureka Sobre Enseñanza y Divulgación de Las Ciencias, 20*(3). https://doi.org/10.25267/rev_eureka_ensen_divulg_cienc.2023.v20.i3.3101

Esquivel-Martín, T., Pérez-Martín, J. M., & Bravo-Torija, B. (2023). Does pollution only affect human health? A scenario for argumentation in the framework of one health education. *Sustainability (Switzerland), 15*, 6984. https://doi.org/10.3390/su15086984

European Commission. (2012). Responsible research and innovation. Europe's ability to respond to societal challenges, 4. https://doi.org/10.2777/11739

Faisal, & Martin, S. N. (2022). Exploring Indonesian biology teachers' perceptions and attitudes towards socio-scientific issues-based instruction. *Asia-Pacific Science Education, 8*, 1–36. https://doi.org/10.1163/23641177-bja10042

Fernández-Oliveras, A., Martín-Gámez, C., Díaz-Moreno, N., & Crujeiras-Pérez, B. (2022). Nuclear cementery, yes or no? A proposal for initial teacher training in critical thinking | Cementerio nuclear, ¿sí o no? Una propuesta para la formación docente inicial en pensamiento crítico. *Estudios Pedagogicos, 48*(1), 11–31.

García-Carmona, A. (2021). Formando a estudiantes de profesorado de Física y Química en naturaleza de la tecnología. In *XI Congreso internacional en investigación en didáctica de las ciencias 2021* (pp. 1379–1383).

García-Carmona, A. (2022). Understanding epistemic aspects of the nature of science in Spain's new curriculum for compulsory-secondary education since the LOMLOE law. *Revista Espanola de Pedagogia, 80*(283), 433–450. https://doi.org/10.22550/REP80-3-2022-01

González-sánchez, E., Acebal Exposito, M. del C., & Brero Peinado, V. (2021). Influencia de una actividad de role-playing y dramatización sobre energía eólica marina en el proceso argumentativo del alumnado de educación secundaria. In *XI Congreso internacional en investigación en didáctica de las ciencias 2021* (pp. 1883–1886).

Hierrezuelo Osorio, J. M., & Franco Mariscal, J. A. (2021). Diseño de tareas para desarrollar pensamiento crítico a través de la argumentación y la toma de decisiones. In *XI Congreso internacional en investigación en didáctica de las ciencias 2021* (pp. 825–828).

Hierrezuelo Osorio, J. M., Franco-Mariscal, A. J., & Blanco López, Á. (2022). Use of socio-scientific dilemmas for the development of critical thinking skills in pre-service teachers. Teachers' perceptions. *Revista Interuniversitaria de Formacion Del Profesorado, 97*(36.1), 99–122. https://doi.org/10.47553/rifop.v97i36.1.92435

Jimenez-Aleixandre, M. P., & Brocos, P. (2017). Processes of negotiation in socio-scientific argumentation about vegetarianism in teacher education. In *Interpersonal argumentation in educational and professional contexts* (pp. 1–223). https://doi.org/10.1007/978-3-319-59084-4

Jiménez-Aleixandre, M. P., & Brocos, P. (2021). Gestión de conflictos entre pruebas, posiciones socialmente establecidas y acción en un dilema sobre dietas. In *XI Congreso internacional en investigación en didáctica de las ciencias 2021* (pp. 1805–1808).

Kinskey, M., & Zeidler, D. (2021). Elementary preservice teachers' challenges in designing and implementing socioscientific issues-based lessons. *Journal of Science Teacher Education, 32*(3), 350–372. https://doi.org/10.1080/1046560X.2020.1826079

Levinson, R. (2006). Towards a theoretical framework for teaching controversial socio-scientific issues. *International Journal of Science Education, 28*(10), 1201–1224. https://doi.org/10.1080/09500690600560753

López Ramos, V., Yuste Tosina, R., Urkidi Elorrieta, P., & Losada Iglesias, D. (2020). Máster de secundaria: análisis de situación en la universidad española. *Educatio. Siglo XXI, 38*(1 Marzo-Ju), 183–208. https://doi.org/10.6018/educatio.413471

López-Fernández, M. del M., González-García, F., & Franco-Mariscal, A. J. (2022). How can socio-scientific issues help develop critical thinking in chemistry education? A reflection on the problem of plastics. *Journal of Chemical Education, 99*(10), 3435–3442. https://doi.org/10.1021/acs.jchemed.2c00223

Manso, J., Matarranz, M., & Valle, J. M. (2019). Supranational and comparative study of the pre-service training teachers in the European Unión. *PRO, 23*(3), 15–33. https://doi.org/10.30827/profesorado.v23i3.9697

Newton, M. H., & Zeidler, D. L. (2020). Developing socioscientific perspective taking. *International Journal of Science Education, 42*(8), 1302–1319. https://doi.org/10.1080/09500693.2020.1756515

Pattier, D., & Olmos Rueda, P. (2021). Administration and teachers: Evidence-based educational practices. *Revista de Educacion, 2021*(392), 33–58. https://doi.org/10.4438/1988-592X-RE-2021-392-478

Portillo-Blanco, A., Díez, J. R., Barrutia, O., Garmendia, M., & Guisasola, J. (2022). Design and evaluation of an educational intervention on the COVID-19 pandemic and prevention measures. *Revista Eureka, 19*(1). https://doi.org/10.25267/REV_EUREKA_ENSEN_DIVULG_CIENC.2022.V19.I1.1302

Pozo-Muñoz, M. P., Martín-Gámez, C., Velasco-Martínez, L. C., & Tójar-Hurtado, J. C. (2023). Research and development of environmental awareness about water in primary education students through their drawings. *Education Sciences, 13*, 119. https://doi.org/10.3390/educsci13020119

Pozuelo Muñoz, J. (2022). Actitud ante el COVID-19 y su vacuna, de estudiantes de ESO y Bachillerato en España. Attitude towards COVID-19 and its vaccine, of ESO and Baccalaureate students in Spain. *REIDOCREA, 11*(1), 1–13.

Puig, B., & Ageitos, N. (2021). Practicar el pensamiento crítico en contextos relevantes como el movimiento antivacunas. In *XI Congreso internacional en investigación en didáctica de las ciencias 2021* (pp. 1793–1796).

Puig, B., Bravo Torrija, B., & Jiménez-Aleixandre, M. P. (2012). Dos unidades de argumentación sobre cuestiones socio-científicas : el determinismo biológico y la gestión de recursos. In *VII Seminario Ibérico/III Seminario Iberoamericano CTS en la enseñanza de las Ciencias "Ciencia, Tecnología y Sociedad en el futuro de la enseñanza de las ciencias"* (pp. 1–7). Retrieved from http://www.oei.es/seminarioctsm/PDF_automatico/F13textocompleto.pdf

Reverte, N., Calero, M., & Vilches, A. (2022). Evolución de las interacciones CTSA en la educación científica. In *Proceedings of the VIII SIACTS—SEMINÁRIO IBERO-AMERICANO CTS* (pp. 931–943).

Romero Ariza, M., Abril, A. M., Quesada, A., & Garcia, F. J. (2014). Bridging inquiry based learning and science education on socio scientific issues: Contributions to the Parrise European Project. *In Inted2014: 8th international technology, education and development conference, 2014* (pp. 2599–2607).

Romero Ariza, M., Abril Gallego, A. M., & Quesada Armenteros, A. (2017). Conectando Los Temas Socio-Científicos, La Naturaleza De La Ciencia Y El Pensamiento Crítico Para Hacer Frente a Los Retos Actuales En La Educación Científica. In *X Congreso internacional sobre investigación en didáctica de las ciencias* (pp. 515–520).

Ruiz González, C., Ayuso Fernández, E., & López-Banet, L. (2021). Una propuesta para mejorar las habilidades comunicativas y de argumentación del alumnado de bachillerato sobre la producción de insulina por biotecnología. In *XI Congreso internacional en investigación en didáctica de las ciencias 2021* (pp. 1947–1950).

Simonneaux, L. (2011). Questions socialement vives and socio-scientific issues: New trends of research to meet the training needs of postmodern society. In C. Bruguière, A. Tiberghien, & P. Clement (Eds.), *Topics and trends in current science education, proceedings of the 9th ESERA conference selected contributions* (pp. 37–54). Springer. https://doi.org/10.1007/978-94-007-7281-6_3

Sjöström, J., Eilks, I., & Zuin, V. G. (2016). Towards eco-reflexive science education: A critical reflection about educational implications of green chemistry. *Science and Education, 25*(3–4), 321–341. https://doi.org/10.1007/s11191-016-9818-6

Solbes, J. (2019). Cuestiones socio-científicas y pensamiento crítico: Una propuesta para cuestionar las pseudociencias. *Tecné, Episteme y Didaxis: TED, 46*, 81–99. https://doi.org/10.17227/ted.num46-10541

Solbes Matarredona, J. (2013a). Contribución de las cuestiones sociocientíficas al desarrollo del pensamiento crítico (I): Introducción. *Revista Eureka Sobre Enseñanza y Divulgación de Las Ciencias, 10*(1), 1–10. https://doi.org/10.25267/rev_eureka_ensen_divulg_cienc.2013.v10.i1.01

Solbes Matarredona, J. (2013b). Contribución de las cuestiones sociocientíficas al desarrollo del pensamiento crítico (II): Ejemplos. *Revista Eureka Sobre Enseñanza y Divulgación de Las Ciencias, 10*(2), 171–181. https://doi.org/10.25267/rev_eureka_ensen_divulg_cienc.2013.v10.i2.03

Solbes, J., & Torres, N. (2012). Análisis de las competencias de pensamiento crítico desde el aborde de las cuestiones sociocientíficas: un estudio en el ámbito universitario. *Didáctica de Las Ciencias Experimentales y Sociales, 0*(26), 247–269. https://doi.org/10.7203/dces.26.1928

Tekin, N., Aslan, O., & Yilmaz, S. (2016). Research trends on socioscientific issues: A content analysis of publications in selected science education journals. *Journal of Education and Training Studies, 4*(9), 16–24. https://doi.org/10.11114/jets.v4i9.1572

Tidemand, S., & Nielsen, J. A. (2017). The role of socioscientific issues in biology teaching: from the perspective of teachers. *International Journal of Science Education, 39*(1), 44–61. https://doi.org/10.1080/09500693.2016.1264644

United Nations. (2017). UNESCO moving forward the 2030 Agenda for Sustainable Development. *The United Nations Educational, Scientific and Cultural Organization*, 22. Retrieved from https://www.un.org/sustainabledevelopment/water-and-sanitation/

Uskola Ibarluzea, A., Burgoa, B., & Maguregi González, G. (2021). Integración del conocimiento científico y de la capacidad argumentativa en tomas de decisión sobre temas sociocientíficos. *Revista Eureka Sobre Enseñanza y Divulgación de Las Ciencias, 18*(1), 1101. https://doi.org/10.25267/Rev

Vázquez-Alonso, Á., Aponte, A., Manassero-Mas, M. A., & Montesano, M. (2016). A teaching–learning sequence on a socio-scientific issue: Analysis and evaluation of its implementation in the classroom*. *International Journal of Science Education, 38*(11), 1727–1746. https://doi.org/10.1080/09500693.2016.1212289

Verdugo Perona, J. J., Gómez Ferragud, C. B., & Pons-Vázquez, S. (2022). Propuesta de intervención para trabajar el pensamiento crítico en torno a los ODS: percepción del alumnado. In *30 Encuentros Internacionales de Didáctica de las Ciencias Experimentales. La enseñanza de las ciencias en un entorno intercultural* (pp. 445–450).

Zeidler, D. L. (2014). Socioscientific issues as a curriculum emphasis: Theory, research and practice. In N. G. Lederman & S. K. Abell (Eds.), *Handbook of research on science education vol. 2*. Routledge.

Zeidler, D. L., & Sadler, T. D. (2023). Exploring and expanding the frontiers of socioscientific issues. In N. G. Lederman, D. L. Zeidler, & J. Lederman (Eds.), *Handbook of research on science education: Volume III* (Vol. 3, pp. 899–929). Routledge. https://doi.org/10.4324/9780367855758-33

Zeidler, D. L., Sadler, T. D., Simmons, M. L., & Howes, E. V. (2005). Beyond STS: A research-based framework for socioscientific issues education. *Science Education, 89*(3), 357–377. https://doi.org/10.1002/sce.20048

Zeidler, D. L., Herman, B. C., & Sadler, T. D. (2019). New directions in socioscientific issues research. *Disciplinary and Interdisciplinary Science Education Research, 1*(11), 1–9. https://doi.org/10.1186/s43031-019-0008-7

Chapter 15
The Socioscientific Issues Framework as a Tool to Enhance Health Education Within the Life Sciences Curriculum in South Africa

Lindelani Mnguni and Velisiwe Gasa

15.1 Introduction

South Africa's standing in global science literacy rankings is notably low, as evidenced by multiple studies, including the Trends in International Mathematics and Science Study (TIMSS) (e.g., Hardman, 2022; Juan et al., 2018; Reddy, 2022). Reddy et al. (2020) report that only 36% of South African Grade 9 learners have mastered the fundamental science knowledge and skills, indicating a substantial deficiency in science education among the majority. In the 2015 iteration of TIMSS, South African Grade 9 learners achieved an average score of 358 points, positioning the country at the bottom of the list among the 39 participating nations (Juan et al., 2018). This underperformance is not isolated to science alone; it aligns with a broader trend observed in international benchmarking assessments where South Africa has consistently shown lower achievement outcomes in literacy, mathematics, and science, as demonstrated by its performance in both TIMSS and the Progress in International Reading Literacy Study (PIRLS) (Hardman, 2022). This pattern underscores the critical need for educational reforms and enhanced focus on these foundational subjects.

Relatedly, South Africa grapples with significant health challenges intricately linked to the broader issue of science literacy (Mnguni et al., 2016, 2023; Wolff & Mnguni, 2015). As reported by the Human Sciences Research Council (2023),

L. Mnguni (✉)
Centre for Health Science Education, Faculty of Health Sciences, University of the Witwatersrand, Johannesburg, South Africa
e-mail: lindelani.mnguni@wits.ac.za

V. Gasa
College of Education, University of South Africa, Pretoria, South Africa
e-mail: gasavg@unisa.ac.za

D. L. Zeidler (ed.), *A Moral Inquiry into Epistemic Insights in Science Education*, Contemporary Trends and Issues in Science Education 61,
https://doi.org/10.1007/978-3-031-63382-9_15

South Africa has the world's highest HIV/AIDS prevalence, affecting 7.8 million individuals, that is, 12.7% of its total population. A striking aspect of this epidemic is the gender disparity in HIV prevalence among adolescents, with females aged 15–19 experiencing a rate twice that of their male counterparts (5.6% vs. 3%). Recently, the country faced a severe COVID-19 crisis, recording over 4 million cases, the highest in Africa and ranking 38th globally. While numerous public health initiatives aim to combat these issues, there is a growing recognition of the need to address the underlying problem of inadequate science literacy, significantly contributing to the challenges in managing and understanding these health crises effectively.

Considering the health and science literacy challenges, this chapter explores the integration of the Socioscientific Issues (SSI) framework (Zeidler et al., 2005) into South Africa's Life Sciences curriculum, mainly focusing on its potential to augment health education through science education and promoting science literacy. The SSI Framework could be used to address health challenges as SSIs that are exacerbated by poor science literacy, pervasive stigma, misinformation, and entrenched negative cultural beliefs. It offers a pathway to deepen scientific understanding of health issues and to cultivate a more nuanced appreciation of the complex interplay between science, society, and ethical considerations. This integration is particularly pertinent given the pressing health issues faced by South Africa, including the HIV/AIDS epidemic and the COVID-19 pandemic.

The chapter also explores how adapting various facets of the curriculum—from specific aims and content to teaching approaches and assessment strategies—could align them with the SSI Framework to foster a more informed, critically engaged, and socially aware youth. This adaptation would not be merely an academic exercise but a crucial step towards empowering learners to tackle real-world health issues effectively with scientific understanding and ethical sensitivity. The chapter also addresses the challenges and opportunities inherent in this transformative educational approach, including the need for targeted resource allocation and comprehensive teacher training. Through this exploration, the chapter aims to contribute to advancing science education in South Africa, preparing learners to navigate and influence an increasingly complex world.

15.2 The SSIs (SSI) Framework

The SSI Framework (Zeidler et al., 2005) informs the development of functional scientific literacy by integrating science education with sociocultural and moral elements. This approach represents a paradigm shift from traditional, decontextualized science education, focusing instead on engaging learners with complex, real-world problems and promoting active participation in democratic decision-making processes (Simonneaux & Simonneaux, 2009; Zeidler et al., 2005). Central to the SSI Framework is the use of controversial, ill-structured social problems that necessitate evidence-based reasoning (Zeidler, 2014). This method involves dialogue, debate, and argumentation, requiring learners to employ scientific knowledge in addressing socially relevant topics. These educational strategies integrate ethical components, emphasizing the formation of virtues and character as long-term pedagogical goals (Zeidler, 2014).

'Science in Context' (SinC) is an umbrella term within the SSI Framework, aiming to promote the well-being of living things in balance with the non-living world (Bencze et al., 2020). According to Bencze et al. (2020), this approach acknowledges the importance of considering anthropocentric and ecocentric factors in socioscientific discourse. Socioscientific reasoning, a vital component of the SSI Framework, comprises complexity, inquiry, perspective-taking, skepticism, and understanding of the affordances and limitations of science. These dimensions foster learners' abilities to engage in the nuanced evaluation of SSI (Sadler et al., 2007).

SSI's role as an educational paradigm extends beyond acquiring scientific content. It includes developing an understanding of the nature of science (NOS), argumentation skills, ethical sensitivity, and reflective judgment (Zeidler & Nichols, 2009). The approach prepares learners to engage with science as a body of knowledge and a dynamic process involving ethical and moral dimensions (Newton & Zeidler, 2020; Sadler, 2004). Socioscientific perspective-taking is a multidimensional construct, blending cognitive and affective elements of moral behavior, values, reasoning, emotion, identity, and meta-moral characteristics (Kahn & Zeidler, 2017).

Recent research on SSI has emphasized real-world applications, highlighting the importance of socioscientific perspective-taking and the integration of informal and place-based contexts (Herman et al., 2019). These elements are crucial in making science more relevant and impactful for learners, connecting them more deeply with the issues they study. The role of place-based and informal learning contexts in SSI has also been increasingly recognized. Herman (2018) suggested that these contexts provide opportunities for learners to engage with SSI in settings that are directly relevant to their lives, thereby enhancing their understanding and responsiveness to these issues.

Ultimately, the SSI Framework's educational significance lies in its holistic approach to science education, blending scientific reasoning with sociocultural awareness and moral sensitivity. It challenges learners to consider science within the broader societal and ethical dimensions, fostering a more informed, empathetic, and responsible citizenry. Integrating socioscientific perspective-taking and place-based learning within the SSI Framework represents a forward-thinking approach to science education, promising to enhance learners' engagement and understanding of the complex interplay between science and society (Herman, 2018; Kahn & Zeidler, 2017).

15.3 The Urgency of the SSI Framework Health Education: COVID-19 and HIV/AIDS

Considering the SSI Framework's potential to enhance science literacy, it is essential to consider its addressing health-related SSIs affecting society. The extent to which science literacy contributes to health crises in South Africa should be explored to determine the potential role of the SSI Framework in addressing these crises and related factors. For example, research has shown that factors such as science illiteracy, misconceptions, stigma, and ill-informed sociocultural beliefs are critical in the spread of HIV/AIDS and have contributed to how society responded to the COVID-19 pandemic. It is, therefore, essential to explore how the SSI Framework could address factors and contribute to improved science and health literacy.

15.3.1 Impact of Poor Science Literacy in the Context of HIV/AIDS

The complex health landscape in South Africa, particularly concerning HIV/AIDS, can be substantially linked to inadequate science literacy among the population. This lack of understanding and interpretation of scientific information has propagated widespread misconceptions, profoundly influencing perceptions and actions regarding disease transmission and prevention (Penn & Ramnarain, 2018). For example, a common misconception is the belief that HIV can be transmitted through casual contact, leading to unnecessary fear and social ostracism of infected individuals. Individuals lacking in scientific literacy are often found to be reluctant to adopt recommended preventive measures, such as the use of condoms, or to seek timely medical intervention for testing and treatment (Kalichman & Simbayi, 2004).

Moreover, the impact of stigma, intertwined with sociocultural beliefs, significantly shapes public responses to HIV/AIDS (Visser et al., 2009). Stigma, fuelled by misinformation and ill-informed cultural myths, poses a substantial barrier to HIV testing, treatment uptake, and open disclosure of one's HIV status. For instance, in some communities, there is a persistent belief that HIV/AIDS is a punishment for moral failings or certain behaviors, leading to a stigmatization of individuals living with HIV/AIDS. This stigma not only perpetuates discrimination but also discourages people from accessing crucial health information or treatment services (Maughan-Brown, 2010).

Additionally, the rampant spread of misinformation, often disseminated through informal networks and unverified sources, exacerbates the situation. Myths regarding the causes of HIV/AIDS, such as false beliefs in supernatural origins or misinformation about potential cures, continue to circulate widely (Dickinson, 2013). This misinformation undermines public health initiatives aimed at both prevention and treatment and significantly impedes progress in controlling the spread of the disease.

Ultimately, the prevalence and management of HIV/AIDS in South Africa are critically hindered by a combination of poor science literacy, pervasive stigma, and sociocultural misconceptions. These factors collectively contribute to a challenging environment for effective public health interventions and necessitate a focused effort to improve science literacy and combat misinformation.

15.3.2 Challenges of Science Literacy and Misinformation in COVID-19

Similar to HIV/AIDS, the COVID-19 pandemic has starkly highlighted the consequences of poor science literacy in South Africa, underscoring its role in exacerbating public health challenges. Misunderstandings about the virus, including its modes of transmission and the efficacy of vaccines, are attributed mainly to a lack of basic scientific understanding among the general populace (Reddy et al., 2020). For instance, early in the pandemic, there was a prevalent misconception that COVID-19 was no more serious than the common flu. This led to underestimating the virus's severity and consequent laxity in following safety protocols (Schmidt et al., 2020). Such misunderstandings have significantly impeded effective efforts to manage the pandemic.

Moreover, the pandemic has seen stigma manifest in numerous forms, notably including discrimination against those infected with the virus and healthcare workers. This stigma, often perpetuated by misinformation, resulted in individuals avoiding testing or treatment for fear of social ostracism (Jarolimova et al., 2021). In some communities, contracting the virus was erroneously associated with personal irresponsibility, leading to heightened stigmatization. Such societal attitudes contribute to a reluctance to adhere to public health guidelines and increase vaccine hesitancy, further complicating public health responses (Schmidt et al., 2020).

The situation was further aggravated by what the World Health Organization (WHO) describes as an 'infodemic'—an overwhelming spread of misinformation (Gagliardone et al., 2021; Zarocostas, 2020). Conspiracy theories regarding the virus's origins, alongside misinformation about the safety and effectiveness of vaccines, have proliferated (Bangani, 2021). For example, false claims linking COVID-19 vaccines to microchip implantation created fear and mistrust among the population (Andrews, 2021; Steenberg et al., 2023). This deluge of misinformation created confusion and eroded public trust in health authorities and scientific experts. The resulting vaccine hesitancy and non-compliance with public health measures posed significant hurdles to controlling the spread of the virus and implementing effective containment and vaccination strategies.

In essence, the COVID-19 pandemic in South Africa has been significantly shaped by a combination of inadequate science literacy, pervasive stigma, and an 'infodemic.' This complex interplay has obstructed public health efforts, highlighting the urgent need for improved science communication and literacy alongside targeted efforts to counteract misinformation and culturally rooted stigmas.

15.3.3 Potential Challenges of Science Literacy and Misinformation for Disease X

The potential emergence of a hypothetical Disease X, predicted to be more devastating than both HIV/AIDS and COVID-19, raises significant concerns, especially when considering the role of poor science literacy and misinformation in disease management (Tahir et al., 2022). "Disease X" is a term the World Health Organization (WHO) uses to describe a hypothetical, unknown pathogen that could cause a future epidemic or pandemic. The concept was included in the WHO's 2018 R&D Blueprint, which outlines strategies for research and development in the face of potential health emergencies (World Health Organization, 2018). The inclusion of Disease X serves as an acknowledgment by health experts that unknown diseases could pose significant threats to global health and that the world needs to prepare for pathogens that have not yet been identified. The idea behind Disease X is to foster flexible preparedness and response plans that are adaptable to various types of threats rather than only preparing for known diseases. It emphasizes the importance of developing systems, such as improved science and health education, to enhance social response to the disease. The experiences with HIV/AIDS and COVID-19, therefore, provide valuable insights into how factors such as science illiteracy, misinformation, stigma, and ill-informed cultural beliefs might impact the handling of Disease X.

Firstly, science literacy, or the lack thereof, has played a critical role in the public's understanding and response to health crises, including HIV/AIDS and COVID-19. If Disease X were to emerge, a similar pattern could be expected. Without a robust understanding of scientific principles among the general populace, misconceptions about Disease X could lead to ineffective or harmful responses. For instance, if Disease X were a highly contagious virus, misunderstandings about transmission modes could result in ineffective or even counterproductive preventive measures similar to how HIV/AIDS spreads in South Africa.

Furthermore, the impact of misinformation, as seen in the 'infodemic' during the COVID-19 pandemic (Zarocostas, 2020), would likely be a significant factor in the response to Disease X. Misinformation could lead to widespread panic, resistance to medical advice, and non-compliance with public health directives. This would be compounded by the rapid spread of false information through social media and other digital platforms, as observed during the COVID-19 pandemic. Misinformation could also fuel stigma associated with Disease X, as seen with HIV/AIDS (Maughan-Brown, 2010), potentially leading to social ostracism of affected individuals and hindering efforts at disease control and treatment.

The lessons from HIV/AIDS and COVID-19 underscore the necessity of enhancing science literacy and combating misinformation in preparing for Disease X. Public health campaigns and education systems need to focus on improving the public's understanding of scientific concepts related to diseases and health. This involves providing factual information and equipping individuals with the skills to evaluate the information they encounter critically. Additionally, proactive measures to counteract misinformation, such as fact-checking initiatives and public awareness campaigns, would be crucial in managing the spread of Disease X.

15.4 Adapting Life Sciences Education to an SSI Framework

The current health challenges in South Africa, characterized by poor science literacy, pervasive stigma, misinformation, and ill-informed cultural beliefs, underscore the need for a transformative educational approach. Incorporating the SSI Framework into the Life Sciences curriculum presents a unique opportunity to address these challenges. By tailoring the Life Sciences curriculum to include SSI, there is potential to significantly enhance science literacy and the practical application of scientific knowledge, particularly within health education. This could include adapting the Life Sciences curriculum's specific aims, content, teaching approaches, and assessment strategies to the SSI Framework to contribute to a more informed, critical, and socially aware learner body equipped to tackle real-world health issues with scientific understanding and ethical sensitivity.

Adapting the specific aims of Life Sciences within the SSI Framework can significantly enhance science literacy and the application of scientific knowledge, especially within the context of health education (Mudaly, 2011). The key is to infuse the curriculum with real-world issues that require both scientific understanding and ethical consideration.

Firstly, developing knowledge of key biological concepts, processes, systems, and theories can be aligned with SSI by incorporating case studies that reflect current health challenges, such as HIV/AIDS and COVID-19. This approach contextualizes learning, making the content more relevant and engaging for learners (Zeidler et al., 2005). For example, discussing the biology of viruses in the context of COVID-19 can illuminate the science of pandemics and their impact on society. Secondly, introducing ethical dilemmas and policy debates can enhance the ability to evaluate and debate scientific issues critically. This could involve discussions on the morality of vaccine distribution and its efficacy, encouraging learners to apply their scientific knowledge to complex societal issues (Sadler, 2004). Teaching approaches should be interactive and inquiry-based, promoting active learning where learners engage in discussions, debates, and problem-solving activities related to real-world health issues. This could include analyzing the impact of misinformation on public health strategies during pandemics or exploring ethical dilemmas in medical biotechnology. Such methods align with the CAPS objectives of fostering critical thinking and evaluative skills in learners. Thirdly, understanding the ways humans impact the environment and living organisms can be explained through the exploration of public health topics. This could include studying environmental factors contributing to disease spread or the impact of human behavior on health outcomes, aligning with the SSI Framework's emphasis on the interconnectedness of science, society, and environment (Kolstø, 2001).

Lastly, adapting assessment approaches in Life Sciences to the SSI Framework involves a shift towards evaluations that assess understanding of scientific knowledge and learners' ability to apply this knowledge in real-world health issues (Mudaly, 2011). The CAPS Life Sciences curriculum already encompasses various forms of assessment, including informal, daily, and formal assessments, along with end-of-year examinations. To align these with the SSI Framework, the focus should be on integrating health-related SSIs into these assessment methods. Teachers could incorporate discussions, reflective journals, or short quizzes focusing on current health issues, their scientific basis, and socioscientific implications for informal and daily assessments. These activities can gauge learners' ongoing understanding and ability to connect scientific concepts with societal issues. Formal assessments and end-of-year examinations could include scenario-based questions or project work where learners analyze case studies on health crises, like the COVID-19 pandemic or HIV/AIDS epidemic. Learners could be assessed on their understanding of the biological aspects and their ability to evaluate the socio-economic, ethical, and environmental factors involved. For instance, an examination question could ask learners to discuss the role of misinformation in public health response to a pandemic, requiring them to apply their scientific knowledge in a socioscientific context.

15.5 Barriers to SSI Implementation

The SSI Framework holds a promise in addressing societal issues through Life Sciences Education; however, several barriers must be addressed to integrate the framework effectively into the curriculum. Firstly, there is a need to consider the gap in specialized teacher training and professional development, particularly acute in rural areas. This deficiency limits educators' ability to effectively integrate SSI into teaching, especially in addressing inclusive education and complex science topics. Secondly, teachers face considerable difficulties incorporating ethical, societal, and cultural elements into the science curriculum. Cultural norms and values, alongside apprehensions about deviating from traditional scientific content, create a challenging teaching environment. This situation often leads to neglecting broader socioscientific reasoning in favor of a narrow focus on factual scientific knowledge. Lastly, the prevalent preference for traditional scientific content over socioscientific reasoning impedes the holistic understanding of SSI. This inclination results in an educational approach that overlooks the complexity and relevance of socioscientific reasoning, limiting learners' capacity for informed decision-making and comprehensive analysis of SSIs. Addressing these barriers necessitates a multifaceted approach, encompassing targeted teacher training, cultural sensitivity, and a balanced curriculum that intertwines scientific and socioscientific content.

15.5.1 Lack of Specialized Teacher Training and Professional Development

Integrating the SSI Framework in Life Sciences education in South Africa faces significant barriers due to the lack of specialized teacher training, professional development, and support. Teachers, particularly those in rural areas, are more likely to encounter challenges such as negative perceptions about the SSI content, limited resources, and inadequate support from educational authorities (Mabasa & Singh, 2020). Nemadziva et al. (2023) report that teaching science in underdeveloped rural South Africa is a serious challenge due to a lack of necessary training and skills acquired through professional development workshops. The effectiveness of teacher training in these contexts is compromised by challenges such as geographical distance, making it difficult for teachers in rural areas to participate in SSI-related teacher development initiatives (Du Plessis & Mestry, 2019; Shikalepo, 2020).

The proficiency of South African science teachers in understanding the nature of science (NOS) is crucial for effectively integrating SSIs into classroom practices. Kurup (2014) highlights the impact of NOS understanding on teaching methodologies, arguing that teachers with a comprehensive grasp of NOS tend to implement more effective teaching strategies incorporating SSIs into the science curriculum. This understanding enables teachers to convey and contextualize scientific concepts within real-world issues, enhancing learners' critical thinking and problem-solving skills. However, Kurup (2014) also notes a significant gap in this area among South African science educators, where many teachers display a limited understanding of NOS, which restricts their ability to integrate SSIs effectively into their teaching. This gap leads to a missed opportunity in science education, where learners could develop a deeper appreciation and understanding of science related to societal, ethical, and environmental issues. Therefore, enhancing teachers' understanding of NOS through professional development initiatives is paramount for fostering an education system where SSIs are effectively integrated, ensuring learners have the knowledge and skills to navigate and understand the complex interplay between science and society.

Furthermore, the proficiency of South African science teachers' understanding of the various domains of science and their linkage to SSIs is another critical element in effective science education. Botha and Reddy (2011) indicate that in-service teachers often perceive pre-service teachers as lacking comprehensive knowledge across different science domains. This shortfall is not just in content knowledge but extends to applying this knowledge to SSIs, which are vital for teaching science in a way that connects with real-world issues (Yilmaz & Salman, 2022). In-service teachers argue that pre-service teachers' limited understanding of science domains can impede their ability to engage learners with SSIs effectively. This gap is concerning because SSIs require an understanding of how various science domains are integrated (Morris, 2014). Botha and Reddy's (2011) observations underscore the need for a more robust training framework for pre-service science teachers in South Africa. Such a framework should enhance content knowledge across different science domains and focus on the pedagogical skills necessary to link this knowledge with SSIs. Therefore, reinforcing the training of pre-service teachers in both the depth of scientific content and the breadth of its application to real-world issues is essential for advancing science education in South Africa.

Integrating SSIs into Life Sciences teaching also requires a high level of pedagogical content knowledge among teachers. Relela and Mavuru (2023) highlight that many South African Life Sciences teachers lack the necessary pedagogical content knowledge for effectively integrating SSIs into their teaching. This gap in pedagogical content knowledge becomes particularly evident when teachers address complex and often controversial topics like evolution (Abrie, 2010; Mpeta et al., 2015). Effective teaching of such issues requires a thorough understanding of the scientific content and the ability to present it in a relevant, understandable, and engaging way for learners. This involves addressing the content's societal, ethical, and personal implications, which are key components of SSIs. Relela and Mavuru (2023) suggest that many teachers struggle with this aspect due to a combination of factors, including inadequate teacher training, personal beliefs, and a lack of resources. The consequence is a missed opportunity to fully engage learners in a deep and meaningful exploration of SSIs, which are crucial for developing critical thinking and informed decision-making skills. To address this issue, there is a need for targeted professional development programs focusing on enhancing teachers' pedagogical content knowledge, specifically in the context of SSIs (Mavhunga & Rollnick, 2016). Such programs should deepen teachers' content knowledge and provide practical strategies and tools for integrating SSIs into their teaching.

Therefore, the lack of specialized teacher training and professional development presents a significant barrier to integrating the SSI framework in science education in South Africa. Addressing this barrier requires targeted professional development programs that focus on enhancing teachers' understanding of the nature of science, strengthening their pedagogical content knowledge, and equipping them with the pedagogical skills needed to integrate SSIs into their teaching effectively.

15.5.2 Challenges Teachers Face in Navigating Ethical, Societal, and Cultural Intricacies

Integrating the SSI Framework into science education in South Africa presents additional challenges when considering the ethical, societal, and cultural intricacies unique to the country. If not addressed, these challenges can hinder the effective implementation of SSIs in science classrooms.

The ethical intricacies faced by South African science teachers primarily revolve around morality, values, and decision-making in science education (Khathi et al., 2022). In a diverse society like South Africa, ethical considerations in science education often intersect with sensitive topics such as genetics, reproduction, and environmental policies. Teachers need to navigate these ethical waters carefully, ensuring that they respect the diverse viewpoints of learners while still conveying scientific principles accurately and objectively. The challenge lies in balancing the need to stimulate critical thinking about ethical issues in science without imposing personal beliefs or values on learners (Khathi et al., 2022; Resnik & Elliott, 2016). This balance is crucial to fostering an environment where learners can openly discuss and evaluate ethical aspects of science, leading to well-rounded scientific literacy (Kim & Roth, 2008).

The societal intricacies involve the broader social context in which learners are taught scientific knowledge. In South Africa, issues like economic disparities, access to education, and historical socio-political dynamics significantly shape the science education landscape (Enaifoghe & Dlamini, 2021; Levy et al., 2019; Oertli, 2023). Teachers often find themselves at the forefront of addressing these societal issues within the classroom. For instance, topics like health disparities are scientific concepts deeply intertwined with the social fabric of the learners' lives. Teachers must, therefore, approach these topics with an understanding of their learners' backgrounds and the broader societal context. This approach ensures that science education is relevant and responsive to the learners' real-world experiences, making SSIs more meaningful and engaging (King, 2016; Taconis & den Brok, 2016).

Cultural intricacies in South African science education pertain to the diverse cultural backgrounds of learners and how these backgrounds impact the teaching and learning of science. South Africa's rich cultural diversity means learners come into the classroom with varied beliefs, traditions, and perspectives, which can influence their understanding and acceptance of scientific concepts (Steenkamp, 2021). For example, in teaching evolutionary biology, teachers might encounter resistance from learners whose cultural or religious beliefs conflict with the scientific explanations of evolution (Relela & Mavuru, 2023). Navigating these cultural intricacies requires teachers to be culturally sensitive and inclusive in their teaching approaches. They need to create a classroom environment where cultural differences are acknowledged and respected and where science is taught in a way that is culturally relevant and accessible to all learners. This inclusivity is key to ensuring that learners from diverse backgrounds feel valued and can engage fully with SSIs in science education.

Therefore, the ethical, societal, and cultural intricacies present significant challenges for South African science teachers in integrating the SSI Framework into science education. Addressing these challenges requires a multifaceted approach that includes professional development for teachers, curriculum adaptations, and a broader shift towards a more inclusive and contextually relevant science education. By acknowledging and addressing these intricacies, Life Sciences education in South Africa can become a powerful tool for imparting scientific knowledge and promoting critical thinking, ethical reasoning, and cultural understanding among learners.

15.5.3 Conflicting Education Priorities: The Tendency to Prioritize Traditional Scientific Content over Socioscientific Reasoning

The integration of the SSI Framework in South African science education is also hampered by a prevailing tendency to prioritize traditional scientific content and exam-oriented teaching over socioscientific reasoning. This inclination towards a more conventional approach to science education, focusing on factual recall and standardized testing, significantly limits the scope for incorporating SSIs essential for developing critical thinking and informed decision-making in learners.

In many South African schools, the emphasis on traditional scientific content—characterized by a focus on basic concepts, theories, and facts—overshadows the integration of SSIs into the curriculum. This conventional approach often overlooks the complex, interdisciplinary nature of contemporary scientific issues that extend beyond the confines of traditional subject boundaries (Hodson, 2003; Presley et al., 2013). The result is a curriculum that is heavily content-driven, leaving little room for exploring SSIs, which require applying scientific knowledge to real-world problems, ethical considerations, and societal impacts.

The predominance of exam-oriented teaching in South African science education further exacerbates the issue. This 'teaching to the test' mentality, where educators focus primarily on preparing learners to succeed in standardized assessments, often leads to a didactic, rote memorization-based teaching style (Jensen et al., 2014; Stefanko, 2021). Such an approach is detrimental to the development of socioscientific reasoning skills, as it prioritizes acquiring factual knowledge over cultivating critical thinking, problem-solving, and decision-making abilities—all of which are integral to understanding and addressing SSIs.

The lack of focus on socioscientific reasoning in the classroom has far-reaching implications. When education is centered around traditional content and exam preparation, learners miss out on opportunities to engage with science in a manner relevant to their lives and societal issues (Jensen et al., 2014). This gap in the curriculum means that learners are often not encouraged or equipped to think critically about how science intersects with ethical, moral, and social dilemmas. As a result, there is a missed opportunity to prepare learners as informed citizens capable of engaging with complex SSIs in a nuanced and thoughtful manner.

Therefore, while the prioritization of traditional scientific content and exam-oriented teaching provides a structured approach to science education, it significantly hinders the integration of the SSI Framework in South African science education. A more balanced approach that includes traditional scientific content and socioscientific reasoning is essential for learners to become well-rounded individuals capable of navigating complex SSIs. This requires a concerted effort from educators, policymakers, and curriculum developers to reimagine and reshape how science is taught in South Africa.

15.6 Conclusion

In conclusion, adapting South Africa's Life Sciences curriculum to include the SSI Framework could mark a significant step towards addressing the complex interplay of poor science literacy, pervasive stigma, misinformation, and cultural beliefs in health education. This transformative educational approach, as detailed in this chapter, underscores the potential of SSI to enhance science literacy and facilitate the practical application of scientific knowledge in addressing real-world health challenges.

Adapting the curriculum's specific aims, content, teaching approaches, and assessment strategies to encompass SSI elements could offer a pathway to develop a more informed, critical, and socially aware learner body. By embedding real-world health issues into the curriculum, the SSI approach ensures that learning transcends traditional academic boundaries, fostering a deeper understanding of science's ethical, societal, and environmental dimensions. Integrating case studies on pandemics, genetic diseases, and public health topics into the Life Sciences curriculum enables learners to grasp the relevance and impact of science in their daily lives and communities. Moreover, the successful implementation of this approach hinges on effective resource allocation and comprehensive teacher training. Providing educators with the necessary resources and ongoing support is crucial for effectively delivering an SSI-infused curriculum. This includes access to contemporary case studies, multimedia content, and professional development opportunities that equip teachers with the skills to navigate the complexities of SSIs.

Ultimately, this chapter advocates for an educational paradigm that not only imparts scientific knowledge but also prepares learners to engage with science's ethical and societal implications. By embracing the SSI Framework, the Life Sciences curriculum in South Africa can produce a generation of learners who are not just scientifically literate but also capable of making informed, responsible decisions in an increasingly complex world. The holistic approach outlined herein promises to elevate the standard of science education, enabling learners to confront and address the multifaceted health challenges of the twenty-first century with scientific understanding and ethical sensitivity.

References

Abrie, A. L. (2010). Student teachers' attitudes towards and willingness to teach evolution in a changing South African environment. *Journal of Biological Education, 44*(3), 102–107. https://doi.org/10.1080/00219266.2010.9656205

Andrews, E. A. (2021). *Combating COVID-19 vaccine conspiracy theories: Debunking misinformation about vaccines, bill gates, 5G, and microchips using enhanced correctives* (Doctoral dissertation).

Bangani, S. (2021). The fake news wave: Academic libraries' battle against misinformation during COVID-19. *Journal of Academic Librarianship, 47*(5), 102390. https://doi.org/10.1016/j.acalib.2021.102390

Bencze, L., Pouliot, C., Pedretti, E., Simonneaux, L., Simonneaux, J., & Zeidler, D. (2020). SAQ, SSI, and STSE education: Defending and extending "science-in-context". *Cultural Studies of Science Education, 15*, 825–851.

Botha, M. L., & Reddy, C. P. S. (2011). In-service teachers' perspectives of pre-service teachers' knowledge domains in science. *South African Journal of Education, 31*(2), 257–274. https://doi.org/10.15700/saje.v31n2a354

Dickinson, D. (2013). Myths or theories? Alternative beliefs about HIV and AIDS in South African working class communities. *African Journal of AIDS Research, 12*(3), 121–130. https://doi.org/10.2989/16085906.2013.863212

Du Plessis, P., & Mestry, R. (2019). Teachers for rural schools-a challenge for South Africa. *South African Journal of Education, 39*(1), 1–9.

Enaifoghe, A., & Dlamini, N. P. (2021). South African political landscape and the quest for youth inclusion: Opportunity for socio-political and economic development. *African Journal of Public Affairs, 12*(1), 207–226.

Gagliardone, I., Diepeveen, S., Findlay, K., Olaniran, S., Pohjonen, M., & Tallam, E. (2021). Demystifying the COVID-19 infodemic: Conspiracies, context, and the agency of users. *Social Media + Society, 7*(3), 205630512110442. https://doi.org/10.1177/20563051211044233

Hardman, J. (2022). A cultural-historical pedagogical model as a potential developmental tool in schools. *European Journal of Education and Pedagogy, 3*(3), 191–198. https://doi.org/10.24018/ejedu.2022.3.3.354

Herman, B. C. (2018). Students' environmental NOS views, compassion, intent, and action: Impact of place-based socioscientific issues instruction. *Journal of Research in Science Teaching, 55*(4), 600–638. https://doi.org/10.1002/tea.21433

Herman, B. C., Owens, D. C., Oertli, R. T., Zangori, L. A., & Newton, M. H. (2019). Exploring the complexity of students' scientific explanations and associated nature of science views within a place-based socioscientific issue context. *Science & Education, 28*(3–5), 329–366. https://doi.org/10.1007/s11191-019-00034-4

Hodson, D. (2003). Time for action: Science education for an alternative future. *International Journal of Science Education, 25*(6), 645–670. https://doi.org/10.1080/09500690305021

Human Sciences Research Council. (2023). New HIV survey highlights progress and ongoing disparities in South Africa's HIV epidemic. https://hsrc.ac.za/press-releases/hsc/new-hiv-survey-highlights-progress-and-ongoing-disparities-in-south-africas-hiv-epidemic/

Jarolimova, J., Yan, J., Govere, S., Ngobese, N., Shazi, Z. M., Khumalo, A. R., Bunda, B. A., Wara, N. J., Zionts, D., Thulare, H., Parker, R. A., Bogart, L. M., & Bassett, I. V. (2021). Medical mistrust and stigma associated with COVID-19 among people living with HIV in South Africa. *AIDS and Behavior, 25*(12), 3967–3977. https://doi.org/10.1007/s10461-021-03307-8

Jensen, J. L., Mcdaniel, M. A., Woodard, S. M., & Kummer, T. A. (2014). Teaching to the test… or testing to teach: Exams requiring higher-order thinking skills encourage greater conceptual understanding. *Educational Psychology Review, 26*, 307–329.

Juan, A., Hannan, S., & Namome, C. (2018). I believe I can do science: Self-efficacy and science achievement of Grade 9 students in South Africa. *South African Journal of Science, 114*(7/8). https://doi.org/10.17159/sajs.2018/20170269

Kahn, S., & Zeidler, D. L. (2017). Using our heads and HARTSS*: Developing perspective-taking skills for socioscientific reasoning (*Humanities, ARTs, and Social Sciences). *Journal of Science Teacher Education, 27*(3), 261–281. https://doi.org/10.1007/s10972-016-9458-3

Kalichman, S. C., & Simbayi, L. (2004). Traditional beliefs about the cause of AIDS and AIDS-related stigma in South Africa. *AIDS Care, 16*(5), 572–580. https://doi.org/10.1080/09540120410001716360

Khathi, J. L., Ajani, O. A., & Govender, S. (2022). Exploring teachers' lived experiences on the integration of values education in South African high schools. *Research in Social Sciences and Technology, 7*(2), 108–128.

Kim, M., & Roth, W.-M. (2008). Rethinking the ethics of scientific knowledge: A case study of teaching the environment in science classrooms. *Asia Pacific Education Review, 9*(4), 516–528. https://doi.org/10.1007/bf03025667

King, D. (2016). Teaching and learning in context-based science classes: A dialectical socio-cultural approach. In *Teachers creating context-based learning environments in science* (pp. 71–85). Brill.

Kolstø, S. D. (2001). Scientific literacy for citizenship: Tools for dealing with the science dimension of controversial socioscientific issues. *Science Education, 85*(3), 291–310. https://doi.org/10.1002/sce.1011

Kurup, R. (2014). The relationship between science teachers' understandings of the nature of science and their classroom practices. *African Journal of Research in Mathematics Science and Technology Education, 18*(1), 52–62. https://doi.org/10.1080/10288457.2014.889791

Levy, B., Cameron, R., Hoadley, U., & Naidoo, V. (2019). Political transformation and education sector performance in South Africa. In *The politics of education in developing countries: From schooling to learning*. Oxford Scholarship Online.

Mabasa, T., & Singh, S. (2020). Decolonising continuing teacher professional development in the teaching of physical science through improvisation in rural areas. *South African Journal of Higher Education, 34*(3). https://doi.org/10.20853/34-3-3455

Maughan-Brown, B. (2010). Stigma rises despite antiretroviral roll-out: A longitudinal analysis in South Africa. *Social Science & Medicine (1982), 70*(3), 368–374. https://doi.org/10.1016/j.socscimed.2009.09.041

Mavhunga, E., & Rollnick, M. (2016). Teacher-or learner-centred? Science teacher beliefs related to topic specific pedagogical content knowledge: A South African case study. *Research in Science Education, 46*, 831–855.

Mnguni, L., Abrie, M., & Ebersohn, L. (2016). The relationship between scientific knowledge and behaviour: An HIV/AIDS case. *Journal of Biological Education, 50*(2), 147–159. https://doi.org/10.1080/00219266.2015.1007888

Mnguni, L., Ramulumo, M., Sari, I. J., & Ahmad Zaky El Islami, R. (2023). The relationship between the covid-19 vaccination status and the reliance on different sources of information among pre-service science teachers from south African and Indonesian universities. *African Journal of Research in Mathematics Science and Technology Education*, 1–15. https://doi.org/10.1080/18117295.2023.2275859

Morris, H. (2014). Socioscientific issues and multidisciplinarity in school science textbooks. *International Journal of Science Education, 36*(7), 1137–1158. https://doi.org/10.1080/09500693.2013.848493

Mpeta, M., de Villiers, J. J. R., & Fraser, W. J. (2015). Secondary school learners' response to the teaching of evolution in Limpopo province, South Africa. *Journal of Biological Education, 49*(2), 150–164. https://doi.org/10.1080/00219266.2014.914555

Mudaly, R. (2011). Risking it: Entering uneven socioscientific spaces in a Life Sciences classroom. *African Journal of Research in Mathematics, 15*(3), 27–40.

Nemadziva, B., Sexton, S., & Cole, C. (2023). Science communication: The link to enable enquiry-based learning in under-resourced schools. *South African Journal of Science, 119*(1/2). https://doi.org/10.17159/sajs.2023/12819

Newton, M. H., & Zeidler, D. L. (2020). Developing socioscientific perspective taking. *International Journal of Science Education, 42*(8), 1302–1319. https://doi.org/10.1080/09500693.2020.1756515

Oertli, R. T. (2023). *Preservice teachers' sociopolitical consciousness in light of the racial disparities highlighted by the COVID-19 pandemic*. University of Missouri Libraries.

Penn, M., & Ramnarain, U. (2018). The effects of scientific literacy on high school science learners' attitudes towards socioscientific issues: The case of Genetically Modified Organisms. https://hdl.handle.net/10210/276497

Presley, M. L., Sickel, A. J., Muslu, N., Merle-Johnson, D., Witzig, S. B., Izci, K., & Sadler, T. D. (2013). A framework for socioscientific issues-based education. *Science Educator, 22*(1), 26–32.

Reddy, V. (2022). *The Gauteng province TIMSS 2019 Grade 9 results: Building achievement and bridging achievement gaps*. HSRC Press.

Reddy, S. P., Sewpaul, R., Mabaso, M., Parker, S., Naidoo, I., Jooste, S., Mokhele, T., Sifunda, S., & Zuma, K. (2020). South Africans' understanding of and response to the COVID-19 outbreak: An online survey. *South African Medical Journal, 110*(9), 894–902.

Relela, M. E., & Mavuru, L. (2023). Life sciences teachers' pedagogical content knowledge when addressing socio scientific issues in the topic evolution. *International Journal of Learning, Teaching and Educational Research, 22*(5), 302–318.

Resnik, D. B., & Elliott, K. C. (2016). The ethical challenges of socially responsible science. *Accountability in Research, 23*(1), 31–46. https://doi.org/10.1080/08989621.2014.1002608

Sadler, T. D. (2004). Moral and ethical dimensions of socioscientific decision-making as integral components of scientific literacy. *Science Educator, 13*, 39.

Sadler, T. D., Barab, S. A., & Scott, B. (2007). What do students gain by engaging in socioscientific inquiry? *Research in Science Education, 37*(4), 371–391. https://doi.org/10.1007/s11165-006-9030-9

Schmidt, T., Cloete, A., Davids, A., Makola, L., Zondi, N., & Jantjies, M. (2020). Myths, misconceptions, othering and stigmatizing responses to COVID-19 in South Africa: A rapid qualitative assessment. *PLoS One, 15*(12), e0244420.

Shikalepo, E. E. (2020). Challenges facing teaching at rural schools: A review of related literature. *International Journal of Research and Innovation in Social Science (IJRISS), 4*(5), 211–218.

Simonneaux, L., & Simonneaux, J. (2009). Students' socioscientific reasoning on controversies from the viewpoint of education for sustainable development. *Cultural Studies of Science Education, 4*, 657–687.

Steenberg, B., Sokani, A., Myburgh, N., Mutevedzi, P., & Madhi, S. A. (2023). COVID-19 vaccination rollout: Aspects of hesitancy in South Africa. *Vaccine, 11*(2). https://doi.org/10.3390/vaccines11020407

Steenkamp, U. (2021). Cultural diversity and its influence on role players in a Full-Service School in Soshanguve: A wellness perspective. *European Journal of Education, 4*(1), 23–39. https://doi.org/10.26417/790dun39d

Stefanko, N. (2021). Negative Impacts of Mandatory Standardized Testing on Teachers and Students. Culminating Experience Projects. 71. Retrieved February 20, 2024 from https://scholarworks.gvsu.edu/gradprojects/71

Taconis, R., & den Brok, P. (Eds.). (2016). *Teachers creating context-based learning environments in science*. Springer.

Tahir, M. J., Sawal, I., Essar, M. Y., Jabbar, A., Ullah, I., & Ahmed, A. (2022). Disease X: A hidden but inevitable creeping danger. *Infection Control & Hospital Epidemiology, 43*(11), 1758–1759.

Visser, M. J., Makin, J. D., Vandormael, A., Sikkema, K. J., & Forsyth, B. W. C. (2009). HIV/AIDS stigma in a South African community. *AIDS Care, 21*(2), 197–206. https://doi.org/10.1080/09540120801932157

Wolff, E., & Mnguni, L. (2015). The integration of HIV and AIDS as a socioscientific issue in the life sciences curriculum. *African Journal of Research in Mathematics, 19*(3), 213–224.

World Health Organization. (2018). *2018 Annual review of diseases prioritized under the Research and Development Blueprint*. http://www.who.int/csr/research-and-development/en

Yilmaz, A., & Salman, M. (2022). Investigation of the relationship between pre-service teachers' critical thinking dispositions and attitudes towards socioscientific issues/Öğretmen Adaylarının Eleştirel Düşünme Eğilimleri ile Sosyobilimsel. *E-Uluslararası Eğitim Araştırmaları Dergisi, 13*, 203–219.

Zarocostas, J. (2020). How to fight an infodemic. *The Lancet, 395*(10225), 676.

Zeidler, D. L. (2014). Socioscientific issues as a curriculum emphasis: Theory, research, and practice. In *Handbook of Research on Science Education, II* (pp. 697–726). Routledge/Taylor & Francis Group.

Zeidler, D. L., & Nichols, B. H. (2009). Socioscientific issues: Theory and practice. *Journal of Elementary Science Education, 21*(2), 49–58. https://doi.org/10.1007/bf03173684

Zeidler, D. L., Sadler, T. D., Simmons, M. L., & Howes, E. V. (2005). Beyond STS: A research-based framework for socioscientific issues education. *Science Education, 89*(3), 357–377. https://doi.org/10.1002/sce.20048

Chapter 16
Evolution and Future Directions of SSI Education in South Korea: A 2-Decade Review and Beyond

Hyunju Lee, Hyunok Lee, and Yeonjoo Ko

16.1 Introduction

The term "socioscientific issues" (SSI) was barely known until mid-2000s. When Zeidler et al. (2005) proposed an SSI education framework, some researchers started to use the term. However, the science, technology, and society (STS) education movement, which emerged in the United States and Europe in the 1980s, began to be introduced from the late 1980s. In the Korean Fifth National Science Curriculum in 1987, one of the key objectives was "to make students recognize the impact of science on technological development and societal progress." (Ministry of Education, 1987, p. 46). Since then, the recognition of the relationship between science, technology, and society has continued to be a goal of every Korean science curriculum reform, and considerable efforts were made to develop STS education programs for elementary, middle, and high school students. Science textbooks also included various examples demonstrating the relationships between science, technology, and society. However, these examples were mostly limited to explaining how scientific knowledge can be applied to natural phenomena or problems in daily life (e.g., kitchen science, seat belts and safety, various types of renewable energy, air and water quality, and land reclamation). Controversial topics involving SSIs were rarely presented.

H. Lee (✉)
Ewha Womans University, Seoul, South Korea

H. Lee
Kangwon National University, Chuncheon-si, South Korea

Y. Ko
Jeju National University, Jeju-si, South Korea

D. L. Zeidler (ed.), *A Moral Inquiry into Epistemic Insights in Science Education*, Contemporary Trends and Issues in Science Education 61,
https://doi.org/10.1007/978-3-031-63382-9_16

Since the mid-2000s, science educators and researchers have begun to focus more on controversial issues, resulting in various SSI studies having been published. Not only has the amount of SSI research increased, but the scope of SSI research has also started to diversify. The distinctive feature here has been that the trends and major emphases of SSI research have also evolved over time with the significant changes in the national curriculum and educational policies because school science in South Korea has been greatly driven by them.

Therefore, this chapter begins by delving into the Korean national science curriculum reforms and education policies which have been connected with SSI research. We then investigate SSI research conducted in Korea over the last 2 decades to understand how SSI studies have been positioned within the Korean academic journal publications. Finally, we explore trends and future directions for SSI research in the context of science education policy and curriculum in South Korea with the aim of providing insights and implications into SSI studies beyond national boundaries.

16.2 Science Curriculum Reforms, Educational Policies, and SSI Research and Education

Science curriculum reform in Korea occurs approximately every 5–8 years. With each iteration of science curriculum reform, there has been an increasing emphasis on students' awareness of the interrelationship between science, technology, and society. It has been also clearly stated that awareness of this interrelationship is important for achieving scientific literacy. Since the National Science Curriculum of 2007 ([NSC 2007], Ministry of Education, Science and Technology, 2007), SSIs have been introduced more prominently in ways that go beyond STS. In subsequent science curricula in 2015 and 2022, SSI content was incorporated more concretely into the curriculum, and SSI research has also expanded widely. In this chapter, we aim to introduce the characteristics of each science curriculum reform and the major educational policies associated with SSI education.

16.2.1 The National Science Curriculum of 2007

Although the field of STS studies had been dealing in depth with the intrinsic nature and potential risks of science and technology and also emphasizing the importance of public awareness of these matters over the previous 2–3 decades, science educators had barely addressed them into the curriculum. When STS was dealt with, the emphasis was more on linking science to everyday life and enhancing students' interest in science learning. However, from the NSC 2007, SSIs began to be explicitly included in the curriculum:

> When selecting content for the science curriculum, it is essential to actively include materials that can connect science concepts with students' real-life experiences to enhance their interest and engagement in science. That means, when choosing content for each science domain, it's vital to use everyday examples, current controversial topics, or areas from other curricula actively to increase awareness of how science is closely related to students' lives. Through science education, students must be accurately made aware of the relationship between science, technology, and society, enabling them to rationally solve social problems related to science and technology in the future. (Ministry of Education, Science, and Technology [MEST], 2007, p. 1)

The NSC 2007 states that it is crucial for students to develop an interest in societal problems related to science and technology advancement and to gain the awareness of the interrelationships between science, technology, and society through SSIs. During this period, SSI education was aligned with the perspective of Vision II scientific literacy (Roberts, 2007). As shown above, it primarily emphasized that individual students should take an interest in SSIs and feel that science is relevant to their lives, and focused on individual moral reasoning or informed decision-making on SSIs. Since SSI education was introduced not long ago, the scope of science education involving SSIs was limited. Most SSI studies focused on analyzing patterns of moral reasoning of individuals, examining values expressed in their reasoning, and identifying factors influencing moral reasoning.

16.2.2 Policies for Character Education: The Basic Plan for Creative and Character Education and the Character Education Promotion Act

In Korea in the late 2000s, the need for character education was raised, and policies and laws were established to officially implement character education in schools. Character education was defined as education that fosters moral values and social responsibility in individuals, enabling them to grow into healthy members of society. The MEST,[1] therefore, announced the Basic Plan for Creative and Character Education in 2010 as part of a policy to simultaneously develop students' creativity and character through education. This plan aimed to support students in acquiring the core competencies needed as society continued to change in the future, focusing on creativity and character at the heart of education across all areas, including the school curriculum, teacher education, and school culture. The plan proposed the development of a curriculum that enhances students' creative thinking and problem-solving skills and provides systematic integration of character education into the subject matter. It also emphasized the establishment of a school culture that allows students to think freely and engage in creative activities and the utilization of various resources inside and outside the school to practice character education. Along these same lines, in 2015 the Character Education Promotion Act was enacted to

[1] The department responsible for education, the Ministry of Education (MOE), was renamed the Ministry of Education, Science, and Technology (MEST) in 2008, but was again given its original designation of the MOE in 2013.

establish a legal foundation for systematically implementing character education (Character Education Promotion Act, 2015). This law aimed to strengthen character education across schools, families, and society at large, officially recognizing the importance of character education.

Since the government encouraged linking character education with each subject, science educators began to explore the ways to connect science education with character development. Some researchers have attempted to address stories of scientists in history as a means to demonstrate their academic integrity, perseverance, and diligence. Efforts have also been made to integrate character education into teaching the nature of science. SSI education had also gained attention as an instructional approach for character development because SSIs could enable students to delve into the social and ethical dimensions of science and technology and to engage in the process of achieving consensus on these controversial issues. It was argued that students could cultivate character and values as citizens (e.g., ecological worldview, social and moral compassion, and socioscientific accountability) through SSI education (Lee et al., 2012; Lee et al., 2013). Schools attempted to develop various programs and materials using SSIs to improve the quality of character education. This indicated that the announcement of the Basic Plan for Creative and Character Education and the Character Education Promotion Act had become a significant catalyst in extending the educational impact of SSIs to character education in science education.

16.2.3 The National Science Curriculum of 2015

The National Science Curriculum of 2015 ([NSC 2015], MOE, 2015) was influenced by the OECD's DeSeCo project (Rychen & Salganik, 2003) and the Next Generation Science Standards ([NGSS], NGSS Lead States, 2013) in United States, bringing to the forefront the importance of competencies and understanding of core ideas. The MOE, in the NSC 2015, defined and explicitly stated the five core competencies to be cultivated through science learning: scientific thinking, scientific inquiry, scientific problem-solving, scientific communication, and the ability to participate in science and engage in lifelong learning.

One of the most noteworthy changes in the NSC 2015 was the gradual expansion of the concept of scientific literacy from Vision II to Vision III. The key features of this transition were as follows: First, among the five core competencies emphasized in the NSC 2015, "scientific engagement and lifelong learning" was newly introduced in South Korea. This refers to:

> the capacity to act rationally and responsibly as a member of the community on social issues related to science and technology, to participate in the decision-making process, and to continuously learn on one's own to adapt to new scientific and technological environments. (MOE, 2015, p. 4)

This officially acknowledges the importance of science education in ensuring that citizens, throughout their lives, not only develop an interest in new scientific discoveries and SSI but also acquire the competence to actively participate in problem solving related to these issues.

Second, in response to the societal and national demands for fostering interdisciplinary talents, the NSC 2015 adopted a thematic approach that integrate physics, chemistry, biology, and earth science. Thus, new units were introduced, for example, The Journey of Water and Energy and Life for elementary schools and Disasters and Safety and Science, Technology, and Human Civilization for middle schools. In such units, SSIs naturally emerged and could be effectively used as material for various learning activities (e.g., case studies, discussions and debates) under the national curriculum.

Last, similar to the NGSS, the NSC 2015 emphasized eight scientific practices. Among these, "evidence-based argumentation" and "communication" were inquiry skills that had not been mentioned in previous national curricula. To implement the new curriculum, science educators and teachers sought various instructional methods to promote students' argumentation and communication skills, and SSIs gained attention as valuable resources for this purpose. Consequently, SSI research published around this period tended to focus on investigating the impact of SSI education programs on enhancing core competencies and scientific practices.

16.2.4 Korean Science Education Standards for the Next Generation and the National Science Curriculum of 2022

Around 2020, in response to climate change and environmental disasters, the MOE (2022) recommended incorporating values such as respect for life, sustainability, and ecological sensitivity into the educational objectives across various subjects with the aim of promoting the harmonious coexistence of humans and the environment, ultimately aiming for a sustainable life. The MOE also stressed the significance of integrating democratic civic education and character education into all subjects. This implies the need to foster students' awareness of their responsibilities in their communities and facilitate collaborative efforts to tackle community issues through critical and reflective thinking. To explicitly implement these ideas, two significant science education documents were published: The Korean Science Education Standards for the Next Generation (KSES) and the National Science Curriculum of 2022 ([NSC 2022], MOE, 2022). These two documents heightened the interest of science educators in Vision III of scientific literacy by explicitly emphasizing "participation and action."

The KSES was jointly announced in 2019 by the MOE and the Korea Foundation for the Advancement of Science and Creativity. In the KSES, the scientific literacy that the next generation should possess was defined as "the attitudes and abilities as democratic citizens to participate in and act for solving personal as well as social problems using science-related competencies and knowledge" (Song et al., 2018, p. 73). The KSES visualized the three dimensions of scientific literacy through a "tree of scientific literacy," with each dimension representing the competences dimension, knowledge dimension, and participation and action dimension, thus signifying that scientific literacy is complete when competences, knowledge, and participation and action are integrated. The participation and action dimension emphasized not only scientific and engineering practices in schools, similar to the NGSS in the United States, but also the enjoyment of scientific culture and engagement in community activities aimed at building a safe and sustainable society as citizens.

The NSC 2022 is based not only on KSES but also on the OECD Future of Education and Skills 2030: Learning Compass 2030 published in 2019 (OECD, 2019). While similar in focus to the KSES, the Learning Compass 2030 sets well-being as the future educational direction and more explicitly emphasizes the need to possess "transformative competencies" and "student agency/co-agency" that go beyond "core foundational competencies." The Learning Compass 2030 defines transformative competencies and student agency/co-agency as follows.

> To meet the challenges of the 21st century, students need to be empowered and feel that they can help shape a world where well-being and sustainability—for themselves, for others and for the planet—are achievable. The OECD Learning Compass 2030 identifies three "transformative competencies" that students need in order to contribute to and thrive in our world, and shape a better future: creating new value, reconciling tensions and dilemmas, and taking responsibility. (p. 13)

> Student agency is defined as the belief that students have the will and the ability to positively influence their own lives and the world around them as well as the capacity to set a goal, reflect and act responsibly to effect change. Student agency relates to the development of an identity and a sense of belonging. When students develop agency, they rely on motivation, hope, self-efficacy and a growth mindset (the understanding that abilities and intelligence can be developed) to navigate towards well-being. This enables them to act with a sense of purpose, which guides them to flourish and thrive in society. Students learn, grow and exercise their agency in social contexts and this is why co-agency is also crucial. Students develops co-agency in an interactive, mutually supportive and enriching relationship with their peers, teachers, parents and communities in an organic way in a larger learning eco-system. (p. 13)

Accordingly, in the NSC 2022, in response to the rapid changes and increasing uncertainties in science and technology, the goal of science education was set to foster students' talents in actively solving problems arising from these changes and uncertainties by having interest and curiosity about natural phenomena and everyday life, understanding phenomena around them through scientific inquiry, and participating and acting as democratic citizens in scientifically and creatively solving personal and social problems. Additionally, for elementary and middle schools, Science and Society has been introduced as a new section of content standards, connecting some areas of science content to climate change, infectious diseases, and career paths, thereby addressing future interdisciplinary problems. Moreover, acknowledging the rapid changes in climate and environmental ecology as an immediate challenge and an international issue, a new high school elective course, Climate Change and Environmental Ecology, has been established. This means that compared to the previous national science curriculum, the emphasis on SSI has become more explicit, and the approach to SSI also more reflects Vision III scientific literacy.

16.3 Research Trends of SSI Education Over 2 Decades

Over the past 2 decades, a diverse range of SSI research has been undertaken in South Korea. Initially, the focus had primarily been on examining the moral reasoning of individual students regarding SSIs. However, in the 2020s, research on SSIs began to emerge with the goal of fostering citizens to be scientifically literate, with an emphasis on collaborative action and co-agency for SSI resolution, as outlined in Vision III (Sjöström & Eilks, 2018; Valladares, 2021). These changes were influenced not only by international SSI research but also by the significant impact of the Korean science curriculum reforms and key policies that were discussed in the previous section.

To illustrate the trends of SSI education in Korea, we have collected journal articles listed in the Korean Citation Index and international articles authored by Koreans that were related to SSI education. Our search strategy incorporated terms such as "socioscientific issues," "science+social+education+issues," and "science+social+education+problems" to ensure a comprehensive inclusion of SSI-related papers. Subsequently, we conducted content analysis to impartially and effectively identify educational research trends, adhering to methodologies proposed by Krippendorff (2004). This began with an initial screening of abstracts and research questions, followed by an analysis of results to distill one to two key themes from each paper. This process facilitated the identification of prevalent research themes across the literature. Moreover, it allowed us to examine the evolution of SSI education research over the years and to determine whether specific events, such as educational reforms, have influenced research directions. We segmented the analysis period into three phases—mid-2000s to early 2010s, early 2010s to late 2010s, and late 2010s to the present—to investigate shifts in research emphasis and the impact of educational and societal changes on visions of scientific literacy. This approach led us to address key questions: (a) What are the primary research themes pursued by Korean SSI educators and researchers? (b) How have these themes evolved over time? and (c) Is there a correlation between educational reforms and the focus of SSI studies?

16.3.1 From STS to SSI Focusing on Moral Engagement (Mid-2000s to Early 2010s)

Although the term "SSI" was introduced around the mid-2000s in South Korea, STS education had been already implemented in various domains. With each iteration of science curriculum reform in South Korea, there has been an increasing emphasis on students' awareness of the interrelationship between science, technology, and society. The curricula have also clearly stated that the awareness of this interrelationship is important for achieving scientific literacy. To achieve these goals, science educators have tried to integrate STS content into science textbooks. Nonetheless, its full integration into the educational curriculum remains incomplete, underscoring a noticeable gap between its theoretical aspirations and practical implementation.

Around mid-2000s, the study of SSIs emerged, enriching the STS educational paradigm by incorporating discussions around the social, ethical, and moral issues associated with scientific research. Though both STS and SSI education have been recognized as innovative approaches within science education, SSI education places greater emphasis on students' moral and ethical development (Zeidler et al., 2005). It signifies a shift in educational focus, encompassing scientific and technological aspects and considering broader ethical implications.

The early studies on SSI were mainly exploratory, focusing on the characteristics of moral reasoning related to SSIs. Although the number of SSI studies published in Korean journals during this period was generally low, these studies delved into different aspects of an individual's moral reasoning, such as exploring patterns or types of moral reasoning, examining the decision-making process regarding specific SSIs and the values that emerge as significant in that process, and investigating factors that influence an individual's SSI moral reasoning, such as personal beliefs, scientific knowledge, contextual factors, and understanding of the nature of science and technology. Additionally, content analyses of SSIs in science curricula and textbooks (Chung et al., 2010; Lee & Chang, 2007), studies investigating teachers' perceptions of SSI teaching (Lee, 2008; Lee et al., 2006; Yoo et al., 2011), and explorations of instructional approaches to SSIs (Cho & Shim, 2009; Shim & Cho, 2009) were performed.

16.3.1.1 Patterns of Moral Reasoning

The primary research topic of this period was the moral reasoning individuals engage in when faced with SSI scenarios. Some studies examined patterns of moral reasoning across various SSI scenarios, utilizing the three types of informal reasoning introduced by Sadler and Zeidler (2005)—rationalistic reasoning, emotive reasoning, and intuitive reasoning—as a conceptual framework. For example, Chang and Chung (2009) presented six hypothetical scenarios on gene therapy and human cloning to 20 college students, exploring their decision-making processes. The results showed that, similar to Sadler and Zeidler (2005), the students primarily exhibited rationalistic reasoning, thinking cognitively and objectively (93%), and showed emotive reasoning (49%) and intuitive reasoning (27%). They also used two or three types of reasoning patterns together. Lee (2008) explored the decision-making processes of pre-service teachers on eight SSIs, analyzing the characteristics of their moral reasoning. For instance, in the topics of designer babies and surrogacy, the teachers demonstrated both rationalistic reasoning and emotive reasoning, but compared to their reasoning on the other six SSIs, they engaged relatively more in emotive reasoning. They often decided their positions based on feelings of care towards those harmed by medical technologies. In contrast, they tended to base their positions on more objective evidence regarding the issues of euthanasia and GMOs, striving to clarify their stance for or against these issues. They also showed a tendency to make their decisions logically and realistically while considering opposing viewpoints.

Another area of research involved exploring individual reactions when encountering SSI scenarios. Some previous studies (Connell et al., 1999; Dreyfus & Roth, 1991) had reported that students often exhibited action paralysis, where they were skeptical about the effectiveness of their actions or felt they could do little about solving the issue being discussed, or they adopted an ambivalent stance, unable to decide their position clearly. For instance, Chang and Lee (2010) identified college students' tendencies when making decisions about SSIs, such as bringing in personal values without further engagement, being confused by incompatible values, being overwhelmed by too many aspects to consider, and trying to be detached from the issues.

16.3.1.2 Values Represented in the Evidence of SSI Reasoning

Studies exploring the evidence used in decision-making processes about SSIs have primarily investigated how individuals constructed evidence for their positions and prioritized values in decisions and explored to what extent they perceived multiple perspectives surrounding SSIs. For example, Ju and Lee (2013) examined the decision-making processes of middle school students on three SSIs—designer babies, prenatal diagnoses, and embryonic stem cell research—and their value application to the issues. The most conflicting values revealed in the evidence on the issues were the therapeutic value achievable through medical technology advancement and the value of human dignity. While acknowledging the importance of treating and preventing incurable and rare diseases, some students argued that life should inherently be valued and not be used as a means for research or other purposes. Some students based their decisions on considering the right to survival of animals or economic burden relief for families with disabilities. Lee and Chang (2012) examined the decision-making process of pre-service science teachers on the nuclear power generation issue. Most teachers supporting nuclear power generation considered economic efficiency most important, mentioning the ability to produce electricity relatively cheaply and meeting Korea's electricity demand. Teachers opposing nuclear power generation primarily mentioned the risks associated with nuclear accidents or the disposal process of nuclear waste. Such studies led to further exploration of how evidence and arguments of individuals are qualitatively enhanced after implementing SSI programs.

16.3.1.3 Factors Influencing Moral Reasoning

In SSI decision-making, various factors influence the process. Personal values, core beliefs, and prior experiences in particular can have a large impact on decision making (Sadler, 2004; Sadler & Zeidler, 2004; Zeidler et al., 2002). Some studies (e.g., Chang & Chui, 2008; Lewis & Leach, 2006; Wu & Tsai, 2011) have suggested that higher understanding of scientific knowledge and the nature of science can lead to more logical and rational decision making, while other studies (e.g., Bell & Lederman, 2003; Means & Voss, 1996; Sadler & Donnelly, 2006) have suggested no significant difference. Additionally, environmental factors such as peers and teachers, culture, and SSI contexts had been shown to lead to various forms of reasoning. Similar research exploring the influence of these factors on decision-making has also been conducted in Korea.

For instance, Lee and Lee (2014) discovered, akin to the findings of Zeidler et al. (2002), that students tended to resist changing their beliefs even when confronted with conflicting evidence in the decision-making process. This reluctance was observed regardless of whether the students were majoring in science or not, suggesting that it might be attributed to a somewhat strong faith in science, particularly among those with a higher level of scientific knowledge. Jho et al. (2013) identified college students' views on the orientation of science (e.g., pragmatic, intrinsic, communal, and ethical) and explored how these views related to decision making concerning SSIs. They observed that students' orientations of science influenced their decision-making processes.

While research outside of Korea has extensively explored the relationship between understanding the nature of science and SSI reasoning, studies in Korea have primarily focused on the relationship between understanding the nature of technology and SSI reasoning. A prominent study in this regard is Lee and Lee (2015), who argued that nature of technology could be more explicitly manifested in SSI reasoning than the nature of science. They pointed out that many SSIs commonly addressed in discussions (e.g., GMOs, nuclear power, environmental pollution, and embryonic cloning) are more closely related to technology and engineering than pure science. Therefore, they explained that a better understanding of the nature of technology and engineering, which have a direct impact on individuals' lives, can more effectively manifest in decision-making regarding SSIs.

16.3.2 SSI Education for Promoting Key Competences and Character as Citizens (Mid-2010s to Late 2010s)

While earlier SSI research primarily focused on examining individuals' moral reasoning and decision-making processes related to SSIs, the scope of SSI studies significantly expanded in this period. By leveraging insights into students' moral reasoning and response patterns, a wide range of SSI instructional models and programs were developed to aid students in making more informed decisions regarding SSIs. Additionally, concerted efforts were made to integrate SSI educational programs with national science curricula, enabling teachers to seamlessly incorporate SSI topics into science classes.

Notably, the NSC 2015 emphasized five core science competencies, and government policies on character education underscored the necessity for teaching methods that bridge science education with character education. As a result, teaching using SSIs was acknowledged as a strategy that nurtures both competencies and character, prompting the creation of various programs aligned with this approach. Additionally, numerous studies were published aiming to measure the outcomes of SSI education across various domains (e.g., scientific knowledge, scientific attitudes, nature of science and technology, and computational thinking), with most reflecting Vision II of scientific literacy.

16.3.2.1 SSI Instructional Models, Strategies, and Program Development

Various instructional models, strategies, and programs have been designed to help students make informed decisions and arguments about SSIs. These resources have aimed to equip students with the knowledge and skills to navigate complex and confusing topics. For instance, Lee et al. (2014) proposed an instructional model to increase students' active engagement in SSIs. This model emphasizes the importance of diversity of opinions, understanding complex issues from multiple perspectives, and facilitating reasoned group decision making. Science teachers who implemented the model reported improved student participation, cooperation, quality of discussions, and evidence. Another study used a creative technique called Six Thinking Hats to develop an SSI discussion program (Jang et al., 2016). The study compared its effects on students' argument types and decision-making abilities across SNS and traditional face-to-face discussions while factoring in personality traits. The study found that SNS discussions, enhanced by this creative thinking method, led to more diverse argument types and were particularly effective in fostering a diversity of perspectives in decision making.

Many studies have investigated the development and effectiveness of SSI programs in terms of enhancing the quality of argumentation (Kang, 2018; Kim & Kang, 2021a; Kim & Kim, 2018; Ko & Lee, 2020; Seo & Wee, 2022). These studies have incorporated writing instruction within the context of SSIs to assess argument quality and monitor changes. When selecting topics for writing, the interests of and relevance to participants were prioritized, particularly choosing SSI topics that were capable of evoking individual emotional responses and fostering empathy. For instance, issues such as abortion, euthanasia, genetic manipulation, and artificial intelligence were introduced to prospective biology teachers in their coursework (Kim & Kim, 2018). This approach acknowledges the emotional impact that may emerge during SSI education on complex social issues, thereby guiding classes toward more profound discussions and problem-solving strategies.

To evaluate educational effectiveness, these argumentation studies employed Toulmin's model of argument elements, developing and utilizing a rubric that considered the presence and extent of refutation. Kim and Kang (2021a) found that pre-service teachers' SSI writings improved as the course in the study progressed. The classes initially focused on simple arguments, while later on, rebuttal elements became more frequent, leading to a diversity of argument structures and enhanced student engagement in conflict management and negotiation.

16.3.2.2 SSI Research Related to South Korea's Science Core Competencies

Since the introduction of the NSC 2015, there has been an increase in research exploring the impact of SSI education on the cultivation of science core competencies (Kim & Kang, 2021b; Kim et al., 2017a; Heo & Kang, 2023). Among the five competencies introduced in the NSC 2015, a significant number of studies have specifically focused on three competencies: communication skills, scientific problem solving, and scientific engagement and lifelong learning. Researchers and educators have highlighted the significance of SSI programs in bolstering students' communication skills and competency to solve scientific problems (Kim & Kang, 2021b; Kim et al., 2017b; Song, 2015). These programs, deeply rooted in critical analysis, evidence-based argumentation, and the practical application of scientific knowledge to address real-world challenges, offer a dynamic platform for interactive learning. Through engaging methodologies like group work, discussions, and debates, SSI programs provide a more collaborative environment for students to express their ideas and listen to differing viewpoints. An example is an SSI class designed for elementary school students aiming to cultivate these abilities that incorporated activities such as brainwriting, fishbone diagrams, and world café discussions, strategically designed to scaffold younger students' engagement in comprehensive discourse (Kim et al., 2017). These activities were aimed at helping students organize their thoughts, visualize and effectively communicate their ideas, and appreciate the value of diverse perspectives in finding solutions. The study demonstrated the effectiveness of SSI education by showing a marked improvement in communication skills and scientific problem-solving competence, affirming its impact on the national curriculum.

In addition, several studies have explored the impact of SSI programs on enhancing scientific engagement and lifelong learning and civic competencies (Kim & Kang, 2021b; Lee & Yoon, 2017; Song, 2015). The researchers in these studies hypothesized that SSI programs sparked students' curiosity about societal issues intertwined with science and technology, motivating them to engage in informed and responsible decision making. They believed that this approach equipped students with the adaptability required to navigate the rapidly changing landscape of science and technology, fostering a culture of continuous learning and empowering them to actively contribute to democratic decision-making processes. The success of these programs, however, was contingent upon their design and execution.

16.3.2.3 SSI Research Related to Character Education

Another distinctive SSI research trend during this period is to examine the effects of SSI education on the development of individuals' character and values as citizens (Kang, 2020; Kim et al., 2016; Kim & Lee, 2017a; Kim et al., 2015; Ko & Lee, 2017; Lee, 2020). These studies assume that SSI education can contribute to not only individuals' moral and psychological development but also the worldview, character, and values necessary to live as responsible citizens in contemporary society. Character education is typically considered a topic mainly addressed in social studies or moral education, so initially, there was some resistance to implementing character education in science classes (Yang et al., 2012). However, as discussed earlier, policies such as the Basic Plan for Creative and Character Education and the Character Education Promotion Act mandated the integration of character education across all subjects. SSI education was reevaluated as a teaching strategy for cultivating character and values as citizens. Many studies have explored the effectiveness of SSI programs in cultivating character and values as citizens.

The definition of character and values as a citizen can vary widely. Lee et al. (2012) argued that individuals living in the modernized society need to develop character and values that enable them to act responsibly toward human life and to demonstrate compassion for others worldwide. They viewed such character and values as serving as a general framework or point of reference to support decision making regarding global issues. Based on the literature review, they identified three components of character and values—ecological worldview, social and moral compassion, and socioscientific accountability—that SSI education might foster and developed a Likert-scale instrument to measure the improvement of the components as an outcome of SSI programs.

In practice, SSI programs integrated with character education have been developed and implemented across diverse target groups, reporting positive outcomes. Researchers tried various instructional strategies that could cultivate character and values as citizens through SSIs. For example, Kim et al. (2016) designed SSI programs for elementary students aimed at encouraging more in-depth engagement in SSIs, choosing issues that students could often encounter in their daily lives. In the program, students were encouraged to consider ways in which humans and the environment could coexist by examining a case where the development of a cable car on a mountain near a school led to the extinction of some animal species. It also offered opportunities of role play to listen to the views of various stakeholders regarding the establishment of a gas plant in their community and to communicate each other to foster understanding and empathy for their viewpoints. Recently, Lee (2020) designed SSI programs for engineering students in college that delved deeply into the interactions between nature and humans by examining efforts to overcome malaria and the historical background and failures associated with DDT usage. It also examined the issue of designer babies, providing students opportunities to explore the diverse perspectives of various stakeholders and to discuss the social responsibilities of engineers in relation to the risks associated with advancements in modern technology. Additionally, the students were encouraged to investigate the potential for a surveillance society due to the proliferation of CCTV and social media, explore the privacy issues this raises for individuals, and seek solutions at the individual, societal, and global levels.

16.3.3 Enlarging the Vision of Scientific Literacy Through SSI Education (Late 2010s to the Present)

SSI research, which primarily focuses on evidence-based argumentation and decision making, has faced criticism for not sufficiently encouraging student participation and action (Bencze & Krstovic, 2017; Roth & Lee, 2004). Some researchers have reported that students became more interested in solving problems (e.g., Kim et al., 2017; Seo, 2021) and enhanced their feeling of responsibility for those affected or marginalized by rapid scientific and technological advancements, but showed little improvement in their willingness to take action (Kim et al., 2015; Lee et al., 2013). However, SSI education has gained increased attention as a way to promote social engagement and action due to the publication of the KSES and the NSC 2022.

16.3.3.1 SSI Education as a Way of Promoting Social Action

The willingness of Korean students to take societal action about SSIs has not appeared to be high. Since they have been more accustomed to being passive consumers of scientific knowledge and artifacts (Jang & Han, 2008; Kang et al., 2016; Lee & Yoon, 2017), they have been likely to lack a deep understanding of the nature of science and technology and its impact on society. Some of them have been shown to believe that SSIs should be resolved by STEM professionals or civic organizations and governmental agencies (Lee & Yoon, 2017). They hold views such as "it is not my place to step in," "it is efficient to solve problems directly by experts in related fields," and "I don't think that ordinary citizens can participate in science and technology-related issues and give reasonable opinions." They also struggle to propose solutions due to a lack of knowledge about practical action strategies. This underscores the need for more explicit educational approaches to enhance societal action in dealing with SSIs.

Even before the publication of the KSES and the NSC 2022, science programs encouraged student actions to resolve SSIs. Science textbooks have often listed simple easy actions (e.g., reducing single-use products, unplugging unused electrical cords, using eco-friendly products, using public transportation, and separate waste collection; Lim et al., 2021), and most SSI education has been limited to proposing personal action plans, with actual implementation being obstructed by time and space constraints within the school environment (Yang & So, 2023). Students still struggled to think of ways to actively engage in society, often perceiving themselves as too powerless to initiate social action due to their age or believing that the social actions planned by their groups would have minimal impact on real social change.

To overcome these limitations, science educators have proposed various teaching methods that emphasize social action, for example, community-based SSI education (Kim et al., 2020; Kim & Lee, 2017a, b). Adopting place-based learning theory, educators have developed programs addressing local community issues (e.g., fine dust, abandoned animals, and recycling) that students encountered in their lives. For example, students made visits to meteorological offices and public health centers, engaged directly with residents affected by fine dust to understand their views, and collaborated on creating and disseminating solutions within their community. Through this initiative, students showed an eagerness and commitment to engage in responsible communication with their peers, family, and the broader community.

Another example is the ENACT program (Lee et al., 2020; http://enactproject.com/). The ENACT program is a project-based approach targeting middle and high school students, STEM college students, and even the general public. It integrates SSI education with problem-solving approaches to develop students' competencies and social responsibility. The program is structured into two cycles and five stages (see Fig. 16.1). Cycle I is dedicated to understanding the nature of science and technology, while Cycle II focuses on the application of scientific, technological, and engineering practices. The aim of Cycle I (Stages 1–3) is for students to understand the nature of science and technology (i.e., it is uncertain, complex, and value laden) by exploring various stakeholders and anticipating the future. In Cycle II (Stages 4–5), students engage in resolving SSIs using scientific and engineering methods and take action based on these solutions for better futures.

Researchers have investigated the effectiveness of the ENACT project across different groups, including university students majoring in natural sciences and engineering, pre-service science and technology teachers, and middle and high school students (Choi et al., 2021; Hwang et al., 2023; Kim et al., 2021; Lee et al., 2022). Middle and high school students realized that even as non-scientists they can solve problems and that solutions should be designed to provide tangible benefits to vari-

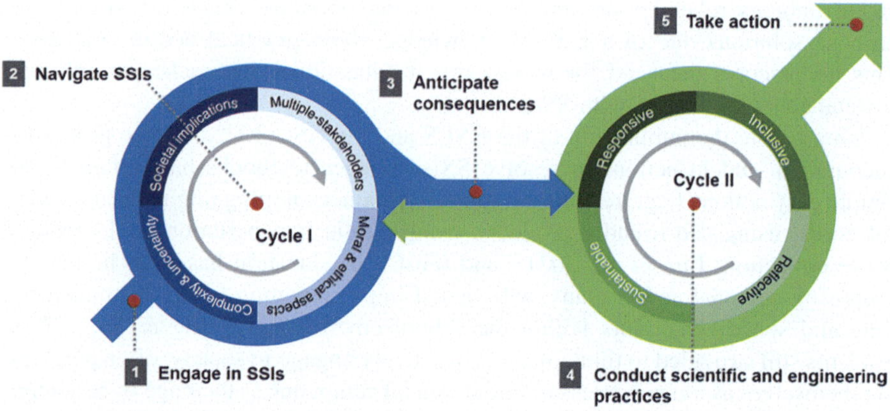

Fig. 16.1 ENACT model (Lee et al., 2020)

ous stakeholders. For example, participating high school students suggested a policy to use color-changing ink to visually display expiration dates on medication packets, motivated by their understanding of the serious issue posed by expired medicines. Although they were unable to make color-changing ink themselves, they felt very proud to contribute a solution from a high school student's perspective. Another example involved middle school students realizing that neonicotinoid, generally regarded as an eco-friendly pesticide, can be fatally harmful to bees and sought to address this issue. They collected photos of various pesticides on the market and developed a pesticide identification program using an artificial intelligence app called Teachable Machine. They believed that this app would greatly assist elderly farmers since it instantly identifies the presence of neonicotinoids when the camera was pointed at the surface of pesticides. However, through conversations with other young farmers, students realized that it could be more challenging for the elderly. They shifted their problem-solving approach to posting posters that inform about the potential dangers of pesticides labeled as eco-friendly. Through these experiences, students realized the significance of their problem-solving and social actions, even if these actions were not large-scale practices to eliminate neonicotinoids or solve the issue entirely. By raising awareness of harmful substances in supposedly environmentally friendly pesticides, they understood the value of their efforts in problem-solving and practice. Engineering students and pre-service science and technology teachers also enhanced their competencies for engagement and action by not only understanding the nature of SSIs, but also examining their responsibilities related to science and technology and how they could actively contribute to solving problems.

16.3.3.2 Enhancing Student Agency

Research has been conducted to enhance student agency. As described in OECD 2030 Learning Compass (OECD, 2019), student agency is the belief that students have the capability to positively influence their own lives and the world around them by responsibly taking actions for changes. A common SSI topic used to promote student agency has been climate change. Some educators have developed climate change programs aimed at providing opportunities to join social and political actions to dealt with the climate crisis. For instance, in Lim et al. (2021), students not only explored the causes and potential solutions for climate change but also took proactive steps to mitigate its effects. They collectively created and uploaded a YouTube video detailing proper actions and solutions, sent letters to Seoul elementary school administrators urging them to amend their policies, and even submitted policy proposals to the Seoul Metropolitan Office of Education. Through these experiences, the students came to understand that their impact extends beyond mere individual efforts. By uniting with their peers to propose policies and actively engaging with municipal authorities through petitions, they recognized their collective agency to effect change and address societal challenges. Another study provided an educational experience highlighting social action to elementary school students through a climate change club (Kim & Kim, 2021). After exploring climate change modeling and seeking solutions, students put their ideas into action by sending emails to beverage manufacturers, making a video on how to separate plastic bottles, and organizing a school campaign of removing all the labels before recycling. Through the cumulative experience of social action and encouraging responses from members of society—including communicating with people through comments on the video, witnessing younger students actively washing bottles for recycling, and receiving responses from educational authorities—students came to realize that they were capable of driving change themselves. Thus, social action and positive feedback led to a transformation in their mind: "We cannot completely change the world, but it seems that we can at least make a small difference." The students started to take initiative, recognizing their ability and responsibility to contribute to sustainable society. This series of studies has shown that focusing on social action within science education can empower students to become democratic citizens.

16.4 Conclusion and Future Directions

This chapter began with a review of the national curriculum reforms and major educational policies in South Korea, and subsequently investigated the changes and features of SSI research over time in response to these curriculum policy changes. The influence of the national curriculum and educational policies has brought significant changes to science textbooks and classroom teaching practice and has been reflected in the trends of SSI research. The journey of Korean SSI education research evolved through three periods. First, the term "SSIs" was introduced around the mid-2000s, although the STS educational philosophy had already been established as an important goal in Korean science education. With the emergence of societal changes and various issues, science educators and teachers have shown increased interest in SSI education as a part of the STS educational movement. Second, SSI research has expanded significantly, with a shift towards developing instructional models and programs aimed at aiding students in making informed decisions regarding SSIs. Korean science education, which has primarily focused on the transmission of scientific knowledge, has paid attention to moral and psychological development and moral reasoning based on diverse values. Third, the continuous reforms and policies over the past 20 years have positioned SSI education as a teaching and learning method that connects character and civic education, leading to an expansion towards education that emphasizes social responsibility, engagement, and action.

In Korea, SSI education has continuously been implemented in connection with education for sustainable development, climate change education, ecological transformation education, and global citizenship education. While each educational movement may have slightly different goals, they share similarities in emphasizing sustainable coexistence between humans and the environment, participation and actions aimed at our wellbeing, and a sense of responsibility towards these goals. The recently revised science education curriculum (NSC 2022) actively reflects these ideas, and the implementation of this curriculum is expected to bring about changes in classroom teaching practices based on these ideas.

SSI education can play a crucial role in realizing education that emphasizes responsible participation and action (i.e., Vision III scientific literacy). To achieve this, various environmental conditions need to be established. It is necessary to break down the thick walls surrounding school science, which primarily delivers value-neutral scientific knowledge and deals with problems that have clear answers. There is a need to transition towards a new culture of science where students, acting as agents, directly tackle real-life, ill-structured problems they encounter in their daily lives (Lee, et al., 2020). Another possible answer is integrating more explicit and robust directions for activism in curricula and educational policies. Current curricula and textbooks deal with participation and action, which are elements that often remain confined to the personal level (Lim et al., 2021). To cultivate socially responsible citizens, it is crucial to include more explicit action-oriented components and seek a paradigm shift toward more engaged and active forms of science education that contribute to societal change. Furthermore, there is a need to develop educational programs that provide students with direct experience of participating in SSI solving as active agents (Hwang et al., 2023; Kim et al., 2020). Finding issues closely related to life within their own communities, engaging in problem solving with peers or experts, and utilizing community resources can play a crucial role in enhancing students' social responsibility.

It is also necessary to change the perspective of science teachers and their understanding of SSI education (Lee, 2022; Lee & Yang, 2019). Although science teachers are fully aware that fostering scientific literacy is an important goal of science education, some of them tend to show their resistance to SSI lessons. This may be partly because, as Zeidler et al. (2011) mentioned, SSI teaching requires a "fundamental reconstruction" of traditional teacher-led science lessons. It is essential that teachers understand what social engagement and action is and why it is important to embrace activism in science education. It is also important to be competent in using a range of instructional strategies that can facilitate activism beyond using such activities as discussion, writing, and role play. Given that social engagement and action are often an unfamiliar topic to in-service teachers relative to traditional science content knowledge and SSI classes, efforts to provide professional development opportunities for teachers would be helpful. Continuous support is needed for science teachers to adopt new epistemological beliefs on scientific knowledge, thereby expanding their vision of the goals of science education. In essence, we highlight the necessity for systemic changes that facilitate a more dynamic, action-oriented, and socially engaged form of science learning and the pivotal role of educators in this transformation.

References

Bell, R. L., & Lederman, N. G. (2003). Understandings of the nature of science and decision making on science and technology based issues. *Science Education, 87*(3), 352–377.

Bencze, L., & Krstovic, M. (2017). Science students' ethical technology designs as solutions to socio-scientific problems. In L. Bencze (Ed.), *Science and technology education promoting wellbeing for individuals, societies and environments* (pp. 201–226). Springer.

Chang, S. N., & Chui, M. H. (2008). Lakatos' scientific research programmes as a framework for analysing informal argumentation about socio-scientific issues. *International Journal of Science Education, 30*(13), 1753–1773.

Chang, H., & Chung, Y. (2009). An analysis of informal reasoning in the context of socioscientific decision-making. *Journal of the Korean Association for Science Education, 29*(2), 253–266.

Chang, H., & Lee, H. (2010). College students' decision-making tendencies in the context of socio-scientific issues (SSI). *Journal of Korean Association in Science Education, 30*(7), 887–900.

Character Education Promotion Act. (2015). No. 26403. https://www.law.go.kr/%EB%B2%95%EB%A0%B9/%EC%9D%B8%EC%84%B1%EA%B5%90%EC%9C%A1%EC%A7%84%ED%9D%A5%EB%B2%95

Cho, H., & Shim, M. (2009). An exploration of the models for teaching/learning bioethics and the bioethical issues appropriate for each model-based teaching/learning. *Biology Education, 37*(2), 199–212.

Choi, Y., Ko, Y., Hong, Y., Lee, H., & Hwang, Y. (2021). Changes of pre-service technology teachers' views and educational needs on social responsibility of science/technology/engineering through the ENACT program. *Journal of Research in Curriculum and Instruction, 25*(2), 1–13.

Chung, Y., Mun, K., & Kim, S. (2010). Exploration of socioscientific issues (SSI) in the science textbook. *Journal of Learner-Centered Curriculum and Instruction, 10*(3), 435–456.

Connell, S., Fien, J., Lee, J., Sykes, H., & Yencken, D. (1999). 'If it doesn't directly affect you, you don't think about it': A qualitative study of young people's environmental attitudes in two Australian cities. *Environmental Education Research, 5*(1), 96–113.

Dreyfus, A., & Roth, Z. (1991). Twelfth-grade biology pupils' opinions on interventions of man in nature: Agreement, indifference and ambivalence. *Journal of Research in Science Teaching, 28*(1), 81–95.

Heo, J., & Kang, N. (2023). The effects of socioscientific issue (SSI)-based instruction on underachieving 9th-grade students: Achievement, attitudes, and scientific participation and lifelong learning competency. *Journal of Science Education, 47*(1), 11–23.

Hwang, Y., Ko, Y., Shim, S. S., Ok, S., & Lee, H. (2023). Promoting engineering students' social responsibility and willingness to act on socioscientific issues. *International Journal of STEM Education, 10*, 11.

Jang, Y., & Han, J. (2008). *Development plan of citizen participatory science and technology policy* (Report No. 2008–04). Science and Technology Policy Institute.

Jang, S., Cha, H., Park, H., & Park, C. (2016). Effectiveness of decision-making skills in SSI class based on debate by utilizing SNS in terms of students' personality traits. *Journal of the Korean Association for Science Education, 36*(5), 757–768.

Jho, H., Song, J., & Levinson, R. (2013). Views on the orientation of science in decision-making revealed in undergraduate students' discussion on socio-scientific issues. *Journal of the Korean Association for Science Education, 33*(3), 581–596.

Ju, I., & Lee, H. (2013). Patterns of middle school students' value-judgment and decision-making on biotechnology-related socioscientific issues. *Journal of Korean Association in Science Education, 33*(1), 79–93.

Kang, K. (2018). Analysis of argument structure presented in pre-service biology teachers' writing on socio-scientific issue. *Biology Education, 46*(1), 55–62.

Kang, K. (2020). The effect of SSI (Socio-Scientific Issues) writing program on pre-service biology teachers' perception for SSI teaching, character and values. *Journal of Education & Culture, 26*(6), 215–231.

Kang, S., Yoon, J., Kim, B., Kim, S., & Lee, G. (2016). *A study on the institutional plan for the contribution of scientists and technologists to social progress*. Korea Foundation for the Advancement of Science and Creativity.

Kim, H., & Kang, K. (2021a). Analysis on changes of argumentation level and structure presented in pre-service biology teachers' SSI (Socio-Scientific Issues) writing. *Biology Education, 49*(2), 195–204.

Kim, H., & Kang, K. (2021b). An analysis of science core competencies presented in pre-service biology teachers' SSI (Socioscientific Issues) writing. *Biology Education, 49*(1), 122–132.

Kim, S., & Kim, S. (2018). Effects of SSI argumentation program based on SEL for preservice biology teachers. *Journal of the Korean Association for Science Education, 38*(2), 259–271.

Kim, J., & Kim, C. (2021). Exploring elementary students' positioning in a context of socio-scientific issues (SSI) education: Focus on an action-oriented climate change club activity. *Journal of the Korean Association for Science Education, 41*(6), 501–517.

Kim, G., & Lee, H. (2017a). Effects of community-based SSI programs on promoting middle school students' understanding of issues and character and values as citizens: Focused on fine dust issues. *Journal of the Korean Association for Science Education, 37*(6), 911–920.

Kim, G., & Lee, H. (2017b). Perceptions of teachers, program instructors, and local experts on implementing community-based socioscientific issues programs. *Journal of the Korean Association for Science Education, 37*(3), 453–464.

Kim, H., Yang, J., & Woo, A. (2015). Effect of socio-scientific issues (SSI) classes on the decision-making and the character and value. *Journal of Research in Curriculum Instruction, 19*(4), 983–1003.

Kim, J., Ko, Y., & Lee, H. (2016). Effects of socioscientific issues instruction on elementary school students' character and values as global citizens. *The Journal of Elementary Education, 29*(3), 1–25.

Kim, J., Lee, H., & Ko, Y. (2017a). Enhancing student key competencies through socioscientific issues instruction. *Journal of Learner-Centered Curriculum and Instruction, 17*(8), 339–362.

Kim, M., Kim, J., Lim, H., Park, D., & Park, S. (2017b). The effect of the program of consensus conference on energy related social science issues (SSI) on the change in consciousness of high school students. *Journal of Energy and Climate Change Education, 7*(2), 171–182.

Kim, G., Ko, Y., & Lee, H. (2020). The effects of community-based socioscientific issues program (SSI-COMM) on promoting students' sense of place and character as citizens. *International Journal of Science and Mathematics Education, 18*, 399–418.

Kim, G., Ok, S., Lee, H., Ko, Y., & Hwang, Y. (2021). A case study of an ENACT model-based engineering design course for fostering social responsibility of engineers. *Journal of Engineering Education Research, 24*(6), 3–19.
Ko, Y., & Lee, H. (2017). Comparison of the effects of socioscientific issues instruction on promoting college students' character and values: Based on idiocentrism and allocentrism. *Journal of the Korean Association for Science Education, 37*(3), 395–405.
Ko, Y., & Lee, H. (2020). Analyzing college students' dialogic argumentation in the context of nanotechnology issues based on idiocentrism and allocentrism. *Journal of the Korean Chemical Society, 64*(5), 291–303.
Krippendorff, K. (2004). *Content analysis: An introduction to its methodology* (2nd ed.). Sage.
Lee, H. (2008). Decision-making patterns of pre-service science teachers on socioscientific issues. *Journal of Research in Curriculum Instruction, 12*(2), 377–395.
Lee, H. (2020). Effects of an SSI-based program for engineering college students on character and values. *Journal of Engineering Education Research, 23*(3), 3–12.
Lee, H. (2022). Pedagogical and epistemological challenges of pre-service science teachers teaching socioscientific issues: Based on the SSI-PCK framework. *Asia-Pacific Science Education, 8*(2), 301–330.
Lee, H., & Chang, H. (2007). The comparison of state-level U.S. science curricula with science teachers' perception regarding teaching socioscientific issues (SSI). *The Journal of Curriculum & Evaluation, 10*(1), 189–209.
Lee, H., & Chang, H. (2012). Patterns of pre-service science teachers' use of evidence in web-based discussions of the nuclear power generation issue. *New Physics: Sae Mulli, 62*(4), 364–373.
Lee, S., & Lee, H. (2014). Pattern of college students' informal reasoning and reactions to anomalous evidence on the controversial nuclear power generation issue. *Journal of Learner-Centered Curriculum and Instruction, 14*(6), 147–168.
Lee, H., & Lee, H. (2015). Analysis of students' socioscientific decision-making from the nature of technology perspectives. *Journal of the Korean Association for Science Education, 35*(1), 169–177.
Lee, H., & Yang, J. (2019). Science teachers taking their first steps toward teaching socioscientific issues through collaborative action research. *Research in Science Education, 49*, 51–71.
Lee, Y., & Yoon, J. (2017). An analysis of undergraduate students' perceptions and practical capabilities on citizen participation in social issues of science and technology. *Journal of the Korean Association for Science Education, 37*(4), 637–650.
Lee, H., Abd-El-Khalick, F., & Choi, K. (2006). Korean science teachers' perceptions of the introduction of socioscientific issues into the science curriculum. *Canadian Journal of Science, Mathematics, and Technology Education, 6*(2), 97–117.
Lee, H., Chang, H., Choi, K., Kim, S., & Zeidler, D. L. (2012). Developing character and values for global citizens: Analysis of pre-service science teachers' moral reasoning on socioscientific issues. *International Journal of Science Education, 34*(6), 925–953.
Lee, H., Yoo, J., Choi, S., Kim, S., Krajcik, S., Herman, B., & Zeidler, D. (2013). Socioscientific issues as a vehicle for promoting character and values for global citizens. *International Journal of Science Education, 35*(12), 2079–2113.
Lee, H., Choi, Y., & Ko, Y. (2014). Designing collective intelligence-based instructional models for teaching socioscientific issues. *Journal of the Korean Association for Science Education, 34*(6), 523–534.
Lee, H., Choi, Y., Nam, C., Ok, S., Shim, S. S., Hwang, Y., & Kim, G. (2020). Development of the ENACT model for cultivating social responsibility of college students in STEM fields. *Journal of Engineering Education Research, 23*(6), 3–16.
Lee, H., Lee, H., & Zeidler, D. L. (2020). Examining tensions in the socioscientific issues classroom: Students' border crossings into a new culture of science. *Journal of Research in Science Teaching, 57*(5), 672–694.

Lee, H., Ko, Y., & Hong, J. (2022). ENACT project: Promoting pre-service science teachers' views on the social responsibility of scientists and engineers. *Journal of the Korean Association for Science Education, 42*(1), 111–125.
Lewis, J., & Leach, J. (2006). Discussion of socio-scientific issue: The role of science knowledge. *International Journal of Science Education, 28*(11), 1267–1287.
Lim, S., Kim, J., & Kim, C. (2021). Exploring elementary teacher's challenges with the perspective of structure and agency when implementing social action-oriented SSI education classes. *Journal of the Korean Association for Science Education, 41*(2), 115–131.
Means, M. L., & Voss, J. F. (1996). Who reasons well? Two studies of informal reasoning among children of different grade, ability, and knowledge levels. *Cognition and Instruction, 14*(2), 139–178.
Ministry of Education [MOE]. (1987). *The 5th national science curriculum*. MOE.
Ministry of Education [MOE]. (2015). *The national science curriculum 2015*. MOE.
Ministry of Education [MOE]. (2022). *The national science curriculum 2022*. MOE.
Ministry of Education, Science and Technology [MEST]. (2007). *The national science curriculum 2007*. MEST.
NGSS Lead States. (2013). *Next generation science standards: For states, by states*. The National Academies Press.
Organization for Economic Co-operation and Development. [OECD]. (2019). *OECD future of education and skills 2030 project background*. OECD Publishing.
Roberts, D. A. (2007). Scientific literacy/science literacy. In S. K. Abell & N. G. Lederman (Eds.), *Handbook of research on science education* (pp. 729–780). Lawrence Erlbaum.
Roth, W. M., & Lee, S. (2004). Science education as/for participation in the community. *Science Education, 88*(2), 263–291.
Rychen, D. S., & Salganik, L. H. (Eds.). (2003). *Key competencies for a successful life and a well-functioning society*. Hogrefe & Huber Publishers.
Sadler, T. D. (2004). Moral sensitivity and its contribution to the resolution of socioscientific issues. *Journal of Moral Education, 33*(3), 339–357.
Sadler, T. D., & Donnelly, L. A. (2006). Socioscientific argumentation: The effects of content knowledge and morality. *International Journal of Science Education, 28*(12), 1463–1488.
Sadler, T. D., & Zeidler, D. L. (2004). The morality of socioscientific issues: Construal and resolution of genetic engineering dilemmas. *Science Education, 88*(1), 4–27.
Sadler, T. D., & Zeidler, D. L. (2005). Patterns of informal reasoning in the context of socioscientific decision making. *Journal of Research in Science Teaching, 42*(1), 112–138.
Seo, S. (2021). *Exploring social environmental identity of students participated in a social action-oriented climate change SSI club* [Unpublished master's thesis]. Seoul National University.
Seo, H., & Wee, S. (2022). Analysis of the argumentation structure shown in writing and earth science teachers' perception for SSI teaching through writing education. *Journal of Learner-Centered Curriculum and Instruction, 22*(11), 125–146.
Shim, M., & Cho, H. (2009). An exploration of the teaching/learning methods of bioethics appropriate for ethical theories and socio-scientific issues in biological sciences. *Journal of the Korean Association for Science Education, 29*(5), 513–530.
Sjöström, J., & Eilks, I. (2018). Reconsidering different visions of scientific literacy and science education based on the concept of Bildung. In Y. Dori (Ed.), *Cognition, Metacognition, and Culture in STEM Education* (pp. 65–88). Springer.
Song, I. (2015). CSI: Conversion class and foundation competences, through operating cases for scientific socio issue utilizing cinema. *Korean Journal of General Education, 9*(1), 265–293.
Song, J., Kang, S., Kwak, Y., Kim, D., Kim, S., Na, J., Do, J., Min, B., Park, S., Bae, S., Son, Y., Son, J., Oh, P., Lee, J., Ihm, H., Jeong, D., Joung, Y., & Joung, J. (2018). *A development of Korean science education standards (KSES) for the next generation*. Korea Foundation for the Advancement of Science and Creativity.
Valladares, L. (2021). Scientific literacy and social transformation: Critical perspectives about science participation and emancipation. *Science & Education, 30*(3), 557–587.

Wu, Y. T., & Tsai, C. C. (2011). High school students' informal reasoning regarding a socio-scientific issue, with relation to scientific epistemological beliefs and cognitive structures. *International Journal of Science Education, 33*(3), 371–400.

Yang, S., & So, K. (2023). Effects of climate change education using the SSI program on climate literacy and decision-making capability of elementary students. *Journal of Energy and Climate Change Education, 13*(2), 137–150.

Yang, J., Kim, H., Gao, L., Kim, E., Kim, S., & Lee, H. (2012). Perceptions of science teachers on socioscientific issues as an instructional tool for creativity and character education. *Journal of the Korean Association for Science Education, 32*(1), 113–128.

Yoo, J., Choi, S., & Lee, H. (2011). Perceptions of science, social studies, and ethics teachers on teaching socio-scientific issues. *Journal of Research in Curriculum Instruction, 15*(2), 415–432.

Zeidler, D. L., Walker, K. A., Ackett, W. A., & Simmons, M. L. (2002). Tangled up in views: Beliefs in the nature of science and responses to socioscientific dilemmas. *Science Education, 86*, 343–367.

Zeidler, D. L., Sadler, T. D., Simmons, M., & Howes, E. (2005). Beyond STS: A research-based framework for socioscientific issues education. *Science Education, 89*, 357–377.

Zeidler, D. L., Applebaum, S. M., & Sadler, T. D. (2011). Enacting a socioscientific issues classroom: Transformative transformations. In T. D. Sadler (Ed.), *Socioscientific issues in the classroom* (pp. 277–305). Springer.

Chapter 17
Science and Moral Inquiry as the Yin and Yang of SSI Education: Two Examples of SSI Research from Sweden

Carl-Johan Rundgren and Shu-Nu Chang Rundgren

17.1 Introduction

Educating for the unknown was the title of a reflective symposium on environmental science education held in 2009 at Linköping University in Sweden (Johansson & Svensson, 2009). Even though education by its very nature always has been an undertaking directed at an unknown future, the current global situation has urgently made educators all over the world keenly aware of the uncertainties of the future and the need to equip young people with adequate awareness and competences to deal with this unknown future. Humanity is currently facing challenges at a global scale in the shape of consequences of climate change, environmental degradation, pandemics, and human conflicts, etc. There is, by and large, an agreement among educators on the goal of fostering, not only knowledgeable citizens, who have knowledge about science and ability to make informed decisions, but also good and empathetic human beings, who have the ability of moral reflection. How to form virtuous citizens through education has been a perpetual issue in history, and each time in history has strived to find solutions to this issue. Different ways to educate people to be able to act with empathy and being capable of moral reflection has been discussed since antiquity. Indeed, Socrates highlighted the value of living "an examined life", as a reflected alternative to a mere adherence to customs and traditions (Wiblom et al., 2020). Aristotle proposed a notion of reflective citizenship, while the Greek and Roman Stoics envisioned an education (in the form of a life-long project), liberating the mind from the bonds of habit and custom, forming *world citizens* (Nussbaum, 1998). The Stoics launched *the humanistic project* as a vision to increase our understanding of what it means to be human through reading different

C.-J. Rundgren (✉) · S.-N. Chang Rundgren
Stockholm University, Stockholm, Sweden
e-mail: carl-johan.rundgren@su.se; shu-nu.chang-rundgren@edu.su.se

D. L. Zeidler (ed.), *A Moral Inquiry into Epistemic Insights in Science Education*, Contemporary Trends and Issues in Science Education 61,
https://doi.org/10.1007/978-3-031-63382-9_17

accounts from different ages and different cultures. The vision of educating world citizens formulated by the Stoics can today be translated to a vision of global citizenship. Undoubtedly, science and science education have important roles to play in contributing to this educational goal in the present science- and technology-dominated society. As this volume shows, research and teaching related to ill-structured and authentic *socio scientific issues* (SSI) has contributed to fostering knowledgeable and virtuous citizens of tomorrow. The SSI tradition has a commitment to a belief that knowing scientific facts alone is not enough, as educators we need to make individuals capable of making moral decisions. Accordingly, there is a need in education to let students tackle actual SSI in school science, developing students' skills such as argumentation, weighing different arguments and perspectives, while learning science content. We believe moral aspects to be at the core of SSI education, and it is thus important to involve moral perspectives in science education, for example in teacher professional development programs.

In this chapter, we discuss the concepts of *science inquiry* and *moral inquiry* and how they are reflected in SSI-based teaching and learning (hereafter called SSI education) in Sweden. There might be more consensus on the meaning of science inquiry, as compared to moral inquiry. Science inquiry can be related to a combination of general science process skills including applying science content knowledge, formal reasoning, creativity, and critical thinking in order to investigate nature (Lederman et al., 2014; Gyllenpalm et al., 2021). Moral inquiry, on the other hand, is a more disputed concept, related to how the concepts of ethics and morality are applied and often function interchangeably in practice, involving informal reasoning and argumentation (Chang & Chiu, 2008). In general, morality is about personal practical decisions about right and wrong, good and bad, while ethics refers to objective and agreed-upon values in communities and societies (e.g., Mattingly & Throop, 2018). In contrast to ethics, morality is subjective and individual. Lenman (2007) relates moral inquiry to the pursuit of individual reflective equilibrium in relation to moral decisions. In SSI education, science inquiry and moral inquiry are both central elements in teaching and learning. We would argue that, in SSI education, science inquiry and moral inquiry are (and ought to be) intertwined as an interconnected entity like the Daoist Chinese philosophical concept of *Yin and Yang,* and that it may be difficult to draw sharp dividing lines between what constitutes *science* inquiry as compared to *moral* inquiry.

Literally, Yin and Yang means the dark side and sunny side, but what Daoism addresses with the concept of Yin and Yang relates to the creation of harmony and balance between different entities in life. A related concept to Yin and Yang is *Dao,* which means 'the way' or even 'the middle way'. In SSI education, the concept of Yin and Yang can be helpful to highlight the different perspectives needed to be considered when we approach a certain SSI. As a basic approach to SSI teaching, Yin and Yang could represent the need to combine science inquiry and moral inquiry in the SSI activities. There are many SSI emerging in the current society, such as how to mitigate effects of climate change while serving human needs for energy, water, foods and materials, which warrant the necessity of equipping citizens with both science inquiry and moral inquiry skills. Echoing the concept of Yin and Yang in SSI education, we argue for the need to educate students, our future citizens, to make informed decisions on SSI based on skills of both science inquiry and moral inquiry, which need to be integrated rather than taught separately in SSI education.

17.1.1 Aim: Science and Moral Inquiry in SSI Research in Sweden

The present chapter reflects, from the perspective of research on SSI in Sweden, on the potential and actual contribution of SSI education to global citizenship education. The relation between what *science inquiry* and *moral inquiry* may entail in SSI education in Sweden is investigated. This is done through a reflection on two studies relating to the SEE-SEP model (Chang Rundgren & Rundgren, 2010), using the concepts of science and moral inquiry as tools for reflection. The reason for this is to understand better what the SEE-SEP model can bring to teaching SSI aiming to embrace both science and moral inquiry for pluralistic and holistic sustainability.

17.1.2 Curricular Traditions: Formation and Education

In many Western countries, there has been a discussion at least since the late nineteenth century about the need to bring about a 'science for the people', and what science content and what aspects of science to include in school (Rudolph, 2005). In Europe, there has been a tradition of different types of science education, one for students that were not supposed to follow science/academic careers, and another in schools for students aiming for academic paths. However, there have also been educational reforms counteracting this tradition. Especially during the decades following WWII, education became increasingly an instrument of social reform and development in many Western European countries, and educational systems were democratized in order to even out socio-economic differences and educational opportunities (Haglund & Hultén, 2017). At the same time in North America, the reforms brought about in many ways related to the need for educating top scientists in the cold war struggle between the West and the Soviet block. However, the post-WWII period was also a time in which a more inclusive conception of science developed in the liberal arts educational tradition (Rudolph, 2002). The cold war period was also the time of birth of the concept *scientific literacy*, which still today in itself holds many of the tensions mentioned above.

Roberts (2007) describes two competing visions of science education. Vision I gives meaning 'by looking inward at the canon of orthodox natural science, that is, the products and processes of science itself' (p. 730), with a focus on matters within science (p. 730). In contrast, 'Vision II derives its meaning from the character of situations with a scientific component, situations that students are likely to encounter as citizens... about science-related situations in which considerations other than science have an important place at the table' (p. 730). Meanwhile, Sjöström and Eilks (2018) argue for a Vision III of scientific literacy in terms of science for transformation, a critical scientific literacy aimed at achieving critical global citizenship, justified by the Central/Northern European *Bildung* tradition.

Traditionally, the mainly Germanic conceptualization of Bildung includes personal and societal development as well as solidarity and responsibility for acting towards positive changes in society (Elmose & Roth, 2005). Klafki (2000) discusses the meaning of the concept of *Allgemeinbildung* as the knowledge required by citizens to take active part of society. Central to a Bildung-related perspective is the notion of individual growth through knowledge, and with this follows aspirations of education to lead to both individual emancipation and contribution to society as a whole. Allgemeinbildung can be related both to the dimension of growth of knowledge among the citizens, and to the individual's cultivation of different skills or knowledge areas (Klafki, 2000).

In Sweden, the academic and the philosophical traditions, as well as the education system has traditionally been closely connected to the German intellectual tradition (by closeness of language and culture since centuries, and for the fact that many Swedish students studied at German Universities), especially since the nineteenth century. However, the establishment of Nazi rule in Germany in 1933 and the subsequent world war resulted in a break in this tradition, and in the school reforms from 1945 and onwards in Sweden became more connected to an Anglo-Saxon tradition. The educational reforms in Post WWII society in Sweden was especially inspired by the progressive education program formulated by John Dewey. In particular, Dewey's ideas about how to organize school education to form and sustain a democratic society became an important inspiration (Dewey, 1903). Swedish students should be 'vaccinated' against authoritarianism and anti-democratic ideas. Actually, there are many parallels between the Bildung tradition and Deweyan progressive education in the strive for character formation and the development of knowledgeable and virtuous citizens. Furthermore, the Bildung tradition has in later years been revitalized in Swedish education and research. In its current form, the Bildung tradition has been related to visions of forming a critical global citizenship education (Sjöström & Eilks, 2018).

We would argue that the Bildung tradition and the influences from Deweyan progressive education both have contributed to a curricular tradition in Sweden in which citizenship education has been important. The SSI tradition seem to have found fertile ground in the Swedish education system, and the national curriculum reforms in Sweden in later decades (1994; 2011; 2022) has resulted in science curricula that are not only compatible with SSI-based science teaching, but moreover increasingly aligned with the view on science education expressed in the SSI tradition. Some of the current goals of the science subjects, especially related to argumentation and decision-making (Swedish National Agency of Education, 2022), can even be said to be hard to achieve without taking in at least some aspects of SSI education.

In the current curriculum in Sweden, education for sustainability constitutes one of the central cross-curricular themes which should be addressed by all subjects (Swedish National Agency of Education, 2022). An important aspect of education for sustainability in later years has been the development of pluralistic teaching (Andersson, 2017). Central to pluralistic teaching is training students to take in different perspectives and to see an issue from different angles. Pluralistic education strives for the development of a multivocal classroom, in which many voices are heard and many different perspectives come into play. These aspects has been, and continues to be, central also in SSI education.

17.1.3 The SEE-SEP Model for Global Citizenship Education and Education for Sustainability

In 2010, based on a literature review work on SSI argumentation, we identified the six subject areas of **s**cience, **e**nvironment, **e**thics/morality, **s**ocial culture, **e**conomy, and **p**olicy (SEE-SEP) as being central parts of SSI. In addition to this, our review showed that students' argumentation not only involved arguments based on knowledge, but also based on values and personal experiences. We came to see knowledge, value and personal experience as *dimensions* of SSI argumentation (Chang Rundgren & Rundgren, 2010). The six subject areas and the three dimensions of knowledge, value and personal experience are holistically intertwined in SSI argumentation in the SEE-SEP model (Fig. 17.1). The SEE-SEP model was developed

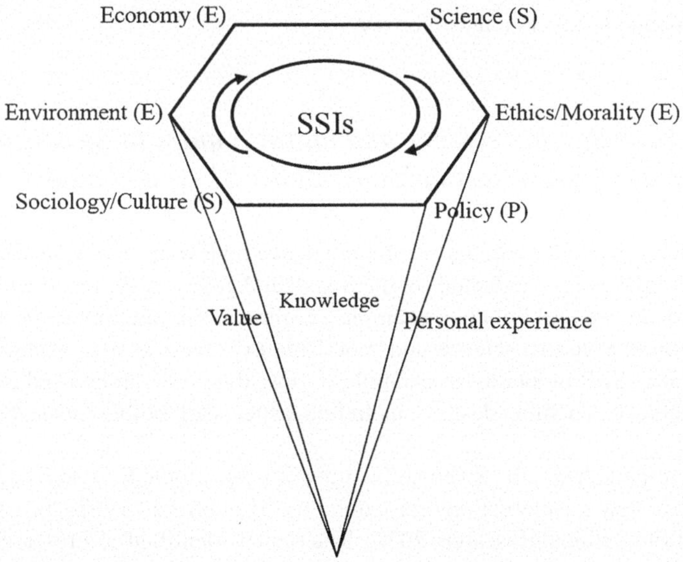

Fig. 17.1 The SEE-SEP model. (Chang Rundgren & Rundgren, 2010, p. 11)

as a tool for analyzing SSI argumentation (e.g. Christenson et al., 2012; Eriksson & Rundgren, 2012), but also aiming to function as a didactic model (Wickman et al., 2020) which can be used for planning and designing SSI teaching. It is perhaps worth to mention that the abbreviation of SEE-SEP was coined not only based on the subject areas, but also the need for people to 'see' SSI from 'separate' perspectives in order to reach a holistic view (Chang Rundgren & Rundgren, 2010). According to Zeidler and Sadler (2023, p. 905), *the SEE-SEP model reflects the inherent interdisciplinary complexity of SSI, and provides a useful lens to examine students' use of justifications embedded in their arguments in a systematic and holistic manner. As an analytical framework, it allows for the examination of multiple perspectives and quality of informal argumentation over a plethora of SSI topics and has implications for how teachers frame and cultivate SSI instruction.* The SEE-SEP model has been further developed for SSI education and a 'Post it!' approach has for example been constructed as a practical example how the model can be used to design SSI teaching (Chang Rundgren, 2011). The approach has been proved feasible in primary through upper secondary school levels and in teacher training programs and workshops (Cebesoy & Chang Rundgren, 2023; Rundgren & Chang Rundgren, 2018). We could say that the SEE-SEP model moves beyond the three ecological, societal and economical dimensions of sustainability by including ethics/morality, science and policy, which also play critical roles for global sustainability. However, including these dimensions in science teaching has proved to be a challenge for many teachers with science education background (Sund & Gericke, 2020; Wang et al., 2019). Our view is that frameworks such as SSI and the SEE-SEP model can function as practical tools helping teachers in their quest to equip future citizens with awareness and competences for sustainability. The SEE-SEP model can help teachers to manage teaching complex issues which include many seemingly divergent aspects in a feasible way.

17.1.4 Investigating science and moral inquiry in Swedish studies—two examples using the SEE-SEP model.

In this chapter, we will not make a comprehensive review of the SSI research, since this has recently been conducted by the Swedish Agency of Educational Research (2022). Instead, we highlight two examples, aiming at illuminating some aspects of SSI education related to science and moral inquiry. We use two research studies using the SEE-SEP model as an example of how this tool can be used and elaborated to achieve teaching designs including aspects of both science and moral inquiry.

As mentioned above, the Swedish Institute of Educational Research conducted a systematic review of international research on SSI published 1997–2021 (Swedish Institute of Educational Research, 2022). The review identified 157 research articles on SSI. No less than 29 of those were published by researchers at Swedish

Universities. Only U.S.-based scholars published more (33) during the period. The third place on the list is divided between the UK and Taiwan, each contributing 14 research articles on SSI. However, there are undoubtedly some biases favoring Swedish research in the review study (and also research from English-speaking countries), since only research published in English or in Scandinavian languages were included in the study. Furthermore, another review by Zeidler and Sadler (2023) identified a considerably higher number of international research articles on SSI, more than 400. Thus, methodological aspects related to data selection may have influenced the result. However, taken as a whole, there are indeed indications that there has been a considerable interest in SSI in Sweden, and that Sweden has made a substantial international contribution to research on SSI education.

Among the SSI research, we choose two examples of SSI studies which have been conducted in Sweden. For detailed descriptions of the studies and their methodologies and results, please see the original publications. Both studies connect regional issues related to environmental (Study A) and in the case of Study B, a combination of environmental and health issues. The two studies are based upon the holistic view of SSI education expressed in the SEE-SEP model. However, the analysis made in the two studies differ, and the SEE-SEP model was used in somewhat different ways in the two studies. While Study A made use of the SEE-SEP model through a quantitative (or mixed methods) analysis of the distribution of students' arguments according to the six different subject areas and the three dimensions of values, knowledge, and personal experiences in the model, Study B conducted a qualitative analysis of the presence of values, knowledge, and personal experiences in the students' arguments. In the following, the concepts of science inquiry and moral inquiry are focused in a reflection on the results from the two studies.

17.1.4.1 Study A: The Wolf Issue

Eriksson and Rundgren (2012) explored Swedish upper secondary school students' attitudes and ways of arguing in relation to the wolf question, relating to the authentic SSI of the management and size of the wolf population in Sweden. The wolf issue can be highlighted from different perspectives, the issue is debated at several levels in society and also affects people on a more personal level. Examples of values described as being in conflict are protecting biological diversity and problems caused by wolves, especially in inhabited areas. Problems that are touched upon include people's concern for their domestic animals and also that the wolf may be perceived as dangerous even for humans. A questionnaire probing attitudes to the existence of wolves in the neighborhood was distributed to 352 upper secondary students. Furthermore, 18 students were interviewed in pairs. In the interviews, an important point of departure was to let students, who initially may have opposing views on the matter, meet in informal conversations to exchange knowledge, put different arguments against each other and test ideas. Prior to the conversations, the students had access to a text and image material that introduced the topic and highlighted some of the conflicts that have been noticed in the matter. The extent to which the students in their arguments referred to factual knowledge, values and personal experiences in the subject areas of society/culture, environment, economics, natural science, ethics/morality and laws/policy was analyzed according to the SEE-SEP model (Chang Rundgren & Rundgren, 2010).

The choice of the wolf issue in this study was made based upon the complex web of scientific, political, social cultural, economic, and ethical aspects interconnecting in this issue. First, it is possible to discern a tension between people living in cities and people living in the countryside related to several aspects of this issue. The wolves are a substantial threat to the livestock of farmers, especially sheep and reindeer are vulnerable to wolf attacks. To livestock owners, a wolf attack on their animals can jeopardize their livelihood and income. People in rural areas may argue that people living in cities, promoting a bigger population of wolves in Sweden, do not face any risk for themselves and their livelihood. Urban citizens, on the other hand, may argue that the current wolf population in Sweden is considerably smaller than it has been in historic times, and that countries in eastern continental Europe, such as Romania, can manage much bigger wolf populations without considerable damage to livestock herding. Most commonly, arguments for an increased wolf population are based on ethical and ecological arguments. Ecological arguments include the positive effect on biodiversity achieved by the presence of large predators such as the wolf. However, the wolf prey on large herbivores, such as elk and raw deer, which rank among the main animals also hunted by human hunters. Hunting is a culturally very important activity in countryside Sweden. One of the students (S1) in the study expressed his personal experience of hunting in the interview:

> S1: Elk and animals like that are as they are. There are a lot of them, I think. I myself have shot elk and such and I see no risk that there will be too few of them.

The experience of hunting elk seemed to make the student see hunting as something natural and positive. Furthermore, this experience made him less anxious that human hunting would reduce the number of elks too much. His argument seemed to be that there are currently enough elk in Sweden for both wolves and humans. However, a higher number of wolves could possibly disrupt this balance. In this case, human needs must be prioritized, and the wolf population needs to be reduced by means of increased wolf hunting. This can be categorized as anthropocentric reasoning, departing from a view of moral reasoning in which humans are the central moral subject.

Another student (S2) expressed an ethical argument for an increase in wolf population:

> S2: They have the same right to live as humans or dogs or pigs or something like that. They were probably here before us humans I guess, so like animals have more right to live than people, I think.

Contrary to S1, S2 takes a more biocentric perspective. Biocentric moral reasoning departs from a view on moral reasoning in which not only humans but also other forms of life, such as animals, are treated as moral subjects. According to her, humans need to make spaces for the wolf to thrive. Actually, wolves may even have a priority over humans, since they may have inhabited the Swedish forests even before the humans settled. S2 does not mention that she has any experience from hunting activities, and does not explicitly reflect upon the problem that an increase in wolf population would reduce the number of available prey for humans to hunt. The difference in life experience seem to lead S1 and S2 to different conclusions.

In this case, the science inquiry in this SSI mostly relates to ecological aspects of the presence of wolves. Especially, the consequences for other species which are preyed upon by the wolves, are in focus in this SSI. The students need to understand the ecological law that the presence of predators will likely give rise to an increased biodiversity, since predators prevent a single species from becoming too dominant and restricting possibilities for other species to live and thrive. If, for instance, the elk become too abundant, causing overgrazing of the forest, the introduction of wolves in the area will likely decrease the elk population to more sustainable levels, thus increasing the amount of resources for other grazing animals and decreasing the damage on the forest, especially on the young trees and seedlings, for the forest owners. However, if the wolf does not exist in the area, or there are too few wolves to make any significant impact on the elk population, human hunters need to take the place of the top predator, and reduce the elk population. The latter is the case in most areas in Sweden. At present, we need to hunt, for example, elk and raw deer to avoid an over-increase in their populations.

The moral inquiry conducted by the students in this issue relates to the moral rights of existence and access to resources. This could be argued from both anthropocentric and biocentric or ecocentric perspectives. Students arguing from an anthropocentric perspective (such as S1) base their argumentation on human needs and human rights. For example, the right to hunt is important from an economic perspective for people in the countryside, providing an important source of meat. Furthermore, hunting is important from a cultural perspective, constituting a central recreational activity with historic roots of such age that we may even argue that it involves cultural aspects of connections to past generations and history. Students arguing from a biocentric or ecocentric perspective (such as S2), on the other hand, argue for the rights of the animals (or even the ecosystems as such) to continue to exist and inhabit the areas they have inhabited for a very long time. S2 argues that, since the wolf has existed in Sweden even before the humans migrated to the areas, the wolf has a right to exist there. Furthermore, the wolf has the right to exist in sustainable populations, not just as a few individuals. This moral right may, according to this line of reasoning, override the interests of people living in the same area.

The quantitative analysis of the total number of arguments identified in the analysis with the SEE-SEP model showed a dominance of value based arguments in the wolf issue (Fig. 17.2). The relation between the different dimensions of the SEE-SEP model has been shown to vary between different SSI (e.g., Christenson et al., 2012, 2014; Karisan & Cebesoy, 2021). For instance, Christenson et al. (2012) investigated upper secondary students' written reflections on four different SSI: GMO, nuclear power, consumption, and global warming. Arguments based on knowledge were most abundant related to the issue of nuclear power, while

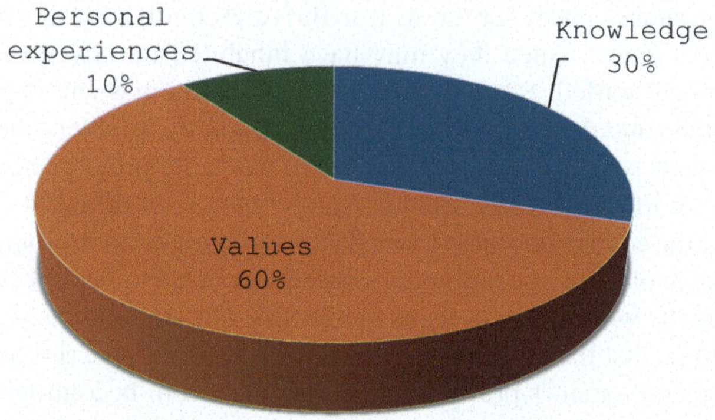

Fig. 17.2 The distribution of students' arguments on the wolf issue according to the three dimensions of knowledge, values and personal experience from the SEE-SEP model. (From Eriksson & Rundgren, 2012)

knowledge-based arguments were least abundant regarding the consumption issue. This may not be surprising, since SSI involving complex aspects of science and technology, such as nuclear power, may create a need for students to consider facts from science and other areas, and to build knowledge-based arguments. While issues no less complex, but with more fuzzy boundaries, such as consumption, may trigger more holistic reasoning where value-based arguments play a greater role.

Reflecting on the outcome of the SSI on the wolf issue, we consider the topic well chosen as a complex SSI, in which knowledge from different subject areas, such as science, policy, ecology, economics, social culture, and morality/ethics are needed to understand different aspects of the issue. Furthermore, the upper secondary students in the study showed an ability to take in different aspects in their reasoning. The interviews in pairs provided an opportunity for the students to listen to the arguments from their peers, which made them consider additional perspectives in their argumentation. The wolf issue is an SSI which can raise strong feelings. In this case, however, the students' argumentation never became agitated. At the same time, it was evident that the wolf issue in some cases touched upon certain strongly held values, which may connect to the students' prior life experiences. The results from the study on the wolf issue gave some interesting indications of the relationship between students' personal experience and values in argumentation. The life experiences of S1 of hunting and the culture of hunting gave him a reason to value this form of life, and subsequently argue from a standpoint of preservation of the possibilities to hunt also in the future. Such personal experiences can be related to the formation of what has been termed the students' *intellectual baggage*. Zeidler (1997) has described intellectual baggage as the previously established cognitive and moral beliefs that students bring to the classroom. These cognitive and moral beliefs can be defined as deeply rooted, abstract motives which guide and motivate a person's attitudes, norms, views, and actions.

17.1.4.2 Study B: Environmental Toxins in Fish from the Baltic Sea

Rundgren et al. (2017) investigated Swedish upper secondary school students' argumentation and decision-making related to an issue on environmental toxins in fish from the Baltic Sea, and food-safety policies related to this. A multi-disciplinary instructional module, designed in order to develop students' skills to argue about complex SSI, was tested in the study. Seven science majors in the final year of their upper secondary studies participated. Their argumentation and decision-making processes were followed closely, and data were collected during multiple stages of the instructional module: group discussions were audio recorded, the participants wrote reports on their decision making, and post exercise interviews were conducted with individual students. The analysis focused on the relationships between the knowledge, values, and personal experiences that they used in their argumentation (based on the three dimensions in the SEE-SEP model).

The accumulation of dioxins in fish from the Baltic Sea is an authentic SSI which relates to environmental toxins, health and sustainable development. The Swedish National Food Agency released a report in 2011, which indicated the presence of high levels of dioxins in certain species of fatty fish in the Baltic Sea, such as Baltic herring and wild salmon. Dioxins are persistent organic compounds, which in concentrations above certain values can pose health risks to humans and other species. Dioxins are among the most potent toxins we know, and can give rise to health risks even in low concentrations. Among other adverse effects, dioxins have been linked to cancer, disorders in the immune- and nervous system, liver damage, and sterility. To ensure the health and safety of EU consumers, the European Commission established regulations in 2002 which imposed an upper limit on the dioxin content in food intended for human consumption. Any food whose dioxin content exceeds this limit may not be marketed in the EU. Therefore, fatty fish from the Baltic Sea can no longer be sold on the European market. Sweden, Finland, and Latvia currently have exemptions from the EU legislation on dioxins for certain fish species (mainly herring and wild salmon) from the Baltic Sea. This permits fish whose dioxin content exceeds the threshold to be marketed within these countries. However, this fish that can be sold in the market within the countries which are part of the exemption, is not allowed to be exported to other European countries. In 2011, Sweden's exemption was made permanent based on the condition that the Swedish National Food Agency would be able to disseminate information regarding the risks of consuming such fish present to certain groups of people (i.e., children, pregnant women and women of fertile age, and presumed high consumers) among the Swedish population (Swedish National Food Agency, 2011). According to the Swedish government, this exemption is of national importance since it provides job opportunities and ensures that rural coastal areas remain livable while also protecting the Swedish cultural tradition of eating fermented herring.

The final task the students were given at the end of the four weeks in which the science lessons were devoted to the study of this SSI was:

*Do you agree or not agree that Sweden should have a permanent exemption from the EU regulations on dioxin levels for fatty fish from the Baltic Sea? Write down and rank the four most important supporting reasons for your decision. Write a brief argumentative text describing your decision and the factors that motivated it.

All seven students who participated in our study had access to the same information but nevertheless weighed this information differently. Even though all of the students agreed on the factual aspects of the issue, the results showed that they came to different decisions. Our interpretation is that their decisions differed depending on their background knowledge, values, and personal experiences. Kolstø (2006) found that certain types of knowledge and values—termed *decisive knowledge* and *decisive values*—are particularly important for individual students' decision-making on SSI. Kolstø further suggests that the main question is not whether certain knowledge is relevant for making policy decisions but what kind of knowledge is regarded as relevant by people with different values. For example, based on the same knowledge base, one individual could argue that nuclear power should be used because of its energy efficiency, while another individual would argue that nuclear power should not be used because of the risk of accidents with far-reaching consequences.

Three out of seven students acknowledged the health risks associated with excessive consumption of fish containing dioxins, but placed a greater weight on the right of individual consumers to freely choose what they would eat—they regarded this as a decisive value that overrode the health risks of eating contaminated fish. All three argued that the health risks of consuming fatty fish should be compared to those presented by other legal products such as alcohol or tobacco, thereby introducing a relativistic view of the risks presented by the consumption of dioxin-contaminated fish. While they acknowledged that some social groups were unaware of the risks presented by the fish, they stated that it is the individual responsibility of the consumer to obtain such information. They also assigned some weight to the economic and cultural arguments supporting the exemption.

S3: The health hazard that dioxin and PCBs represent is obviously not as intimidating compared to many other threats, but can anyway of course affect a few unfortunate individuals. The damage caused when taking a schnapps to the food at the Swedish midsummer celebration is probably much greater than those caused by eating one or two sandwiches with fermented herring.......There is considerably worse health hazards than dioxins and PCBs, and they are only limited by an age limit, e.g. tobacco and alcohol. As long as legal drugs are sold by the state, the exemption should not be abolished.....Parents of young children have the same responsibility to choose not to eat fish with high dioxin levels as women should abstain from consuming alcohol and tobacco during their pregnancy. What you consume is highly your own active choice and the dangers of consuming the wrong goods can easily be avoided.

S4: However, to just make prohibitions are rarely good solutions. Instead, the government could introduce regulations about labeling dangerous fish, similar to the marking of products that contain tobacco and alcohol: A label that points out the dangers and how often the fish can be consumed without being harmful to the human body. Citizens may then decide for themselves whether they want to eat that kind of fish and how often. Today, similar regulations are applied to other hazardous products sold on the Swedish market.

The other group consisted of the remaining four students who came to the decision that the exemption is wrong. They weighted the health-risk argument heavily enough for it to override all other concerns. They did not consider economic and cultural arguments to have the same weight as the health risks. The students in this group seemed to embrace the view that the government authorities have a special responsibility to inform citizens about health risks and protect them from harm. We may identify, especially in the argumentation of S6, the idea of acting with extra caution when the long-term risks are relatively unknown.

> S5: The long term effects of dioxins are relatively unknown, and we do not have much knowledge about the subject. Dioxins are suspected to cause some forms of cancer and affect the immune system, endocrine system, reproductive ability, and higher doses can affect the brain and nervous system…….'Besides affecting us today with not yet fully known consequences, all future generations will also be affected. A woman exposed to toxins such as dioxins, accumulates them and transfers the dioxins via the placenta and breast milk to her child. Therefore, the levels of dioxins in individuals will increase for each generation. It is known that we can tolerate a certain level of intake, but how will this affect us, if we already have dioxins from our mothers and perhaps simultaneously are exposed to other toxins.

> S6: Dioxins constitute a threat. Dioxins are found in small invisible amounts. However, it is toxic even in small quantities. This means that dioxins represent an invisible threat. Dioxins must be met with caution. We must not be soft, and because the threat is invisible we must take the utmost precautions……there are other toxins in the environment as well, similar to dioxins but coming from completely different sources. Nobody knows what happens when they are mixed together.

In the case of this issue, the science inquiry in this SSI mainly relates to understanding the spread and effect of environmental toxins in food webs. The students need to understand the concepts of bioaccumulation (that persistent environmental toxins such as dioxins can be stored in the fat tissues of animals) and biomagnification (meaning that concentrations of environmental toxins become higher the further up in the food pyramid you come, which means that apex predators such as eagles, seals, and humans are more vulnerable). Furthermore, the students need to understand the physiological and possible medical consequences for humans of regular consumption of fish containing dioxins. The consequences of bioaccumulation will have the result that, even if there are no immediate toxic effects of consuming such fish, the dioxins are stored in our fatty tissue, and adverse health effects may be visible after many years of consumption. The consequences of biomagnification place humans among the more vulnerable species, since we consume animals higher up in the food pyramid, which have already accumulated toxins in their body. This would not have been the case if we, for example, would follow a strictly vegetarian diet and only consume algae from the same waters. The algae are on a low level of the food pyramid, and accumulate less environmental toxins as compared to salmon and herring.

The moral inquiry conducted by the students on this issue mainly relates to aspects of risk in relation to food safety, and values of freedom of choice. The issue has a strong component of policy-making. Here, the students are confronted with a situation in which they need to balance between public food safety with potential little-known long-term risks, and freedom of choice to eat foods that have been eaten for centuries, and has a strong place in Swedish food culture, especially in the coastal regions in Northern Sweden. Interestingly, S3 and S4 who came to the decision that the exemption was right, both discussed risk in terms of relative risk. They compared the risk of eating fatty fish from the Baltic with the use of other products with known risks, such as alcohol and tobacco. For them, the decisive value was freedom of choice, and that the government should avoid banning products from the market as far as possible. On the contrary, S5 and S6 highlight the need for caution in a case with little known risks. They came to the decision that the exemption was wrong, and argued that the government should ban Baltic wild salmon and herring from the market. S5 and S6 seem to argue that it is morally wrong to sell the same fish in Sweden which is forbidden to export to other European countries such as France. In her reasoning, S5 further highlights the possible consequences of the exemption for coming generations. She argues that it is morally wrong that an unborn fetus should be exposed to environmental toxins already in the womb.

Reflecting on the outcome of the issue on environmental toxins in fish from the Baltic Sea, we consider the topic work well as a complex SSI, in which knowledge from different subject areas, such as ecology, policy, economics, culture, ethics, chemistry and medicine are needed to understand different aspects of the issue. Furthermore, the upper secondary students in the study showed ability to take in different aspects in their reasoning. It is interesting to see how they could come to different decisions based on the same factual content. The design of the task made the students focus on policy aspects of the issue, whether the exemption from the EU regulations was right or wrong. A consequence of this was that some of the ecological or environmental aspects of the spread of environmental toxins were less discussed than may have been the case with a different framing of the task. It would, for example, have been possible to frame the issue in a way which more clearly highlighted the consequences of the spread of dioxins for other top predators than humans, such as seals and sea eagles. Such aspects were indeed briefly touched upon by the students in their discussions, but it was in no way prominent in their reflections. In contrast to the wolf issue, this SSI did not seem to promote biocentric or ecocentric moral reasoning in the same way. The design of the SSI seemed to lead the students into more or less exclusive anthropocentric moral reasoning.

17.1.5 Reflection on the Intertwinement of Science and Moral Inquiry in the Two Studies

From the above-given examples, we can conclude that in order for the students to conduct a moral reflection, they at the same time need to understand certain aspects of the science related to the issue. In the case of the wolf issue, there was a need for understanding some aspects of ecology in order to come to a moral standpoint on the management of the wolf population. In his argumentation, S1 needed to show that he grasped the ecological consequences of an over-increase in the elk population, and that human hunters need to take the place of the wolf as top predator in the forest ecosystems. He showed awareness that a substantial increase in the wolf population may disrupt this balance. In his moral reasoning, he valued this as something which should be avoided, and that we should adopt a policy of increased wolf hunting in order to avoid this situation. In the moral reasoning of S2, this situation was not seen as something to be avoided, but as something natural and positive. A substantially increased wolf population would be a situation closer to the long-term historical situation as she saw it.

In the case of the fish issue, the students needed to make a risk assessment of the consumption of fish from the Baltic Sea. They needed to have an adequate understanding of science concepts such as bioaccumulation and biomagnification in order to make a sound risk assessment. In the case of S5, her science understanding of the possible passing of lipophilic substances from the placenta to the fetus, makes her come to the moral conclusion that the sales of herring and wild salmon from the Baltic is wrong, since it may have negative consequences for future generations. However, an adequate understanding of the science related to the fish issue does not necessarily make it easy to come to a conclusion regarding the question whether the exemption is right or not. An adequate understanding of the risks involved may very well be that the long-term risks are relatively unknown. We may, like S5 and S6, conclude that caution is a good principle when the long-term risks are relatively unknown. However, without the exemption we risk the income of coastal fishermen who may have fished for many generations, and we also put an effective end to the tradition of producing and eating fermented herring in Sweden. Which decision that in the end is to be preferred relates to how we balance estimation of long-term risk, the possibility to disseminate information about risks and how to handle those, as compared to economic and cultural values.

17.2 Reflective Discussion

In the light of the current pressing global sustainability issues, the development of global citizenship education has become imperative. However, this can be done in many different ways. One way is through SSI education. Below we present a didactic model as a way to design SSI education aiming to train science and moral inquiry skills to promote global citizenship competences. The two examples of SSI education in Sweden given above show how science and moral inquiry can be intertwined in SSI tasks. In both cases, care was taken to make students take in opposing views and arguments from their peers. We could also see how they, based on this, revised their arguments during the teaching unit. Still, many students seem to hold certain decisive values or had certain decisive experiences which at an early stage in the discussion seem to decide their basic standpoint. From a pluralistic standpoint, it is important to let many voices be heard in the classroom, and to design the teaching and the tasks in a way that makes it necessary for the students to take in different perspectives in their reasoning. In the case of the wolf issue, the interviews were performed in pairs where one of the students had expressed that s/he was negative to the presence of wolves in the neighborhood, while the other had expressed positive attitudes to this in the questionnaire. In the case of the SSI about fish from the Baltic Sea, the students were assigned to have a discussion in group before writing down his or her individual decision.

17.2.1 A Didactic Model for SSI Education and Sustainability Embracing Science and Moral Inquiry

Over the past 30 years, SSI-related educational research has become one of the main trends in the field of science education and many researchers have appreciated the significance of SSI, emphasizing the need to design issue-based curricula as part of the school science canon (Chang Rundgren & Rundgren, 2010; Driver et al., 2000) and for education for sustainability (Chang Rundgren, 2023; Levrini et al., 2021). The teacher's crucial role for ensuring the sustainability of our world has been addressed in several studies (e.g., Bourn et al., 2017; Montebon, 2018; UNESCO, 2014, 2020). However, just as many teachers were found to be positive about but ill-prepared to integrate education for sustainability into science lessons, teachers has also been found to be positive about yet ill-prepared for SSI-based teaching (Amos & Levinson, 2019; Rundgren & Chang Rundgren, 2018). There have been ongoing discussions about how to integrate education for sustainability into science education programs and science teacher education to form responsible citizens for a sustainable future (e.g., Feldman & Nation, 2015; Stratton et al., 2015). It has been suggested that school education in the twenty-first century should embrace a broader perspective to prepare citizens to understand the various components of sustainability—including social, environmental, and economic aspects—and enable them to make social, political, and environmental decisions for themselves and their communities (Feldman & Nation, 2015). All the above-mentioned ESD and twenty-first century science education initiatives are embraced in the earlier mentioned critical scientific literacy or Vision III of scientific literacy, which discusses scientific engagement and socio-political action (Sjöström & Eilks, 2018). Based on the importance and complexity of education for sustainability, the need for developing didactic models has been recognized (Wickman et al., 2020; Herranen et al., 2021) to allow for the implementation of teaching designs and analyses in new teaching contexts for sustainability. Here, we want to suggest an integrated didactic model for SSI education and sustainability.

A didactic model called the Inquiry- and Context-based (IC-based) three-step model was developed in an EU FP7 project, PROFILES (Walan & Chang Rundgren, 2015; Chang Rundgren, 2023). As shown in Fig. 17.3, the IC-based three-step

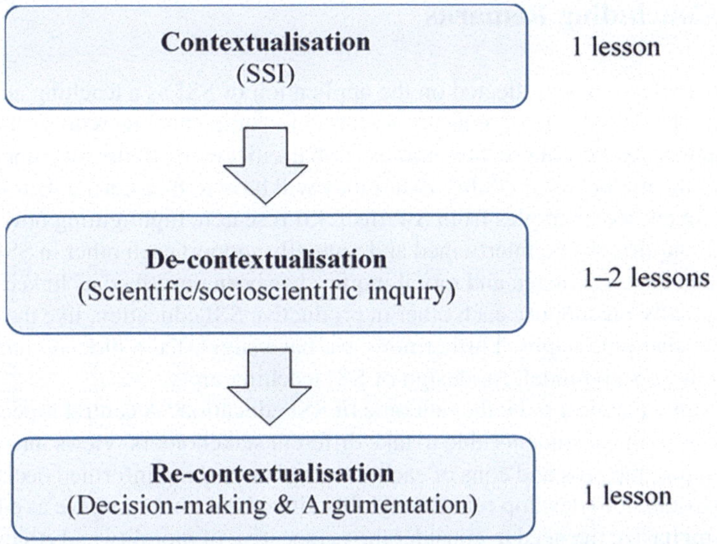

Fig. 17.3 The IC-based three-step model. (Chang Rundgren, 2023, p. 25)

model includes the steps of contextualization (introducing SSI), de-contextualization (conducting scientific or socio-scientific inquiry) and re-contextualization (making decisions and argumentation), and this teaching activity can be structured as a minimum three-hour activity in a school lesson or longer one-month course or even a one-term program.

In our opinion, the second step of de-contextualization provides many possibilities for conducting science and moral inquiry. For example, teachers can use the SEE-SEP model to enable students to find multi-disciplinary, multi-stakeholder, and multi-community solutions (based, for instance, on the SEE-SEP model's six subject areas of science, economy, ethics/morality, social culture, environment, and policy) by searching information and conducting inquiry via the Internet or interviewing local, national, and international stakeholders. Here, ethics/morality perspective can be highlighted, thus making students focus on moral and ethical perspectives simultaneously with science perspectives in the same inquiry process. It is certainly possible to apply ICT in this step, for example, by using video link for interviews or searching for viewpoints via social media. Then, in the step of re-contextualization, ICT, youth-led activities, policy, and strategy aspects can be engaged in decisions and solutions, feeding back into step 1 of the SSI context. Much depends on the available time and how teachers design the activities. It is important to address that the IC-based three-step model with SSI context can also be applied purely for science inquiry in the step two while laboratory work for example is involved.

17.3 Concluding Remarks

In this chapter, we have reflected on the application of SSI as a teaching and learning context in school science education aiming to equip students with competences and awareness to be able to function as global citizens of tomorrow, not least in relation to the major sustainability issues they will have to face during their lifetime. We have given two examples from Swedish SSI research, highlighting how science and moral inquiry can be intertwined and mutually support each other in SSI education. In the chapter, science and moral inquiry has been described as linked entities which mutually presuppose each other in productive SSI education, like the Yin and Yang of Daoist philosophy. Furthermore, we have suggested a didactic model, the IC-based three-step model, for design of SSI teaching units.

What can we expect to be the outcome of SSI education? A central aspect of SSI education is to make students able to take different stakeholders' views into account and evaluating the pros and cons of each view before making informed decisions on SSI. In particular, to develop reflective global citizens of tomorrow, we as educators need to emphasize the need to consider the perspective of morality and ethics. In our view, morals and ethics are critical components of SSI education and sustainability. The moral and value discussion related to SSI may not be easy to handle for science teachers, at least partly due to the lack of a tradition of educational training in many teacher education programs. Therefore, we also want to argue for the need in teacher professional development to target the management of moral and value aspects of science education in general and of SSI education in particular. We all need to learn from the diversity of human values, showing respect to others and their values and views, and how we can find a balance of the Yin and Yang of science and moral inquiry in SSI education.

The visions of education laid out already by the Stoics during antiquity aiming to educate world citizens who could liberate their thinking from limiting traditions of mind and rise above the confined, and perhaps narrow, interests of their local community, to exercise awareness and care for humanity as a whole is just as valid today. The Bildung tradition has kept this vision alive through the past centuries. Today, the vision has become even more inclusive, including awareness, knowledge of and care for the whole planet and its nonhuman inhabitants. Whether we chose to explicitly acknowledge those ancient roots or not, SSI education continues this tradition today.

References

Amos, R., & Levinson, R. (2019). Socio-scientific inquiry-based learning: An approach for engaging with the 2030 Sustainable Development Goals through school science. *International Journal of Development Education and Global Learning, 11*(1), 29–49.

Andersson, K. (2017). Starting the pluralistic tradition of teaching? Effects of education for sustainable development (ESD) on pre-service teachers' views on teaching about sustainable development. *Environmental Education Research, 23*(3), 436–449.

Bourn, D., Hunt, F., & Bamber, P. (2017). *A review of education for sustainable development and global citizenship education in teacher education*. UNESCO Press.

Cebesoy, U. B., & Chang Rundgren, S.-N. (2023). Embracing socioscientific issues-based teaching and decision-making in teacher professional development. *Educational Review, 75*(3), 507–534.

Chang, S.-N., & Chiu, M. H. (2008). Lakatos' scientific research programmes as a framework for analysing informal argumentation about socioscientific issues. *International Journal of Science Education, 30*(13), 1753–1773.

Chang Rundgren, S.-N. (2011). Post it!—A cross-disciplinary approach to teach socioscientific issues. *Teaching Science, 5*(3), 25–28.

Chang Rundgren, S.-N. (2023). Demonstrating didactic models for ESD and Bildung in school education. *Multidisciplinary Journal of School Education, 12*(2(24)), 15–31.

Chang Rundgren, S.-N., & Rundgren, C.-J. (2010). SEE-SEP: From a separate to a holistic view of socioscientific issues. *Asia-Pacific Forum on Science Learning and Teaching, 11*(1), Article 2.

Christenson, N., Chang Rundgren, S. N., & Höglund, H. O. (2012). Using the SEE-SEP model to analyze upper secondary students' use of supporting reasons in arguing socioscientific issues. *Journal of Science Education and Technology, 21*, 342–352.

Christenson, N., Chang Rundgren, S. N., & Zeidler, D. L. (2014). The relationship of discipline background to upper secondary students' argumentation on socioscientific issues. *Research in Science Education, 44*, 581–601.

Dewey, J. (1903). Democracy in education. *The Elementary School Teacher, 4*(4), 193–204.

Driver, R., Newton, P., & Osborne, J. (2000). Establishing the norms of scientific argumentation in classrooms. *Science Education, 84*, 287–312.

Elmose, S., & Roth, W.-M. (2005). Allgemeinbildung: Readiness for living in risk society. *Journal of Curriculum Studies, 37*(1), 11–34.

Eriksson, M., & Rundgren, C.-J. (2012). Vargfrågan—Gymnasieelevers argumentation kring ett sociovetenskapligt dilemma. *NorDiNa (Nordic Studies in Science Education), 8*(1), 26–41.

Feldman, A., & Nation, M. (2015). Theorizing sustainability: An introduction to science teacher education for sustainability. In S. K. Stratton, R. Hagevik, A. Feldman, & M. Bloom (Eds.), *Educating science teachers for sustainability* (pp. 3–13). Springer.

Gyllenpalm, J., Rundgren, C.-J., Lederman, J., & Lederman, N. (2021). Views about scientific inquiry: A study of students' understanding of scientific inquiry in grade 7 and 12 in Sweden. *Scandinavian Journal of Educational Research, 66*(2), 336–354.

Haglund, J., & Hultén, M. (2017). Tension between visions of science education: The case of energy quality in Swedish secondary science curricula. *Science & Education, 26*, 323–344.

Herranen, J., Yavuzkaya, M., & Sjöström, J. (2021). Embedding chemistry education into environmental and sustainability education: Developing of a didactic model based on an eco-reflexive approach. *Sustainability, 13*(1746), 1–15.

Johansson, M., & Svensson, T. (Eds.). (2009). Att lära för det okända: Miljövetarprogrammets jubileumsskrift. Linköping University, The Tema Institute, Department of Water and Environmental Studies. Linköping University, Faculty of Arts and Sciences. Retrieved from https://urn.kb.se/resolve?urn=urn:nbn:se:liu:diva-59049

Karisan, D., & Cebesoy, U. B. (2021). Use of the SEE-SEP model in preservice science teacher education: The case of genetics dilemmas. In W. A. Powell (Ed.), *Socioscientific issues-based instruction for scientific literacy development* (pp. 223–254). IGI Global.

Klafki, W. (2000). The significance of classical theories of Bildung for a contemporary concept of Allgemeinbildung. In I. Westbury, S. Hopmann, & K. Riquarts (Eds.), *Teaching as a reflective practice: The German Didaktik tradition* (pp. 85–108). Lawrence Erlbaum.

Kolstø, S. D. (2006). Patterns in students' argumentation confronted with a risk-focused socioscientific issue. *International Journal of Science Education, 28*(14), 1689–1716.

Lederman, J. S., Lederman, N. G., Bartos, S. A., Bartels, S. L., Meyer, A. A., & Schwartz, R. S. (2014). Meaningful assessment of learners' understandings about scientific inquiry—The views about scientific inquiry (VASI) questionnaire. *Journal of Research in Science Teaching, 51*(1), 65–83.

Lenman, J. (2007). What is moral inquiry? *Proceedings of the Aristotelian Society, Supplementary Volumes, 81*, 63–81. http://www.jstor.org/stable/20619102

Levrini, O., Tasquier, G., Barelli, E., Laherto, A., Palmgren, E., Branchetti, L., & Wilson, C. (2021). Recognition and operationalization of Future-Scaffolding Skills: Results from an empirical study of a teaching-learning module on climate change and future. *Science Education, 105*, 281–308.

Mattingly, C., & Throop, J. (2018). The anthropology of ethics and morality. *Annual Review of Anthropology, 47*, 475–492.

Montebon, D. R. T. (2018). Pre-service teachers' concept of sustainable development and its integration in science lessons. *Journal Pendidikan Humaniora, 6*(1), 1–8.

Nussbaum, M. C. (1998). *Cultivating humanity*. Harvard University Press.

Roberts, D. A. (2007). Scientific literacy/science literacy. In S. K. Abell & N. G. Lederman (Eds.), *Handbook of research on science education* (pp. 729–780). Lawrence Erlbaum Associates.

Rudolph, J. L. (2002). *Scientists in the classroom: The cold war reconstruction of American science education*. Palgrave.

Rudolph, J. L. (2005). Epistemology for the masses: The origins of the scientific method in American schools. *History of Education Quarterly, 45*(3), 341–376.

Rundgren, C.-J., & Chang Rundgren, S.-N. (2018). Aiming for responsible and competent citizenship through teacher professional development on teaching socioscientific inquiry-based learning (SSIBL). *Asia-Pacific Forum on Science Learning and Teaching, 19*(2), article 2. http://www.eduhk.hk/apfslt/v19_issue2/changsn/index.htm#con

Rundgren, C.-J., Eriksson, M., & Chang Rundgren, S.-N. (2017). Investigating the intertwinement of knowledge, value and experience of upper secondary students' argumentation concerning socioscientific issues. *Science & Education, 25*(9), 1049–1071.

Sjöström, J., & Eilks, I. (2018). Reconsidering different visions of scientific literacy and science education based on the concept of Bildung. In Y. J. Dori, Z. Mevaresh, & D. Baker (Eds.), *Cognition, metacognition, and culture in STEM education—Learning, teaching and assessment* (pp. 55–67). Springer.

Stratton, S., Hagevik, R., Feldman, A., & Bloom, M. A. (2015). Toward a sustainable future: The practice of science teacher education for sustainability. In S. Stratton, R. Hagevik, A. Feldman, & M. A. Bloom (Eds.), *Educating science teachers for sustainability* (pp. 445–458). Springer.

Sund, P., & Gericke, N. (2020). Teaching contributions from secondary school subject areas to education for sustainable development—A comparative study of science, social science and language teachers. *Environmental Education Research, 26*(6), 772–794.

Swedish Institute of Educational Research (2022). *Samhällsfrågor med naturvetenskapligt innehåll—en kartläggning av under visningsmöjligheter* [Socio-scientific Issues—A systematic mapping review]. Systematisk forskningssammanställning 2022:01. Solna: Skolforskningsinstitutet.

Swedish National Agency of Education. (2022). *Curriculum for the compulsory school, preschool class and the recreational center 2022*. Skolverket.

Swedish National Food Agency. (2011). *Redovisning av regeringsuppdrag rörande gränsvärden för långlivade miljöföroreningar i fisk från Östersjöområdet* [Description relating to Limit Values for Persistent Environmental Contaminants in Fish from the Baltic Sea Region.] Report. Swedish National Food Agency, Dnr 115/2010. http://www.slv.se/upload/dokument/remisser/regeringsuppdrag_2011/rapport_regeringsuppdrag_dioxinundantag_2011_02_28.pdf (10.07.2013).

UNESCO. (2014). *UNESCO roadmap for implementing the Global Action Programme on education for sustainable development.* Retrieved from https://sustainabledevelopment.un.org/content/documents/1674unescoroadmap.pdf

UNESCO. (2020). *Education for sustainable development; a roadmap*. UNESCO.

Walan, S., & Chang Rundgren, S.-N. (2015). Student responses to context- and inquiry-based three-step teaching model. *Teaching Science, 61*(2), 33–39.

Wang, G., Zhou, X., & Cui, H. (2019). Exploring education for sustainable development in a Chinese kindergarten: An action research. *ECNU Review of Education, 2*(4), 497–514.

Wiblom, J., Andrée, M., & Rundgren, C.-J. (2020). Navigating alarming media messages about nutrition and health: How students engage in critical examination of science in news media. *Science & Education, 29*(1), 75–100.

Wickman, P.-O., Hamza, K., & Lundegård, I. (2020). Didactics and didactic models in science education. In P. J. White, R. Tytler, J. Ferguson, & J. C. Clark (Eds.), *Methodological approaches to STEM education* (Vol. 1, pp. 34–49). Cambridge Scholars Publishing.

Zeidler, D. (1997). The central role of fallacious thinking in science education. *Science Education, 81*(4), 483–496.

Zeidler, D. L., & Sadler, T. D. (2023). Exploring and expanding the frontiers of socioscientific issues. In N. G. Lederman, D. Zeidler, & J. Lederman (Eds.), *Handbook of research on science education* (pp. 899–929). Routledge.

Chapter 18
Comprehensive Exploration of Science Education Research within Socioscientific Contexts in Taiwan: Examining Curriculum Reform, Research Status, Reflections and Implications

Shu-Sheng Lin, Shiang-Yao Liu, and Ying-Shao Hsu

18.1 Science Education and Socioscientific Issues

After more than two decades of efforts by numerous science education scholars worldwide, the term 'Socioscientific Issues' (SSI) has become a widely recognized term in the field of science education. It represents a conceptual framework that fosters and guides the development of science education theory, research, and practice (Zeidler, 2015), gradually giving rise to the 'SSI movement' in the research domain of science education (Zeidler et al., 2005). Following the prevalence of "Science, Technology, and Society (STS) movement" in the 1990s, this signifies another wave of research trends in the global science education community. In relative terms, SSI places a stronger emphasis on epistemology, higher-order thinking skills, character development, and emotional and moral reasoning in students, in contrast to the focus of STS teaching and research (Zeidler & Keefer, 2003; Berkowitz & Simmons, 2003). Even today, many scholars continue to use SSI as a context or teaching material, publishing scientific teaching or related research in various journals. Taiwanese scholars, as members of the global community, are no exception to this trend.

In essence, SSI involves social controversies arising from the development and progress of science and technology (Stradling, 1984). The occurrence of controversy is precisely because any interest group involved in the issue will propose

S.-S. Lin (✉)
Department of Education, National Chiayi University, Chiayi City, Taiwan
e-mail: lin-s-s@mail.ncyu.edu.tw

S.-Y. Liu · Y.-S. Hsu
Graduate Institute of Science Education, National Taiwan Normal University, Taipei City, Taiwan

D. L. Zeidler (ed.), *A Moral Inquiry into Epistemic Insights in Science Education*, Contemporary Trends and Issues in Science Education 61,
https://doi.org/10.1007/978-3-031-63382-9_18

solutions that align with its own position and values. Due to the lack of clear evidence to prove which solution is superior and the absence of consensus on solutions, issues remain unresolved in a situation where each party holds its own views without compromise. Millar (1997) pointed out that technologies causing social controversies typically exhibit the following characteristics: (1) Experts lack consensus on whether the application of this science and technology in society will result in predominantly positive or negative outcomes. Predictions about the technology's consequences are probabilistic rather than absolute certainty. (2) Disputes over judgment involve incomplete and unclear evidence and data, leading to a lack of public consensus in discussions and debates.

In our lives, what kinds of SSI have occurred or are currently taking place? Issues such as climate change and global warming, the emergence of cloned individuals, the cultivation of genetically modified crops, embryonic stem cell research, the discharge of nuclear wastewater, energy utility, the construction of reservoirs, the invasion of exotic species, and more. These topics encompass global issues, transnational regional issues, as well as local issues specific to a particular country or region. Zeidler et al. (2003) pointed out that regardless of the type of SSI, each can serve as a case study, guiding students in exploration led by teachers. The entire teaching process can promote personal development in knowledge, ethics, and emotions related to the issues. It also stimulates interaction between teachers and students or peers, fostering an understanding of diverse perspectives and positions for resolving controversies. This, in turn, connects to dialogues across different disciplinary areas.

SSI occurs in our surroundings, and we cannot ignore their presence or disregard their impact on our lives and society. Therefore, strengthening students' abilities to face SSI through science education is a crucial goal in education (Kolsto, 2001) and a significant aspect in cultivating students' functional scientific literacy (Zeidler et al., 2019). In essence, we aspire to train students to approach SSI with rationality, engage in evidence-based discussions and negotiations, and embrace the social responsibility of future citizens by making decisions that benefit the public and are environmentally friendly to the Earth's ecology. However, engaging in rational thinking within the context of SSI poses a highly challenging task. We will further elaborate on this point in the upcoming section, "The Difficulties and Challenges in SSI-based Teaching."

Compared to science problem-solving with standardized answers and value-free science instruction, SSI shows characteristics such as ill-structured, open-ended questions, interdisciplinary nature, involvement of diverse and conflicting perspectives, and influence of emotions and values (Oulton et al., 2004; Levinson, 2006). Simultaneously, SSI aligns with the concept of situated learning, enabling students to confront real-life issues and connect school science with problems happening out of school, and apply acquired scientific knowledge and skills (Klosterman & Sadler, 2010; Sadler, 2009). Therefore, integrating SSI into science teaching and curriculum yields effects distinct from purely concept-based science instruction and learning, achieving the goal of fostering scientific literacy and cultivating responsible citizens in the broader sense.

The mentioned argument also align with the curriculum advocated by social reconstructionism in education. Stanley (1992) notes that this theory emphasizes that schools should guide students in understanding their connection to society, enabling them to contribute to societal improvement. It puts emphasis on the importance of continuously staying attentive to social issues, taking action to engage in social participation, actively contributing to problem-solving, and ultimately working towards creating a more equitable society. In other words, education in schools goes beyond assisting students in adapting to social life; it must also foster students' active discussion and participation in societal matters, prompting them to shoulder their social responsibilities (Groenke, 2009). Consequently, under the guidance of school teachers, building upon students' existing knowledge and abilities, engaging in collaborative inquiries into SSI within the classroom becomes a significant experience in teaching and learning. This approach aligns seamlessly with the advocated practices in the field of science education.

Previous studies have shown that using SSI as teaching topic and materials or contexts for science instruction can achieve various instructional and learning outcomes, which includes the construction of science knowledge and concepts related to the issue (e.g. Kiryak & Çalik, 2018; Sadler et al., 2016), attitudes toward science (eg. Pelch & McConnell, 2017), inquiry skills (e.g. Eastwood et al., 2013; Lee & Brown, 2018), argumentation skills (Su & Lin, 2012; Evagorou & Osborne, 2013; Lin & Mintzes, 2010), decision-making and its quality (Gresch et al., 2013; Hsu & Lin, 2017), reflective judgment (Zeidler et al., 2009), moral or ethical reasoning (Lu & Lin, 2019), moral sensitivity (Fowler et al., 2009), informal reasoning (e.g. Chin & Wu, 2017a; Karpudewan & Roth, 2018), understanding of the nature of science (e.g. Eastwood et al., 2012; Khishfe, 2013; Lederman et al., 2014; Khishfe et al., 2017).

An important observation is that, among the aforementioned learning outcomes, there has been relatively limited scholarly research on improving the concept or variable of 'character'. Despite being identified by some scholars as a teaching goal achievable through SSI-based instruction (Choi et al., 2011; Lee et al., 2012a; Zeidler et al., 2019), it appears challenging to attain. Lee et al. (2013) demonstrated significant enhancement in the character and values of 132 ninth-grade students using the issue of Genetically Modified technology and seven class sessions (each lasting 45 min). they confined the assessment of character and values to three dimensions: ecological worldview, social and moral compassion, and socioscientific accountability. However, the cultivation of students' character is a long-term process influenced by various factors such as cultural influences and societal norms from different countries. It requires prolonged observation of students' discourse and behavior to comprehend their character development. While it might be plausible to improve students' character through one or two SSI teaching experiences, the sustainability of such improvements is a critical aspect that the education community is concerned about. Furthermore, the assessment of character not only involves the scope measurable by tools but also includes challenges related to setting assessment indicators and determining whether the assessment results genuinely reflect students' character. Therefore, the hypothesis that SSI promotes students' character development still requires further extensive research for a deeper understanding.

As the SSI movement spreads and influences the research and teaching practices in science education worldwide, Taiwanese scholars are naturally not exempt from its impact. In the following sections, we explore the reforms in Taiwan's science education, the contributions of Taiwanese scholars to the global research on teaching with SSI as materials or contexts, and offer some reflections on the subject.

18.2 Socioscientific Issues and Curriculum Reform in Taiwan

Examining the status of SSI in science curriculum documents is crucial and necessary, reflecting opportunities for implementing issue-based teaching in science classrooms at the compulsory education stage. Taiwan's science curriculum has seen two waves of reform this century. The first reform refers to the curriculum guidelines for grades 1–9 promulgated in 2001, which subsequently triggered the reform of high school science subject curricula with the provisional version and the newly revised version released between 2006 and 2010. The second reform, the "Curriculum Guidelines of 12-year Basic Education–Natural Sciences," emerged in 2018. Understanding the evolution of SSI in these documents is essential for fostering issue-based teaching in science classrooms during compulsory education.

The 1–9 grades curriculum guidelines integrating natural sciences and life technology began to include the scientific literacy attributes as educational goals. Under the "Scientific Applications" attribute, for students in grades 7–9, one of the competence indicators is designated as incorporating "analysis of social issues related to science" into the learning outcomes. When detailed learning contents are listed in the appendixes of the curriculum document for references in lesson plans, there is a dedicated statement under the sub-topic "Ethics in Science" that emphasizes teaching students to think critically about science-related social issues through data collection or group discussions.

The high school science curriculum revised in 2010 was considered more in line with the concept and scope of basic education curriculum reform at that time (Li, 2010). Among science subjects, only biology curriculum guidelines explicitly required the inclusion of social, legal, and ethical issues related to the biological sciences in the teaching materials and proposed issue-based teaching strategies. High school compulsory biology was divided into basic biology and applied biology. The applied biology curriculum document addressed specific SSI topics, including genetically modified organisms, genetic screening, and biomass energy, which served as the basis for textbook development. In addition to biology, a teaching unit "complexities and dilemmas of environmental issues" was listed in the earth science elective course, but there was no description of relevant learning objectives and teaching strategies.

The latest science curriculum, spanning grades 3–12, underscores scientific inquiry at all learning stage, extending to the creation of a mandatory high school course, “Natural Science Inquiry and Practice.” In the curriculum guidelines document, the specific term “socioscientific issues” has appeared on page one, in the first chapter “fundamental rationale.” The integration of science-related social issues into learning content and learning performance has become evident throughout the document. Especially in the descriptions of the “Natural Science Inquiry and Practice” course, it is emphasized that students should learn systematic knowledge, interdisciplinary dialogue and critical thinking through these daily-life oriented issues. This reflects Taiwan’s ongoing emphasis on issue-based teaching in science education at the compulsory education stage, and teachers are thus compelled to create relevant teaching materials, marking a significant evolution to align with the changing educational landscape.

18.3 The Contributions of Taiwanese Scholars to Research on Socioscientific Instruction or Contexts

Taiwanese science education scholars not only publish papers in English in international journals but also use their native language, Chinese, as a medium for journal article publication. Similar to the publication practices in various countries worldwide, these articles, after undergoing anonymous review and acceptance, are published in academic journals by Taiwan’s academic institutions, associations, or publishing companies. Examples include journals like the “Contemporary Journal of Science Education,” “Curriculum and Instruction,” and “Journal of Research in Education Sciences”….etc. The papers published in these journals are included in the traditional Chinese literature database “Hwa-I.”

Using the Chinese term “社會性科學議題” (Socioscientific Issues) as a keyword, we searched the “Hwa-I” database for journal articles published from January 2000 to September 2023. The search yielded a total of 40 articles. After thorough reading and screening, four non-empirical studies were excluded. The remaining 36 articles were empirical studies related to SSI-based instruction or SSI context. The authors then employed content analysis and inductive methods to categorize these empirical articles based on their research objectives, resulting in five major categories (Fig. 18.1).

The first category is “Teachers’ instruction,” focusing on describing and interpreting the professional states and changes of teachers in SSI instruction. This includes the adjustment and change of professional knowledge and strategies among secondary school teachers during SSI teaching (Genetically modified food; GMO) (Lin, 2006), the transformation of roles among an elementary school teacher in SSI teaching (GMO) (Lin, 2007), the differences between expert and novice instructors at the college level guiding students in SSI argumentation (The use of cosmetics) (Hung et al., 2010), and the similarities and differences in elementary school

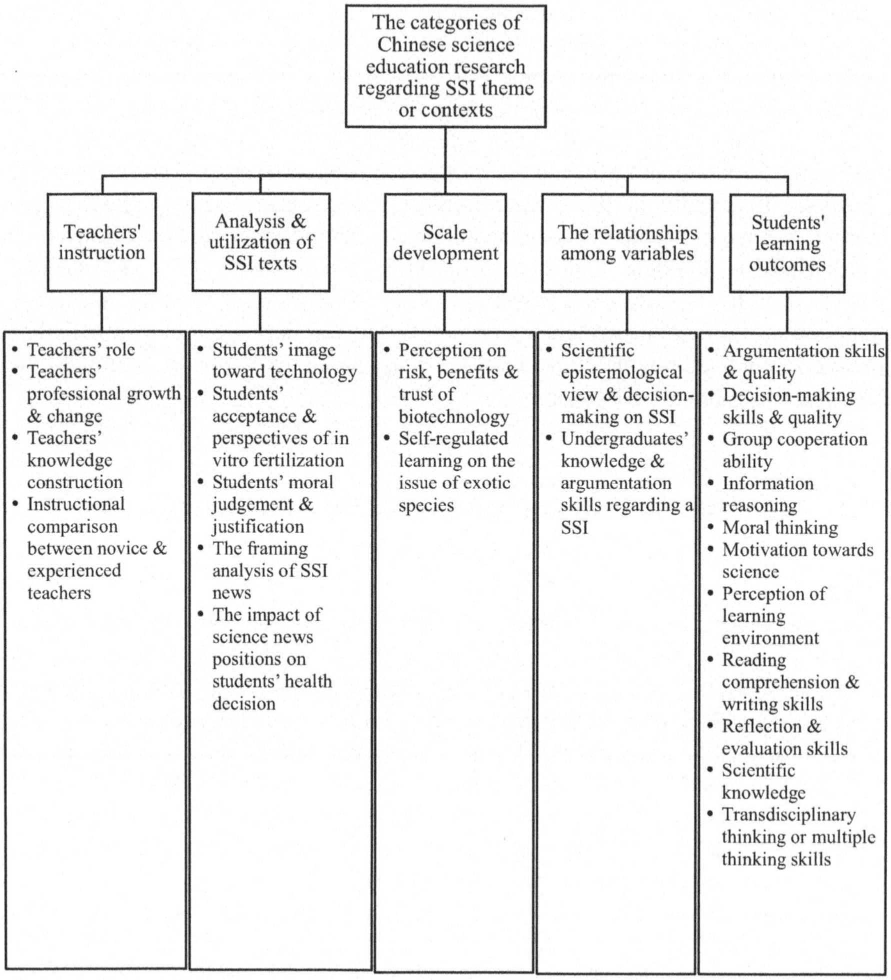

Fig. 18.1 Classification of Chinese science education research regarding SSI theme or context in Taiwan

teachers' construction of subject-specific teaching knowledge for SSI instruction (The establishment of Ma-Gao national park) (Lin & Chin, 2012).

The second category is "Analysis and utilization of SSI texts." Researchers explore into how news represents SSI (Chen & Lee, 2017), investigate university students' construction of impressions toward technology through SSI news (Huang & Jian, 2008), examine the differences between students and experts in monitoring science news (Huang, 2008), survey university students' understanding and monitoring of SSI news output (Huang, 2017), and explore changes in the stance and decision-making of female university students before and after reading SSI news on hormone-replacement therapy (Lin et al., 2014). Additionally, studies investigate the viewpoints and thoughts of non-science major university students facing a SSI (Undergraduates' acceptance and perspectives of in vitro fertilization) (Lin & Hsin, 2017), as well as the moral judgment and justifications of high school students facing SSIs (Regenerative medicine, GMO, and animal experiments) (Lu & Lin, 2019).

In the third category, "SSI scale development," a scale named "Perception on Risks, Benefits and Trust of Biotechnology" (Chin, 2015) for high school students consists of 24 items distributed across six dimensions, covering knowledge perception, relevance to daily life, issue importance, benefits, trust in risk management, and risk assessment. The scale effectively explains 65.414% of the variance, with high reliability ($\alpha > 0.70$) in each dimension. The overall alpha coefficient for the scale is .89 ($N = 450$), demonstrating its robustness in assessing students' perceptions of biotechnology.

Another scale in this category is the "A Scale for Self-Regulated Learning on the Issue of Exotic Species" (Lin, 2021), administered to elementary school students. This scale consists of 15 items distributed across three dimensions: "Driving force for SSI learning" (4 items), "Action strategies" (8 items), and "Reflection on learning gains" (3 items). The overall internal consistency (Cronbach's α) for the scale was 0.86 ($N = 406$), with individual subscales ranging from 0.63 to 0.85. The complete set of items accounts for 52.38% of the variance. Confirmatory factor analysis, based on responses from an additional 666 elementary students, confirms the data's good fit to the three-factor structure of the scale.

"The relationships among variables" in SSI contexts, providing insights into the direction and strength of relationships between different variables. This type of research helps measure whether changes in one variable are associated with changes in another. Examples include examining the correlation between college students' scientific epistemological views and their decision-making on socio-scientific issue choices (The Establishment of a mountain cable car & global warming) (Liu et al., 2007), and exploring understandings and argumentation on genetically modified organisms among college students and their correlation (Lin et al., 2010).

The fifth category primarily focuses on exploring" Students' learning outcomes" after SSI-based Instruction, such as knowledge, emotions, high-order thinking abilities, or skills among students after exposure to different SSI-based instructions. This category constitutes the largest number of studies, totaling 22 articles. In addition to enhancing students' knowledge about the issues, the examined learning outcomes include argumentation ability and quality (Lin & Huang, 2009; Lee et al., 2012a, b; Su & Lin, 2012; Yang et al., 2017), moral reasoning (Lin, 2012), decision-making ability and quality (Fang & Hsu, 2017; Lin & Hsu, 2017; Cheng et al., 2021; Wu et al., 2021; Zhang et al., 2023), reflection and evaluation skills (Zhang & Hsu, 2021a), decision-making sub-skills (issue identification, criterion establishment, evidence collection) (Wu et al., 2017), informal reasoning (Chin & Wu, 2017a), transdisciplinary or multiple thinking skills (Lin & Li, 2018; Chang et al., 2022), changes in perspective and positions (Lin & Chin, 2018; Chin & Hu, 2018), reading comprehension and writing (Xia & Chin, 2021; Chin et al., 2010), learning motivation and group collaboration abilities (Chin & Hu, 2016), and classroom environment awareness (Chin & Wu, 2017b).

Regarding studies on "Students' Learning Outcomes" after SSI-based instruction in the fifth category, we further summarize relevant findings, insights, and reflections.

Firstly, some studies adopt SSI related to transnational issues, involving controversies such as international cooperation between nations, political negotiation, economic competition, ecological destruction and biological survival, ethical norms, and the formation of laws and regulations. Examples include global warming (Lin & Chin, 2018) and cloning (Chin et al., 2010). Other studies select local SSIs, addressing disputes involving the preservation of traditional culture, local economic development, challenges in environmental conservation, and the labor rights of marginalized communities. Examples of local SSIs include the construction of high-speed railways (Lin & Huang, 2009), the celebration of Ma Zu's birthday (Su & Lin, 2012), reservoir site selection (Fang & Hsu, 2017), hillside development (Chin & Hu, 2016), Hakka community traditional holy pig competition (Xia & Chin, 2021), and mining in indigenous territories (Chang et al., 2022), among others. Compared to global or international issues, these local topics are more closely connected to students' lives, aligning with their existing life knowledge and experiences, and are more likely to engage students in classroom participation and discussions.

Secondly, various teaching strategies have been employed to guide students in learning within SSI contexts or through SSI-based instruction in these studies. These strategies include questioning, demonstration, small group or whole-class discussions, cooperative learning, debates, role-playing, public hearings, case analysis, and more. In addition to instructional strategies, teachers often provide teaching scaffoldings for student learning, guiding students in their learning process and assisting them in realizing their potential to accomplish various learning tasks (Lin et al., 2012, 2022). All kinds of instructional or learning scaffolding also play a crucial role in SSI-based instruction.

These studies have explored different scaffolding approaches to enhance students' learning experiences in diverse contexts. Chin et al. (2010) introduced the "Guided TAPing (Toulmin's Argument Pattern) cell" to improve elementary students' expression of arguments in a cloning context, revealing improved informal reasoning but still weaknesses in providing backings. Su and Lin's (2012) study demonstrated that elementary students, instructed with both oral and writing scaffoldings in the context of "Mazu's birthday celebration, promoting heavy industry, and replacing manual labor with robots," generated at least two additional valid rebuttals compared to those receiving only oral scaffoldings. However, both groups demonstrated enhanced argumentation skills. Lin and Chin (2018) employed 'Reading-and-Writing worksheets' as scaffolding to facilitate elementary students' discussions and revisions of proposed plans for polar bear protection, highlighting group collaborative efforts that led to consensus and improved decision-making quality. Wu et al. (2021) focused on high school students facing the context of utilization of nuclear power, discovering that open-ended question prompts significantly improved decision-making performance in a computer-supported collaborative learning environment. Zhang and Hsu (2021a) developed an SSI-based decision-making curriculum (coastline protection) with metacognition prompts, engaging high school students in deep reflection, especially regarding trade-off analysis. Xia and Chin (2021) successfully applied the "Question-Answer-Relationship" strategy as a teaching scaffold, improving reading comprehension and argumentation ability in elementary students across contexts of animal experiments and traditional "holy pig" competition ritual.

Together, these studies provide valuable insights into the effectiveness of various scaffolding methods across different educational levels and subjects within the context of SSI-based instruction.

Thirdly, the experimental teaching durations in these studies ranged from 4 to 18 h. Almost all studies indicated significant improvement in learning outcomes for the experimental group, either significantly outperforming the control group or comparison group. For studies with relatively short teaching interventions (e.g., 2 weeks with 2 h each week), we believe that incorporating delayed assessments into the evaluation could better demonstrate the sustained and persistent impact of instructional interventions on students' learning outcomes. This approach could provide stronger support for the depth of the instructional impact on students, showcasing learning retention and transfer, rather than just the effects of short-term teaching.

Fourthly, implications and reflection on these studies, aspects that future research can continue to explore include:

1. Attention to affective assessment

   Compared to assessments in the cognitive and skills domains, fewer studies assess students' changes in the affective domain within SSI contexts. After undergoing SSI-based instruction, do students' attitudes and sense of responsibility in participating in discussions or solving SSIs change? Does their level of concern and attention to SSIs alter? Can SSI-based instruction sustain students' motivation and interest in science learning continuously? What kind of technological integration into SSI teaching and curriculum design is needed to further enhance students' willingness and motivation to learn?

2. Further enhancement of many higher-order thinking skills

   Since SSIs involve complex and ill-structured problems, and most of them exist in a science-in-the-making state (Kolsto, 2001), evidence for problem resolution is often insufficient or undetermined. Therefore, when students engage in SSI exploration, teachers need to guide them in collecting and analyzing various data, interpreting, evaluating, weighing, and utilizing diverse scientific evidence, and ultimately making informed decisions. In other words, the process of engaging students in SSI argumentation or decision-making requires a considerable range of epistemic abilities related to evidence interpretation, assessing the credibility and validity of evidence, weighing evidence, and other factors. These aspects are less explored and merit further in-depth investigation in the future.

3. The lack of tracking students' learning progression

The current limitation lies in the lack of tracking students' learning progression in most SSI-based instructions. Typically, only one or two SSIs are utilized to assess students' learning outcomes. However, the development of higher-order thinking skills related to facing and discussing SSIs, such as argumentation, decision-making, evidence evaluation, and the ability to support or counter arguments using evidence, requires continuous cultivation and repeated practice. These skills need time to internalize as part of students' capabilities, enabling them to demonstrate learning transfer in various contexts. To address this, it is crucial to provide students with diverse SSIs for learning while concurrently undertaking long-term tracking of their changes in higher-order thinking skills.

18.4 Two Cases of SSI-Based Instruction

The problem-solving within SSI is complex, encompassing discussions of diverse perspectives and involving dialogue and negotiation within the democratic process. Therefore, when teachers implement SSI-based instruction, they often design instructional scaffolds to guide student learning (Erman et al., 2022; Zhang & Hsu, 2021b). Simultaneously, adopting strategies of deliberative democracy and dialogue allows students to simulate experiencing future SSI decision-making scenarios in society. The following are two specific cases of SSI-based instruction respectively developed by the second and the third author in this chapter in their research: one integrates a technology-assisted learning scaffold, while the other employs the method of deliberative democracy dialogue.

18.4.1 Case 1: Guardians of the Coastline

In the first case, the researchers designed e-learning SSI activities called as "Guardians of the Coastline: focus on designing e-learning SSI activities (called as "Guardians of the Coastline: Planning for the Future") which guided students to collaboratively choose an appropriate option through three phases of the decision-making process summarized from the literature (Fang et al., 2019; Lee & Grace, 2012). The SSI scenario required students to solve a SSI about sand lose along with a coastline by choosing coastal engineering. Coastline management is a SSI because people with diverse perspectives on sustainability, human safety, or economic development can often barely reach a consensus on what coastal engineering the government should adopt. The first phase was required students to recognize the issues to formulate the decision-making space and to collect the necessary information and evidence to identify the possible perspectives about the issue. Then, they were guided to formulate at least five criteria for coastline protection. The second phase helped students weigh the criteria, analyze each given option's pros and cons, and recognize the dilemma to stimulate students to consider the decision via trade-off analysis and make their decision with justification. Seeking a consensus might evoke students' negotiation on the inconsistencies to reach the complementarity convincing those who disagreed with their decision in the group. The third phase was related to continually differentiating the features via multiple methods, especially monitoring and regulating the quality of the decision-making process. To promote students' regulation learning in their decision-making process, a peer-review task was designed to guide students to critique the decision made by another group, and debate or reflect on the quality of the decision. Through the processes of debate and reflection, students had more opportunities to monitor and regulate their decision-making and then to simultaneously improve their decision-making ability.

The study used a single-group pre-and post-test design to explore the influence of SSI-based learning on students' decision-making ability. The SSI-based learning was designed with scaffolds such as an animation for evidence-based decision-making (Fig. 18.2), peer collaboration, and e-learning prompts. A total of 172 10th-grade students participated in the study. The remarkable findings revealed that students' decision-making ability significantly improved in these three phases. Students could consider at least two aspects to make their decision, use a compensatory method to weigh four coastline-protected engineering techniques, and monitor and evaluate the quality of their own decision. However, the content analysis of students' responses to worksheets in the groups indicated that students' understanding of the compensatory method was not satisfactory. Although they reported a preference for employing the compensatory method in decision-making tasks, 15 groups made mistakes while applying the method, such as failing to weigh the criteria or confusing the compensatory method with the deletion method. This finding suggests that more learning supports for students to develop their decision making strategies are needed.

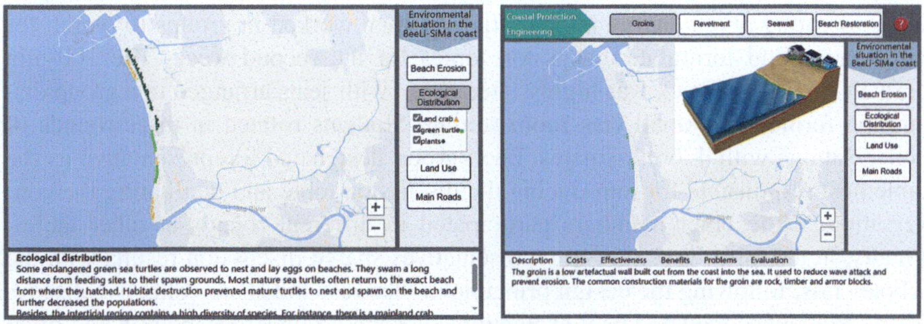

Fig. 18.2 An animation for evidence-based decision making

18.4.2 Case 2: Facilitating Students' Deliberative Dialogues on Energy-Related Issues

The energy-related topics were selected as teaching materials for the second SSI-based teaching case. The issue originated from a referendum on four initiatives held in Taiwan in December 2021, two of which were related to the construction of power plants. The referendum seems to have fueled public demand for scientific evidence on the governance of social issues, especially energy issues related to people's livelihood.

Drawing on prior research (e.g., Dawson & Carson, 2020; Suwono et al., 2021), the goal of this SSI teaching unit was set to engage students in developing thinking skills and citizenship values through argumentation and decision-making on SSI. World Café is a well-known strategy for democratic deliberation allowing participants to share opinions through dialogues for social change (Carson, 2011). Design principles by founders Juanita Brown and David Isaacs include: (1) set the context; (2) create hospitable space; (3) explore meaningful questions; (4) encourage everyone's contribution; (5) cross-pollinate and connect diverse perspectives; (6) listen together for patterns, insights, and deeper questions; and (7) harvest and share collective discoveries (Brown & Isaacs, 2005, p. 40). World Café adapts for learning activities in various settings, including small classrooms, designed as highly interactive events of varying durations (Cassidy & Fox, 2013). Only a few recent studies in the field of science education have utilized the World Café approach to socioscientific inquiry-based learning (Baek et al., 2022) and to practice negotiation of science-related ethical dilemmas (Kali et al., 2019). Therefore, this case study integrating World Café activities into the SSI teaching model, using energy-related topics for deliberative dialogues.

This teaching unit lasted 3 weeks (2 lesson hours per week). The first week's lesson allowed the teacher to clarify the context, and students explored questions of interest. Using the Kahoot platform, the teacher engaged students in an energy knowledge quiz and led them to explore Taiwan's power system and energy sources through review quiz questions, focusing on rising electricity prices and the importance of caring about energy issues. Students then worked in groups to search for information and formulate discussion topics for the second week. The 100-min class in the second week was highly interactive, with seats arranged in a group discussion format in a small classroom setting. Students rotated in three rounds of conversations with 4–6 classmates. Each group designated a representative as the table host responsible for introducing the discussion topic and facilitating the conversation, while other members participated in different rounds at other tables. Finally, in the third week, group representatives shared discussion results with the whole class, following the design principle of "harvest and share collective discoveries." Students conducted peer evaluations on each group's presentation and filled out the argumentation questionnaire.

This SSI teaching unit, incorporating World Café design principles, was tested with three 10th-grade classes (15–16 years old). Effectiveness was assessed by comparing students' argumentation performance before and after teaching.

Qualitative data analysis of group discussions suggested the first week's lesson was a critical stage that affected students' argumentation performance. The teacher utilized the knowledge quiz to lead students to review energy background concepts and problematic situations. Only with sufficient background knowledge of energy issues can students engage in meaningful group discussions and ask deeper questions for follow-up deliberative dialogues. For example, in class A, students proposed six topics, covering a variety of energy sources such as nuclear, wind, solar, and even human power (biking). Two other topics were related to policy, including raising electricity rates and limiting the number of automobiles in a household. In class B, one group asked a question about why we care about energy issues, while other groups explored different opinions on nuclear power, green energy facilities, biomass energy, and electric vehicle tax exemption. In contrast, most groups in class C struggled with complete questions, probably because about one-third of the students missed the knowledge quiz. Further dialogue analysis in each round of group discussion may provide concrete evidence for integrating SSI instruction with World Café.

18.5 The Challenges of SSI-Based Instruction

Numerous empirical studies published in journals on SSI-based instruction or science teaching within SSI contexts, including the research conducted in Taiwan mentioned above, consistently show positive teaching outcomes. Teachers have successfully facilitated student learning, and instances of failure are rare. However, these findings do not necessarily guarantee the success of in-service science teachers in their classroom practices. After all, each science teacher possesses unique professional abilities and skills and faces different students, situations, teaching environments, and resources. Therefore, the implementation of SSI-based instruction in practice can give rise to various potential difficulties and challenges. Overcoming these challenges is a crucial aspect that must be addressed in the promotion of SSI-based instruction. Only by doing so can the practical implementation of SSI-based science teaching become more likely and have a broader impact on students' science learning.

18.5.1 Challenge 1: Duration of Mutual Support Partnerships in Science Teaching Research

In the field of science teaching research, the researcher themselves or teams composed of scholars and teachers form the backbone supporting science teachers in implementing SSI-based instruction. The research team, led by the researcher, not only provides timely professional advice to science teachers but also offers encouragement both spiritually and practically. These aspects are made possible due to financial support from institutional grants for research. This kind of "mutual support partnership" exists because of this support (Lin, 2022).

Although we know professional growth communities are crucial and essential for individual teacher development (e.g. Akerson et al., 2009; Clair et al., 2023; Vangrieken et al., 2017), most science teachers are often left to work independently in educational settings, without specific scholars or teacher groups to assist them. Therefore, when practicing unfamiliar teaching methods for the first time, teachers often experience anxiety and uncertainty, fearing that they may not perform correctly or effectively. Thus, they must engage in continuous trial and error, making adjustments to eventually achieve better outcomes. During this trial-and-error process, having a team as a supportive backbone can enhance teachers' confidence in implementing SSI-based instruction. The success of promoting SSI-based instruction becomes more feasible. However, when scholars are unable to secure continued financial support, such mutual support partnerships may disappear, potentially leading to a decrease in teachers' willingness to promote SSI-based instruction.

18.5.2 Challenge 2: Limited Integration of SSIs in Science Textbooks

In SSI-based instruction research, the SSIs commonly used are often not included in the original school curriculum. This means that the science textbooks adopted by schools do not cover these SSIs, making them almost exclusively supplementary materials outside the school curriculum or extended materials related to specific units. For science education in Taiwan, the problem lies in the fact that the national curriculum guidelines for the natural sciences explicitly allow schools to implement issue-based teaching, and the actual implementation of SSI-based instruction is legal. However, due to the heavy reliance on textbooks in school teaching, if SSIs are not included in the textbooks, the opportunities for science teachers to practice SSI-based instruction are reduced. Moreover, if teachers face pressures related to the curriculum pace, limited class time, and the student-centered nature of SSI-based instruction requiring more time, the likelihood and willingness of science teachers to incorporate SSIs in the classroom become even lower.

Furthermore, while activity design and development are undoubtedly part of a science teacher's professional performance, it is an acknowledged time-consuming and mentally consuming task. Many science teachers have additional responsibilities beyond teaching, such as mentoring, administrative work, and collaboration with national or local education initiatives, collaboration with national or local education initiatives, which encompass specific programs, projects, or actions aimed at addressing and promoting certain educational goals or improvements. These additional responsibilities demand extra effort from science teachers. Requesting them to implement SSI-based instruction under such circumstances creates a dilemma in their willingness to engage in teaching.

18.5.3 Challenge 3: Time Needed for Teachers and Students to Adapt to Teaching, Learning, and Open-Ended Assessment

For both teachers and students, SSI-based instruction, learning, and assessment differ significantly from the predominant focus on knowledge, concepts, and problem-solving in traditional teaching methods. This is especially true during activities involving SSI argumentation or decision-making processes. In essence, SSI-based instruction brings about a fundamental shift in science teaching and learning practices (Zeidler et al., 2011), moving away from the 'comfort zone' of familiar teaching habits and approaches.

Teachers must develop their beliefs, value and pedagogical content knowledge (PCK) specific to SSI-based instruction. Moreover, through the implementation of SSI-based instruction, they need to adjust and refine their PCK to address previously unencountered teaching challenges. This includes navigating open-ended assessments that require reasoned arguments without standard answers, and transitioning roles to facilitate students in engaging in value-oriented dialogue, argumentation, or decision-making activities (Bossér et al., 2015).

Similarly, students' learning experiences diverge from traditional approaches. They must take a more active role in learning, moving beyond passive listening, pursuing standard answers, and focusing only on individual learning. Lee et al. (2019) emphasize that learning SSI often demands students to transition across different scientific learning cultures. This involves a shift in learning styles from passive engagement to actively participating in discussions, collaborative work, evidence-seeking, and argumentation with peers.

Particularly in Taiwanese classrooms, where student participation may not be as prominent, teachers need to design effective strategies, such as incorporating oral encouragement or providing incentives, to overcome the "culture of silence" among students. This facilitates meaningful dialogue and interaction between teacher and students or among students in SSI-based instruction. Adapting to these changes for both teachers and students is a gradual process, requiring not only time but also consistent practice to become proficient and familiar.

18.5.4 Challenge 4: Navigating the Complex Rational Thinking in SSI Decision Making

In the process of conducting SSI-based instruction in science teaching, we often train students to first understand the controversies in the issues, identify the stakeholders involved, and explore perspectives on resolving disputes. Then, they proceed to collect data and find evidence. Afterward, we require them to develop criteria and evaluating strategies, rely on evidence, weigh the importance and appropriateness of different perspectives, and finally make decisions. This is what is known as a systematic and rational thinking process (Hsu & Lin, 2017; Gresch et al., 2013; Gresch & Bögeholz, 2013).

However, rational thinking is not an easy task, especially in the context of SSI. Firstly, as mentioned earlier in this article, many SSI involve scientific and technological developments that are still in the science-in-the-making stage. Therefore, the evidence provided for us to base our conclusions on is still insufficient, incomplete or even lacking. Secondly, prior studies have already told us that SSI argumentation, decision-making, and moral thinking are influenced by many factors, such as students' knowledge of the issues, communication and expressive abilities, and the SSI scenarios (Topcu et al., 2010). One of important factors is personal emotions and values (Powell et al., 2021; Rundgren et al., 2016).

In other words, rational thinking cannot work perfectly when facing SSI, especially when making decisions related to the survival, life, medical and treatment experiences. Emotional factors, particularly, cannot be ignored or excluded (Lin, 2012). In other words, the consideration of making decisions in SSI does not solely rely on evidence and personal rational thinking; it must also consider personal psychological feelings, projection and attachment of emotions, empathy, and the impact of personal values. Therefore, while science teaching put emphasis on students' rational thinking, if emotional considerations can also be taken into consideration, more humane decisions should be made. This increases the complexity of educating students in rational thinking. Whether teachers can grasp this in the classroom becomes another challenge.

18.6 Conclusion

In summary, empirical studies by Taiwanese scholars on SSI-based instruction or SSI contexts, published in Chinese journals, encompass five categories: "Teachers' instruction, Analysis and utilization of SSI texts, SSI scale development, The relationships among variables, and Students' learning outcomes," all of which are noteworthy. These findings enhance our understanding of research related to SSI themes or contexts, suggesting directions for further exploration. They also shed light on the potential for diverse designs and scaffolding approaches in SSI-based instruction, influencing the learning outcomes of Chinese-speaking students. For future research, exploration can be continued in areas such as strengthening students' affective assessment, focusing on higher-order thinking skills related to epistemology, and tracking students' learning progression. When considering teaching practices, the recognition that SSI-based instruction can improve students' cognition, affect, and abilities brings to light the importance of overcoming the aforementioned challenges. Collaborative efforts from scholars and teachers are essential to make further progress in the practical implementation and promotion of SSI-based instruction.

References[1]

Akerson, V. L., Cullen, T. A., & Hanson, D. L. (2009). Fostering a community of practice through a professional development program to improve elementary teachers' views of nature of science and teaching practice. *Journal of Research in Science Teaching, 46*(10), 1090–1113. https://doi.org/10.1002/tea.20303

Baek, S., Shin, H., & Kim, C. J. (2022). Development of a climate change SSIBL-STEAM program aligned to the national curriculum for SSI elementary school in Korea. *Asia-Pacific Science Education, 8*(1), 109–148.

Berkowitz, M. W., & Simmons, P. (2003). Integrating science education and character education. In D. L. Zeidler (Ed.), *The role of moral reasoning on socioscientific issues and discourse in science education* (pp. 117–138). Kluwer Academic Publishers.

Bossér, U., Lundin, M., Lindahl, M., & Linder, C. (2015). Challenges faced by teachers implementing socio-scientific issues as core elements in their classroom practices. *European Journal of Science and Mathematics Education, 3*(2), 159–176.

Brown, J., & Isaacs, D. (2005). *The world café*. Berrett-Koehler Press.

Carson, L. (2011). Designing a public conversation using the World Café method. *Social Alternatives, 30*(1), 10–14.

Cassidy, A., & Fox, J. (2013). Modified world café discussion model for conference and course settings. In *Collected essays on learning and teaching Vol. 6: Learning without boundaries* (pp. 99–104). Society for Teaching and Learning in Higher Education.

*Chang, H.-C., Lowsing, L., & Lee, H. (2022). Place-based socioscientific teaching and learning practice in indigenous elementary schools. *Contemporary Journal of Science Education, 30*(S), 403–424. https://doi.org/10.6173/CJSE.202212/SP_30.0002

*Chen, Y.-C., & Lee, S.-T (2017). The framing analysis of SSI news-examples of gene-modification news. *Contemporary Journal of Science Education, 25*(S), 439–462. https://doi.org/10.6173/CJSE.2017.25S.02

*Cheng, T.-S., Lin, J.-W., Lee, Y.-Y., & Lin, C.-H. (2021). The impacts of socioscientific issue modeling curriculum for preservice teachers' decision making on highway route selection. *Contemporary Journal of Science Education, 29*(3), 219–243. https://doi.org/10.6173/CJSE.202109_29(3).0002

*Chin, C.-C. (2015). High school students' perception on risks, benefits and trust of biotechnology: A scale development study. *Contemporary Journal of Science Education, 23*(3), 241–264. https://doi.org/10.6173/CJSE.2015.2303.02

*Chin, C.-C., & Hu, F.-J. (2016). If learning science is like this- Elementary students' change in knowledge, motivation, and cooperative skills through socioscientific issues-based learning. *Education Journal, 44*(2), 101–126.

*Chin, C.-C., & Hu, F.-J. (2018). Action research on mudslide-module instruction in grade-five level-students' change in decision making, reasoning, and environmental ethic belief. *Contemporary Journal of Science Education, 26*(1), 51–70. https://doi.org/10.6173/CJSE.201803_26(1).0003

*Chin, C.-C., & Wu, C.-Y. (2017a). Elementary students' change in informal reasoning through the SSI Instruction: Their decisions on energy choice under the different conditions. *Contemporary Journal of Science Education, 25*(1), 21–46. https://doi.org/10.6173/CJSE.2017.2501.02

*Chin, C.-C., & Wu, C.-Y. (2017b). Elementary students' perception of socioscientific issue-based learning environment- Perspectives of personal learning motivation, group cooperative ability, and the classroom environment. *Education Journal, 45*(1), 71–97.

*Chin, C.-C., Yang, W.-C., & Tuan, H.-L (2010). A study on informal reasoning of sixth graders through guided TAPing integrated with clone issue. *Curriculum & Instruction Quarterly, 13*(1), 209–232. https://doi.org/10.6384/CIQ.201001.0209

[1] * in Chinese with English abstract.

Choi, K., Lee, H., Shin, N., Kim, S., & Krajcik, J. (2011). Re-conceptualization of scientific literacy in South Korea for the 21st century. *Journal of Research in Science Teaching, 48*(6), 670–697. https://doi.org/10.1002/tea.20424

Clair, T. L. S., Wheeler, L. B., Maeng, J. L., & Bell, R. L. (2023). Mixed-methods analysis of science teacher educator professional development: a five-year longitudinal study. *Professional Development in Education, 49*(2), 273–296. https://doi.org/10.1080/19415257.2020.1787191

Dawson, V., & Carson, K. (2020). Introducing argumentation about climate change socioscientific issues in a disadvantaged school. *Research in Science Education, 50*, 863–883. https://doi.org/10.1007/s11165-018-9715-x

Eastwood, J. L., Sadler, T. D., Zeidler, D. L., Lewis, A., Amiri, L., & Applebaum, S. (2012). Contextualizing nature of science instruction in socioscientific issues. *International Journal of Science Education, 34*(15), 2289–2315. https://doi.org/10.1080/09500693.2012.667582

Eastwood, J. L., Sadler, T. D., Sherwood, R. D., & Schlegel, W. N. (2013). Students' participation in an interdisciplinary, socioscientific issues based undergraduate human biology major and their understanding of scientific inquiry. *Research in Science Education, 43*, 1051–1078. https://doi.org/10.1007/s11165-012-9298-x

Erman, E., Pare, B., Susiyawati, E., Martini, M., & Subekti, H. (2022). Using scaffolding set to help student addressing socio-scientific issues in biochemistry classes. *International Journal of Instruction, 15*(4), 871–888. https://doi.org/10.29333/iji.2022.15447a

Evagorou, M., & Osborne, J. (2013). Exploring young students' collaborative argumentation within a socioscientific issue. *Journal of Research in Science Teaching, 50*(2), 209–237. https://doi.org/10.1002/tea.21076

Fang, S.-C., & Hsu, Y.-S. (2017). Exploring university students' group and individual socioscientific decision making with web-based instruction. *Contemporary Journal of Science Education, 25*(4), 391–412. https://doi.org/10.6173/CJSE.2017.2504.04

Fang, S.-C., Hsu, Y.-S., & Lin, S.-S. (2019). Conceptualizing socioscientific decision making from a review of research in science education. *International Journal of Science and Mathematics Education, 17*(3), 427–448. https://doi.org/10.1007/s10763-018-9890-2

Fowler, S. R., Zeidler, D. L., & Sadler, T. D. (2009). Moral sensitivity in the context of socioscientific issues in high school science students. *International Journal of Science Teacher Education, 31*(2), 279–296. https://doi.org/10.1080/09500690701787909

Gresch, H., & Bögeholz, S. (2013). Identifying non-Sustainable courses of action: A prerequisite for decision-making in education for sustainable development. *Research in Science Education, 43*, 733–754. https://doi.org/10.1007/s11165-012-9287-0

Gresch, H., Hasselhorn, M., & Bögeholz, S. (2013). Training in decision-making strategies: An approach to enhance students' competence to deal with socio-scientific issues. *International Journal of Science Education, 35*(15), 2587–2607. https://doi.org/10.1080/09500693.2011.617789

Groenke, S. L. (2009). Social reconstructionism and the roots of critical pedagogy: Implications for teacher education in the neoliberal era. In S. L. Groenke & J. A. Hatch (Eds.), *Critical pedagogy and teacher education in the neoliberal era*. Springer. https://doi.org/10.1007/978-1-4020-9588-7_1

Hsu, Y.-S., & Lin, S.-S. (2017). Prompting students to make socioscientific decisions: Embedding metacognitive guidance in an e-Learning environment. *International Journal of Science Education, 39*(7), 964–979. https://doi.org/10.1080/09500693.2017.1312036

*Huang, C.-J. (2008). Trying to construct real-time science learning in a technical society: A study of the differences between students and experts when monitoring science news. *Contemporary Journal of Science Education, 16*(1), 105–124. https://doi.org/10.6173/CJSE.2008.1601.03

*Huang, C.-J. (2017). The general education as a way to cultivate a civil society: A media literacy course related to science news as an example. *Curriculum & Instruction Quarterly, 20*(1), 45–72. https://doi.org/10.6384/CIQ.201701_20(1).0003

*Huang, C.-J., & Jian, M.-J. (2008). "What have scientists created?"-students' image toward technology when they read science news reports. *Contemporary Journal of Science Education, 16*(4), 415–438. https://doi.org/10.6173/CJSE.2008.1604.05

*Hung, J.-F., Lin, U.-L., & Wei, T.-T (2010). The analysis and comparison of students' argumentation in inexperienced and experienced teachers' classroom. *Contemporary Journal of Science Education, 18*(3), 205–227. https://doi.org/10.6173/CJSE.2010.1803.02

Kali, Y. Van, Horne, K., Watkins, D., Jacobs, J. K., Penuel, W. R., & Sumner, T. (2019, April). *Ethical dilemmas and deliberative dialogue as means for increasing students' science capital: A World Café inspired design.* Paper presented at the annual meeting of the American Educational Research Association, Toronto.

Karpudewan, M., & Roth, W. M. (2018). Changes in primary students' informal reasoning during an environment-related curriculum on socioscientific issues. *International Journal of Science and Mathematics, 16*(3), 401–419. https://doi.org/10.1007/s10763-016-9787-x

Khishfe, R. (2013). Transfer of nature of science understandings into similar contexts: Promises and possibilities of an explicit reflective approach. *International Journal of Science Education, 35*(17), 2928–2953. https://doi.org/10.1080/09500693.2012.672774

Khishfe, R., Alshaya, F. S., BouJaoude, S., Mansour, N., & Alrudiyan, K. I. (2017). Students' understandings of nature of science and their arguments in the context of four socioscientific issues. *International Journal of Science Education, 39*(3), 299–334. https://doi.org/10.1080/09500693.2017.1280741

Kiryak, Z., & Çalik, M. (2018). Improving grade 7 students' conceptual understanding of water pollution via common knowledge construction model. *International Journal of Science and Mathematics, 16*(6), 1025–1046. https://doi.org/10.1007/s10763-017-9820-8

Klosterman, M. L., & Sadler, T. D. (2010). Multi-level assessment of scientific content knowledge gains associated with socioscientific issues-based instruction. *International Journal of Science Education, 32*(8), 1017–1043. https://doi.org/10.1080/09500690902894512

Kolsto, S. D. (2001). Scientific literacy for citizenship: Tools for dealing with the science dimension of controversial socioscientific issues. *Science Education, 85*(3), 291–310. https://doi.org/10.1002/sce.1011

Lederman, N. G., Antink, A., & Bartos, S. (2014). Nature of science, scientific inquiry, socioscientific issues arising from genetics: A pathway to developing a scientifically literate citizenry. *Science & Education, 23*(2), 285–302. https://doi.org/10.1007/s11191-012-9503-3

Lee, E. A., & Brown, M. J. (2018). Connecting inquiry and values in science education. *Science & Education, 27*(1–2), 63–79. https://doi.org/10.1007/s11191-017-9952-9

Lee, Y. C., & Grace, M. (2012). Students' reasoning and decision making about a socioscientific issue: A cross-context comparison. *Science Education, 96*(5), 787–807. https://doi.org/10.1002/sce.21021

Lee, H., Chang, H., Choi, K., Kim, S.-W., & Zeidler, D. L. (2012a). Developing character and values for global citizens: Analysis of pre-service science teachers' moral reasoning on socioscientific issues. *International Journal of Science Education, 34*(6), 925–953. https://doi.org/10.1080/09500693.2011.625505

*Lee, J.-W., Su, M.-C., Huang, P.-H., Liu, C.-C., & Kao, H.-L. (2012b). A study of adopting the scientific reading and writing model to promote elementary pupils' argumentation ability. *Contemporary Journal of Science Education, 20*(6), 483–514. https://doi.org/10.6173/CJSE.2012.2006.01.

Lee, H., Yoo, J., Choi, K., Kim, S.-W., Krajcik, J., Herman, B. C., & Zeidler, D. L. (2013). Socioscientific issues as a vehicle for promoting character and values for global citizens. *International Journal of Science Education, 35*(12), 2079–2113. https://doi.org/10.1080/09500693.2012.749546

Lee, H., Lee, H., & Zeidler, D. L. (2019). Examining tensions in the socioscientific issues classroom: Students' border crossings into a new culture of science. *Journal of Research in Science Teaching, 57*(5), 672–694. https://doi.org/10.1002/tea.21600

Levinson, R. (2006). Towards a theoretical framework for teaching controversial socio-scientific issues. *International Journal of Science Education, 28*(10), 1201–1224. https://doi.org/10.1080/09500690600560753

*Li, K.-C. (2010). On the 95 guidelines and the 99 guidelines of high school. *Educational Resources and Research, 92*, 1–24. https://doi.org/10.6209/JORIES.202203_67(1).0010.

*Lin, S.-S. (2006). Interpretative research on the roles of the science teacher in instruction of a controversial issue in science and technology-A case study. *Contemporary Journal of Science Education, 14*(3), 237–255. https://doi.org/10.6173/CJSE.2006.1403.01.

*Lin, S.-S. (2007). The professional change of an experienced science teacher: A case study of the instruction of the issue of genetically modified food. *Contemporary Journal of Science Education, 15*(3), 241–264. https://doi.org/10.6173/CJSE.2007.1503.01.

*Lin, S.-S (2012). Fostering the sixth grade students' moral thinking through the instruction of controversial issues in science classroom. *Contemporary Journal of Science Education, 20*(5), 435–459. https://doi.org/10.6173/CJSE.2012.2005.03.

*Lin, S.-S. (2021). A scale for self-regulated learning on the issue of exotic species. *Contemporary Journal of Science Education, 29*(4), 351–373. https://doi.org/10.6173/CJSE.202112_29(4).0003.

*Lin, S.-S. (2022). Preparing science teachers to design and implement socioscientific decision making instruction: Researcher's and teachers' experiences. In Y.-S. Hsu, R. Tytler, & P. J. White (eds.), *Innovative approaches to socioscientific issues and sustainability education- Linking research to practice* (pp. 159–178). Springer. https://doi.org/10.1007/978-981-19-1840-7_10.

*Lin, S.-S., & Chin, C.-C. (2012). The comparison of the novice and experienced teachers' knowledge construction about socioscientific instruction before and after actual implementation. *Contemporary Journal of Science Education, 20*(1), 41–68. https://doi.org/10.6173/CJSE.2012.2001.03.

*Lin, T.-W., & Chin, C.-C. (2018). The change of sixth graders' recognition and position after socioscientific issues-based instruction-The case of global warming. *Contemporary Journal of Science Education, 26*(4), 283–303. https://doi.org/10.6173/CJSE.201812_26(4).0001.

*Lin, S.-Y., & Hsin, M.-C. (2017). A study on undergraduates' acceptance and perspectives of in vitro fertilization. *Contemporary Journal of Science Education, 25*(S), 463–484. https://doi.org/10.6173/CJSE.2017.25S.03.

*Lin, S.-P., & Hsu, Y.-S. (2017). A teaching research on improvements of eleventh graders' decision-making abilities and climate change adaptability literacy. *Contemporary Journal of Science Education, 25*(S), 413–437. https://doi.org/10.6173/CJSE.2017.25S.01.

*Lin, S.-S., & Huang, P.-H. (2009). Students' constructing argumentation about a socioscientific issue-The differences between sixth graders with different levels of academic achievement. *Contemporary Journal of Science Education, 17*(2), 111–133. https://doi.org/10.6173/CJSE.2009.1702.02.

*Lin, C.-I., & Li, Y.-Y (2018). Sustainability interdisciplinary education for facilitating undergraduate students' holistic thinking and conflict resolution competencies: A study on education for ocean sustainability. *Contemporary Journal of Science Education, 26*(1), 1–27. https://doi.org/10.6173/CJSE.201803_26(1).0001.

Lin, S.-S., & Mintzes, J. J. (2010). Learning argumentation skills through instruction in socioscientific issues: The effect of ability level. *International Journal of Science and Mathematics Education, 8*(6), 993–1017. https://doi.org/10.1007/s10763-010-9215-6

*Lin, T.-J., Lin, S.-S., & Chen, Y.-C. (2010). University students' knowledge and argumentation skills concerning a socioscientific issue. *Contemporary Journal of Science Education, 18*(3), 229–252. https://doi.org/10.6173/CJSE.2010.1803.03.

Lin, T.-C., Hsu, Y.-S., Lin, S.-S., Changlai, M. L., Yang, K.-Y., & Lai, T.-L. (2012). A review of empirical evidence on scaffolding for science education. *International Journal of Science and Mathematics Education, 10*(2), 437–455. https://doi.org/10.1007/s10763-011-9322-z

*Lin, J.-W., Lin, C.-H., & Chen, M.-R. (2014). The impact of science news positions on different age levels women's health decision-An example of hormone-replacement therapy. *Contemporary Journal of Science Education, 22*(3), 281–306. https://doi.org/10.6173/CJSE.2014.2203.03.

Lin, T.-C., Tang, K.-Y., Lin, S.-S., Changlai, M. L., & Hsu, Y.-S. (2022). A co-word analysis of selected science education literature: Identifying research trends of scaffolding in two decades (2000–2019). *Frontiers in Psychology, 13*, 844425–844439. https://doi.org/10.3389/fpsyg.2022.844425

*Liu, S.-Y., Lee, L.-C., & Tsai, C.-C. (2007). Scientific epistemological view and decision-making on socioscientific issues. *Contemporary Journal of Science Education, 15*(3), 335–356. https://doi.org/10.6173/CJSE.2007.1503.03.

*Lu, J.-D., & Lin, S.-S. (2019). Exploring high school students' moral judgment and justifications in socioscientific contexts. *Contemporary Journal of Science Education, 27*(3), 147–166. https://doi.org/10.6173/CJSE.201909_27(3).0001.

Millar, R. (1997). Science education for democracy: What can the school curriculum achieve? In R. Levinson & J. Thomas (Eds.), *Science today—Problem or crisis?* (pp. 87–101). Routledge.

Oulton, C., Dillon, J., & Grace, M. M. (2004). Reconceptualizing the teaching of controversial issues. *International Journal of Science Education, 26*(4), 411–423. https://doi.org/10.1080/0950069032000072746

Pelch, M. A., & McConnell, D. A. (2017). How does adding an emphasis on socioscientific issues influence student attitudes about science, its relevance, and their interpretations of sustainability? *Journal of Geoscience Education, 65*, 203–214. https://doi.org/10.5408/16-173.1

Powell, W. A., Newton, M. H., & Zeidler, D. L. (2021). Impact of socioscientific issues on middle school students' character and values for global citizenship. In *Socioscientific issues-based instruction for scientific literacy development* (pp. 56–91). https://doi.org/10.4018/978-1-7998-4558-4.ch003

Rundgren, C. J., Eriksson, M., & Rundgren, S.-N. C. (2016). Investigating the intertwinement of knowledge, value, and experience of upper secondary students' argumentation concerning socioscientific issues. *Science & Education, 25*, 1049–1071. https://doi.org/10.1007/s11191-016-9859-x

Sadler, T. D. (2009). Situated learning in science education: socio-scientific issues as contexts for practice. *Studies in Science Education, 45*, 1–42. https://doi.org/10.1080/03057260802681839

Sadler, T. D., Romine, W. L., & Topçu, M. S. (2016). Learning science content through socio-scientific issues-based instruction: a multi-level assessment study. *International Journal of Science Education, 38*(10), 1622–1635. https://doi.org/10.1080/09500693.2016.1204481

Stanley, W. B. (1992). *Curriculum for utopia: Social reconstructionism and critical pedagogy in the postmodern era*. State University of New York.

Stradling, R. (1984). The teaching of controversial issues: an evaluation. *Educational Reviews, 36*(2), 121–129. https://doi.org/10.1080/0013191840360202

*Su, Y.-C., & Lin, S.-S. (2012). Improving sixth graders' argumentation skills through scaffolding instruction in socioscientific contexts. *Contemporary Journal of Science Education, 20*(4), 343–366. https://doi.org/10.6173/CJSE.2012.2004.03

Suwono, H., Rofi'Ah, N. L., Saefi, M., & Fachrunnisa, R. (2021). Interactive socio-scientific inquiry for promoting scientific literacy, enhancing biological knowledge, and developing critical thinking. *Journal of Biological Education, 57*(1), 1–16. https://doi.org/10.1080/00219266.2021.2006270

Topcu, M. S., Sadler, T. D., & Yilmaz-Tuzun, O. (2010). Preservice science teachers' informal reasoning about socioscientific issues: The influence of issue context. *International Journal of Science Education, 32*(18), 2475–2495. https://doi.org/10.1080/09500690903524779

Vangrieken, K., Meredith, C., Packer, T., & Kyndt, E. (2017). Teacher communities as a context for professional development: A systematic review. *Teaching and Teacher Education, 61*, 47–59. https://doi.org/10.1016/j.tate.2016.10.001

*Wu, J.-I., Hsu, Y.-S., Zhang, W.-X., & Fang, S.-C. (2017). The influence of the online learning module on the multi-perspective decision-making of high school students. *International Journal on Digital Learning Technology, 9*(3), 69–94. https://doi.org/10.3966/2071260X2017070903004

*Wu, J.-I., Zhang, W.-X., & Hsu, Y.-S. (2021). The effects of question-prompt scaffolding designs on students' decision-making on socioscientific issues. *Curriculum & Instruction Quarterly, 24*(4), 199–227. https://doi.org/10.6384/CIQ.202110_24(4).0008

*Xia, Y.-L., & Chin, C.-C. (2021). The effect of question-answer-relationship strategy on the argumentation and reading comprehension abilities of the sixth graders through infusing socioscientific issue texts. *Contemporary Journal of Science Education, 29*(3), 191–217. https://doi.org/10.6173/CJSE.202109_29(3).0001

*Yang, C.-S., Tung, Y.-Y., Chen, S.-R., & Wang, K.-H. (2017). The effect of argumentation-based inquiry instruction and course in socioscientific context on 7th grade students' achievements, argumentation skills and scientific literacy. *Contemporary Journal of Science Education, 25*(S), 485–500. https://doi.org/10.6173/CJSE.2017.25S.04

Zeidler, D. (2015). Socioscientific issues. In R. Gunstone (Ed.), *Encyclopedia of Science Education*. Springer. https://doi.org/10.1007/978-94-007-2150-0_314

Zeidler, D. L., & Keefer, M. (2003). The role of moral reasoning and the status of socioscientific issues in science education. In D. L. Zeidler (Ed.), *The role of moral reasoning on socioscientific issues and discourse in science education* (pp. 7–40). Kluwer Academic Publishers.

Zeidler, D. L., Osborne, J., Erduran, S., Simon, S., & Monk, M. (2003). The role of argument during discourse about socioscientific issues. In D. L. Zeidler (Ed.), *The role of moral reasoning on socioscientific issues and discourse in science education* (pp. 97–116). Kluwer Academic Publishers.

Zeidler, D. L., Sadler, T. D., Simmons, M. L., & Howe, E. V. (2005). Beyond STS: A research-based framework for socioscientific issues education. *Science Education, 89*(3), 357–377. https://doi.org/10.1002/sce.20048

Zeidler, D. L., Sadler, T. D., Applebaum, S., & Callahan, B. E. (2009). Advancing reflective judgment through socioscientific issues. *Journal of Research in Science Teaching, 46*(1), 74–101. https://doi.org/10.1002/tea.20281

Zeidler, D. L., Applebaum, S. M., & Sadler, T. D. (2011). Enacting a socioscientific issues classroom: Transformative transformations. In T. D. Sadler (Ed.), *Socioscientific issues in the classroom* (pp. 277–305). Springer. https://doi.org/10.1007/978-94-007-1159-4_16

Zeidler, D. L., Herman, B. C., & Sadler, T. D. (2019). New directions in socioscientific issues research. *Disciplinary Interdisciplinary Science Education Research, 1*, 11. https://doi.org/10.1186/s43031-019-0008-7

*Zhang, W.-X., & Hsu, Y-S. (2021a). Senior high school students' reflection and evaluation of decision-making process regarding a socioscientific issue. *Contemporary Journal of Science Education, 29*(2), 113–135. https://doi.org/10.6173/CJSE.202106_29(2).0002

Zhang, W.-X., & Hsu, Y.-S. (2021b). The interplay of students' regulation learning and their collective decision-making performance in a SSI context. *International Journal of Science Education, 43*(11), 1746–1778. https://doi.org/10.1080/09500693.2021.1933250

*Zhang, W.-X., Hsu, Y.-S., Ho, Y.-T., & Liu, C.-C. (2023). Effects of a SSI-based learning module on students' decision-making abilities. *Journal of Research in Education Sciences, 68*(1), 167–198. https://doi.org/10.6209/JORIES.202303_68(1).0006

Chapter 19
A Moral Inquiry into Epistemic Insight through Socioscientific Issues: A Turkish Perspective

Özgül Yılmaz-Tüzün and Mustafa Sami Topçu

19.1 Introduction

Humans continuously transform their physical, biological, social and informational environments and transmit them to the next generations. Transformation of these environments includes complex processes due to complex interaction between people and these environments. It is like playing social chess to compete over limited resources. Playing this chess is getting more difficult because of having complex environments and their complex issues over generations (Sterelyn, 2007). These complex issues are generally called as worldwide crisis in economy, environment, climate, health, education, migration etc. generally resulting in having complex interaction of humans with their natural, artificial and social environments. COVID 19 pandemic, climate change, sustainable development and some scientific and technological practices such as genetically modifying organisms, artificial intelligence, nuclear energy are good examples of these issues. Most of the cases these issues do not have easy solutions and create societal dilemmas. Thus, these dilemmas include a close link between society and science and are called socioscientific issues (SSI). SSI "display a unique degree of societal interest, effect, and consequent" of scientific enterprises (Sadler & Zeidler, 2003, p. 5).

All around the world during the last two decades SSI has gained both education researchers' and policy makers' attention and diffused into different subject matter curriculum at different levels from early childhood to university. Researchers

Ö. Yılmaz-Tüzün (✉)
Middle East Technical University, Ankara, Türkiye
e-mail: ozgul@metu.edu.tr

M. S. Topçu
Yıldız Technical University, Istanbul, Türkiye
e-mail: mstopcu@yildiz.edu.tr

D. L. Zeidler (ed.), *A Moral Inquiry into Epistemic Insights in Science Education*, Contemporary Trends and Issues in Science Education 61,
https://doi.org/10.1007/978-3-031-63382-9_19

realized the importance of understanding how learners perceive, negotiate, resolve these issues and develop informed decisions. Developing informed decision necessitates the discussion of SSI from a variety of perspectives (social, economic, politic, etc) by using informal reasoning (cognitive, affective intuitive) and considering morality and ethics. Each SSI has a unique situation and students need to consider morality and ethics in their discussion to attain decision making. Thus, in their decisions students may consider their own personal experiences and universally recognized conceptions of human welfare, justice, and rights along with what is right to do and what should be done. In other words, some extent of ethical ramification should be used in moral reasoning by students. If students form their decisions through this process then they will recognize the messiness of the social and moral contexts of science. In this case development of morality, perspective taking, civic engagement, and responsibility during SSI discussions may help students to link scientific knowledge to understand social situations. Evidently formation of virtue and character development at the individual level and having genuine community at the societal level can be attained as a prolonged goal of the education (Sadler & Zeidler, 2003). With all these characteristics SSI discussion also helps attain the three key competencies defined in OECD's (2018) report of The Future of Education and Skills: Education 2030. These competencies address creating new value, reconciling tensions and dilemmas, and taking responsibility. All these competencies emphasize that students should learn to cope with uncertainties in different contexts by evaluating the issues' past, present and future situation by considering their impacts on local environments such as family and community, at global scales, and in digital space. Especially in the third competency the importance of developing "…a sense of responsibility, and moral and intellectual maturity, with which a person can reflect upon and evaluate his or her actions in light of his or her experiences, and personal and societal goals, what they have been taught and told, and what is right or wrong" is emphasized (OECD, 2018, p. 6).

In the present chapter, the SSI framework developed by Zeidler and Sadler (2023) used as a base to assess and synthesize SSI research in Türkiye context. In this framework, SSI can be defined as ill-structured, contentious, and complicated problems with theoretical, conceptual, and practical ties to science (Sadler, 2004). These issues may include scientific, political, environmental, and social impacts at the local, national, and global levels. Thus, in this chapter a literature synthesis about SSI based research in Turkish context is given in several aspects. First part includes the synthesis of the SSI based research trend in Türkiye. The synthesis highlighted that most of the SSI research supported the idea that science teaching is the focal point to raise functional scientific literacy by using scientific knowledge to generate evidence based reasoning and informed decision about SSI. Also, in terms of pedagogical approach, argumentation, and debate has been preferred widely to engage students in science learning through SSI based teaching. The second part addresses how scientific inquiry and moral inquiry have been conceptualized and compared in light of the SSI research conducted in the Turkish context. The synthesis mainly includes how the current Turkish SSI literature on epistemic insights (moral reasoning, character education, and socioscientific reasoning) through SSI were investigated by Turkish scholars. In the last part of the chapter, the state of SSI research in Türkiye in terms of research, practice, and policy were criticized and how these dimensions can be advocated through the synthesis given in the first and second parts.

In this chapter in order to present and synthesize SSI research in Türkiye we searched for articles. All these articles were used in the following sections of the chapter. Thus, here we would like to share brief information about the selection of the articles. For the purpose of this chapter, our search included the research base covering the last two decades, beginning with research from approximately 2008 through 2023. The reason for choosing the starting year as 2008 was actually shaped by the first SSI thesis in Türkiye that was completed in 2008 and the articles that were published in 2010. Our comprehensive review included research and review articles indexed in Scopus and ULAKBIM Scholar (Turkish National Academic Network and Information Center). Scopus was chosen because of being recognized and promoted by most of the universities in Turkiye. ULAKBIM was selected to reach articles published in local journals and used Turkish language. We used a selection criterion in which the keywords were searched in the title, abstract, and keywords. We sought empirical, conceptual, and review research studies. First two databases were searched by using "socioscientific issues and Turkey or Turk", "socioscientific issues and moral or morality or Turkey or Turk" as keywords and in total 76 articles were reached. These keywords helped us select all the research focus of SSI and morality in any perspective. All of the articles were then examined in terms of as being empirically and/or conceptually sound and whether they fit the themes structured in this chapter. These themes included comprehensive research areas: curriculum studies, pedagogical implementations, SSI as a teaching context, argumentation used to negotiate SSI, perspectives taking in SSI, character development, morality, moral reasoning, values, ethics, and citizenship education. We believed that our search was successful in capturing all the published studies in Turkiye SSI literature; however, due to limitations in keywords some of the related studies dealing with these subjects might be omitted. Thus, with the articles that we reached and our own expertise in the field we tried to present how recent works have shaped the themes in Türkiye and will influence the future research interest in SSI and morality research both in Türkiye and other countries as well.

19.2 Türkiye's SSI Research Agenda: A General Overview (Scientific Literacy and SSI)

There is currently no consensus among science educators over the definition of scientific literacy. Various parties offered varying interpretations of what scientific literacy entails. Three visions for scientific literacy were put forth by Roberts (2007): In Vision I, scientific literacy mostly reflects science contents and procedures and generally ignores moral and social aspects of science topics. Vision II concentrates on contextualizing scientific knowledge to provide its application in life and society significance. Vision III focuses on politicization, philosophical values, and critical global citizenship education, and this vision necessitates an understanding of how our culture, including our norms, values, and worldviews, and the society we live in, influence the content we choose to teach and how we choose to select it (Sjöström & Eilks, 2018). Roberts's (2007) scientific literacy Visions II and III reflect our understanding about science literacy we accepted both visions in this book chapter. According to Zeidler and Sadler (2023), on the one hand Vision II seeks to understand science in common sociocultural contexts. On the other hand, Vision III (Haglund & Hultén, 2017) concentrates on sociopolitical issues, social justice and activism-related concerns, and an ethic of care in order to establish the necessary links between the context of science and moral reasoning. Since we believe that relationships among character education, moral reasoning, and social aspects of science are important to understand, our SSI research framework in the present book chapter is based on Roberts's Vision II and Vision III of scientific literacy.

Recently there is much SSI-based research around the world exploring the effects of SSI-based learning environments on students' science literacy. Parallel to this research trend around the World, Türkiye's SSI research agenda over the past 20 years revealed that interest in SSI-based education has grown steadily in terms of practice, research, and policy-making. SSI is given the necessary weight in Türkiye's curriculum documents and science education research because it offers a highly productive learning environment for students in a research context that helps students' science literacy. For example, in the Middle School Science Program published in Türkiye in 2013, SSI were directly included, and these issues have become one of the basic components of the program. The program is based on four basic learning areas (MoNE, 2013): (1) Knowledge (Living Things and Life, Matter and Change, Physical Phenomena, World and Universe) (2) Skills (Scientific Process Skills, Life Skills) (3) Affect (Attitude, Motivation, Values, Responsibility) (4) Science-Technology-Society-Environment (Socioscientific Issues, Nature of Science, Relationship between Science and Technology, Social Contribution of Science, Sustainable Development Awareness, Science and Career Awareness). Within the scope of Science-Technology-Society-Environment, socio-scientific issues as well as the nature of science, the relationship between science and technology, the social contribution of science, sustainable development awareness, and science and career awareness have found a place under this heading.

In the middle school science program updated in 2018, it is seen that SSI is not one of the basic components of the program, but is among the specific objectives of the program. SSI is included as a specific purpose in the program as follows (MoNE, 2018a, p. 9): "To develop reasoning ability, scientific thinking habits and decision-making skills by using socioscientific issues." In addition, SSI have been tried to be included in many objectives. For example, an objective is included as follows regarding the health of systems (MoNE, 2018a, p. 37): "Understands the importance of organ donation in terms of social solidarity".

In the Turkish High School Physics Program, SSI are mentioned as follows in the general objectives of the program: "It is aimed to make inferences about socioscientific events regarding the use of different energy resources" (MoNE, 2018b, p. 11). In this case, we see that SSI are discussed in the context of energy resources in the Physics Curriculum. This situation was also reflected in the objectives. The objective regarding Energy Resources is as follows: "9.4.5.1. It evaluates renewable and non-renewable energy sources in terms of their advantages and disadvantages" (MoNE, 2018b, p. 19). Similarly, there is an objective regarding Energy Transmission Paths and Energy Transmission Rate: "9.5.4.5. Develops projects for measures to be taken against global warming. c) The necessity of being sensitive to the environment and the contributions that can be made individually are discussed" (MoNE, 2018b, p. 20).

In the High School Biology Program, SSI are mentioned as follows in the specific objectives of the program: "It is aimed to enable students to make conscious evaluations about SSI (controversial social issues related to science)" (MoNE, 2018c, p. 11). In this case, we see that SSI are discussed in the context of specific objectives in the curriculum. Considering all science disciplines, it stands out that the High School Biology Program is the program that gives the most importance and place to SSI. This situation is also reflected in the objectives that include SSI in the High School Biology Program. The objectives regarding Ecosystem Ecology and Current Environmental Problems are as follows: "10.3.1.3. Analyzes the flow of matter and energy in the ecosystem. c) It is ensured that the negative effects of biological accumulation on human health and other living things are investigated and discussed" (MoNE, 2018c, p. 21). There are two objectives regarding the Protection of Natural Resources and Biological Diversity: "10.3.3.2. Explores the importance of biological diversity for life. (a) Socioscientific Issues and Teaching are discussed about the factors that make Türkiye rich in terms of biological diversity." "10.3.3.3. It suggests solutions for the protection of biological diversity. (b) Solution proposals for the protection of biological diversity and the prevention of bio smuggling are discussed" (MoNE, 2018c, p. 22).

Above mentioned program efforts helped accelerate Turkish researchers' attention to SSI. Number of master and Ph.D thesis and published articles dramatically increased especially after curriculum revisions in 2013. This trend also shows a similar pattern with other countries. Delen and Yılmaz-Tüzün (2024) conducted a bibliometric analysis of global trends in science education from 2000 to 2021 revealed four main themes: attitude, achievement and technology; science, science education, and knowledge; argumentation and scientific literacy; inquiry, beliefs and curriculum. Under the third theme researchers more frequently investigated SSI in between 2010 and 2021. In order to address the SSI research trend in Türkiye, we examined the bibliometric and systemic research conducted on SSI. In this sense a total of eight different synthesis research on SSI was found (e.g., Başkan-Takaoğlu, 2023; Sönmez-Eryaşar, 2021; Sevgili et al., 2022; Tatar & Adıgüzel, 2019; Topçu et al., 2014). When considering that the first unpublished thesis was conducted in 2008 (Topçu, 2008) and first articles published in 2010 (Sürmeli & Şahin, 2010; Topçu, 2010; Topçu et al., 2010) this number shows that Turkish researchers are highly interested in synthesis studies on SSI. The most recent synthesis was conducted by Başkan-Takaoğlu (2023). In this synthesis Turkish database was searched and a total of 95 master and PhD theses were obtained between 2008 and 2021. It was seen that the number of theses on SSI reached a peak value in 2019 with 28 theses in total. Out of 95 theses 74 of them were conducted in science education departments of the education faculties. Quantitative (31 theses), qualitative (35 theses), and mixed studies (27 theses) were used as research approaches. When this data is compared with the earlier synthesis done by Topçu et al. (2014), there has been a shift from the use of quantitative approach to qualitative and mixed research approaches. This tendency shows that researchers are interested in more in-depth analysis of the SSI teaching practices. Of the studies analyzed by Başkan-Takaoğlu (2023) 54 of them studied with preservice teachers, 37 of them worked with preservice science teachers, 41 of thesis participants were students, and 13 of them studied with teachers. This data shows that most of the theses' participants were preservice teachers and almost majority of them worked with middle school students. This study and Sönmez-Eryaşar's (2021) study revealed the theses were commonly focused on perceptions for teaching SSI, argumentation skills, reasoning skills, attitudes toward and knowledge about SSI, epistemological beliefs, self-efficacy beliefs, and argumentation-based learning. The factors affecting the effectiveness of SSI teaching are listed as affective characteristics, argumentation, content knowledge, teaching methods and techniques, reasoning, and academic achievement. Teacher participants of the studies acknowledged that they did not have the necessary knowledge for SSI teaching, did not feel competent in SSI teaching, and had low self-efficacy beliefs to teach SSI. Therefore, they were not eager to include SSI argumentation and give much space to teaching SSI in the classroom. Besides lack of knowledge and experience, time constraints to attain requirement of the curriculum, and physical difficulties in school and classroom were also indicated as the possible reasons for teachers and preservice teachers to hesitate teaching SSI.

When the teaching context of the master and PhD theses included in above mentioned research synthesis was examined, SSI research in Türkiye serves two distinct purposes. These purposes were also stated in Topçu et al.'s (2014) systematic research. In one of the purposes, SSI was used as ends in teaching environments, and in the second one, it was used as means. It was also seen that different SSI were included in studies. Tatar and Adıgüzel's (2019) synthesis revealed that the issues addressed in master and doctoral theses were mostly genetically modified foods/organisms, biodiversity, global warming, air pollution, food additives substances, alternative energy sources, climate change, industrial revolution, cloning, nuclear power plants, environmental problems, and genetic screening test. Başkan-Takaoğlu's (2023) synthesis revealed that only four studies emphasized etic and values in SSI teaching and researchers studied ethics and values only with bioethics topics. These findings revealed that values, ethics, moral perspectives have been rarely addressed by Turkish scholars.

Compared to the master/PhD synthesis, a few researchers analyzed published articles in their synthesis regarding SSI teaching (Gül et al., 2023; Yapıcıoğlu, 2021). Yapıcıoğlu (2021) conducted a bibliometric analysis and search for the articles that were published on "socioscientific issues" from 1994 to 2019 all around the world. The publications indexed in the Web of Science (WoS) database used as the study's data source and a total of 372 research articles were obtained. Analysis of the articles revealed that when compared with other countries the United States of America (USA), Sweden, and Türkiye have published most of the articles. It was discovered that the USA and Türkiye were leading nations in terms of both strong collaboration and the quantity of publications. Furthermore, it was discovered that reasoning, argumentation, science education, nature of science, decision-making, and environmental education are the keywords most commonly utilized in the papers. Through a systematic literature review, Gül et al. (2023) investigated articles published by Turkish scholars in Türkiye. A total of 35 works, comprising 31 articles and 4 theses, were reached and analyzed. Analysis showed that studies typically look into how teaching on local socioscientific issues affects students' understanding of scientific concepts, critical thinking, and perceptions of the nature of science. The sample group was primarily composed of high school and middle school students, as well as teacher candidates. The majority of the research focus was on environmental based issues, including nuclear and hydroelectric power plants, global climate change, nuclear power plants, hydroelectric power plants, river pollution and biofuel energy sources were selected as local SSI. All these research syntheses revealed that Turkish scholars closely follow SSI research trends all around the world, even conduct lead studies in the field and are highly interested in SSI teaching and using SSI as a teaching context for students' science learning.

Zeidler and Sadler (2023) made an extensive research synthesis to show how that research has expanded the frontiers of SSI and examined these recent works under four broad themes and provided near-future trends in SSI research. In this chapter, we have synthesized the studies carried out in Türkiye with the same objectives by considering comprehensive structures of these themes. Thus, we had the chance to evaluate the studies carried out in the field of SSI in Türkiye in a parallel infrastructure and to reveal its place in world literature more clearly. These four themes can be shortly summarized as (1) SSI as engagement of curriculum practice and teachers' pedagogical beliefs, (2) SSI teaching and learning relates to the cultivation of epistemic and sensemaking practices, (3) SSI outside of conventional classrooms in terms of situating sociocultural factors that contribute to the activity of science, and (4) SSI as character development, citizenship and socioscientific perspective taking—considers how SSI serves to promote the development of morality, perspective taking, civic engagement, ethics, and responsibility in relation to science teaching. The first part of this chapter mainly focused on the first three themes and the second part focused on the fourth theme. Since the fourth theme addressing how scientific inquiry and moral inquiry have been conceptualized and compared in light of the SSI research, we provided a brief conceptual understanding about morality, moral education, and moral inquiry before addressing research synthesis on how the current Turkish SSI literature on epistemic insights (moral reasoning, character education, and socioscientific reasoning) through SSI were investigated by Turkish scholars.

19.3 Part 1: SSI, Informal Reasoning, SSI Teaching Practices, and Place-Based Education

19.3.1 Theme 1: Teachers' Beliefs, Experiences and Teaching Practices about SSI

In the Turkish SSI research agenda, investigating teachers' beliefs, experiences and teaching practices about SSI has played an important place in the last decade. While some researchers particularly focused on teachers' *beliefs and/or belief systems* about teaching SSI (Kılınç et al., 2013; Kılınç et al., 2017a), some researchers particularly studied teachers' *experiences and/or practices* about SSI-based instruction (Genel & Topçu, 2016; Karahan & Roehrig, 2017; Kılınç et al., 2017b; Topçu et al., 2022). Kılınç and his research group (2013, 2017a) studied two different groups of preservice science teachers *teaching efficacy beliefs and belief system* about genetically modified food as a SSI. In their first study (Kılınç et al., 2013) a self-efficacy instrument was developed and administered to the Preservice Science Teachers (PST) and found that PST showed moderately high *teaching efficacy beliefs*. As a sub dimension of the self-efficacy scale teaching efficacy scores were predicted by content knowledge and risk perceptions. In their second study (Kılınç et al., 2017a) with another group of PST, they investigated PST belief system about teaching genetically modified food as SSI. PST had relatively knowledgeable, high risk perceptions and certain moral beliefs about genetically modified foods, and high teaching efficacy beliefs. Both studies show that Turkish PST had high efficacy beliefs about teaching GM foods and risk perceptions and moral beliefs were important predictor variables in their decision making.

Kılınç and his research group (Kılınç et al., 2017b) also studied another group of PST *experiences* with use of dialogic discourse for SSI teaching. Researchers argued that science teachers have resistance to use dialogic discourse in SSI teaching even when they receive professional training. In this study the nature of this resistance was investigated. For this purpose, the preservice teacher's yearlong SSI teaching experience was followed after receiving a professional development program. The teacher implemented SSI lesson plans in real classrooms during teaching practicum course and induction year. It was found that an argumentation-based workshop encouraged the teachers to use dialogic discourse. But, when they had negative experiences during their teaching, it enhanced their preexisting beliefs about resistance to use dialogic discourse in SSI teaching. Researchers concluded that resistance to change is a complex process and difficult to overcome by nature of discourse, SSI and science education. The researchers suggested that in order to eliminate these factors to form barriers for teachers to use dialogic discourse in SSI teaching, teachers should be supported during their first year of teaching experiences in schools after graduation. With this research, the researchers presented the barriers teachers faced when they teach SSI in real science classrooms.

Another research exploring teachers' experiences while they implemented SSI-based instruction was conducted by Turkish scholar, Karahan and Roehrig (2017). They studied a social and a science studies high school teacher's experiences regarding collaboratively designing and teaching an SSI class. The researchers found that both participants criticized the science curriculum for not being able to address different SSI dimensions. Therefore, they included social, economic, and environmental perspectives in their environmental ethics class. The river-related issues were presented to students from social, economic, and cultural contexts and students were expected to develop critical and informed decisions by considering multiple perspectives. Both studies of Kılınç et al. (2017b) and Karahan and Roehrig (2017) showed that teachers have some barriers and difficulties about teaching SSI in their classrooms, so they need professional support from their peers or researchers.

Considering this necessity in the field, some professional development programs were developed in order to support science teachers' professional development in SSI-based teaching. Topçu et al. (2022) investigated science teachers' SSI-based teaching practices while creating and implementing a professional development (PD) program for them. Then, teachers performed their SSI units in their own classrooms. Using the Interconnected Model of Professional Growth theoretical framework (personal domain, domain of practice, external domain, domain of consequence), the data were analyzed to examine how the teachers implemented SSI, how they interacted with the researchers during professional development, how they believed in teaching and learning, and how they perceived the results of SSI. All teachers showed significant gains in terms of IMPG components. For example, Teacher A stated that she learned about the characteristics and outputs of SSI-based instruction during the professional development process. According to Teacher B, the SSI-based teaching methodology helps students learn in a way that is significant, long-lasting, and useful. It also helps them to discuss, draw conclusions, and locate supporting data for the positions they take. Therefore, this study showed that after the professional development program about SSI-based instruction, teachers developed their beliefs and practices about teaching SSI. Another study conducted by Genel and Topçu (2016) investigated ten Turkish PST SSI-based teaching practices in middle school science classrooms (for students ages 11–14). The five primary categories were found in this study to describe Turkish PST SSI-based teaching practices: media, argumentation, risk analysis, SSI selection and presentation, and moral perspective. As a summary of this section, in the Turkish SSI research context, we need much more studies exploring the effectiveness of different SSI-based PD programs on teachers' professional development.

Han-Tosunoğlu and İrez (2019) made a research synthesis about SSI teaching and proposed a comprehensive and holistic pedagogical model and named the model as SSI-PCK model. According to this model, the basic knowledge areas that teachers must have for effective SSI teaching are determined as the curriculum knowledge required for SSI (knowledge of concepts, objectives and units), pedagogical knowledge for SSI teaching (knowledge of teaching methods, learning approaches, assessment strategies), subject matter knowledge (subject matter self-efficacy, importance of subject matter), knowledge of students' understandings (knowledge of learning difficulties and misconceptions), and knowledge of school culture (knowledge of school culture and social environment). However, when considering the characteristics of SSI the suggested model has deficiency because it fails to include moral and ethical considerations.

19.3.2 Theme 2. Informal Reasoning and Argumentation about SSI and Related Constructs

In Turkish SSI research literature, we can claim that practitioners and researchers mostly used informal reasoning and argumentation as a sense-making and epistemic practice in their SSI-based research. One of the most studied topics on SSI in Turkiye is the examination of teachers' and students' informal reasoning and decision-making processes about SSI. We also determined that one of the most studied groups on informal reasoning and decision-making process is pre-service science teachers. For example, Han-Tosunoğlu and Özer (2021) studied Turkish preservice biology teachers' informal reasoning and decision-making modes for COVID-19 pandemic. A case was given to the participants and the questions were directed to them to reveal their argumentation. In the study informal reasoning was addressed through decision making modes, self-reasoning modes, and persuasive means. The findings revealed that decision making modes were coded as rationalistic, emotive, intuitive, and moral based, among them rationalistic mode was obtained with the highest frequency and the moral based mode was obtained with the second highest mode. Highly used self-reasoning modes were found as risk perception, moral based, and economic oriented and for persuasive means moral-based, scientific oriented authority-based, evidence-based, and experience-based were found as frequently used codes. Another example of the study was conducted by Cebesoy (2021) and she investigated the pre-service science teachers' risk perceptions and their informal reasoning patterns for genetic engineering cases. It was found that preservice science teachers made decisions by using at least one rationalistic, emotive, or intuitive but frequently displayed multiple reasoning patterns in their decisions. In this study risk perceptions and moral reasoning were determined through preservice teachers' evaluation of the pros and cons of related cases. As seen from these two examples of the studies, moral reasoning and risk perceptions were key modes characterizing teacher candidates' informal reasoning and decision-making process.

Topçu et al. (2011) also investigated Turkish preservice science teachers' informal reasoning about SSI and the factors impacting their informal reasoning. Three informal reasoning patterns were explored: rationalistic, emotive, and intuitive reasoning. The factors that influenced informal reasoning were technological concerns, personal experiences, moral and ethical considerations, and social considerations. Similar to Topçu et al. (2011), Cebesoy and Chang-Rundgren (2021) investigated the factors influencing PST decision making on other SSI, abortion. They explored the possibility of PST considering a variety of topic areas when deciding on SSI in relation to abortion. In three scenarios pertaining to genetics, the forty-two PST were asked to make abortion-related decisions and provide written reports outlining their reasoning. To investigate the rationale behind the pre-service teachers' choices, the SEE-SEP (science, environment, ethics/morality, sociology/culture, economy, and policy) model was used. The findings indicated that fields of science and ethics/morality had the biggest influence on their decisions. The findings also showed that the individuals who were in favor of abortion made their choices based on both their emotions and scientific evidence. Conversely, the pre-service teachers who opposed abortion did so on the basis of uncertainty and scientific evidence. Another study using the SEE-SEP model was conducted by Türköz and Öztürk (2020) and they explored pre-service science teachers' decisions and reasons about sugar loading during pregnancy, raw/open milk, processed/packaged milk, and nuclear power plants. The SEE-SEP model was used for identifying which perspectives they used in their decisions. It was found that participants changed their prior con claims to pro claims after the implementation except the nuclear power plant case. It was concluded that even though different dimensions of SEE-SEP model were used, for all cases pro evidence mostly given from society/culture and ethic/moral perspective was used to provide evidence for nuclear power plant and sugar loading during pregnancy. As inferred from these three studies, although they used different analytical frameworks in order to determine teacher candidates' informal reasoning and influencing factors, moral and ethical considerations were one of the important factors for teacher candidates' informal reasoning and decision-making process about SSI.

Another emerging research trend on SSI in Türkiye is the examination of the relationship between the informal reasoning and decision-making process of teachers or students about SSI and other variables that are important in science education, such as the nature of science. For example, Karakaya and İrez (2022) studied with preservice biology teachers and investigated a possible relationship between decision-making and reasoning in SSI and understanding the nature of science and scientific knowledge. Participants were grouped as naive, eclectic, and conscious-informed in terms of their understanding of nature of science and scientific knowledge. Findings revealed that due to the complex characteristics of socioscientific reasoning, factors such as moral concerns and personal experiences, were used by participants. Similar to previous studies, this study also showed that moral concerns are an important factor shaping SSI-decision making of science teacher candidates.

Another study that explored the relationship patterns between teacher candidates' informal reasoning about SSI and epistemological beliefs about science was conducted by Ozturk and Yilmaz-Tuzun (2017), and they explored PST informal reasoning regarding SSI, their epistemological beliefs, and the relationship between informal reasoning and epistemological beliefs. Nuclear power usage was selected as a SSI. They found negative correlations between epistemological beliefs and argumentation construction. The findings revealed that PST preferred to generate evidence-based arguments and mostly generated supportive argument construction. Researchers suggested that teacher education programs should develop preservice teachers' pedagogical skills about SSI teaching. Similarly, Adal and Cakiroglu (2022) investigated how decision-making processes on a SSI show differences between preservice science teachers who had unsophisticated and sophisticated nature of science understandings. Authors concluded that when participants hold different perspectives regarding SSI they were better in using these perspectives in their decision making.

Irmak (2021) studied socioscientific reasoning and another important concept for the SSI literature, attitude towards SSI of preservice teachers. These two constructs were measured as a pre-test and post-test when preservice teachers participated to an elective course containing. Statistical analysis revealed that socioscientific reasoning competencies except for the inquiry sub-dimension and attitude towards SSI except for the anxiety sub-dimension were improved. As another example of investigating the relationship between informal reasoning about SSI and an important variable for SSI research literature, media literacy was studied by Irmak et al. (2023) and they investigated the relationships between preservice teachers' (PT) informal reasoning regarding COVID-19 vaccination and new media literacy. Results revealed that preservice teachers mostly generated high quality arguments for scientific and risk-oriented modes and low-quality arguments for economy and ecology-oriented reasoning modes.

As one example of the Turkish SSI literature, Özden (2020) only studied elementary school students' informal reasoning patterns related to three SSI scenarios. It was observed that the participants mostly created intuitive reasoning patterns. This reasoning pattern is followed by the emotional reasoning pattern then logical reasoning pattern. Most of the participants only developed claims regarding SSI scenarios and justified these claims. A few participants used high quality reasoning by creating claims, justifying them, identifying possible opposing views and refuting these views. One of the most important findings of this study is that elementary students mostly used intuitive informal reasoning patterns and a few students revealed sophisticated argumentation. These findings may suggest more future research on elementary or children's reasoning about SSI instead of teacher candidates and/or adults.

As a last example from the Turkish SSI literature for this section, like Özden (2020), Topçu et al. (2010) studied PST (but not elementary students) informal reasoning and argumentation across different SSI contexts. They investigated to what extent PST informal reasoning processes are influenced by the issue contexts in which they are engaged. A qualitative research interpretive approach was used to analyze the data. Strong consistency was observed in the quality of informal reasoning at the sample level across different socioscientific scenarios. Finer-grained analyses, however, showed that there was more variation in each PST informal reasoning. These findings corroborate earlier findings that imply context dependence in informal reasoning about SSI.

19.3.3 Theme 3. SSI Outside of Conventional Classrooms

Turkish scholars also used place-based SSI interventions and authentic investigations to improve students' content knowledge and values as global citizens. Compared with the above mentioned two themes the studies conducted in this theme were very limited in number. We reached two studies on this theme. In one of these studies Avşar-Erumit et al. (2023) explored changes in preservice science teachers' nature of science views and their values as global citizens after a place-based intervention augmented with role-play activities and discussions of variety of SSI (e.g., hydroelectric power plants, industrial wastes, plastics) concerning climate change. Field trips were organized during the course that took six weeks and various locations (tea factory, recycling center etc.) within and surrounding the town were visited. Value data were collected by using the 'Character and Values as Global Citizens Assessment' scale developed by Lee et al. (2013) and adapted into Turkish by Karisan and Yilmaz-Tuzun (2017). The scale used in the study included four dimensions with accompanying subdimensions namely ecological worldview (sustainable development), social and moral compassion (moral and ethical sensitivity, empathic concerns), and socioscientific accountability (willingness to act). The results showed that after engaging in authentic activities both preservice science teachers' contextualized NOS views and willingness to act at personal, local, and global levels developed. However, they made limited efforts to carry out these actions in their daily lives due to economic difficulties. Participating in SSI in a real environment also enabled participants to be aware of multiple perspectives and sociocultural conditions due to contextual features of each SSI. In another study Atabey and Topçu (2017) developed and implemented a place-based SSI-based unit, which included visiting thermal, wind, and hydroelectric power plants. Researchers interested in how visiting power plants help developing 7th grade students' argumentation skills (proving claim, evidence, and reasoning). Written argumentation forms were used for data collection tools as pre and post applications. The authors concluded that the place-based SSI intervention improved middle school students' argumentation skills. Most of the students developed claims, used appropriate and sufficient evidence, and provided high quality reasoning. Authors argued that seeing socioscientific problems in situ, making observations about different dimensions of the problem, gaining concrete experiences, obtaining scientific information from experts during the trips developed students' argumentation skills. Researchers observed that most of the information that students obtained from the scientists in the field were used in their reasoning. Moreover, the students had a chance to learn by experiencing and gain first-hand experience. These studies supported that teaching science in informal and nonformal learning settings was seen as important and alternative ways in helping students to learn science.

19.4 Part 2: Moral Inquiry and Moral Reasoning in the Practice of SSI

Being an informative citizen in a society requires an educational practice in which students have a chance to link school subjects to societal dynamics (such as societies' concerns, interest, needs, problems) rather than learning concepts in disciplines to understand certain facts and processes unique to each discipline. This practice has a high potential to help students realize the functional link between schools and societies. At the same token, societies also need schools to educate future citizens as intelligent and informed decision makers and problem solvers. As Zeidler and Sadler (2023) argued if schools functioned as a part of the societies, responsible future citizens will be better in being aware of, understanding, and taking part in societal dynamics during and after their education in schools. The need for citizens donated with these characteristics have been recently supported with school-based research studies by many researchers all around the world (Guérin, 2019; Kahn, 2021; Karisan & Yilmaz-Tuzun, 2017; Kim et al., 2020; Powell et al., 2021; Rundgren & Chang Rundgren, 2018). Scholars have begun to conceptualize all these issues as the goals of moral education and argue that these goals can be attained by considering moral education as it crosscuts all school disciplines at all levels of the education system. In terms of science education most of the SSI necessitates students to make informed decisions and solve problems regarding science related societal issues. Thus, not only the science concepts and processes (nonnormative components) but also moral reasoning, values and ethics (normative components) should be considered by science educators. This can be attained by considering moral intelligence and social responsibility as the center of the science curriculum (Zeidler & Sadler, 2023). Thus, Zeidler and Sadler (2023) support the idea of progressive science education where the SSI teaching framework focuses on SSI and socioscientific reasoning in a philosophically, sociologically, and psychologically consistent and complementary manner in science education as well as across other disciplines.

Morality has been seen as a higher mental function by scholars such as Socrates and Aristotle (cited in LaBarge, 2005), Piaget (1932), Dewey (1959), Kohlberg (1969) and Kristjánsson (2015). Over the many years accumulation of theoretical, educational and philosophical understanding about moral education has supported the systemic inclusion of moral education into the educational system. Profound researchers in many countries have been providing scholarly and persuasive arguments about moral education with respect to theory, practice, and current implemented policies based on their empirical and theoretical research. Walker and Frimer (2007) provided a research synthesis on moral education from theoretical and practical perspectives. In this synthesis they stressed that moral education regains the world-wide attention in inclusion to the school curriculum as a meta-objective of all educational aims. This success was attuned by well supported educational and psychological discourses, but there still room for practical implementation. Zeidler and Sadler (2023) stressed that researchers and educators omitted this issue and left it to politicians. Having less practical implementation was explained through two problematic issues. The first one was seeing moral education as "old-fashioned, religious, paternalistic, conservative and situation-dependent" (p. 81). The second one was not getting enough attention from teacher education programs and policy makers.

In order to overcome the first problematic issue, the definition and characterization of moral education should be understood from the historical perspectives. Since the readers of this chapter are considered mostly science education researchers, we believe that presenting a short historical background about moral education could be beneficial to better communicate our ideas about morality and moral education. Currently moral education is defined as "cultivation of young people's moral characters and virtues" (p. 80) and studied in terms of philosophical, educational, and psychological considerations. For the scope of this chapter educational consideration is focused widely, however to make our ideas clear for the readers philosophical and psychological considerations are presented to some degree with the historical presentation. Walker et al. (2015) presented moral education in their historical synthesis from the eighteenth century to today. In the eighteenth-century teachers needed to discipline students' sinful nature and let them have good character according to religious conception of goodness. In the nineteenth century moral education focused on raising polite pupils by controlling their sin socially through effective class-based socialization. In this era religious and psychological doctrines still influenced the social control in the background. Early twentieth century schools gained the main power to educate the morally adjusted pupils. Schools act as a moral community to let students experience morally valuable activities and develop positive habits to better adjust society. During the late twentieth century moral adjustment of pupils by schooling resulted in having context dependent virtue and character qualities and fear for homogenization in moral development. This homogenization was seen as problematic, especially experiencing moral depressions among the young generation in societies after World War II. Thus, Kohlberg's moral development theory suggested cognitive development in pupils' morality and gained huge attention. Kohlberg's theory of moral education was focused on developing students' moral reasoning through six sequential stages of development and also acknowledged that not all students can pass through these stages at the same speed and most of them stay at the stage 4. In this rationalistic stage-bound cognitive development students developed skills to raise moral claims, justify their claims by developing logical and rational strategies, and generate correct inferences to deal with morally dilemmatic issues. Thus, moral education aimed to help students understand how to behave morally well (Darnell et al., 2019; Walker et al., 2015). Kohlberg's theory mainly influenced moral education throughout the 1960s and 1970s and it is still very influential in both curriculum development efforts (Schuitema et al., 2008; Ye, 2014) and curriculum in different disciplines besides education such as engineering education (LaPatin et al., 2023), medical education (Kumar & Choudhury, 2023), information technology (Sullins, 2023).

At the beginning of the twenty-first century emotionally vulnerable pupils gained much more attention. It was also supported by the research that Kohlberg's stages of moral reasoning do not always lead to actual moral behavior, some empirical research supported that moral reasoning only accounts for 10% of the variance in moral behavior (Walker et al., 2015). Darnell et al. (2019) argued that even though Kohlberg theory was accepted as the milestone in moral psychology, he missed the complex relations between cognition, emotion, identity, and motivation in moral education. At this era Goleman's Emotional intelligence gained pedagogical attention by accepting that moral action needs emotions more than rationalistic reasoning. Thus, schools help emotionally vulnerable students in developing their self-esteem for being resilient. So, they can better adapt to the challenging issues in society (Walker et al., 2015). However, later in this century research showed that emotionally intelligent people of high self-esteem cannot be resulted in performing morally right. Baumeister et al. (2003) carried out a research synthesis on self-esteem and found limited benefits of self-esteem. Correlational analysis of self-esteem with other variables indicated that high self-esteem may increase both prosocial and antisocial tendencies. Moreover, in some cases high self-esteem might be related to showing bad behaviors such as bullying and drug-taking. The authors concluded that accurate self-knowledge would be more useful than high self-esteem. Thus, more recently researchers started to argue that not only the individual students change but also institutional change is necessary for flourishing pupils as the focus of moral education. Moral education is not only focusing on an individualistic base but also considering students' socio-cultural context by considering a variety of issues such as minorities, class differences, power relations etc. With this approach complete virtue can be attained by students through being reflective and aware of the limitations of the behaviors. Performing this virtue over time and across different situations resulted in having virtuous habits. Virtuous pupils act morally by reflecting on the right reasons from the right motives (intrinsically), weighing the behaviors as either worthwhile or not, and considering the ethical perspectives of given situation by performing ethically salient considerations (Darnell et al., 2019; Kristjánsson, 2013; Walker et al., 2015). Thus, today it is accepted that a person's moral development occurs at both individual and societal levels because it has been affected by the specific social, cultural, and historical background in which it is shaped (Chen et al., 2023).

In Türkiye, it was seen that globally dominant moral education philosophies also influenced curriculum development. Sözer's (2023) study analyzed the educational system in Türkiye by considering the National Education Councils because these councils determine the educational policies by providing a philosophical foundation for curriculum development and its implementation in schools. In the early years of the Republic education principles focused on being national, republican, secular, science oriented, and modern according to National Education Law number 1739 (LePage et al., 2011). Thus, raising ideal citizens with national consciousness and eliminating social stratification and a sociopolitical role for religion formed the basis of the educational system. Thus, in moral terms religiously free (Salmoni, 2004) and virtuous citizens were also addressed in this time period (LePage et al., 2011). The period after the 1960s traditional and conservative values gained importance in education rather than imparting universal values (Sözer, 2023). Between 2000 and 2023 years the core values (justice, friendship, honesty, self-control, patience, respect, love, responsibility, patriotism, helpfulness) were placed at the center of the education programs (LePage et al., 2011). Moreover, all subject matters' curriculum aimed to develop students' national, moral, ethical, historical, cultural, social, and esthetical values throughout the formal education.

Above mentioned summary on morality and moral education from historical and philosophical perspectives reached to the point that preparing students to participate in societies by considering social, cultural, and historical background was the main goal (Chen et al., 2023). In this preparation students have to deal with dilemmatic SSI due to controversial characteristics of these issues that people face today and will face in the near future. Thus, having responsible citizenship has been advocated by many scholars in SSI research. Moreover, it has been accepted that use of moral reasoning by performing ethically salient considerations and acknowledging sociocultural context in dealing with dilemmatic situations is the central focus of moral development. According to Zeidler and Sadler (2023), any given SSI for science education purposes should have the following components, even if it is on a local and smaller scale: Incorporate implicit and/or explicit ethical elements that call for a certain amount of moral reasoning and place an emphasis on the development of virtue and character as long-term educational objectives. An essential component of SSI-based education has been the moral perspective, and a great deal of research has examined how moral reasoning influences students' ability to make well-informed decisions regarding SSI. SSI framework also addresses the use of moral reasoning through addressing macromorality and micromorality. SSI framework emphasizes intense discussion of issues through argumentation in which students develop moral reasoning in a face-to-face sociocultural context. In terms of macromorality students can develop moral reasoning by considering formal society-wide considerations such as laws, duty, formal societal norms without indicating their positions. In micromorality students consider their everyday experiences through face-to-face interaction in their social circles such as school, family, friends (Thoma & Dong, 2014). Thus, when students engage with SSI based argumentation, they gain an experience in understanding and responding to other perspectives and consider them in their moral judgment. In other words when compared with macromorality micromorality is context dependent and influenced by multiple perspectives. SSI framework also helps students' moral growth through providing opportunity to them to develop macromorality as well as micromorality.

Many researchers in the field suggested that morality is an indispensable part of the informed decision-making process of SSI. According to Sadler and Zeidler (2003), SSI are typically controversial in nature and generally touch on morals and ethics. Researchers (e.g., Pedretti, 1999; Sadler & Zeidler, 2003; Zeidler, 1984) have emphasized that morality and ethics need to be considered when making decisions in SSI, especially when dealing with concerns like genetic engineering. The study by Sadler and Zeidler (2003) suggested that scientific curricula should not only incorporate SSI, but also explicitly attend to (and not deny or overlook) the moral aspects of these issues, since students' (at least college students') decision-making regarding SSI (in this case, genetic engineering dilemmas) is largely determined by moral considerations.

19.4.1 Theme 4. Exploring Ethics and Moral Reasoning in SSI-Based Research

Literature search revealed that many Turkish scholars studied the role of morality and ethics in Turkish students' socioscientific decision making about different SSI. Keskin-Samanci et al. (2014) developed "Bioethical Values Inventory" to determine secondary school students' use of ethical values in debates regarding the biological sciences. Inventory included scenarios for the ethical dilemmas related to use of animals in experiments, prenatal genetic diagnosis and abortion, deciding the gender or physical appearance of unborn babies (designer babies), genetically modified organisms, genetic screening tests, and therapeutic cloning. This study revealed that high school students held a variety of moral, ethical, and civic engagement perspectives in responding to the dilemmatic issues given in the scenarios. These responses were considered as ethical values and categorized under utilitarian, rights, justice, virtue, normative, theological, nature preference, science and technology based, and believing in humans' superiority.

Studies considered under this theme mostly focus on implementing an intervention by using SSI for teaching context and investigating participants' perspective taking in their decision making. Interventions generally used SSI as context for students' discussion and were implemented throughout a course or as a part of a course. During discussions participants were encouraged to use moral, values, ethical, civic engagement, and responsibility perspectives in the decision-making process. In one of these Namdar and Namdar (2021) developed and implemented an SSI unit focusing on nuclear energy to foster high school students' values and decision making. An SSI teaching model guided their SSI-based instruction design and they utilized role playing as a teaching/learning approach. They used Choi et al.'s (2011) value framework to determine students' values. Their research concluded that role playing engages students to use emotive, moral, and ethical reasonings in their decision making. Moreover, in another study, Namdar and Namdar (2022) further supported that role reversal enabled students to see others' points of views and develop empathy and sympathy for others, which are known as the important aspects of social and moral compassion. The researchers argued that in order to elicit less articulated ethical, moral, and responsibility perspectives possessed by SSI teaching strategies should be developed to focus on these perspectives. Keskin-Samanci et al.'s (2014) study revealed that even though they were given a variety of perspectives they used a limited number of perspectives in addressing given SSI during discussions.

Some researchers studied relationships between SSI based teaching and value responsibility at middle school level. Yaman and Anilan (2021) studied on developing sixth grade students' value responsibility by implementing activity-based values education for matter and health subjects. The students who were instructed with an enriched content and learning activities provided more diverse reasons in their responses to dilemmatic cases when compared with the control group students. These students were also better in addressing the value responsibilities of citizens' in a society and had the necessary knowledge to fulfill these responsibilities. Researchers also found that before the instruction the students tended to perform certain behaviors just because they were approved by other people (parent, society, teacher etc.). However, after the treatment they realized that performing these behaviors were useful and they internalized their responsibilities at the individual, societal and environmental levels to solve the problems. This study also revealed that interdisciplinary science learning was achieved by students regarding energy saving, thermal insulation, and renewable energy resources. They also corrected their erroneous ideas. The authors took readers' attention to the importance of the use of real-life problems and having students to engage with solving these problems in their implementation. By this way the students even at the middle school were able to address SSI by considering ethical, moral, and responsibility perspectives.

A few Turkish scholars investigated the relationship between moral reasoning and other related variables with preservice science teachers. Özturk (2018) developed a human rights curriculum based on SSI as a part of the environmental education course and implemented this curriculum by using a controlled experimental design. Preservice elementary school teachers participated in the study and during the course they discussed several environmental SSI from a variety of perspectives. Special attention was given to draw the participants' attention towards human rights. All of the discussions were centered around real events/problems from daily life. Results revealed that participants who took this newly designed course had higher knowledge and attitudes regarding human rights. It was concluded that SSI provided a suitable context for human rights education within the environmental education course. Encouraging students to consider human rights aspects in discussions improved participants' awareness about violation of human rights. Observing significant differences between experimental and control groups supported that SSI teaching enabled teachers to give special attention to civic engagement and responsibility while enhancing their knowledge and attitudes regarding human rights.

In another study Tuncay-Yüksel et al. (2023) worked with 1524 preservice science teachers and determined how their environmental moral reasoning about moral dilemma scenarios (hiking, picnicking, fishing, camping) was predicted by epistemological beliefs and values. Moral reasoning was identified as ecocentric, anthropocentric, and egocentric based on the responses provided for scenarios. Values were categorized as openness to change versus traditional (conservation) values and self-transcendence (being helpful and protecting the environment) versus self-enhancement (considering social authority, power, and wealth). Results revealed that path analysis for each scenario provided positively significant relationships between environmental moral reasoning categories and self-transcendence and tradition values. It was supported that self-transcendence and (altruistic/anthropocentric and biospheric/egocentric) value orientations acted together with traditional values (feelings of moral obligation to act in pro-environmental ways by considering self, other people, and nonhuman species) and also individuals' traditional values are affected by contextual constructs shaped by cultures. The participants preferred to present more egocentric and anthropocentric value orientations (e.g., authority, influential, wealth, ambitious, social power) when they were confronted with humility, tolerance, and empathy values. Researchers pointed out the importance of implementing well-designed (e.g., scaffolding of experiences and questions requiring socioscientific perspective taking) pedagogical approaches to develop more ecocentric values by developing students' awareness about the reciprocal interconnected relationships in nature and elements of sustainable development.

In another study preservice science teachers' moral reasoning patterns and moral decisions for local and global environmental dilemmas and the factors underlying these reasoning patterns were investigated (Tuncay et al., 2012). Preservice science teachers provided responses on local and global environmental dilemmas for deforestation, e-waste, oil spills, and global warming. Moral reasoning patterns were categorized as ecocentric, anthropocentric, and non-environmental reasoning. The participants' moral reasoning mostly reflected ecocentric and anthropocentric orientations, with a few displaying non-environmental reasoning. On the contrary to Tuncay-Yüksel et al.'s (2023) study, in this study ecocentric moral reasoning was observed with the highest values across for all local or global dilemmas, which means that the participants had highest concern for nature aspects of the dilemmas. The moral decision-making interview protocol was used to understand the participants' moral decisions. It was found that different values such as effect on human life, formal principles, notion of rights, moral emotions, potential harm to others, and popular culture played a role behind the participants' moral reasoning. When we compared the patterns of moral reasoning used by preservice science teachers from 2012 to 2023, in ten years preservice science teachers valued more of individual comfort, success, and self-esteem over tolerance, empathy, sympathy in addressing moral reasoning in environmental dilemmas. As it was reported in Tuncay-Yüksel et al.'s (2023) study, this change can be seen as alarming signals for targeting the aims of functional scientific literacy proposed in the SSI framework by Zeidler et al. (2005).

Tuncay-Yüksel et al. (2023) also investigated relationships among epistemological beliefs, values, and environmental moral reasoning and found that epistemological beliefs and values were more predictive of nature centered moral considerations when compared to human-centered ones. Quick learning and omniscient authority dimension of epistemological belief and self-transcendence and traditional values had statistically significant path coefficients with all of the environmental moral reasoning categories of the four scenarios. All these relationships landed itself to a common pattern in which importance of family and obeying laws for security reasons. Researchers argued that the preservice science teachers valued both laws and parents as significant figures for protecting humans and the environment. For instance, participants who were more concerned about the potential harm caused by humans to the environment also believed that damaging actions and problems described in the scenarios should be legitimate by those in authority. In terms of quick learning it was found that believing learning as quick or not at all related to having higher levels of moral concerns about egocentric and anthropocentric aspects of the scenarios. Researchers concluded that less sophisticated epistemological beliefs in this dimension might be resulted in spending less time to analyze and evaluate information given in the scenarios. Even though this study revealed some significant relationships, Topçu's (2011) study found no significant relationship between elementary preservice teachers' epistemological beliefs and moral reasoning. In this study moral reasoning was determined by stories including moral dilemmas and each dilemma followed by 12 statements with a 5-point Likert scale. The instrument developed by Rest (1986) by considering Kohlberg's moral reasoning theory, which includes six stages of moral development. The researcher concluded that sophisticated epistemological beliefs can be related to well-developed moral reasoning. Thus, rather than presenting cases to students, teachers should use SSI to provide a learning environment with enriched argumentation to let students be aware of the moral and ethical perspectives and use them in their arguments to negotiate ill-structured problems. Theoretical work on Turkish students' challenges with socio-scientific argumentation and decision-making research findings was produced by Acar et al. (2009) also supported research findings of Tuncay-Yüksel et al. (2023) and Topçu (2011). Acar et al.'s (2009) synthesis pointed out that high-quality student arguments on SSI can be developed as a result of in-depth argumentation in which they can evaluate various options and think about their values regarding given SSI.

19.5 Conclusion and Future Directions in SSI and Moral Reasoning Research

19.5.1 Barriers and Difficulties about Teaching SSI in Classrooms and PD Programs

Zeidler and Sadler (2023) reported that while profitable, implementing SSI-based instructions is not without its difficulties. Teachers are facing many tensions when they attempt to implement SSI strategies in their classrooms. The biggest reason for this tension comes from the difference between traditional and progressive cultures of school science (Zeidler & Sadler, 2023). As expected, it is not easy to implement an SSI-based instruction (including inquiry-based and student-centered activities) in traditionally created learning environments. Kinskey and Zeidler (2020) stated that teachers had reluctance to implement SSI instruction because teachers had a lack of confidence in both content knowledge and pedagogy. Consistent with the existing international literature, Turkish scholars (Karahan & Roehrig, 2017; Kılınç et al., 2017b) reported that Turkish teachers have some barriers and difficulties about teaching SSI in their classrooms. The first order barriers (limited educational opportunities, an unsupportive environment, and gap between teacher education and the school) and the second order barriers (naive epistemologies, an argumentation-avoider personality, low self-efficacy, limited pedagogical content knowledge and pedagogical abilities) resulted in using monological SSI teaching. In order to address these barriers and tensions based on the different school culture and learning environments, we need much more studies in a highly developing SSI research field.

One of the solutions in response to these barriers can be increasing the number and quality of PD programs for SSI-based instruction. Zeidler and Sadler (2023) highlighted the importance of supporting teachers in the design and implementation of SSI-based instruction. Friedrichsen et al. (2020) implemented a PD program for effective SSI-based instruction, including active participation and engagement of teachers in preparing SSI units and implementation of these units. Consistent with this international trend, Turkish scholars (e.g., Topçu et al., 2022) also investigated science teachers' SSI teaching practices while creating and implementing a professional development (PD) program for them using a case study methodology. Both international and Turkish scholars reported positive effects of these PD programs on teachers' professional development for effective SSI-based instruction. However, we need much more studies exploring the effectiveness of the SSI-based PD programs on teachers' professional development since the number of research studies claiming tensions and barriers for effective SSI-based instruction is increasing day by day (Karahan & Roehrig, 2017; Kılınç et al., 2017b; Kinskey & Zeidler, 2020; Topçu et al., 2022; Zeidler & Sadler, 2023).

19.5.2 SSI and STEM Education Integration

Zeidler and Sadler (2023) claimed that during the past ten years, widespread STEM-centric programs across the world have highlighted the dangers of teaching science without more liberating and balanced sociocultural methods. Kahn (2015) suggested that "M" as a moral issue related to SSI should be integrated into STEM education. She also suggested that any standards for science education in the world should include moral growth by emphasizing argumentation that supports moral reasoning and discourse in STEM decision-making.

Thanks to Roehrig et al. (2021) and Zeidler (2020), they recognized the benefits of integrated STEM and SSI. They criticized the existing state of global STEM projects and advocated for a shift in emphasis toward STEM literacy and awareness as a means of promoting fair education and reducing poverty. Specifically, Roehrig et al. (2021) showed how SSI instruction can be integrated into the engineering design process (STEM Education). They considered the role of ethical and moral considerations in decisions on engineering and SSI integrated cases. Although there are some initiatives in integrating moral reasoning with SSI and STEM education in the field, we can suggest that the field needs much more research highlighting the role of human flourishing and moral and ethical considerations in highly developing areas of STEM education across the world. Without necessary integration of moral and ethical values in STEM education the world can be a more dangerous place for everyone. Without considering moral and ethical values, technological developments like artificial intelligence may create big risks for creating a more liberate, democratic and equitable society. For example, selling our data to third parties without our personal permission seems an important risk not only for Turkish society, but also the whole world. Therefore, as educational researchers, we need to take the attention of policy-makers, curriculum makers and policy makers into these controversial issues like responsible artificial intelligence use in education as a context in order to show the importance of SSI integration into STEM education, accordingly as an opportunity for growing more democratic and liberal societies.

19.5.3 SSI and Pedagogical Content Knowledge (PCK)

The connection between SSI instructional strategies and teachers PCK was rarely addressed by Turkish scholars. The same trend was observed in other countries as well. Han-Tosunoğlu and İrez (2019) developed a SSI-PCK model in their theoretical study. There are a few empirical studies conducted to understand how characteristics of SSI teaching and learning can enhance teachers' teaching repertoire by considering the PCK. As it was stated earlier PCK generally address teachers' content specific expertise in domains specified by it. SSI framework however has other specific areas such as ethical/moral, political, economic, sociocultural issues beyond domain specific content knowledge. Han-Tosunoğlu and Lederman (2021) developed a model and in their model, they added teachers' understanding about SSI and Teachers' background knowledge to Shulmans' PCK domains. Lee (2016) developed a SSI-PCK model by adapting the Magnusson PCK (Magnusson et al., 1999) approach to SSI oriented teaching. In this model the researcher added knowledge of learning context as a new domain. Kutluca (2021) developed another model and called it as PCK-SSI for argumentation teaching strategy and adapted the domains from Magnusson PCK approach (Magnusson et al., 1999). When all these models were analyzed it was seen that students' ethical and moral development and teachers' knowledge regarding these issues were not considered. When we consider the importance of moral and ethical reasoning and perspective taking for informed decision making and performing desirable actions in societies, future PCK studies should consider these issues as either a separate domain or cross-cutting theme of all the domains.

19.5.4 Challenging Students Learning through SSI Based Authentic Learning Environment

Above mentioned studies revealed that in Turkish culture at the middle school and university levels students and preservice science teachers had benefited from placed-based education and their science learning along with nature of science understanding and argumentation skills were enhanced. However, similar to other countries, our research resulted in a few studies focused on these areas (Herman et al., 2021). Thus, SSI by their own nature provide suitable context to teach science outside of conventional classrooms. Not only learning science but also experiencing how sociocultural factors contribute to the activity of science and understanding the complexities and perspectives embedded in scientific processes can be attained by students through contextualization SSI in science teaching. Increasing these practices both at schools and universities necessitates encouraging teachers and university instructors to adapt informal learning environments in their teaching repertoire through contextualization SSI in science teaching.

Realizing placed based learning through contextualization SSI in science teaching supports student learning in two ways. On the one hand, it enabled students to realize in-depth learning. Both younger and older students look for the rigor in learning activities they engage with. In other words, they are eager to see how participating in certain activities may challenge and motivate them to learn and improve the quality of their learning. On the other hand, authentic learning experiences enable students to apply their cognitive and affective skills in different contexts. In other words, knowledge, skills, attitudes and values gained in a school learning environment can be transferred to real world contexts and shared with practitioners in the field and vice versa. In parallel to that the quality of argumentation held by students can be improved because they gain experiences to see how different disciplines contribute to each other to operate complex systems and how these systems are important for their lives. For example, when they visit a hydroelectric power plant they will better capture the importance of energy for societies and the complex system operated in this plant. For ensuring successful operation of this complex system students can see the collaboration among different parties' (mechanical, civic and civil engineers, administrators, workers, environmentalists) perspectives and knowledge shared by them as well as rules and regulations used in the complex.

Developing students' socioscientific reasoning (SSR) competencies is also critical in helping students to understand the complex systems addressed by SSI. Through effective use of SSR competencies students will be able to perceive and reason of complexity inherent to SSI, incorporate multiple perspectives, recognize ongoing, open-ended nature of scientific inquiry, demonstrating skepticism, and relate scientific knowledge, processes as well as its limitations and affordances in their arguments (Zeidler et al., 2019). Other researchers also consider the importance of scientific competencies in SSI argumentation (Tsai, 2018) and using decision making competencies in understanding complex scientific evidence, normative considerations and personal values for constructing moral judgments (Gresch et al., 2013). Research on competencies and SSI direct researchers and practitioners' attention to the wide range of knowledge, skills, attitudes as well as moral, ethical and value considerations in negotiation and resolution of SSI. Thus, the responsibility of developing these competencies cannot be achieved through science education curriculum, these competencies should be addressed across different subject matter curriculum from early childhood to university.

19.5.5 Filling the Gap Between Knowledge (Thought) and Behavior (Action) in Dilemmas Bring by SSI

Synthesis of the research on using moral reasoning, ethics, perspective taking and values in SSI context supported that character and moral development starting from the early age of schooling to university level for having future citizens who can make informed decisions regarding SSI. Research synthesis also revealed that at different education levels, students had problems in developing their moral and ethical reasoning to decide a moral value weighted as the most important among alternatives and considering different perspectives possessed by a given dilemma in their decision-making process. Even studies conducted with preservice science teachers in different time intervals showed that these participants' moral values shaping their moral reasoning fails because their developed reasoning abilities found in 2012 were found as undeveloped in 2023. These findings suggest an urgent need for school curriculum across all disciplines to consider components of functional scientific literacy. Functional scientific literacy can be attained by SSI based teaching and learning and includes cognitive reasoning, moral reasoning, emotive belief systems, and moral and character education. Main goal here is to attain students' psychological, epistemological, moral and character development while connecting science and society (Kahn & Zeidler, 2019). Thus, functional scientific literacy has unique characteristics from conventional scientific literacy as having ethical and moral consideration in its structure. The research synthesis presented in this chapter regarding students and preservice teachers' use of ethics and moral reasoning in negotiation and resolution of different SSI revealed that these reasoning types were challenging for participants to develop and use them in their argumentation one hand and act morally and ethically right on the other hand. Moral education aims to serve the development of society by cultivating students' prosocial behaviors through attaining moral development at individual and societal levels (Chen et al., 2023). Moving from individual level to societal level puts emphasis on considering multiple perspectives in moral reasoning and using them in discourse and argumentation. This is called moral excellence and guided by knowing what good and virtuous people can do right with this type of reasoning. Cultivating prosocial behavior however deals with doing what is good for the good reason and called as phronesis guided by practical wisdom. Thus, functional scientific literacy aims to help students to "consider practical ways to practice morally mature judgment and moral action in real-life situations" (Shim, 2023, p. 4).

However, the gap between knowing the good (thought) and doing the good (action) is a well-known challenge in moral education (Darnell et al., 2019). We believe that SSI has a strong potential to fill this gap by helping students to develop phronesis through considering the basis of functional scientific literacy. Darnell et al. (2019) stated that theoretical and practical understanding in moral psychology and moral education supported that phronesis is gained through education and supported with daily life experiences and it includes virtuous habits formation by using intellectual abilities, experiences, and reflective judgements. Thus, virtuous people rely on their habits over time across different situations. Throughout their formal education students go through the developmental stages and acquire rational capacities to gain these habits if pedagogical implications are well suited for this aim. Development of phronesis depends on considering four components and Darnell et al. (2019) defined a four-component model for developing phronesis by adding some additional characteristics to the neo-Kohlbergian four component model of moral functioning. Narvaez and Rest (1995) developed a neo-Kohlbergian moral functioning model and defined four psychological components of acting morally as moral sensitivity, moral judgment, moral motivation, and implementation.

Individuals need to use their cognitive and affective skills within each component as well as personality traits. Darnell et al. (2019) models' four components are constitutive function, integrative function, blueprint, and emotional regulations. First components of these models include the ability to identify, attend, notice, or perceive moral issues in a given situation. The second components of both models consider the ability to weigh, decide, and justify the morally ideal course of action among alternatives through reflection. The two models' third components focus on the ability to determine priority among moral values for a given situation. According to Darnell et al. (2019), the fourth component of the two models differs substantially in that implementation refers to ability to act morally and stay on moral task (conception of character) and emotion regulation refers to regulating emotions to perform virtuous action (conception of both character and personality traits or virtue). Being unsuccessful in bringing moral action can result in the failure to attend any of these four components as a consequence of the individual or collective efforts.

When considering the characteristics of SSI and its pedagogical importance to science education (engaging students in discussions and dilemmas) the four components of phronesis can be attained by students. The first component can be attained by bringing the science related societal dilemmas to the classroom. Teachers encourage students to identify the moral issues and let them think about how others might be affected by considering the feelings of others (emotions of empathy and sympathy). Each SSI dilemma has alternative solutions and possess multiple perspectives; thus, students need to weigh their own solutions/decisions/reasoning against alternative ones and justify them scientific evidence. This characteristic of SSI is very helpful to attain the second and the third components of the phronesis. For the last component, once students develop their informed decision as a result of high quality of argumentation then these students will want to put it into practice of their decisions without resistance. At this stage teachers can be role models to help them to satisfy and regulate their actions both emotionally and cognitively. For the focus of this book the studies conducted all over the world provided theoretical, conceptual, and practical evidence in terms of how moral and character development along with cognitive, affective and nature of science understanding can be attained through SSI based teaching and learning to reach the long-range pedagogical goals of functional scientific literacy as a result of educational practices in schools.

References

Acar, O., Turkmen, L., & Roychoudhury, A. (2009). Student difficulties in socio-scientific argumentation and decision-making research findings: Crossing the borders of two research lines. *International Journal of Science Education, 32*(9), 1191–1206. https://doi.org/10.1080/09500690902991805

Adal, E. E., & Cakiroglu, J. (2022). Investigation of preservice science teachers' nature of science understanding and decision making on socioscientific issue through the fractal model. *Science & Education, 32*(2), 529–565. https://doi.org/10.1007/s11191-022-00319-1

Atabey, N., & Topçu, M. S. (2017). The effects of socioscientific issues based instruction on middle school students' argumentation quality. *Journal of Education and Practice, 8*(36), 61–71.
Avşar-Erumit, B., Namdar, B., & Oğuz Namdar, A. (2023). Promoting preservice teachers' global citizenship and contextualised NOS views through role-play activities integrated into place-based SSI instruction on climate issues. *International Journal of Science Education*, 1–30. https://doi.org/10.1080/09500693.2023.2251189
Başkan-Takaoğlu, Z. (2023). Sosyobilimsel konulara yönelik yürütülen lisansüstü tezlerin incelenmesi: sistematik bir analiz. *Milli Eğitim Dergisi, 52*(237), 547–576. https://doi.org/10.37669/milliegitim.1075185
Baumeister, R. F., Campbell, J. D., Krueger, J. I., & Vohs, K. D. (2003). Does high self-esteem cause better performance, interpersonal success, happiness, or healthier lifestyles? *Psychological Science in the Public Interest, 4*(1), 1–44.
Cebesoy, U. B. (2021). Pre-service science teachers' informal reasoning patterns and risk perceptions in SSI: Case of gene therapy. *European Journal of Science and Mathematics Education, 9*(4), 211–229. https://doi.org/10.30935/scimath/11237
Cebesoy, U. B., & Chang-Rundgren, S. N. (2021). Embracing socioscientific issues-based teaching and decision-making in teacher professional development. *Educational Review, 75*(3), 507–534. https://doi.org/10.1080/00131911.2021.1931037
Chen, J. Y., Liu, Y. D., Dai, J., & Wang, C. L. (2023). Development and status of moral education research: Visual analysis based on knowledge graph. *Frontiers in Psychology, 13*, 1079955. https://doi.org/10.3389/fpsyg.2022.1079955
Choi, K., Lee, H., Shin, N., Kim, S. W., & Krajcik, J. (2011). Re-conceptualization of scientific literacy in South Korea for the 21st century. *Journal of Research in Science Teaching, 48*(6), 670–679. https://doi.org/10.1002/tea.20424
Darnell, C., Gulliford, L., Kristjánssona, K., & Paris, P. (2019). Phronesis and the knowledge-action gap in moral psychology and moral education: A new synthesis? *Human Development, 62*, 101–129. https://doi.org/10.1159/000496136
Delen, B., & Yılmaz-Tüzün, Ö. (2024). Farklı ülkelerde fen eğitimi: Türkiye'de ve farklı bölgelerde gözlenen eğilimler arasındaki farklılıklar. *TED Eğitim ve Bilim*. https://doi.org/10.15390/EB.2024.12578
Dewey, J. (1959). *Moral principles in education*. Philosophical Society.
Friedrichsen, P., Sadler, T. D., & Zangori, L. (2020). Supporting teachers in the design and enactment of socio-scientific issue-based teaching in the USA. In M. Evagorou, J. A. Nielsen, & J. Dillon (Eds.), *Science teacher education for responsible citizenship: Towards a pedagogy for relevance through socioscientific issues* (pp. 85–99). Springer Nature Switzerland.
Genel, A., & Topçu, M. S. (2016). Turkish preservice science teachers' socioscientific issues-based teaching practices in middle school science classrooms. *Research in Science & Technological Education, 34*(1), 105–123.
Gresch, H., Hasselhorn, M., & Bögeholz, S. (2013). Training in decision-making strategies: An approach to enhance students' competence to deal with socio-scientific issues. *International Journal of Science Education, 35*(15), 2587–2607. https://doi.org/10.1080/09500693.2011.617789
Guérin, L. (2019). Socio-scientific issues and citizenship education: From theory to the classroom. *Inter-UndTransdisziplinäre Bildung, 1*, 57–64.
Gül, E., Sezen-Vekli, G., & Çapkınoğlu, E. (2023). Yerel sosyobilimsel konular üzerine yapılan çalışmaların genel eğiliminin incelenmesi: Bir sistematik derleme. *Nevşehir Hacı Bektaş Veli Üniversitesi SBE Dergisi, 13*(3), 1345–1364.
Haglund, J., & Hultén, M. (2017). Tension between visions of science education. *Science & Education, 26*(3–4), 323–344. https://doi.org/10.1007/s11191-017-9895-1
Han-Tosunoğlu, Ç., & İrez, S. (2019). Sosyobilimsel konuların öğretimi için pedagojik bir model. *Yükseköğretim ve Bilim Dergisi/Journal of Higher Education and Science, 9*(3), 384–401. https://doi.org/10.5961/jhes.2019.340

Han-Tosunoğlu, Ç., & Lederman, N. G. (2021). Developing an instrument to assess pedagogical content knowledge for biological socioscientific issues. *Teaching and Teacher Education, 97*, 103217. https://doi.org/10.1016/j.tate.2020.103217

Han-Tosunoğlu, Ç., & Özer, F. (2021). Exploring pre-service biology teachers' informal reasoning and decision-making about COVID-19. *Science & Education, 31*(2), 325–355. https://doi.org/10.1007/s11191-021-00272-5

Herman, B. C., Newton, M. H., & Zeidler, D. L. (2021). Impact of place-based socioscientific issues instruction on students' contextualization of socioscientific orientations. *Science Education, 105*(4), 585–627. https://doi.org/10.1002/sce.21618

Irmak, M. (2021). Lisans öğrencilerinin sosyobilimsel konularla ilgili muhakeme yeterliklerinin ve tutumlarının geliştirilmesi. [Developing undergraduate students' reasoning competencies and attitudes related to socioscientific issues]. *Gazi Üniversitesi Gazi Eğitim Fakültesi Dergisi, 41*(3), 1801–1838.

Irmak, M., Ozturk, N., Tuncay Yüksel, B., Çakır Yıldırım, B., & Karaarslan Semiz, G. (2023). Reasoning in the Era of COVID-19 pandemic. *Science & Education*. https://doi.org/10.1007/s11191-023-00467-y

Kahn, S. (2015). Another "M" for STEM? Moral considerations for advancing STEM literacy. *K-12 STEM Education, 1*(4), 149–156.

Kahn, S. (2021). No child too young: A teacher research study of socioscientifc issues implementation at the elementary level. In W. A. Powell (Ed.), *Socioscientifc issues-based instruction for scientifc literacy development* (pp. 1–30). IGI Global.

Kahn, S., & Zeidler, D. L. (2019). A conceptual analysis of perspective taking in support of socioscientific reasoning. *Science & Education, 28*, 605–638. https://doi.org/10.1007/s11191-019-00044-2

Karahan, E., & Roehrig, G. (2017). Case study of science and social studies teachers co-teaching socioscientific issues-based instruction. *Eurasian Journal of Educational Research, 17*(72), 1–1. https://doi.org/10.14689/ejer.2017.72.4

Karakaya, E., & İrez, O. S. (2022). The relationship between understanding the nature of scientific knowledge and reasoning and decision making in socioscientific issues. *Hacettepe University Journal of Education, 37*(4), 1329–1358. https://doi.org/10.16986/huje.2022.452

Karisan, D., & Yilmaz-Tuzun, O. (2017). Adaptation of character and values as global citizens assessment questionnaire into Turkish: Validity and reliability study. *Pamukkale University Journal of Education, 42*, 74–85.

Keskin-Samanci, N., Özer-Keskin, M., & Arslan, O. (2014). Development of 'Bioethical Values Inventory' for pupils in secondary education within the scope of bioethical education. *Eurasia Journal of Mathematics, Science and Technology Education, 10*(2), 69–76. https://doi.org/10.12973/eurasia.2014.1029a

Kılınç, A., Kartal, T., Eroğlu, B., Demiral, M., Afacan, Z., Polat, D., Demirci Guler, M. P., & Görgülü, Z. (2013). Preservice science teachers' efficacy regarding a socioscientific issue: A belief system approach. *Research in Science Education, 43*(6), 2455–2475. https://doi.org/10.1007/s11165-013-9368-8

Kılınç, A., Kelly, T., Eroglu, B., Demiral, U., Kartal, T., Sonmez, A., & Demirbag, M. (2017a). Stickers to facts, imposers, democracy advocators, and committed impartialists: Preservice science teachers' beliefs about teacher's roles in socioscientific discourses. *International Journal of Science & Mathematics Education, 15*(2), 195–213. https://doi.org/10.1007/s10763-015-9682-x

Kılınç, A., Demiral, U., & Kartal, T. (2017b). Resistance to dialogic discourse in SSI teaching: The effects of an argumentation-based workshop, teaching practicum, and induction on a preservice science teacher. *Journal of Research in Science Teaching, 54*(6), 764–789. https://doi.org/10.1002/tea.21385

Kim, G., Ko, Y., & Lee, H. (2020). The effects of community-based socioscientific issues program (SSI-COMM) on promoting students' sense of place and character as citizens. *International Journal of Science and Mathematics Education, 18*(3), 399–418.

Kinskey, M., & Zeidler, D. (2020). Elementary preservice teachers' challenges in designing and implementing socioscientific issues-based lessons. *Journal of Science Teacher Education, 32*(3), 350–372. https://doi.org/10.1080/1046560x.2020.1826079

Kohlberg, L. (1969). Stage and sequence: the cognitive developmental approach to socialization. In D. Goslin (Ed.), *Handbook of socialization theory and research* (pp. 347–480). Rand McNally.

Kristjánsson, K. (2013). Ten myths about character, virtue and virtue education – and three well-founded misgivings. *British Journal of Educational Studies, 61*(3), 269–287.

Kristjánsson, K. (2015). *Aristotelian character education*. Routledge/Taylor & Francis Group. https://doi.org/10.4324/9781315752747

Kumar, S., & Choudhury, S. (2023). Cognitive morality and artificial intelligence (AI): A proposed classification of AI systems using Kohlberg's theory of cognitive ethics. *Technological Sustainability, 2*(3), 259–273. https://doi.org/10.1108/TECHS-12-2022-0047

Kutluca, A. Y. (2021). An investigation of elementary teachers' pedagogical content knowledge for socioscientific argumentation: The effect of a learning and teaching experience. *Science Education, 105*, 743–775. https://doi.org/10.1002/sce.21624

Labarge, S. (2005). Socrates and moral expertise. In L. Rasmussen (Ed.), *Ethics expertise. philosophy and medicine* (Vol. 87). Springer. https://doi.org/10.1007/1-4020-3820-8_2

LaPatin, M., Roy, A., Poleacovschi, C., Padgett-Walsh, K., Feinstein, S., Rutherford, C., Nguyen, L., & Faust, K. M. (2023). Measuring ethical development of engineering students across universities and class years. *International Journal of Ethics Education, 8*, 49–65. https://doi.org/10.1007/s40889-022-00150-w

Lee, H. (2016). Conceptualization of an SSI-PCK framework for teaching socioscientific Issues. *Journal of the Korean Association for Science Education, 36*(4), 539–550.

Lee, H., Yoo, J., Choi, K., Kim, S. W., Krajcik, J., & Herman, B. C. (2013). Socioscientific issues as a vehicle for promoting character and values for global citizens. *International Journal of Science Education, 35*(12), 2079–2113.

LePage, P., Akar, H., Temli, Y., Şen, D., Hasser, N., & Ivins, I. (2011). Comparing teachers' views on morality and moral education, a comparative study in Turkey and the United States. *Teaching and Teacher Education, 27*(2), 366–375. https://doi.org/10.1016/j.tate.2010.09.005

Magnusson, S., Krajcik, J., & Borko, H. (1999). Nature, sources, and development of pedagogical content knowledge for science teaching. In J. Gess-Newsome & N. G. Lederman (Eds.), *Examining pedagogical content knowledge* (pp. 95–132). Springer.

Ministry of National Education [MoNE]. (2013). *Fen bilimleri dersi öğretim programı* [Turkish elementary and middle school science curriculum]. MEB Publishing.

Ministry of National Education [MoNE]. (2018a). *Fen bilimleri dersi öğretim programı* [Turkish elementary and middle school science curriculum]. MEB Publishing.

Ministry of National Education [MoNE]. (2018b). *Lise Fizik dersi öğretim programı* [Turkish high school physics curriculum]. MEB Publishing.

Ministry of National Education [MoNE]. (2018c). *Lise Biyoloji dersi öğretim programı* [Turkish high school biology curriculum]. MEB Publishing.

Namdar, B., & Namdar, A. O. (2021). Fostering students' values through role play about socioscientific issues. *The Physics Teacher, 59*(6), 497–499. https://doi.org/10.1119/5.0019320

Namdar, B., & Namdar, A. O. (2022). Role reversal to facilitate social & moral compassion. *The American Biology Teacher, 84*(4), 242–245. https://doi.org/10.1525/abt.2022.84.4.242

Narvaez, D., & Rest, J. (1995). The four components of acting morally. In W. Kurtines & J. Gewirtz (Eds.), *Moral behavior and moral development: An introduction* (pp. 385–400). McGraw-Hill.

OECD. (2018). *The future of education and skills: Education 2030. The future we want*. OECD Publishing.

Özden, M. (2020). Elementary school students' informal reasoning and its' quality regarding socio-scientific issues. *Eurasian Journal of Educational Research, 20*(86), 1–24. https://doi.org/10.14689/ejer.2020.86.4

Özturk, A. (2018). Human rights education with socioscientific issues through the environmental education courses. *Eurasian Journal of Educational Research, 18*(77), 1–30. https://doi.org/10.14689/ejer.2018.77.3

Ozturk, N., & Yilmaz-Tuzun, O. (2017). Preservice science teachers' epistemological beliefs and informal reasoning regarding socioscientific issues. *Research in Science Education, 47*(6), 1275–1304. https://doi.org/10.1007/s11165-016-9548-4

Pedretti, E. (1999). Decision making and STS education: Exploring scientific knowledge and social responsibility in schools and science centers through an issues-based approach. *School Science and Mathematics, 99*, 174–181.

Piaget, J. (1932). *The moral judgment of the child*. Harcourt, Brace.

Powell, W. A., Newton, M. H., & Zeidler, D. L. (2021). Impact of socioscientific issues on middle school students' character and values for global citizenship. In W. Powell (Ed.), *An introduction to socioscientific issues-based instruction for scientific literacy development* (pp. 56–91). IGI Global.

Rest, J. (1986). *Moral development: Advances in research and theory*. Praeger.

Roberts, D. A. (2007). Scientific literacy/science literacy. In S. K. Abell & N. G. Lederman (Eds.), *Handbook of research on science education* (pp. 729–780). Lawrence Erlbaum.

Roehrig, G., Keratithamkul, K., & Hiwatig, B. M. R. (2021). Intersections of integrated STEM and socioscientific issues. In W. Powell (Ed.), *An introduction to socioscientific issues-based instruction for scientific literacy development* (pp. 256–278). IGI Global.

Rundgren, C.-J., & Chang Rundgren, S.-N. (2018). Aiming for responsible and competent citizenship through teacher professional development on teaching socioscientific inquiry-based learning (SSIBL). *Asia-Pacific Forum on Science Learning and Teaching, 19*(2).

Sadler, T. D. (2004). Informal reasoning regarding socioscientific issues: A critical review of research. *Journal of Research in Science Teaching, 41*, 513–536.

Sadler, T. D., & Zeidler, D. L. (2003). The morality of socioscientific issues: Construal and resolution of genetic engineering dilemmas. *Science Education, 88*(1), 4–27. https://doi.org/10.1002/sce.10101

Salmoni, B. A. (2004). Ordered liberty and disciplined freedom: Turkish education and republican democracy. *Middle Eastern Studies, 40*(2), 80–108.

Schuitema, J., Dam, G., & Veugelers, W. (2008). Teaching strategies for moral education: A review. *Journal of Curriculum Studies, 40*(1), 69–89. https://doi.org/10.1080/00220270701294210

Sevgili, C., Selvi, M., & Irmak, M. (2022). 2008–2020 Yılları arasında fen bilgisi eğitiminde sosyobilimsel konular üzerine yayınlanmış lisansüstü tezlerin değerlendirilmesi. *Eğitim ve Toplum Araştırmaları Dergisi, 9*(2), 430–446. https://doi.org/10.51725/etad.1210246

Shim, J. (2023). Investigating the effectiveness of introducing virtual reality to elementary school students' moral education. *Computers & Education: X Reality, 2*, 100010.

Sjöström, J., & Eilks, I. (2018). Reconsidering different visions of scientific literacy and science education based on the concept of *Bildung*. In Y. J. Dori, Z. R. Mevarech, & D. R. Baker (Eds.), *Cognition, metacognition, and culture in STEM education. Innovations in science education and technology* (Vol. 24). Springer. https://doi.org/10.1007/978-3-319-66659-4_4

Sönmez-Eryaşar, A. (2021). Sosyobilimsel konular ve öğretmenler: türkiye'deki tezlere yönelik bir sistematik literatür incelemesi. *Ahi Evran Üniversitesi Kırşehir Eğitim Fakültesi Dergisi, 22*(3), 1604–1640. https://doi.org/10.29299/kefad.855194/10.15390/eb.2024.12578

Sözer, Y. (2023). Türkiye Millî Eğitim Şuraları eğitim programlarına ilişkin kararlarının felsefi yönden incelenmesi: Cumhuriyet'in ilk yüzyılına ilişkin dönemsel bir analiz. *Gaziantep University Journal of Social Sciences*, 55–85.

Sterelyn, K. (2007). Social intelligence, human intelligence and niche construction. *Philosophical Transactions of the Royal Society, 362*(1480), 719–730. https://doi.org/10.1098/rstb.2006.2006

Sullins, J. (2023). Information technology and moral values. In E. N. Zalta & U. Nodelman (Eds.), *The Stanford encyclopedia of philosophy*. https://plato.stanford.edu/archives/fall2023/entries/it-moral-values/

Sürmeli, H., & Şahin, F. (2010). Üniversite öğrencilerinin genetik mühendisliği ile ilgili biyoetik görüşleri: genetik testler ve genetik tanı. *Türk Fen Eğitimi Dergisi, 7*(2), 119–132.

Tatar, Ş., & Adıgüzel, O. (2019). Türkiye'de tartışmalı ve sosyobilimsel konular üzerine yazılan lisansüstü tezlerin eğitim bilimleri perspektifinden incelenmesi [Özel Sayı]. *Eskişehir Osman Gazi Üniversitesi Sosyal Bilimler Dergisi, 20*, 305–325. https://doi.org/10.17494/ogusbd.548368

Thoma, S. J., & Dong, Y. (2014). The defining issues test of moral judgment development. *Behavioral Development Bulletin, 19*(3), 55–61.

Topçu, M. S. (2008). *Pre-service science teachers' informal reasoning regarding socioscientific issues and the factors influencing their informal reasoning*. Unpublished doctoral thesis, Middle East Technical University, Ankara, Turkiye.

Topçu, M. S. (2010). Development of attitudes towards socioscientific issues scale for undergraduate students. *Evaluation & Research in Education, 23*(1), 51–67.

Topçu, M. S. (2011). Turkish elementary student teachers' epistemological beliefs and moral reasoning. *European Journal of Teacher Education, 34*(1), 99–125. https://doi.org/10.1080/02619768.2010.534132

Topçu, M. S., Sadler, T. D., & Yilmaz-Tuzun, O. (2010). Preservice science teachers' informal reasoning about socioscientific issues: The influence of issue context. *International Journal of Science Education, 32*(18), 2475–2495.

Topçu, M. S., Yilmaz-Tuzun, Ö., & Sadler, T. D. (2011). Turkish preservice science teachers' informal reasoning regarding socioscientific issues and the factors influencing their informal reasoning. *Journal of Science Teacher Education, 22*(4), 313–332.

Topçu, M. S., Muğaloğlu, E. Z., & Güven, D. (2014). Fen eğitiminde sosyobilimsel konular: Türkiye örneği. *Educational Sciences: Theory & Practice, 14*(6), 2327–2348.

Topçu, M. S., Atabey, N., & Çiftçi, A. (2022). Supporting science teachers' professional development and teaching practices: A case study of socioscientific issue-based instruction. In Y. S. Hsu, R. Tytler, & P. J. White (Eds.), *Innovative approaches to socioscientific issues and sustainability education. Learning sciences for higher education*. Springer.

Tsai, C. (2018). The effect of online argumentation of socio-scientific issues on students' scientific competencies and sustainability attitudes. *Computers & Education, 116*, 14–27. https://doi.org/10.1016/j.compedu.2017.08.009

Tuncay, B., Yılmaz-Tüzün, Ö., & Teksoz, G. T. (2012). Moral reasoning patterns and influential factors in the context of environmental problems. *Environmental Education Research, 18*(4), 485–505. https://doi.org/10.1080/13504622.2011.630576

Tuncay-Yüksel, B., Yılmaz-Tüzün, O., & Zeidler, D. L. (2023). Epistemological beliefs and values as predictors of preservice science teachers' environmental moral reasoning. *Journal of Research in Science Teaching, 60*(9), 2111–2144. https://doi.org/10.1002/tea.21889

Türköz, G., & Öztürk, N. (2020). Fen bilgisi öğretmen adaylarının bazı sosyo-bilimsel konularla ilgili kararlarının çok boyutlu bakış açısı ile incelenmesi. *Cumhuriyet International Journal of Education, 9*(1), 175–197.

Walker, L. J., & Frimer, J. A. (2007). Moral personality of brave and caring exemplars. *Journal of Personality and Social Psychology, 93*, 845–860.

Walker, D. I., Roberts, M. P., & Kristjánsson, K. (2015). Towards a new era of character education in theory and in practice. *Educational Review, 67*(1), 79–96.

Yaman, H., & Anilan, B. (2021). Values education in science lessons with activities: responsibility value. *Science Education International, 32*(3), 237–247. https://doi.org/10.33828/sei.v32.i3.7

Yapıcıoğlu, A. E. (2021). Investigation of the bibliometric features of the articles on socioscientific issues. *OPUS–International Journal of Society Researches, 17*(36), 2402–2428. https://doi.org/10.26466/opus.841772

Ye, W. (2014). Power and moral education in China: Three examples of school-based curriculum development. .

Zeidler, D. L. (1984). Moral issues and social policy in science education: Closing the literacy gap. *Science Education, 68*, 411–419.

Zeidler, D. L. (2020). STEM: Unrequited dreams in a material world. In V. Ackerson & G. A. Buck (Eds.), *Critical questions in STEM education* (pp. 257–260). Springer.

Zeidler, D. L., & Sadler, T. D. (2023). Exploring and expanding the frontiers of socioscientific issues. In N. G. Lederman, D. L. Zeidler, & J. S. Lederman (Eds.), *Handbook of research on science education* (pp. 899–929). https://doi.org/10.4324/9780367855758-33

Zeidler, D. L., Sadler, T. D., Simmons, M. L., & Howes, E. V. (2005). Beyond STS: A research-based framework for socioscientific issues education. *Science Education, 89*(3), 357–377.

Zeidler, D. L., Herman, B. C., & Sadler, T. (2019). New directions in socioscientific issues research. *Disciplinary and Interdisciplinary Science Education Research, 1*(11), 1–9. https://doi.org/10.1186/s43031-019-0008-7

Chapter 20
The Contribution of Socioscientific Issues to Science Education: A United Kingdom Perspective

Alex Craig Jeffries, Amanda McCrory, and Michael J. Reiss

20.1 Introduction

There are a number of arguments in favor of a greater inclusion of socioscientific issues (SSI) in science education. In particular, such issues can be motivating for learners, give them deeper understanding of how science is interdisciplinary, and help them appreciate the complexities of applying the science learnt in classrooms and labs to the real world. In the UK though, SSI are under-used in science education, whether at primary, secondary, or tertiary level. This chapter looks at the utility of SSI from a range of perspectives and examines the reasons why SSI are not used more widely in science education, with a particular focus on neoliberal understandings of education that are increasingly prevalent. We discuss how SSI might be used to contribute to the formation of scientific literacy, epistemic insight and human flourishing at all three levels of formal education. We begin with an unusual argument as to the value of SSI, namely one derived from Stoicism.

20.2 Stoicism and SSI

Most people know of Stoicism only through the injunction 'to be stoical', meaning that one should endure hardship without showing one's feelings or complaining. Somewhat unexpectedly, there has recently been an upsurge of influencers on

A. C. Jeffries (✉)
Milner Centre for Evolution, Department of Life Sciences, University of Bath, Bath, UK
e-mail: bssacj@bath.ac.uk

A. McCrory · M. J. Reiss
UCL Institute of Education, University College London, London, UK

D. L. Zeidler (ed.), *A Moral Inquiry into Epistemic Insights in Science Education*, Contemporary Trends and Issues in Science Education 61,
https://doi.org/10.1007/978-3-031-63382-9_20

TikTok, Instagram, YouTube and other platforms enjoining us to draw on Stoic philosophy to overcome destructive emotions. This sudden popularity has surprised philosophers, and it has been pointed out that these short videos and the general recent enthusiastic advocacy of Stoicism, promising us productivity and happiness, ignore its full scope and underpinnings, in particular its relationship to science (e.g., Duncombe, 2023).

The philosophy of the ancient Stoics is best known today for its focus on practical ethics. However, historically it was grounded in a tripartite framework, prioritizing logic, physics, and ethics, in that order. Logic, for the Stoics, was foundational, serving as a tool for reasoning and rationality and therefore the discernment of truth and fallacy. 'Physics', in Stoic philosophy, was a much broader concept than the word represents today. For the Stoics, physics encompassed not just the workings of the natural world, but also metaphysical elements, psychology, and even theology (Menn, 1997). Ethics, forming the third part of this framework, provided normative principles and practical guidance to achieve and sustain *eudaimonia*, a state of flourishing or well-being. For a comprehensive review of ancient Stoic history and philosophy, see Sellars (2006).

A crucial aspect of Stoic philosophy, particularly relevant to SSI, is the Stoic view of physics as a necessary foundation or precursor for ethics. They believed that making informed ethical decisions required a deep understanding of the world in which these ethics were applied. This is because ethical principles are anchored in value judgments about what is significant and meaningful, but confined within the realm of what is realistically achievable. The practice of making normative propositions and assigning values in human behavior and psychology are an integral part of this process and cannot be known *a priori* but require knowledge of the natural world by a suitably literate society.

While the fundamental nature of the human mind remains consistent from ancient times to the present, the practices and contexts of the modern world are heavily influenced by scientific and technological developments and so differ from those of our Stoic predecessors in many significant ways. These differences manifest in new possibilities and concerns, especially in fields that are influenced by scientific knowledge and understanding. Though ancient philosophy laid the groundwork for principles of argumentation and reasoning, there have been significant developments since, particularly in probabilistic and statistical thinking, largely spurred by the development of scientific and mathematical knowledge. But despite the huge advances in the volume and complexity of our knowledge and understanding, the basic Stoic perspective of physics informing ethics is still a valid one. Perhaps even more so today, ethical judgements need to be informed by scientific realities, which requires a scientifically literate society.

20.3 The Aims of the UK Education System

In the modern world, secular education has become a nearly ubiquitously practiced system in many countries for the structured transmission of knowledge, skills, values, and beliefs from one generation to another. It is largely the vehicle for developing a scientifically literate society. The United Kingdom (UK), as one would expect from it being a democratic, industrially developed country with a relatively high standard of living and technological infrastructure, has a well-established, diverse, and (to some extent) decentralized education system that ranges from early years through to tertiary level.

The aim of this system is to 'produce' citizens who can engage in a diverse range of activities that ultimately help towards the maintenance and enhancement of society's wellbeing. The potential activities that could contribute is vast, and extremely difficult to rank by order of impact. However, for better or worse, the primary aim of the UK education system is the generation of productive and constructive citizens who can contribute, directly or indirectly to the country's economic growth. It is largely to this end that science education is directed, and the concept of a 'physics'-guided ethical society is either completely unacknowledged or at best peripherally alluded to. For example, the UK government's latest policy statement on the future of science and technology speaks almost entirely to economic impact, with just a nod towards human health and sustainability, and never mentions wider societal concerns such as ethics, or character development (Department for Science, Innovation, & Technology and Prime Minister's Office, 2024). Similarly, policy advisories on education's contributions focus on economics (e.g., Grant, 2017). In an occasional exception, Power et al. (2015), in a non-governmental policy advice initiative by the Health & Education Advice & Resource Team (HEART), draw attention in an annex to Unesco's 1998 *World Declaration on Higher Education for the Twenty-first Century: Vision and Action*, but do not discuss it.

20.4 Tertiary Education

The tertiary education system in the UK is decentralized in terms of not having a truly universal curriculum and institutions having considerable scope for autonomy, albeit less now than in the past as successive governments have exercised greater control. University status is granted through a rigorous process *via* either royal charter, Orders of Council, and/or other approvals under the Further and Higher Education Act 1992 (The National Archives, 2024). Each of these approaches require the submission, scrutiny, and approval of documents that outline aspects such as the institution's powers, internal governance structure and regulations, purpose, and charitable status, and are judged on criteria such as academic quality, financial sustainability, and public benefit. The public benefit criterion is usually generic, but a central role is typically now given to the economic concept of return on investment for graduates and society. Overall, the university creation process provides high level oversight of the number and quality of higher education institutions (HEIs) in the country and details of an HEI's remit is largely left to the individual institutions.

Universities, once created, develop their teaching program portfolios, faculty, and infrastructure with a large degree of autonomy. Curricula are not set out directly by a central authority but are monitored by an independent body, the Quality Assurance Agency for Higher Education (QAA; https://www.qaa.ac.uk/). The QAA monitors and advises on the standards and quality of UK HEIs through periodic inspections of institutions and reporting of its findings and recommendations to those institutions, in addition to providing overarching advice to various concerned government bodies about the whole UK higher education sector. In addition, the QAA develops and maintains the UK Quality Code for Higher Education which sets out the expectations for HEIs. Key components of this are a set of Subject Benchmarks that specify the expected standards of graduates from various subject degrees (QAA, 2024). At the time of writing there are 28 Subject Benchmarks developed, of which arguably 19 are within the 'science' umbrella. None of the QAA's outputs have statutory or regulatory authority; instead the system relies on self-regulation within UK universities (statutory powers over HEIs are held by the Office for Students (OfS), established in 2018; https://www.officeforstudents.org.uk/). Nevertheless, UK HEIs take QAA inspections and recommendations very seriously and the Subject Benchmark Statements are typically consulted during the design and development of new courses.

As an example, the Subject Benchmark Statement for Biosciences (QAA, 2023) makes no mention of morals, or character development. It does however concern itself with ethics from a range of perspectives, though mainly focused on professional practice (such as scientific integrity), the use of humans and other animals in research, sustainability, and associated legislative and regulatory frameworks. Section 1.14, within a wider equity, diversity and inclusion section, suggests that the historical role of biological science in social injustice could be critically explored as part of a program. Similarly, both historical and current ethical issues and perspectives in the practice of science are suggested as exploratory topics, and from a perspective ranging from the individual student through to the whole of society. Section 1.15 suggests that individual students explore the biosciences from a perspective of their personal and cultural backgrounds, hinting at moral or ethical self-development, but an explicit justification for this perspectival approach is not given. The section concludes with a succinct steer for curriculum development stating "Curricula should facilitate the development of culturally aware graduates, with ethical and open approaches to thinking and practising the Biosciences." (QAA, 2023, 5). Each Subject Benchmark Statement is specific to its discipline in terms of its coverage and the Statements vary in their formats and extent. For instance, the Subject Benchmark Statement for Physics, Astronomy and Astrophysics (QAA, 2019) is much shorter than the one for Biosciences and makes only two bullet-point references to ethics, and only in the context of "professional behaviour."

The biological sciences contain within them numerous issues, past and present, that have direct ethical implications, perhaps more so than any other scientific discipline. These range from historical topics, such as eugenics, through to current technologies, such as genetically modified organisms, cloning, assisted reproductive technologies, and the ownership of biological sequence data (e.g., Miller & Selgelid, 2007). All are examples of contentious, often ill-structured ('wicked') topics with significant societal impacts; they require sophisticated scientific knowledge and evidence-based reasoning to fully comprehend and evaluate. This makes them ripe for SSI-based teaching and learning, and the combination of reasoning, science and ethics illustrates that a Stoic approach, whether explicitly acknowledged as such or not, not only has merit but is essential.

As the biological sciences often raise ethical questions, it is not uncommon to find a biological ethics module offered to students on programs across the biological sciences subdisciplines and there now exist a number of textbooks to help bioscience students to reason ethically about their subject (e.g., Bryant & la Velle, 2018). A full module in this area is typically an optional choice for students, but some addressing of professional practice (e.g., scientific integrity, plagiarism, etc.) will likely be a mandatory component of key skills training. As with the Subject Benchmark Statement for Biosciences, the focus of such teaching will largely be on professional practice and the ethical implications of technologies and ethical approval frameworks. A more introspective approach to personal moral/ethical/character development will most likely be a byproduct of this teaching rather than an explicitly taught and/or facilitated educational experience.

Another influence on curriculum design and implementation in UK HEIs, and common worldwide, comes from external program accreditation organizations. Certain degrees are required to meet industry or professional standards and use independent verification of curricula, faculty expertise, teaching facilities and graduate outcomes to regulate subsequent registration and employment of graduates. For instance, the General Pharmaceutical Council (GPhC) and the General Medical Council (GMC) rigorously control pharmacists and medical doctors, respectively. These accreditations have statutory status. For example, the Pharmacy Order 2012 mandated the GPhC accreditation for pharmacists and pharmacy technicians to practice, whilst the Medical Act 1983 established the GMC as the independent regulator of medical practice.

In addition to these statutory accreditation organizations, a number of non-professional and non-statutory accreditation organizations also exist for various disciplines and have a similar influence on curriculum development and teaching practice. For instance, the Royal Society of Biology (RSB; https://www.rsb.org.uk/) exists to quality assure the curricula, faculty competence and graduate attainment for a range of biosciences programs at UK HEIs, from the undergraduate to doctoral level. Such accreditation is voluntary for the institution but is typically sought to help with employability of graduates, the fostering of industry contacts, and recruitment of students.

Similar to the Subject Benchmark Statement for Biosciences discussed above, the RSB's Accreditation Handbook (RSB, 2023), which lays out the details of its accreditation process, makes no mention of morals, moral character or development, or personal character development in general. However, it does make mention of ethics in multiple places but mostly in a generic manner and again with an emphasis on professional practice. It briefly recognizes that the teaching of ethics will provide learning outcomes consistent with those of SSI. For example, it states that, "The study of ethics helps students to develop widely applicable skills in communication, reasoning and reflection, as well as an introduction to codes of conduct and work as a professional scientist." (RSB, 2023, 23). The Accreditation Handbook further advises that, "Regulatory and ethical issues, including environmental and social aspects, are considered and addressed by students at appropriate times throughout their programme of study." (RSB, 2023, 29).

Within the RSB Accreditation Handbook, the strongest statement that aligns with the ethos of SSI are guidelines for the "Development of creativity and innovation relevant to the work place" (RSB, 2023, 26). This section advises that, "Institutions should provide evidence that they encourage students to be creative by 'thinking differently." (RSB, 2023, 26). It further details that, "Students should be encouraged to: be inquisitive and open-minded; welcome the unexpected; challenge assumptions and (from time-to-time) defy convention; think beyond their own discipline and make interdisciplinary connections; and consider problems from the perspective of non-biologists." (RSB, 2023, 26). The Accreditation Handbook also gives some general advice about teaching and assessment in this domain:

> HEIs should make it clear how they promote creativity and creative problem solving, using techniques designed to develop individual and group creativity. For group sessions there should be evidence that institutions offer structured, constructive and inclusive approaches to creative problem solving. Where these activities are assessed, emphasis should be placed on students demonstrating how they have engaged with techniques designed to promote creativity in individuals, and the extent of their participation in group sessions. As an example for the former, students could be asked how they have utilized a specific technique during creative problem solving. Students should not be awarded marks solely on the basis of coming up with novel ideas, as this is frequently an unrealistic expectation. (RSB, 2023, 26)

Such guidance facilitates teaching and learning about SSI. However, while UK tertiary bioscience courses do often include a certain amount of ethics, they are less likely to draw on other non-science disciplines. History is likely to be used simply to contextualize (e.g., Darwin when teaching evolution, Mendel when teaching genetics—and that less now than was the case) and, in our experience, some disciplines—e.g., politics, economics and sociology—seem rarely to feature, though we are unaware of any systematic research on this issue.

In conclusion, although an explicit articulation of SSI principles is absent in the various guidelines and oversight mechanisms governing UK HEIs by both governmental and independent entities, the ethos of SSI is evident in instances pertaining to ethical considerations and student skills such as creativity and problem-solving. The UK HEI pedagogic culture is quite uniform which facilitates the dissemination and integration of SSI-related principles implicit in the above-mentioned sources, with adjustments made to accommodate the distinctive requirements of each academic discipline. However, as discussed above, the predominant focus is on fostering professional practice, upholding scientific integrity, and adhering to pertinent legal frameworks and monitoring mechanisms. Within the realm of the sciences in higher education, the incorporation of extended moral or personal ethical development, as well as broader character development, seems to be significantly scarcer. The curricula for scientific disciplines typically allocate minimal time for such pursuits, with faculty members generally, in our experience, exhibiting a limited inclination and proficiency for facilitating in-depth exploration of moral or personal ethical dimensions. The emphasis in character development within the sciences, when there is any, predominantly revolves around cultivating traits such as originality, curiosity, creativity, personal responsibility, conscientiousness, and tenacity, and all within the context of carrying out scientific research. This gap is a potential opportunity for more explicit SSI approaches to teaching at HEIs as it pertains to moral and character development. However, this will face the challenges of forming a coherent justification for the value of such teaching and the formation of advocacy and dissemination that would be needed to generate what is, in essence, a cultural change. Arguably, the vehicles best suited to initiate this would seem to be the governmental and independent entities that in their own ways have been influencing the development of program content and structures at UK HEIs over many years. Otherwise, the inclusion of SSI in UK HEIs is likely to depend on local enthusiasms, leading to, at best, a patchwork coverage.

20.5 School Education

To a greater extent than at tertiary level, primary and secondary school education depends on which of four UK nations (England, Northern Ireland, Scotland, Wales) one is talking about. Scotland, in particular, has an education system that is, and has long been, very different in structure and curriculum from that of the other three UK nations.

In England, Northern Ireland and Wales, a particularly important date for science education was the introduction, in 1989, of the National Curriculum. Initially, this applied to all state schools, with only the 7% or so of students in independent (fee-paying) schools exempt. However, the reality has been that the examinations sat at age 16 (GCSEs, General Certificates of Secondary Education), normally after 2 years of study, need to be aligned with the requirements of the National Curriculum. As a result, the various subject curricula taught to 14–16 year-olds in state and independent schools are very similar, and this has had a certain amount of 'washback', so that the similarities at early ages are greater than might have been expected.

The UK school education system is generally held to be one of the most complicated to explain, not just because of the differences between the four UK nations but because of the existence not only of a state and independent system but also of a plethora of types of school within the state system. For example, academies were introduced by the Labour government in 2002 and free schools by the Conservative-Liberal Democrat coalition following the 2010 general election. The relevance of this is that both academies and free schools are not obliged to follow the National Curriculum. Indeed, now that the majority of state schools are academies or free schools, it might be thought that government control over what is taught in school has weakened. In reality, partly because of GCSEs, as mentioned above, and other elements of the examination system, this has not proved to be the case.

Another important lever controlling what is taught in schools is the inspection regime. This occurs in all four UK nations with Ofsted (the Office for Standards in Education, Children's Services and Skills) being the non-ministerial department responsible for school inspection (amid other responsibilities) in England. Ofsted was established in 1992, under a Conservative government, amidst a quite widespread presumption that school standards were not improving. Prior to this, HMI (His Majesty's Inspectorate) undertook a certain number of inspections of schools in England but these were not systematic and HMI, which was smaller than Ofsted, had a particular responsibility to inform government in relation to policy matters (Lee & Fitz, 1997).

In Ofsted's first Annual Report, Professor Sutherland, the first chief Inspector of Ofsted, wrote that the purpose of Ofsted, "is to make a contribution, through these inspections, to raising standards and improving the quality of educational experience and provision" (Ofsted, 1993) and this remains its main aim. However, since the introduction of the National Curriculum (in 1989) and Ofsted (in 1992), central government (of whatever political persuasion) has increasingly been accused of micromanaging the teaching and administration of schools, and in pillorying schools that do not conform. Ofsted has, in effect, become a government enforcement agency with less independence than its HMI predecessor.

20.5.1 Neoliberalism in UK School Education

Since the 1980s, neoliberalism has increasingly been the dominant ideology shaping school education policy in the UK. It has been argued that the school education system should be managed according to the principles of a free market. Thus, neoliberals believe that education should be privatized both endogenously, promoting competition between schools to incentivize them to raise standards, and exogenously, enabling private companies to take over education services which had been previously managed by the public sector.

In 1988, the Education Reform Act, which heralded the introduction of the National Curriculum, was enacted by the Conservative government; it was strongly influenced by neoliberal ideals. With the introduction of a statutory, and some would assert restrictive, National Curriculum and the introduction of new, standardized assessment practices in schools, it was maintained by the government that both primary and secondary schools were given the tools needed to raise standards of achievement for the students they served (MacBeath & McGlynn, 2004). At the same time, school league tables were introduced, with the promise that this would empower parents to judge and compare schools in the best interest of their child's education.

In doing this, parental choice was highlighted as an important factor of an educational market, but the reality was that schools were forced to engage in hyper-competitive practices, such as single-mindedly focusing on getting the best possible grades to the detriment of a more holistic education (Ball, 2016). And so, from this emerged a pervasive culture of performativity in schools, producing a high stakes environment with the tendency for teachers to 'teach to the test' to ensure student grades and measures of school performance aligned with what was, and still is, seen as acceptable assessment benchmarks. Formula funding for schools was also introduced, meaning that schools then needed to compete to a greater extent for students. This policy, inevitably, led to polarization across schools with those deemed 'best' being oversubscribed and those deemed 'worst' being under-funded, leading to significant wealth inequalities whilst widening the education inequality gap (Gray et al., 2018).

Integral to a neoliberal approach to education is the argument that both parents and students are given more agency, choice and voice whilst standards in schools are monitored *via* a top-down approach of surveillance and performance management. This has been and continues to be achieved *via* a focus on teachers' performativity which can be defined as, "a mode of regulation that employs judgements, comparisons and displays as a means of incentive, control, attrition and change: (Ball, 2003, 216). There can be a set of rigid conventions to which teachers feel they must adhere in order to be considered 'good' teachers. Neoliberalism argues that all learners will perform and succeed in standardized testing if taught as deemed appropriate, thus meaning that performance management provides a sure-fire way to raise educational standards.

Critics of neoliberalism have consistently pointed out that education is not an industry; rather, teachers are professionals and not technicians (Ball, 2003). Classrooms are environments of teaching and learning and not production lines. Schools are a microcosm of society, not factories. In schools, students and teachers (should) work and live together in a learning community, working collaboratively to nurture and promote students' interests and abilities by engaging in the scholarship of learning; but the neoliberal agenda prevails (Gray et al., 2018). This is despite the fact that neoliberalism tends to perpetuate, even accentuate, educational inequalities by prioritizing market-based solutions over and above the common good. Free markets tend to advantage those with capital and resources. Exposing schools to market forces can therefore be detrimental to students who come from disadvantaged backgrounds (Ingleby, 2021) as they often lack sufficient resources or navigational capital (Yosso, 2005) to get into the best schools in the open educational market. If schools are to narrow the pervasive attainment gap for children from disadvantaged backgrounds, which evidence suggests is already present when children come to school, then teaching for engagement is key. This includes building on the ideas that students bring to lessons and helping children to build self-regulation for intrinsic motivation.

A different and more recent criticism of neoliberal approaches to education has come, particularly post-COVID, from an increasing acknowledgement that the present system of education in England is unhealthy for many students and teachers. It is generally accepted (Glazzard, 2018) that children's mental health has deteriorated in recent decades in the UK (Pitchforth et al., 2019) and that teachers have high levels of stress and low and deteriorating levels of mental health (Kidger et al., 2016). Much has been written in the popular press and professional literature about the impact of school inspections and other aspects of the neoliberal apparatus on the mental health and well-being of school staff, including those in leadership. The recent tragedy of primary school headteacher Ruth Perry, who took her own life in 2023 after an Ofsted inspection downgraded her school from the top grade ('outstanding') to be bottom one ('inadequate') is one such example which hit the headlines. An independent coroner's inspection concluded that Perry's cause of death was "Suicide, contributed to by an Ofsted inspection carried out in November 2022" (Connor, 2023).

There has understandably therefore been a backlash against the neoliberal culture in which schools currently operate. Critics of neoliberalism argue that a neoliberal focus on school education omits much of value that schools can and should do. Critics also maintain that to raise standards and ensure effective student progress, schools must become more autonomous and collegial entities, responsible for their own development and for the demographic they serve. In order to achieve this, instead of an excessive focus on external inspection, as though schools are only concerned with the cognitive output of students, as measured *via* an obsessive focus on examination outcomes, internal accountability and self-evaluation is needed *via* moderation in and across schools and governing bodies so that the school itself can answer the fundamental question: 'Are standards as good as they can be?' (Ball, 2016).

20.5.2 The Contribution of SSI to School Science Education

The neoliberal approach to education fails to take into account the complexity of the learning needs present in any science classroom and the diverse demographic of students that science teachers serve. The irony here, which we are sure is not lost on the reader, is that this approach can prevent creative and innovative approaches to teaching, which are likely ultimately to help the UK's economy, given that the UK is increasingly a knowledge and service economy. A narrow neoliberal approach to education, including STEM (Science, Technology, Engineering and Mathematics) education, means that learners are more likely to be disengaged, with deeper, critical thinking skills not being sufficiently developed (cf. Bencze et al., 2018). Student engagement and interest are key to school students "learning a body of knowledge relating to the products and practices of science … and developing a sense of excitement and curiosity about natural phenomena" (Ofsted, 2021). One way in which this can be achieved is when science courses personalize and localize science learning by emphasizing the links between science and everyday life, as this has been shown to foster more positive attitudes to science and school science (Bennett et al., 2007).

A high stakes neoliberal environment, an overburdened science curriculum, and the constant pressure for students to perform at their very best in standardized tests is not a conducive environment for the teaching and learning of SSI. Engaging in interdisciplinary learning, inquiry-based learning, problem-based projects, real-life problem solving, critical thinking, discussion, and perspective taking does not translate well into a neoliberal perspective which is built on a narrow conceptualization of teaching and learning. This, we argue, goes a long way to accounting for why SSI approaches to science education are still underutilized in schools in the UK. Nonetheless, there has been a gradual turn to utilizing SSI in school science education, particularly at secondary level but also in primary schools where the existence of a culture of cross-curricular teaching and project-based learning provides fertile ground to employ SSI (McCrory & Reiss, 2023).

For science teachers, a neoliberal environment can also have a profound impact on their agency as well as on the development of their professional identity, pedagogy, and views regarding the teaching of science. This can cause conflict between personal and professional values in relation to the aims of science education. Am I teaching this because it is important, worthwhile and aligns with my own teaching ethos or simply because ultimately it will help my students and my school to perform better when their performance is assessed and compared with others? We are not, of course, against learners and schools doing well in assessment; rather, our point is two-fold: first, some of the best teaching can result in valuable learning that cannot be assessed in national tests and examinations; secondly, the present system of national tests and examinations in the UK, perhaps especially in England, is based on a much narrower conceptualization of learning than could (should) be the case. The reality of a neoliberal approach in the science classroom can mean that there is the potential not only to impact negatively on the educational achievement of some learners, but also on the development of their scientific literacy, epistemic insight and knowledge, as well as their aspirations to become scientists or work in scientific fields (McCrory & Reiss, 2023).

We live in a highly connected and globalized world, but one that remains in some ways fragmented. Science operates within a landscape of rapid technological change with uncertainty about the impact of this and what the future might look like, including concerns that humanity is facing threats to its continued existence (Reiss, 2023). If we take, for example, the current debate surrounding Artificial Intelligence, a clear example of technological advancement which is challenging to keep up with and is beginning to impact education and our daily lives in unprecedented ways, we can see the complexity of the issues and sympathize with the concerns voiced. This is an example of a socially relevant, multi-disciplinary issue, based in science, and which has the potential to have a significant impact on society—an SSI.

Science teachers in England are now expected to address matters which have previously been the province of teachers of history, PSHE (Personal, Social, Health and Economic) education, and even Religious Education. Ofsted inspection requirements stipulate that a school's intention to provide for the personal development of all students must be a whole school issue. Schools are required to demonstrate how their programs of personal development aim to enable their students to develop into responsible, respectful, and active citizens who are able to play their part and become actively involved in public life as adults. This should be achieved by developing students' character, dispositions, and virtues that inform their motivation and guide their conduct so that they reflect wisely, learn eagerly, behave with integrity, and cooperate with others. In 2008, changes to the science National Curriculum (for 11–16-year-olds) for England and Wales encouraged the recognition of the role of controversy in science. Subsequent changes to the National Curriculum for science in 2015 require 14–16-year-old students to learn about practical and ethical considerations of modern biotechnology (DfE, 2015, 10). Aims across the whole of the 5–16 age range include ensuring that students are "equipped with the scientific knowledge required to understand the uses and implications of science, today and for the future" (DfE, 2015, 2).

Specific references to SSI and scientific debate are lacking from the current (DfE, 2015) primary science National Curriculum (for 5–11-year-olds), with guidance stating that the "social and economic implications of science are important but, generally, they are taught more appropriately within the wider school curriculum" (DfE, 2015, 2). However, topics within the primary science National Curriculum include adaptation, evolution, deforestation, diet, drugs, nutrition, materials, micro-organisms, exploring human impact on ecology, and studying scientists and their work. Clearly there is room here for discussion and debate in the primary science classroom, and the primary science National Curriculum also states that teachers will, "wish to use different contexts to maximize their pupils' engagement with and motivation to study science" (DfE, 2015, 3). We take the view that this does in fact give schools the freedom and agency to teach in ways that they see best, so long as the statutory requirements are adhered to.

Teaching in primary schools in the UK is not normally compartmentalized by subject in the way that it typically is in secondary schools. By and large (there can be exceptions, such as music and Physical Education), in primary schools, the one teacher teaches all subjects to a class. Take, for example, the suggestion in the primary science programme of study (DfE, 2015) that when learning about materials, students might wish to learn about people who have developed useful materials, for example John Dunlop and the invention of pneumatic tyres. Using their scientific research skills, it should hopefully not take long for a student to raise the ethical issue of the impact of rubber plantations on endangered species and the broader ecology in Southern and Eastern Asia (Fig. 20.1). In certain areas of the curriculum, e.g., PSHE education and citizenship, teachers are encouraged to undertake activities such as 'circle time' to discuss various issues such as bullying, and, with older students, drugs and relationships education. Citizenship, which advocates the teaching of controversial issues, can provide approaches and strategies for teachers when planning to teach SSI, since engaging in citizenship learning involves clarifying values, evaluating information and viewpoints, discussion, and decision making

Fig. 20.1 Rubber plantation in Kerala, India. Rubber plantations damage native forests and cause other ecological problems. On the other hand, they can support local economies. (https://commons.wikimedia.org/wiki/File:Rubber_plantation,_Kerala_01.jpg)

(Woolley, 2010). Fortunately, whole school approaches to cross-curricular teaching in primary schools enable students to make links between subjects and for teachers to deliver a curriculum that is enriched.

20.5.3 The Contribution of SSI to Visions of Science Education

Roberts and Bybee (2014) who build on Roberts' (2007) give a helpful conceptualization of the two contrasting visions for science education—Vision I and Vision II:

> Vision I looks inward at science, to build curriculum from its rich and well established array of techniques and methods, habits of mind, and well-tested explanations for the events and objects of the natural world. Literacy, in this view, is within science – general familiarity and fluency within the discipline, based on mastering a sampling of the language, products, processes, and traditions of science itself.
>
> Vision II, developed later in the history of school science, begins by looking outside science to build curriculum that illuminates how science permeates and interacts with many areas of human endeavor and life situations. These societal issues and individual life situations usually include political, economic, and ethical considerations. This view is sometimes called science for citizenship, concentrating on matters of more obvious personal and social relevance to students than preparing to grasp more demanding science they might or might not study. The slogan "science for all" represents a viewpoint that all students need

> some introduction to citizen science, including such matters as environmental quality, resource use, personal health, and decision making about complex socioscientific issues. (Roberts & Bybee, 2014, 546)

More recently, Sjöström and Eilks (2018) have conceptualized what they term 'Vision III' to include 'critical scientific literacy' and 'socio-political action'. This builds on earlier ideas of such authors as Hodson (1994), Roth and Barton (2004) and Aikenhead (2007). Vision III therefore has various elements to it. It builds on Visions I and II in that it requires a good understanding of science (Vision I) and is a science education for all (Vision II) but it extends these to integrate both cognitive and affective domains including politicization to address complex SSI.

Increasingly, school science courses attempt to combine Visions I and II in order to serve the majority of students and the minority who will go on to be scientists, or at least use science to a certain extent in their paid employment. However, utilizing SSI in school science courses actually has the potential to address all three Visions in a meaningful way.

20.6 How Can SSI Be Utilized to Promote Scientific Literacy, Epistemic Insight and Human Flourishing in Science Education?

Fostering students' appreciation of the epistemology of science is an important goal of science education, one that has only grown in importance in recent years as misinformation in science has increased and trust in science has decreased (Reiss, 2022; Osborne & Pimentel, 2023). Understanding the nature of science and developing scientific literacy so that students can not only apply epistemological knowledge to school science, but also participate effectively in civic and cultural affairs is important. At a personal level, it is desirable for students to be able to make informed decisions and judgements regarding what matters to them using their understanding of scientific concepts and processes (Archer et al., 2013).

Indeed, it has been argued that science education can contribute to human flourishing (Reiss & White, 2014). In writing about this, Reiss and White (2014, 83) note, "how difficult it is to pigeon-hole knowledge into discrete school subjects", indicating that an approach to school science education that makes good use of SSI is more likely to contribute to human flourishing. One way therefore for educators to promote scientific literacy, epistemic insight and human flourishing is to empower students through the study of SSI that connect with their interests, values and experiences so that they can become reflective citizens who engage in, and are a part of, a society that neither disparages nor worships science.

Perhaps one of the most important aspects of teaching is student motivation (Silvia, 2006); for many students learning science, whether at primary, secondary or tertiary levels, engagement in contemporary issues can help stimulate interest. As

Reiss (2009) notes, consideration of ethical issues in school science can enhance student motivation; however, as he points out, "it is perhaps unsurprising that for many young people it is the ethical issues raised by science that too often seem to be lacking from their science lessons" (p. 137). If the social, cultural and ethical aspects of science were included more fully in school science, many students might be encouraged to study science longer, as the humanistic side of science appeals to many learners, particularly girls and young women (Davila Dos Santos et al., 2022). It is therefore encouraging that in a bid to enable more students to enjoy studying science as well as to help them to become more ethically sensitive, some schools have been including more SSI within their science curricula (Hancock et al., 2019).

Science teaching is complex, and the measure of success is not simply about ensuring good summative results in standardized tests such as those at the end of secondary schooling, although, given the structure of the education system in the UK, this is an important step to continuing to study science once it is no longer compulsory. There is now a considerable body of evidence that effective teaching in science must start from the beginning of a learner's time in school by ensuring their engagement and developing their commitment to learn (Bennett et al., 2023). We are sure that science teachers across all age ranges can identify with the moments when curiosity and interest in science have been sparked, and the sense of satisfaction this brings to both student and teacher alike (e.g., Pugh et al., 2010). The quality of science teaching is therefore a major determinant of student engagement and attainment, whilst high quality teaching can support and develop positive attitudes to science. Zeidler et al. (2011) have shown that an SSI approach to science education, when fully enacted, can become a transformative process for participating students and their teachers. In addition to the reasons that we have already considered, this is in part because SSI teaching entails a shift to a more student-focused pedagogy, which itself tends to engage learners.

In the UK, there is an engagement problem in school science with declining attitudes beginning to emerge by the end of primary school (9–11-year-olds). Many students by the age of 14 feel that science is 'important, but not for me' and this is worrying, given that pursuing further study in science is largely determined by this age (Sheldrake et al., 2019). Why is this when students do in fact recognize the value of science? Furthermore, we are living at a time when significant numbers of young people are realizing the power they have to effect social change, using social media to demonstrate their agency and voice so as to have an impact (Fullam, 2017). In 2019, a series of school strikes organized by young people, inspired by a then 16-year-old Greta Thunberg's protests regarding inaction on climate change, saw more than a million young people take part in demonstrations across the world. It is therefore evident that students understand that science is powerful but there is too often a disconnect between school science and the real world. For many students, science is not relevant to their daily lives and school science does not focus on the issues for which they are impassioned and matter to them. This is one of the principal reasons why SSI have a central place in school science.

20.7 Learning Progressions in SSI

Finally, if SSI are to play a more prominent role in science, whether at primary, secondary or tertiary level, it is important that students are enabled to progress in their learning. If one thinks of a topic like the theory of evolution, there is a degree of consensus about what one might expect learners at different stages of their education to be studying (Wyner & Doherty, 2017; Moore & Graham, 2021). However, consensus about progression in SSI—what we would expect learners to know at roughly what age, and in what order, and how this might be assessed—are less advanced. We report now how one in-depth study looked at the understandings of SSI among primary children in England and at what was found about their learning progressions.

McCrory (2014) explored the capabilities of 7–8-year-olds, in relation to their perceptions, knowledge and understanding of SSI in tandem with their ability to think morally and scientifically. Three schemes of work—'Morals', 'Energy' and 'Genetics'—were taught to 21 Year 3 students, one topic per term, across one academic year. Lessons were carefully planned to take into account the students' cognitive capabilities, especially important when considering topics such as genetics. SSI were used that were raised by the children themselves which either affected them or their families or arose from their engagement with film and media. For example, topics such as cloning and genetic inheritance arose due to watching *Star Wars* films, reading newspaper reports about Dolly the Sheep, and talking about inherited genetic diseases, e.g., sickle-cell anemia which one of the children in the class had and openly shared. A discussion of the merits of luminous zebra fish which glow green when put in polluted water was a focus after being reported *via* Children's BBC Newsround television program. The effects of climate change (for example, flooding) on countries such as Bangladesh (a country where many children in the class had relatives) were also a focus.

In line with thinking about pedagogical content knowledge (Pollard et al., 2023) and relevant research on teaching SSI at primary level, lessons incorporated a variety of pedagogies, including argumentation, role-play, pictorial representations, mind-maps, concept cartoons, dialogic talk, and reporting. Outcomes indicated that students' moral thinking when considering SSI was underpinned by their belief in the following values:

- *Being fair*. For example, when discussing renewable energy, one student commented:

> It's very unfair when we [humans] pollute habitats, like when there was that ship that spilt loads of oil into the sea. The birds were covered in oil and they couldn't fly and the fish died, lots were washed up on the beach. It made me feel really angry and upset, the poor birds and fish were suffering. If we used more renewable energy, like wind turbines, then that probably wouldn't happen as much because we would be using less oil.

When considering the possibility of human cloning, one student stated: "Only the rich will be able to afford a clone and that's not fair to the poor."

- *Honesty and truthfulness*. For instance, when considering recycling, one student said:

> Sometimes people don't recycle because they can't be bothered even though they know that they should ... and then some people say that they recycle even when they don't, so they lie which is silly. Maybe it's because they are embarrassed?

In another example, when thinking about a dilemma based on inherited diseases, one student said of one of the protagonists: "She has to tell her partner [about having the faulty gene] ... the partner will not get angry with her because she has told the truth".

- *Respect*. For example, when thinking about their local environment, one student said:

> When we throw rubbish on the ground and litter in the parks we are disrespecting it [the environment] and that's wrong. It looks disgusting, birds can choke on food wrappings, it makes the water dirty ... Once, I saw a bike in a pond!

When considering whether or not to use a human clone for their own future benefit, one student said: "We are not respecting the clone ... We'd be saying that we're better than the clone and that it is not worth anything which is really terrible, because it would be a person just like me."

- *Being responsible and considering the welfare of others*. For instance, when thinking about the behavior of some of the engineers at the Chernobyl plant, one student argued: "They didn't think about anyone but themselves. They didn't care that other people might get hurt or die."

In terms of learning progression, the aim of the assessment frameworks in Tables 20.1 and 20.2 (McCrory, 2014), is to promote the knowledge, understanding, and skills needed when teaching scientific and moral decision making to children of primary age. The work that led to these two tables entailed initially presenting the students with moral dilemmas, followed by presenting them with SSI which demanded both moral and scientific decision-making skills. The children were given repeated opportunities to become aware of and identify with their own moral values, to promote moral awareness and development, since clarification of values is integral to moral education. The lessons aimed to help the children to develop a critical appreciation of issues of right and wrong, fairness, rights and obligations in society, and to promote social awareness and development through helping them to begin to acquire relevant knowledge, understanding and skills.

Studies suggest that children apply multiple forms of reasoning, namely moral (fairness, justice, rights), conventional (customs, etiquette) and psychological (autonomy, identity), simultaneously when evaluating transgressions or social dilemmas. The frameworks in Tables 20.1 and 20.2 were constructed to help analyze the children's statements, not to neatly categorize the children's responses into separate sub-levels as can be the norm in primary school assessment in the UK.

It is important to distinguish between what is desirable and what is possible when evaluating learning of SSI. Authentic evaluation is difficult to achieve due to

Table 20.1 Framework for the determination of moral thinking and perspective taking when reasoning about SSI in the primary school

Level	Criterion	Example(s)
MTPT0	No use of moral thinking or perspective taking	When asked to consider whether anything might be wrong about choosing the hair color or sex of a child when designing babies Sh replied, "If you choose too many boys, then we won't be able to make as many babies [in the future] because you need a boy and a girl to make a baby"
MTPT1	Some use of moral thinking or perspective taking but manifests moral misconception(s)	When asked "Should we clone humans?", V replied, "There will be a population increase which means that there will not be enough homes. So, poverty will rise and only people who are rich will be able to pay which is unfair to the poor. Also, more food will be needed so species of animals may die out"
MTPT2	Student identifies a moral concern/ dimension in a SSI but is unable to reason further (for example, to explore intentions and consequences or take more than one perspective)	When thinking about why the two engineers responsible for the Chernobyl disaster carried on with their experiment, R wrote, "They were selfish, they didn't care". R was unable to explain this any further
MTPT3	Student analyses the morality of a specific SSI using intentions and consequences and/or demonstrating perspective taking	When deciding whether or not Jane [a character who has inherited Huntington's disease] should tell her partner, M wrote, "She has to tell her partner [about having the faulty gene] because then her partner will know. And when she says, 'I might die', the partner will not get angry because she told the truth"
MTPT4	Student analyses the morality of a specific SSI using intentions and consequences and/or demonstrating perspective taking and makes a decision about what to do using moral language and thinking	When asked to consider "What if, in the future, we could make a clone of ourselves and keep him/her to use in case we became ill?", V noted, "We are not respecting the clone", followed by Sh adding "We'd be saying that we're better than the clone, that it's worthless and that it doesn't have any human rights"
MTPT5	Student uses moral language and understanding to ask abstract questions to further his/her knowledge and understanding of SSI	When asked "What if, in the future, we could make a clone of ourselves and keep him/her to use in case we became ill?", S questioned, "But what if the clone refuses to give away their heart [if needed] because they know that they won't live? Then, that's not right and I'd feel guilty"

the relationship between "education and behavioural outcomes … and the practicality of authentic assessment" in the classroom (Ratcliffe & Grace, 2003, 41). In addition, in a neoliberal environment where accountability of students' attainment and progress is high on the agenda, it is important to present science teachers with

Table 20.2 Framework for the determination of scientific knowledge and understanding when reasoning about SSI

Level	Criterion	Example(s)
SKU0	No use of scientific knowledge or understanding	When asked, "Do you know what climate change is?", J responded "No"
SKU1	Some use of scientific knowledge or understanding but there is evidence of scientific misconception(s) or a lack of scientific knowledge	V asked, "If you are not married and have a baby on your own, does it mean that it will only have 23 chromosomes?"
SKU2	Student identifies a cause or an effect of a problem using scientific knowledge, or understanding, but is unable to reason further (for example, to link causes and effects)	K stated, "A genetic disease is a disease which is passed on to you by your mother and father, in your genes" but was unable to elaborate further about this
SKU3	Student gives an advantage or disadvantage of a specific SSI or identifies its risk(s) and makes a decision about what to do using scientific reasoning	When deciding whether or not Jane [a character who has inherited Huntington's disease] should have children of her own, E includes risk analysis in her explanation: "No, she should not have children because if she has children then they might inherit Huntington's disease [so] she is taking a risk"
SKU4	Student identifies and links causes and effects of a SSI using the correct scientific knowledge and understanding	When considering the impact of the Chernobyl disaster, S identifies the cause of the explosion as well as multiple consequences: "Caused by engineers playing with the safety valves. Impacts: radioactive clouds, massive fire for 10 days, people lost their lives, people got cancer, livestock died, farmers lost their money, bad relations between countries, people were unhappy"
SKU5	Student proposes possible solutions and gives reasons for these	When asked "What can we do to combat climate change?", M replied, "Fuel-efficient transport means using less fuel and producing less carbon dioxide. Using transports like moped, bicycle, train, and bus uses less energy. If you use a bicycle it uses far less energy than a car"
SKU6	Student uses scientific knowledge and understanding to ask abstract questions to further his/her knowledge and understanding of science without prompting or cues from the teacher	When discussing renewable energy, S asked, "Do people use man-made chlorophyll to work solar panels?"

assessment frameworks which 'work'. In the case of the frameworks and approach devised by McCrory (2014), primary classroom science teachers would be able to apply the frameworks not only for summative assessment that links to the National Curriculum but also, and which is perhaps more pertinent in relation to primary teaching pedagogy of science, for in-the-moment formative assessment.

References

Aikenhead, G. S. (2007). Expanding the research agenda for scientific literacy. In C. Linder, L. Östman, & P.-O. Wickman (Eds.), *Promoting scientific literacy: Science education research in transaction* (pp. 64–71). Geotryckeriet.

Archer, L., Osborne, J., DeWitt, J., Dillon, J., Wong, B., & Willis, B. (2013). *ASPIRES: Young people's science and career aspirations, age 10–14*. King's College London.

Ball, S. J. (2003). The teacher's soul and the terrors of performativity. *Journal of Education Policy, 18*(2), 215–228.

Ball, S. J. (2016). Neoliberal education? Confronting the slouching beast. *Policy Futures in Education, 14*(8), 1046–1059.

Bencze, L., Reiss, M. J., Sharma, A., & Weinstein, M. (2018). STEM education as "Trojan Horse": Deconstructed and reinvented for all. In L. A. Bryan & K. Tobin (Eds.), *13 Questions: Reframing education's conversation—Science* (pp. 69–87). Peter Lang.

Bennett, J., Lubben, F., & Hogarth, S. (2007). Bringing science to life: A synthesis of the research evidence on the effects of context-based and STS approaches to science teaching. *Science Education, 91*(3), 347–370.

Bennett, J. M., Dunlop, L., Atkinson, L., Compton, S., Glasspoole-Bird, H., Lubben, F., Reiss, M. J., & Turkenburg-van Diepen, M. (2023). A systematic review of approaches to primary science teaching, Endowment Education Foundation. Available at https://d2tic4wvo1iusb.cloudfront.net/production/documents/guidance/systematic-review-of-approaches-to-primary-science-teaching.pdf?v=1701173359

Bryant, J. A., & la Velle, L. B. (2018). *Introduction to bioethics* (2nd ed.). Wiley.

Connor, H. (2023). *Regulation 28: Report to prevent future deaths.* https://www.judiciary.uk/wp-content/uploads/2023/12/Ruth-Perry-Prevention-of-future-deaths-report-2023-0524_Published-1.pdf

Davila Dos Santos, E., Albahari, A., Díaz, S., & De Freitas, E. C. (2022). 'Science and Technology as Feminine': Raising awareness about and reducing the gender gap in STEM careers. *Journal of Gender Studies, 31*(4), 505–518.

DfE (Department for Education). (2015). *National curriculum in England: Science programmes of study*. Department for Education. https://www.gov.uk/government/publications/national-curriculum-in-england-science-programmes-of-study

Department for Science, Innovation & Technology and Prime Minister's Office. (2024). *Policy paper: The UK science and technology Framework*. https://www.gov.uk/government/publications/uk-science-and-technology-framework/the-uk-science-and-technology-framework

Duncombe, M. (2023, November 16). Stoicism on Tiktok promises happiness—But the ancient philosophers who came up with it had something very different in mind. *The Conversation*. https://theconversation.com/stoicism-on-tiktok-promises-happiness-but-the-ancient-philosophers-who-came-up-with-it-had-something-very-different-in-mind-216693

Fullam, J. (2017). Becoming a youth activist in the internet age: A case study on social media activism and identity development. *International Journal of Qualitative Studies in Education, 30*(4), 406–422.

Glazzard, J. (2018). The role of schools in supporting children and young people's mental health. *Education and Health, 36*(3), 83–88.

Grant, C. (2017). *The contribution of education to economic growth*. K4D Helpdesk Service. https://assets.publishing.service.gov.uk/media/5b9b87f340f0b67896977bae/K4D_HDR_The_Contribution_of_Education_to_Economic_Growth_Final.pdf

Gray, J., O'Regan, J. P., & Wallace, C. (2018). Education and the discourse of global neoliberalism. *Language and Intercultural Communication, 18*(5), 471–477.

Hancock, T. S., Friedrichsen, P. J., Kinslow, A. T., & Sadler, T. D. (2019). Selecting socio-scientific issues for teaching: A grounded theory study of how science teachers collaboratively design SSI-based curricula. *Science & Education, 28*, 639–667.

Hodson, D. (1994). Seeking directions for change: The personalisation and politicisation of science education. *Curriculum Studies, 2*(1), 71–98.

Ingleby, E. (2021). *Neoliberalism across education*. Springer International.

Kidger, J., Brockman, R., Tilling, K., Campbell, R., Ford, T., Araya, R., et al. (2016). Teachers' wellbeing and depressive symptoms, and associated risk factors: A large cross sectional study in English secondary schools. *Journal of Affective Disorders, 192*, 76–82.

Lee, J., & Fitz, J. (1997). HMI and OFSTED: Evolution or revolution in school inspection. *British Journal of Educational Studies, 45*(1), 39–52.

MacBeath, J., & McGlynn, A. (2004). *Self-evaluation: What's in it for schools?* Routledge.

McCrory, A. (2014). *Investigating the moral and scientific thinking of 7–8 year olds when taught socioscientific issues related to energy and genetics*. Unpublished PhD thesis, University of London, Institute of Education.

McCrory, A., & Reiss, M. J. (2023). *The place of ethics in science education: Implications for practice*. Bloomsbury.

Menn, S. (1997). Physics as a virtue. In J. J. Cleary & W. C. Wians (Eds.), *Proceedings of the Boston area colloquium in ancient philosophy* (Vol. XI, pp. 1–34). University Press of America.

Miller, S., & Selgelid, M. J. (2007). Ethical and philosophical consideration of the dual-use dilemma in the biological sciences. *Science and Engineering Ethics, 13*(4), 523–580.

Moore, A., & Graham, C. (2021). Evolution. In M. J. Reiss & M. Winterbottom (Eds.), *Teaching secondary biology* (3rd ed., pp. 207–227). Hodder Education.

Ofsted. (1993). *Standards and quality in education 1992–93*. HMSO.

Ofsted. (2021). *Research review series: Science*. https://www.gov.uk/government/publications/research-review-series-science

Osborne, J., & Pimentel, D. (2023). Science education in an age of misinformation. *Science Education, 107*(3), 553–571.

Pitchforth, J., Fahy, K., Ford, T., Wolpert, M., Viner, R. M., & Hargreaves, D. S. (2019). Mental health and well-being trends among children and young people in the UK, 1995–2014: Analysis of repeated cross-sectional national health surveys. *Psychological Medicine, 49*(8), 1275–1285.

Pollard, A., Wyse, D., Craig, A., Daly, C., Seleznyov, S., Harley, S., Hayward, L., Higgins, S., & McCrory, A. (2023). *Reflective teaching in primary schools* (6th ed.). Bloomsbury.

Power, L., Millington, K. A., & Bengtsson, S. (2015). *Building capacity in higher education topic guide*. The Health & Education Advice & Resource Team (HEART). https://assets.publishing.service.gov.uk/media/57a08984e5274a27b2000109/Capacity-Building-in-Higher-Education-Topic-Guide.pdf

Pugh, K. J., Linnenbrink-Garciam, L., Koskey, K. L. K., Stewart, V. C., & Manzey, C. (2010). Motivation, learning, and transformative experience: A study of deep engagement in science. *Science Education, 94*, 1–28.

QAA. (2019). *Subject benchmark statement: Physics, astronomy and astrophysics*. https://www.qaa.ac.uk/docs/qaa/subject-benchmark-statements/subject-benchmark-statement-physics-astronomy-and-astrophysics.pdf?sfvrsn=eff3c881_4

QAA. (2023). *Subject benchmark statement: Biosciences*. https://www.qaa.ac.uk/docs/qaa/sbs/sbs-biosciences-23.pdf?sfvrsn=b570a881_6

QAA. (2024). *Subject benchmark statements*. https://www.qaa.ac.uk/the-quality-code/subject-benchmark-statements

Ratcliffe, M., & Grace, M. (2003). *Science education for citizenship: Teaching socioscientific issues*. McGraw-Hill Education.

Reiss, M. J. (2009). Ethical reasoning and action in STSE education. In A. Jones & M. de Vries (Eds.), *International handbook on research and development in technology education* (pp. 307–318). Sense.

Reiss, M. J. (2022). Trust, science education and vaccines. *Science & Education, 31*(5), 1263–1280.

Reiss, M. J. (2023). Science education at a time of existential risk. *School Science Review in Depth, 105*(389), 5–10.

Reiss, M. J., & White, J. (2014). An aims-based curriculum illustrated by the teaching of science in schools. *The Curriculum Journal, 25*(1), 76–89.

Roberts, D. A. (2007). Scientific literacy/science literacy. In S. K. Abell & N. G. Lederman (Eds.), *Handbook of research on science education* (pp. 729–780). Lawrence Erlbaum.

Roberts, D. A., & Bybee, R. W. (2014). Scientific literacy, science literacy, and science education. In N. G. Lederman & S. K. Abell (Eds.), *Handbook of research on science education, Volume II* (pp. 559–572). Routledge.

Roth, W. M., & Barton, A. C. (2004). *Rethinking scientific literacy*. RoutledgeFalmer.

RSB. (2023). *The accreditation handbook*. RSB. https://www.rsb.org.uk/images/accreditation_home/RSB_Accreditation_Handbook_2024.pdf

Sellars, J. (2006). *Stoicism*. Routledge.

Sheldrake, R., Mujtaba, T., & Reiss, M. J. (2019). Students' changing attitudes and aspirations towards physics during secondary school. *Research in Science Education, 49*, 1809–1834.

Silvia, P. J. (2006). *Exploring the psychology of interest*. Oxford University Press.

Sjöström, J., & Eilks, I. (2018). Reconsidering different visions of scientific literacy and science education based on the concept of Bildung. In Y. J. Dori, Z. R. Mevarech, & D. R. Baker (Eds.), *Cognition, metacognition, and culture in STEM education* (pp. 65–88). Springer.

The National Archives. (2024). *Further and Higher Education Act 1992*. https://www.legislation.gov.uk/ukpga/1992/13/contents

Woolley, R. (2010). *Tackling controversial issues in the primary school: Facing life's challenges with your learners*. Routledge.

Wyner, Y., & Doherty, J. H. (2017). Developing a learning progression for three-dimensional learning of the patterns of evolution. *Science Education, 101*(5), 787–817.

Yosso, T. J. (2005). Whose culture has capital? A critical race theory discussion of community cultural wealth. *Race, Ethnicity and Education, 8*(1), 69–91.

Zeidler, D. L., Applebaum, S. M., & Sadler, T. D. (2011). Enacting a socioscientific issues classroom: Transformative transformations. In T. D. Sadler (Ed.), *Socio-scientific issues in the classroom: Teaching, learning and research* (pp. 277–305). Springer.

Chapter 21
Next Generation Socioscientific Issues Education: Incorporating Practices in SSI Learning Experiences

Troy D. Sadler, Li Ke, and Amanda Peel

In 1986, Reg Fleming introduced the phrase "socio-scientific issues" to the field of science education in a pair of articles published in the *Journal of Research in Science Teaching* (Fleming, 1986a, b). The phrase had been used previously in other fields (e.g., Currie, 1980), but Fleming's use of it is the first that reflects a sense of the concept as it is considered today. Over the next decade and a half, science educators did not take up the phrase in a widespread or systematic way. In the US during this time, the mid-1980s and early 1990s, the Science-Technology-Society movement was popular and provided the dominant framing for how science educators considered the integration of societal issues in science learning contexts (Yager, 1991). In the early 2000s, Dana Zeidler's work altered this trajectory by reintroducing "socioscientific issues" as the descriptor for an approach to science education that engaged learners in the negotiation of societal issues with a focus on moral and ethical reasoning (Zeidler, 2003; Zeidler et al., 2002). The early socioscientific issues (SSI) scholarship simultaneously built on and challenged the Science-Technology-Society (STS) approach (Sadler, 2004; Zeidler et al., 2005). By the late 2010s, SSI had supplanted STS as the dominant frame for issues-based science education in the US.

T. D. Sadler (✉)
University of North Carolina – Chapel Hill, Chapel Hill, NC, USA
e-mail: tsadler@unc.edu

L. Ke
University of Nevada, Reno, Reno, NV, USA
e-mail: lke@unr.edu

A. Peel
New Mexico State University, Las Cruces, NM, USA
e-mail: apeel@nmsu.edu

D. L. Zeidler (ed.), *A Moral Inquiry into Epistemic Insights in Science Education*, Contemporary Trends and Issues in Science Education 61,
https://doi.org/10.1007/978-3-031-63382-9_21

The emergence of the SSI movement in the early 2000s was happening in the midst of the first wave of national standards for science education in the US. The *National Science Education Standards* (NSES) were published in 1996 and served as the dominant policy document for US science education. The NSES advanced a vision of science education that emphasized science as a body of knowledge and a process of inquiry, and "Scientific Inquiry" became the central theme of NSES-informed reform. The NSES offered some statements that were consistent with goals of SSI such as its characterization of a scientifically literate person as someone who is able to "use appropriate scientific processes and principles in making personal decisions" and "engage intelligently in public discourse and debate about matters of scientific and technological concern" (National Research Council, 1996, p. 13). The early SSI work drew from NSES language like this as a part of its justification, but the standards did little to promote priorities directly aligned with SSI. However, there is evidence of SSI advocates taking up ideas championed by the NSES to advance SSI work. The chief example of this was applying principles of inquiry, from the NSES, to student experiences with SSI (e.g., Taylor et al., 2006).

The NSES continued to substantively shape science education in the US until uptake of the *Next Generation Science Standards* (NGSS; NGSS Lead States, 2013). The NGSS and the *Framework for K-12 Science Education* (NRC, 2012), which was the foundation for creation of the NGSS, precipitated a second wave of science standards in the US. The NGSS moved away from prioritizing inquiry and, instead, made a case for framing science learning as the necessary integration of Disciplinary Core Ideas, Science and Engineering Practices, and Crosscutting Concepts. These organizing features of the NGSS were different as compared to the NSES, but failing to attend to SSI in a substantive way was a point of consistency with the NSES. SSI as a research thread had clearly been established by the time groups were putting together the *Framework* and NGSS, and good examples of SSI curricula and implementation cases were available (e.g., Ratcliffe & Grace, 2003), but the new standards did not incorporate SSI as a focus. Like the NSES, the NGSS presents a few sweeping statements that are consistent with the goals of the SSI movement, but competencies such as reasoning about complex issues, SSI decision-making, and negotiating the ethical dimensions of societal issues are not a part of the expectations specified in the standards.

Standards in the US context are somewhat paradoxical. On one hand, national standards such as NGSS are not binding, mandated, or even national in scope. States have control over education standards. This results in a patchwork combination of state science standards in which many states adopt NGSS, some states have standards informed by NGSS, and others have standards altogether dissimilar from NGSS. On the other hand, standards like NGSS exert enormous influence on the broader science education community. The NGSS have shaped K-12 science textbooks and other curricular materials, directions for professional organizations such as the National Science Teachers Association (NSTA), and funding priorities for agencies like the National Science Foundation (NSF). Therefore, national standards are consequential for researchers, curriculum designers, teacher educators, and professional development providers even if the standards are not adopted in their local or state communities and even if the standards ignore themes most significant to these scholars' work. In the case of NGSS, SSI was not adequately featured in the document, but US educators committed to advancing the SSI movement are well-advised to explore ways in which the NGSS could be used to advance SSI teaching and learning.

Considering how NGSS connects with and supports any science education efforts, including SSI work, begins with an analysis of the standards' tripartite framing: Disciplinary Core Ideas, Science and Engineering Practices, and Crosscutting Concepts. Disciplinary Core Ideas correspond to science content knowledge. Science content knowledge has always been a part of science education, and from an SSI perspective, teaching science content is necessary but not sufficient for SSI approaches. In this sense, the NGSS focus on Disciplinary Core Ideas has little impact on the SSI movement. Positioning Crosscutting Concepts as a central element of science education is a more novel direction for the standards, although not a completely new development. (For example, Project 2061 standards presented "Common Themes" that correspond to several of the NGSS Crosscutting Concepts [AAAS, 1993].) There is some potential to situate SSI learning with Crosscutting Concepts such as *Systems* and *Cause and Effect*. However, Crosscutting Concepts as one of the three constituent dimensions of NGSS is a muddled group (Osborne et al., 2018). The individual concepts make sense, and reasonable arguments can be made about their individual importance, but it is not entirely clear how or why this group of concepts should necessarily be linked together. As a result, there has been less attention in terms of curricular innovations, assessment development, and research on the Crosscutting Concepts. Further, it is challenging to make a case for a systematic linkage between Crosscutting Concepts as a dimension of science education to SSI teaching and learning.

Science and Engineering Practices (SEP), the third dimension of the tripartite NGSS framework, represent the range of activities necessary for the production of scientific knowledge and engineering solutions (NGSS Lead States, 2013). The full list of the NGSS SEP is presented below.

- Asking questions (for science) and defining problems (for engineering)
- Developing and using models
- Planning and carrying out investigations
- Analyzing and interpreting data
- Using mathematics and computational thinking
- Constructing explanations (for science) and designing solutions (for engineering)
- Engaging in argument from evidence
- Obtaining, evaluating, and communicating information

The SEP are unique in terms of their framing for standards in at least a couple ways. First, engaging in practice is necessarily dependent on context and knowledge associated with that context. For example, the way in which a learner (or a scientist) engages in the first practice listed, asking questions, depends on what they are asking questions about. How one asks questions about an ecological problem is necessarily different than how questions are asked about theoretical physics.

This characteristic of practice distinguishes SEP from other learning goals such as skills. (In the NSES era, the teaching of "process skills" which do not necessarily require context and knowledge were common.) Second, practices are epistemic in nature; that is, they are associated with knowledge building. The practitioners of science engage in all sorts of activities, but the practice label is reserved for those activities directly related to knowledge building. For example, in a modern biology laboratory, it is common to find technicians and students making solutions, pipetting samples, and running equipment like DNA sequencers. These isolated activities can be conducted with little regard for knowledge products and, therefore, are not practices in and of themselves. In contrast, the SEP highlight activities like modeling, argumentation, and developing evidence-based explanations that have knowledge creation, revision, and sharing as central components of the work.

Practices represent a way of thinking about the knowledge-based and knowledge-generating activities that describe the kinds of work in which members of scientific communities engage. The NGSS suggest that these practices also represent the kinds of activities that science learners should take up in the process of learning science. Although the NGSS practices were not designed to connect with SSI teaching and learning, several SSI researchers have incorporated the NGSS focus on scientific practices into SSI focused work. Exploring possible connections between the SSI approach and SEP is reasonable prima facie for at least two reasons. First, the epistemic nature of practice aligns with SSI goals to support learners in building and using knowledge. Second, some of the practices featured in the SEP list were important aspects of the SSI approach prior to the existence of NGSS. For example, *Engaging in argument from evidence* and *Obtaining, evaluating, and communicating information* were central to SSI approaches since the earliest SSI frameworks (Zeidler et al., 2005). As the NGSS has spread and its underlying commitments have become a more influential part of the expectations and assumptions of the science education community, integration of science practices with SSI teaching and learning has become a significant direction for the SSI movement.

In the remainder of this chapter, we present two different projects which exemplify SSI teaching, learning, and research that prioritize learner engagement in scientific practices. In our work including the projects described below, we tend to focus on the practices that are most likely to support students' generation, use, and evaluation of ideas and knowledge. The NGSS present all of the SEP as epistemic, and while we acknowledge that realizing epistemic outcomes is possible across the full set of SEP, we also have found that some of the SEP are more likely to support true knowledge building among learners than others. For example, in a typical classroom context, students can engage in the practice of *planning and carrying out investigations,* and they may do so in ways that do not encourage deep thinking and knowledge generation. In contrast, we have found that opportunities for learners to engage in practices such as *developing and using models*, *constructing explanations*, and *using mathematics and computational thinking* tend to more consistently lead to epistemic outcomes. As a result, in our work we focus on practices such as modeling, argumentation, explanation, and computational thinking.

In our work, we have also found that practices are often interconnected. The standards documentation presents eight discrete practices, and it is certainly possible to conceptualize the practices independently. However, when individuals and communities (whether professionals or students) actually engage in epistemic work, the theoretical boundaries among some practices fade. For example, some forms of modeling have a goal of generating explanations, so the NGSS SEP of *developing and using models* and *constructing explanations* merge in the formation of model-based explanations (Zangori & Forbes, 2015). The first project that we present demonstrates this interconnectedness of practices—it highlights work that we did with learners to engage in both modeling and computational thinking. The second project focuses on modeling and student engagement with different kinds of models. As students work with models that take different forms and have different purposes, natural opportunities arise for engagement with companion practices. For example, learners develop mechanistic models that have the goal of explaining a scientific phenomenon (i.e., *constructing explanations*) central to an SSI. In other modeling activities, students work with a computational model, which presents opportunities for computational thinking.

We have also found that modeling is a particularly productive practice in terms of the kinds of learning that we want to support in SSI learning environments. Models are useful pedagogical tools for learners to make sense of scientific phenomena or systems (Schwarz et al., 2009; Windschitl et al., 2008). Furthermore, in SSI contexts, models can help learners gain the scientific understanding needed to inform decision-making about the underlying issues (Lewis & Leach, 2006; Sadler & Donnelly, 2006). Over the past decade, an emerging body of research has shown that model-based instruction is an effective way to promote students' SSI learning across grade levels (Evagorou & Puig-Mauriz, 2017; Zangori et al., 2017). In this chapter, we advance a case for incorporating epistemic practices, as emphasized in the NGSS, in SSI education. In the project examples that we share next, multiple practices are featured, but modeling is a consistent thread because we have found this particular practice to be so supportive of the kinds of learning that we strive to support.

21.1 Project 1: Exploring Antibiotic Resistance Through Computational Thinking and Modeling

In this project we combined computational thinking (CT) with modeling to help high school biology students reason about natural selection within an SSI unit that focused on antibiotic resistance in bacteria. In an initial design-based research study, modeling was used as a central practice as students created hand-drawn models that explained antibiotic resistance. Students were faced with the central SSI question: How do we mitigate the development of antibiotic resistance? Natural Selection is a key biology concept, but many students struggle to fully understand it (Gregory, 2009). Antibiotic resistance was selected as the SSI topic because it is an important natural selection health issue. After students conducted a lab experiment with bacteria, they created models to account for how antibiotic resistance occurs (Williams et al., 2018). After learning about natural selection, students worked in teams to create a general, or context-free, model of natural selection that did not include any specific organism, selective pressure, etc. Findings indicated students explained natural selection differently in bacteria than in their general models (Peel et al., 2019b). Additionally, students showed different natural selection misconceptions in each of their models (the bacterial model for antibiotic resistance and the general model).

These findings prompted revisions to help target natural selection misconceptions and help students reason about natural selection as a process in different organismal contexts. CT is a practice that centers around generalization, decomposition of complex processes, and sequencing ideas and concepts (Selby & Woollard, 2013), making it a good fit for addressing student's natural selection challenges. In this revision of the antibiotic resistance unit, CT was used in addition to modeling in different contexts to help students understand and generalize natural selection across contexts. This unit used CT through Algorithmic Explanations (CT-AE), which is an unplugged, or computer-free, approach that shifts CT from programming to solve a problem to sense-making explanations using algorithmic logic (Peel et al., 2022).

21.1.1 Overview of the Revised Unit

The antibiotic resistance unit included three investigations of natural selection (Peel et al., 2021). The first was the same bacteria lab used in the prior implementation, and this was followed by a modeling activity about changes in horn size in a big horned sheep population due to hunting. The third investigation was about changes in field mustard flowering time with drought conditions. After each investigation, students created a hand-written algorithmic explanation using algorithm concepts and engaging in CT practices. Figure 21.1 presents a transcribed version of an algorithmic explanation created by one of the students. Students used algorithm concepts: *branching*, *iteration*, *methods*, and *variables*, as they engaged in CT practices: *sequencing steps*, *abstracting*, *generalizing*, *recognizing patterns*, *decomposing*, and *evaluating*.

Next, students worked in groups to create a general natural selection algorithmic explanation with a focus on using variables to generalize across biological contexts. For example, instead of "bacteria" or "sheep", they could use "organism" or "population." In groups, students shared their context-specific algorithmic explanations with each other and compared each context to identify where variables could be used to generalize. Groups posted their algorithmic explanations and performed a gallery walk where they received and provided feedback to other students. The instructors also provided feedback and prompted students to clarify their algorithms and to apply additional algorithm concepts where helpful. As a final step, groups revised their general algorithmic explanations; an example can be seen in Fig. 21.2.

1. Mt. sheep horns are one size
2. Mutations produce alleles that carry the genetic code for different horn size
3. Varying horn sizes → variable
4. Hunters introduced to population
5. If sheep has bigger horns will get shot → branching
   If not, survives
6. Sheep reproduce
   If sheep has longer horns, offspring will have larger horns → branching/
   If sheep has smaller horns, smaller horns will be had by offspring variable
7. Repeat steps 2-6 until, and if ever, the population of sheep die out → iteration

Fig. 21.1 Student example of a big horned sheep algorithmic explanation

Variable	1. Population has <u>single trait</u>
	2. Population reproduces
Variable	3. Genetic mutations occur to create different variation of the original <u>trait</u>
Variable	4. Population introduced to <u>selective pressure</u>
Branching	→ survives & reproduces
Variable	5. If individual in population has <u>adaptation</u> → else dies
	6. Population continues to reproduce
	7. Over time, population has one variation of original trait occurring more frequently than other variations due to differential survival
Iteration	8. Repeat steps 5,6,7 till <u>selective pressure</u> abates
Variable	

Fig. 21.2 Student example of a general algorithmic explanation

21.1.2 Findings

Empirical investigations were conducted and focused on students' algorithmic explanations and pre- to post-unit natural selection scenario explanations to determine the effectiveness of this cycle of design-based research. Students showed significant improvements in natural selection conceptions from pre- to post-unit as determined through their use of natural selection factors (initial variation, mutation, favorable trait, selection pressure, differential survival, reproduction, and population shift; Peel et al., 2019a). Students also used the factors in the correct sequencing more often and showed significant decreases in presentation of natural selection misconceptions (Peel et al., 2019a, b). As students created algorithmic explanations across the unit, they gradually increased use of algorithm concepts (branching, iteration, variables, and methods; Peel, 2019). The general algorithms displayed a significant increase in algorithm concepts and students indicated in interviews that this lesson helped them learn about natural selection and algorithmic explanations because they worked with peers and received feedback on their ideas (Peel et al., 2019a, b). We investigated students' understanding of natural selection through a comparison of pre- to post-unit performance on a natural selection concept inventory assessment (Opfer et al., 2012). Students showed significant differences in their understandings when prompted with different organismal contexts (plant, animal, insect, microorganism) prior to the unit. After the unit, students did not demonstrate these differences suggesting more coherent understandings of the target science concepts (Peel & Arastoopour Irgens, 2020).

21.1.3 Discussion

The design-based nature of this project facilitated iterative design, implementation, assessment, and revision of a practice-oriented high school biology unit about natural selection through an SSI related to antibiotic resistance. The first modeling iteration highlighted struggles that students had with generalization, which limited the understandings they were building about natural selection. These findings prompted incorporation of CT as a sense-making practice. Combining models, algorithmic explanations (an unplugged form of CT), and multiple organismal contexts provided a learning experience that helped students learn about natural selection and generalize the process across contexts. As students created algorithmic explanations, the algorithm concepts provided a structure for representing their ideas that helped them make sense of natural selection steps, the order of steps, and how the steps interacted with each other. The generalized lesson provided experiences that helped students integrate variables in their algorithmic explanations to engage in generalization of the natural selection process. After these experiences, students created proposals to address the SSI question: How can we mitigate the development of antibiotic resistance? Students were able to produce evidence-based proposals for steps they could take to mitigate antibiotic resistance. Some students focused on steps they could take in their day-to-day lives and others focused on approaches that could be taken at higher levels of society, such as in agriculture. Interviews with students and investigations of their classroom artifacts indicate the practices of modeling and CT were essential in students' learning about natural selection and provided avenues for sense-making that allowed students to propose solutions to the SSI.

Interviewer: Do you think computational thinking helped you think about the issue?
Student: Yes, because it helped me rationalize it in my head. If people have less antibiotics, then their bacteria aren't able to become as resistant. If they have more antibiotics they don't need that, their bacteria are going to be able to adapt and then adapt, and adapt. I was able to mentally put it into the natural selection algorithm, and see if that stuff was taken out, it wouldn't adapt into something that's resistant. (Student post-unit interview)

This excerpt it from one of many students who engaged in computational thinking to identify which step in the natural selection algorithm needed to be eliminated to mitigate the development of antibiotic resistance. This example shows how students can engage in SSI problem-solving with CT practices.

21.2 Project 2: Learning About Viral Epistemics Through Multiple Models

The second project that we highlight focuses on students' modeling practice as they explore the issue of viral epidemics. A key feature of the project is that students navigate among different kinds of models, rather than a single model, to explore the issue of viral epidemics. In professional settings, scientists routinely explain, predict, and cross-validate their findings with multiple models to gain a comprehensive understanding of complex phenomena or systems. In a similar vein, we want learners to take up this multidimensional element of the practice of modeling in their own investigations of the world. This approach is innovative as little work has been done around student interactions with multiple models. Yet, understanding the merits and limitations of different models, as well as being able to navigate among these models, is a key aspect of modeling practice emphasized in the NGSS (NRC, 2012).

The use of multiple models can be especially advantageous in the context of SSI because SSIs are multifaceted and can encompass various science disciplines and their social implications. Engaging with different models, each designed to address specific facets of the underlying issue, enables learners to fully grasp the issue's complexity, thereby informing and improving their decision-making processes regarding the issue. Consider the case of COVID-19: to develop an informed stance on the pandemic and related policies, one must understand the virus' reproduction within the human body, its transmission dynamics through populations, the effectiveness of preventive measures such as masks and vaccinations, and the broader societal impacts, including the disproportionate effects on marginalized communities. Modeling can provide learning supports for exploring each of these aspects. Next, we introduce the three models we use in our curriculum materials to help students make sense of viral epidemics.

Models can take many forms, including drawings, computer simulations, and mathematical equations, each with unique affordances and constraints. Different models serve different epistemic goals. It is essential for learners to recognize the values and interconnectedness of different models, as each provides a partial representation of the target phenomena or system. Learners also need to be aware of the trade-offs and limitations inherent to each model. In our project, we use design-based research methodology (Cobb et al., 2003) to develop SSI-based curriculum materials that engage students in three types of models: (1) mechanistic models of viral reproduction, (2) computational models of viral transmission, and (3) system models of social implications. These three models have unique epistemic aims and address different aspects of viral epidemics. For example, students create their own mechanistic models (see Fig. 21.3 for an example from a group of middle school students) as they make sense of and synthesize their understanding of spread of viruses. In this example, students created a model to help them explain how a person infected with a virus could infect a healthy individual. Here, the primary epistemic goal is to explore the phenomenon for scientific explanation. In contrast, for the computational model (see Fig. 21.4), students manipulate variables and compare

Fig. 21.3 An example of a student-generated mechanistic model

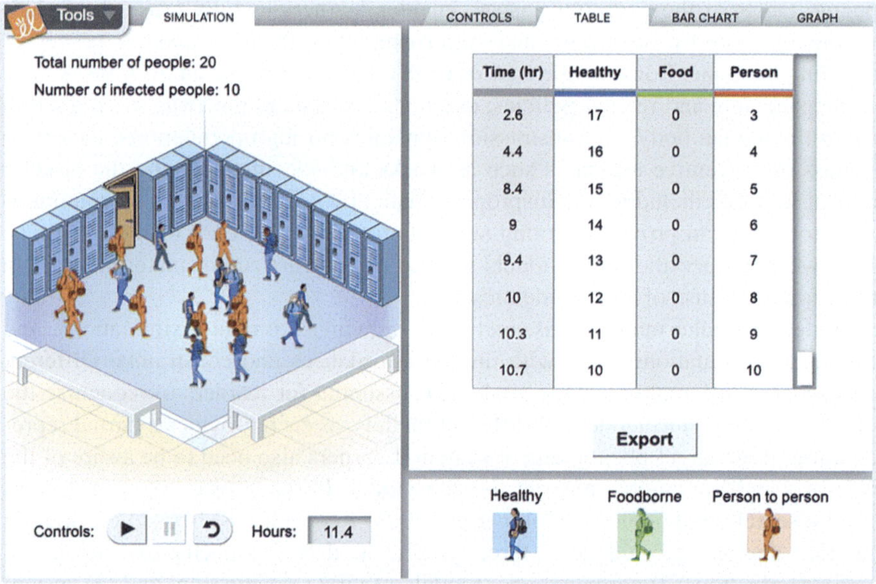

Fig. 21.4 An example of computational model

outcomes of an already programmed simulation. This opportunity makes it possible for students to test their hypothesis and make predictions about the impacts of different measures (e.g., social distancing, masks) on viral spread in a semi-quantitative way. In a system model (see Fig. 21.5), students conceptualize and represent multidirectional causal interactions among many dimensions (e.g., infection rates, economy, policy) of an overall viral disease system. The epistemic goal is to develop

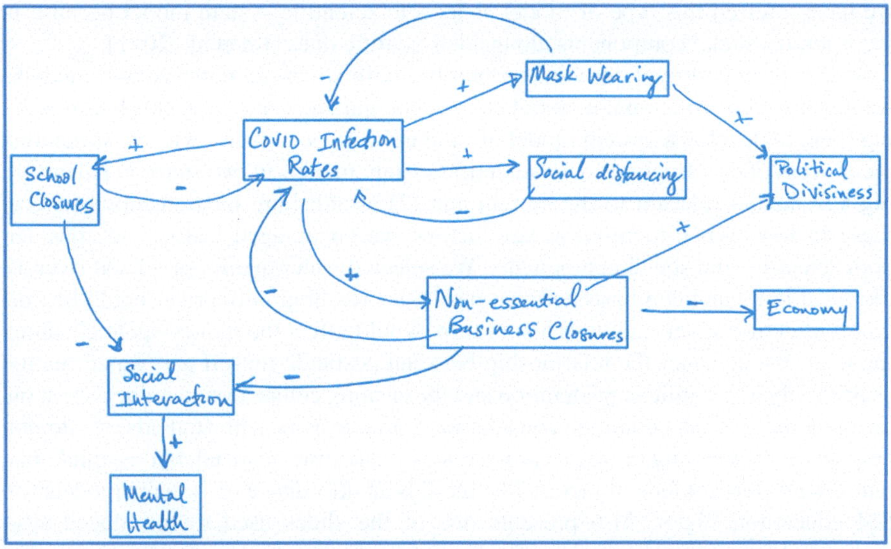

Fig. 21.5 An example of system model

a deep understanding of the issue, taking into consideration both the scientific and social dimensions. In this project, we are particularly interested in how the epistemic dimensions of different models afford or constrain students' negotiations of the issue of viral epidemics.

21.2.1 Promoting Systems Thinking About Issues Through System Models

SSI can be viewed as complex social systems as they encompass components and relationships spanning both scientific and social dimensions interacting at different levels (Ke et al., 2023). Developing systems thinking skills is critical in SSI-based instruction. Prior research on SSI has shown that learners often overlook the complexity of issues and propose solutions indicative of simple causal reasoning (Hogan, 2002; Sadler et al., 2007). Yet, developing an informed and agentic stance towards SSI requires learners to view issues from a systems perspective that considers both scientific and social dimensions of the issues (Ke et al., 2020). System models that integrate a broad social science perspective are often unfamiliar to K-16 learners because they differ from traditional models that primarily focus on scientific knowledge and theories. In the context of SSI, it is crucial for learners to also consider the social ramifications of science knowledge and how values and decision-making at various levels can impact their daily lives. In our previous work, we have referred this type of model as a socio-scientific system model because it incorporates social components alongside scientific ones (Ke et al., 2021).

In our project, to get participants familiar with socio-scientific system models, we developed instructional materials to provide guidance and scaffolds before asking them to develop a system model of viral epidemics on their own. We structured the process of constructing socio-scientific system models in two steps: (1) identifying key factors relevant to the system and (2) establishing relationships between these factors. When identifying key factors, we encouraged learners to consider both scientific and social components. We selected water quality of a local river as the focal issue and demonstrated that pesticides washing into a river could be one factor affecting water quality. This, in turn, would affect the money spent cleaning the river. We consider the relationship between pesticide run-off and water quality as primarily a scientific consideration and the ensuing connection to money spent on cleaning the river an economic consideration. In our work with students, we do not emphasize the labeling of factors as necessarily scientific or social, but we think that featuring different kinds of factors like these is an affordance of systems models for SSI education. Figure 21.6 presents one of the slides used to introduced system models.

We then illustrated that arrows could be used to represent causal relationships. To prompt learners to consider the system dynamics of the underlying issue, we introduced conventions of "+" and "-" signs to represent the direction of the relationships. For example, a negative sign between pesticides washing into a river and water quality indicates that an increase in pesticide use will lead to decreased water quality. For the COVID-19 socio-scientific system model, we gave participants the driving question, "how has COVID-19 impacted your life?" In the following section, we present findings regarding learners' engagement in socio-scientific system models when making sense of COVID-19.

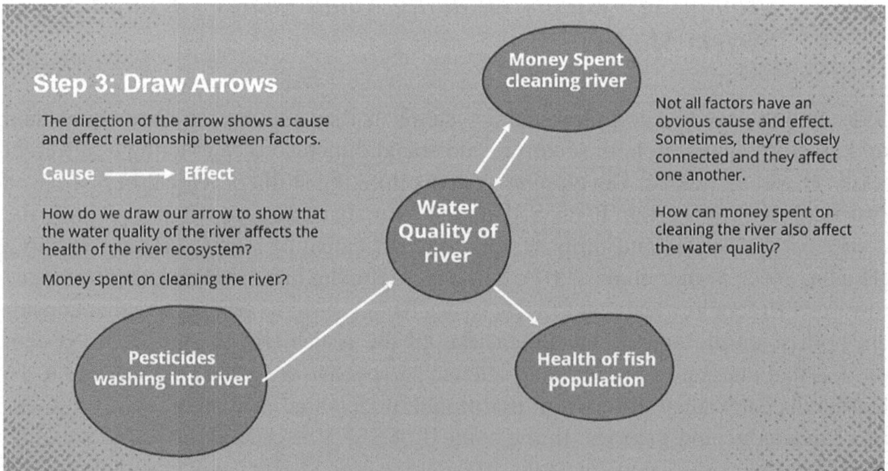

Fig. 21.6 The slide used to introduce system models

21.2.2 Findings

We have conducted multiple rounds of empirical studies with different age groups ranging from middle school to undergraduate students, in which participants created system maps about COVID-19 in pairs or small groups. Participants were given the central component, COVID-19 infection rates. They were asked to create the model with factors affecting or affected by infection rates. Below we highlight patterns we observed regarding learners' modeling experiences with socio-scientific system models.

First, we found that learners tended to include social components that were personally relevant to them, such as remote teaching and learning, labor shortages, and mental health. We believe that inviting students to consider social dimensions is critical, allowing them to bring in their own personal values and experiences. It makes the system modeling practice expansive and personally meaningful (Schwarz et al., 2022). One participant, for instance, included employment in her system model because her mom lost her job during the pandemic. In addition, system modeling can also serve as a useful tool to interrogate social justice issues. Specifically, the interconnectedness of different components within an SSI system makes visible the underlying systemic injustice faced by underrepresented communities (Lesnefsky et al., 2023).

Another pattern observed in our data, which includes student-generated artifacts (i.e., the systems models they created), interviews, and writing prompts, is that students see systems mapping as a tool that helped their thinking. In reflecting on their engagement in a system modeling activity, students often indicated that they thought the exercise pushed them to think about factors associated with the SSI and relationships among these factors that they would not have considered on their own. The excerpts below are taken from a writing assignment that undergraduate students completed after completing a systems modeling assignment related to the Covid pandemic.

> Thinking through the systems modeling approach really widened my perspective as to what other factors and affected sectors that need to be taken into consideration…My previous response [written before creating a systems model] was slightly more black and white and felt that the only necessary thing in a pandemic was to protect the health of citizens, however, politicians and decision-makers must consider significantly more nuanced scenarios when making large decisions.
>
> After learning about systems modeling, I now understand that there are many more aspects to managing an emerging pandemic… The systems model perspective is an effective way to create a layout of how the pandemic can be amplified or deescalated and therefore is a good way for researchers to analyze how well an approach slowed down the spread of COVID-19 or even the complications that arose from each approach.

An important goal of SSI education is to support students in conceptualizing the complexity of issues and encouraging expansive thinking about multidimensional problems (Sadler et al., 2007). Students being able to recognize how systems modeling helped them to think about an issue more expansively is an indication that this form of modeling can help to achieve this goal.

Finally, our work with student creation of systems models suggests that learners need further support when justifying the relationships among components in their models. It was particularly challenging when it came to defining relationships among social components. Learners mainly relied on their personal experiences or common sense when establishing the relationships in their models. However, for those relationships that were not as relevant, learners need additional information to determine the relationships. For example, one participant noted that she was not confident if her relationship between air travel and mortality was correct in her model. She said that she did not do extensive research on the topic and only used information from social media to help her decision-making.

21.2.3 Discussion

This project provides us the opportunity to design various modeling learning environments to examine student learning regarding multiple models in the context of SSI. It was evident from data that learners tend to focus on different epistemic goals as they were engaged in different modeling activities. In particular, in the context of socio-scientific models, we found learners constantly contemplating the system dynamics related to the issue of COVID-19 through identifying relevant components and relationships. We argue that the practice of system modeling makes the epistemic goal of appreciating the complexity of underlying issue meaningful and accessible. Furthermore, preliminary findings suggest that learners are able to leverage different modeling experiences to reason about viral epidemics and potential solutions, outcomes that align very well with the goals for SSI education.

21.3 Conclusions

When this volume's editor asked us to write the U.S. chapter for the book, we thought carefully about what a U.S. chapter should present. We are U.S. based scholars, but we do not necessarily conceptualize our national context as a defining feature of our SSI scholarship. We also do not think that there is a single U.S. approach to SSI education. The variation in the ways in which SSI teaching and learning in the U.S. is planned for and implemented likely rivals the variation in these factors among national contexts. As we pondered the challenge to somehow represent SSI in the U.S., we tried to think about factors that are, in fact, unique to or help to define American education. This line of thinking brought us to standards because standards have certainly exerted substantial influence on education in the U.S. for three decades. Standards movements are not unique to the U.S., but the NGSS, for better or worse, are unique to our national context. As soon as we started thinking about the NGSS, discussing the connections between practices and SSI education became a clear choice for us. The practice-orientation to science education is not happening only in the U.S., but the influence of the NGSS and its prioritization of scientific practices has certainly shaped the work of U.S. based SSI educators and researchers.

A significant line of justification for the prioritization of practices in the NGSS is that they aptly describe what scientists do. We do not necessarily find this argument, on its own, all that compelling—we are not convinced that having K-12 learners replicate the activity of scientists should guide our field. However, epistemic practices like those included in the NGSS support sensemaking, and this is a compelling justification for why educators ought to consider practice-centered teaching approaches as a part of science education. This kind of thinking has shaped some of our work particularly in our development and research efforts associated with SSI.

In our early work to integrate SSI and practices, we created opportunities for learners to engage in scientific practices, that is, epistemic activity related to the scientific elements of SSI. This is evident in the first project presented earlier in the chapter. Students engage in modeling and computational thinking in order to develop more sophisticated understandings about science ideas including bacterial resistance and natural selection. Within the broader SSI unit, it was critical for students to understand how bacteria develop resistance to antibiotics and how this process reflects natural selection. In order for the students to engage thoughtfully in discourse about the broader issue of how communities and society should respond to the emergence and spread of antibiotic resistant bacteria, they need to understand these key scientific points.

As our work in these areas has progressed, we have begun to expand the objects of practice. The NGSS practices are scientific practices—they are derived from science as a disciplinary community and the objects of these practices are science ideas and understandings. We are interested in how to help learners think about issues that extend beyond the boundaries of science. We want learners to consider scientific ideas in concert with factors such as economics, politics, and ethics. The multidimensional nature of SSI presents opportunities to consider ways that practices extend beyond science. We use the phrase socio-scientific practice, as opposed to scientific practice, to denote this expansion. The second project that we describe in the chapter presents examples of socio-scientific modeling, that is, modeling activities in which students engage in sensemaking about scientific and social dimensions of the issue they are exploring. There are other examples of this extension of practice to go beyond science; for example, several scholars have explored socio-scientific argumentation (e.g., Evagorou & Osborne, 2013). However, most science education scholarship on practice focuses on scientific practice.

One of the challenges of expanding practice from scientific to socioscientific is a shift in the disciplinary norms that typically provide frameworks for defining and evaluating practice (Kelly & Licona, 2018). These disciplinary norms are often implicit or completely unrecognized, but they give shape to the expectations and criteria for practices. For example, there is general recognition that scientific argumentation requires the use of evidence in support of claims advanced in an argument. Furthermore, evidence may be stronger or weaker based on the data that underly it and the theory used to interpret the data. From where does the requirement for evidence come, and how do we know what counts as stronger or weaker evidence? These questions are addressed by the disciplinary norms of the communities from which practice is framed. We can distinguish an argument based on empirical evidence that has been systematically collected and analyzed from an argument based on a personal anecdote because the scientific community has collectively negotiated norms for what counts as strong evidence.

When learners engage in scientific practices, in the context of SSI (like project 1) or other science learning opportunities, teachers and researchers can rely on scientific norms to help navigate criteria for competency. However, when the nature of practice is extended beyond science, as in socioscientific modeling or socioscientific argumentation, teachers and researchers have to reconsider criteria for competency. If a student makes a socioscientific argument and uses an ethical stance to support their claim, it would be inappropriate to evaluate the argument based on criteria for evidence from science. This shift in criteria can present a challenge for teachers and researchers—we cannot just rely on relatively uncontested norms from science—but the challenge is not insurmountable. One potential avenue is to look toward other disciplinary communities (e.g., ethicists, economists, and political scientists) to help inform standards of socioscientific practices when the practices incorporate ideas from those disciplines. A different strategy is to employ more general norms that appeal to a broad range of disciplinary traditions such as logic, coherence, and consistency. Another option is to turn to groups that actively engage with SSI and explore their norms and criteria for practice (e.g., Feinstein, 2014). At this point, these are just suggestions; more work will need to be done to better understand how to create and enact criteria for socioscientific practices which are necessarily interdisciplinary.

We think that there is a lot more to learn about how to incorporate practices in SSI education. We have highlighted some work that focuses on computational thinking and modeling, and there are several other practices that might provide unique affordances for SSI education. There are also interesting opportunities to explore the role of scientific practices in SSI education as compared to socioscientific practices and how we as a community of SSI scholars should conceptualize the norms and criteria for both scientific and socioscientific practice in support of SSI teaching and learning.

References

American Association for the Advancement of Science. (1993). *Benchmarks for science literacy*. Oxford University Press.

Cobb, P., Confrey, J., diSessa, A., Lehrer, R., & Schauble, L. (2003). Design experiments in educational research. *Educational Researcher, 32*(1), 9–13.

Currie, L. A. (1980). Scientific uncertainty and societal decisions: The challenge to the analytical chemist. *Analytical Letters, 13*(A1), 1–31.

Evagorou, M., & Osborne, J. (2013). Exploring young students' collaborative argumentation within a socioscientific issue. *Journal of Research in Science Teaching, 50*, 209–237. https://doi.org/10.1002/tea.21076

Evagorou, M., & Puig-Mauriz, B. (2017). Engaging elementary school pre-service teachers in modeling a socioscientific issue as a way to help them appreciate the social aspects of science. *International Journal of Education in Mathematics, Science and Technology, 5*(2), 113–123.

Feinstein, N. W. (2014). Making sense of Autism: Progressive engagement with science among parents of young, recently diagnosed autistic children. *Public Understanding of Science, 23*, 592–609.

Fleming, R. (1986a). Adolescent reasoning in socio-scientific issues: II. Nonsocial cognition. *Journal of Research in Science Teaching, 23*(8), 689–698. https://doi.org/10.1002/tea.3660230804

Fleming, R. (1986b). Adolescent reasoning in socio-scientific issues, part I: Social cognition. *Journal of Research in Science Teaching, 23*(8), 677–687. https://doi.org/10.1002/tea.3660230803

Gregory, T. R. (2009). Understanding natural selection: Essential concepts and common misconceptions. *Evolution: Education and Outreach, 2*(2), 156–175.

Hogan, K. (2002). A sociocultural analysis of school and community settings as sites for developing environmental practitioners. *Environmental Education Research, 8*(4), 413–437.

Ke, L., Sadler, T. D., Zangori, L., & Friedrichsen, P. J. (2020). Students' perceptions of socio-scientific issue-based learning and their appropriation of epistemic tools for systems thinking. *International Journal of Science Education, 42*(8), 1339–1361.

Ke, L., Sadler, T. D., Zangori, L., & Friedrichsen, P. J. (2021). Developing and using multiple models to promote scientific literacy in the context of socio-scientific issues. *Science & Education, 30*(3), 589–607.

Ke, L., Kirk, E., Lesnefsky, R., & Sadler, T. (2023). Exploring system dynamics of complex societal issues through socio-scientific models. *Frontiers in Education, 8*, 1219224.

Kelly, G. J., & Licona, P. (2018). Epistemic practices and science education. In M. R. Matthews (Ed.), *History, philosophy and science teaching* (pp. 139–187). Springer.

Lesnefsky, R., Kirk, E., Yeldell, J., Sadler, T., & Ke, L. (2023). Socio-scientific modeling as an approach towards justice-centered science pedagogy. *London Review of Education, 21*(1), 30.

Lewis, J., & Leach, J. (2006). Discussion of socio-scientific issues: The role of science knowledge. *International Journal of Science Education, 28*, 1267–1287.

National Research Council. (1996). *National science education standards*. The National Academies Press. https://doi.org/10.17226/4962

National Research Council. (2012). *A framework for K-12 science education: Practices, crosscutting concepts, and core ideas*. National Academies Press.

National Research Council. (2013). *Next Generation Science Standards: For states, by states*. The National Academies Press.

Opfer, J. E., Nehm, R. H., & Ha, M. (2012). Cognitive foundations for science assessment design: Knowing what students know about evolution. *Journal of Research in Science Teaching, 49*(6), 744–777.

Osborne, J., Rafanelli, S., & Kind, P. (2018). Toward a more coherent model for science education than the crosscutting concepts of the next generation science standards: The affordances of styles of reasoning. *Journal of Research in Science Teaching, 55*, 962–981.

Peel, A. (2019). *Developing computational thinking competencies and natural selection understanding thorough unplugged algorithmic explanations* (Doctoral dissertation, University of Missouri--Columbia).

Peel, A., & Arastoopour Irgens, G. (2020). *Scaffolding secondary students' natural selection transfer through computational thinking*. NARST Annual International Conference (NARST 2020), Virtual.

Peel, A., Sadler, T. D., & Friedrichsen, P. (2019a). Learning natural selection through computational thinking: Unplugged design of algorithmic explanations. *Journal of Research in Science Teaching, 56*(7), 983–1007.

Peel, A., Zangori, L., Friedrichsen, P., Hayes, E., & Sadler, T. (2019b). Students' model-based explanations about natural selection and antibiotic resistance through socio-scientific issues-based learning. *International Journal of Science Education, 41*(4), 510–532.

Peel, A., Sadler, T. D., & Friedrichsen, P. J. (2021). Using unplugged computational thinking to scaffold natural selection learning. *The American Biology Teacher, 83*(2), 112–117.

Peel, A., Sadler, T. D., & Friedrichsen, P. J. (2022). Algorithmic explanations: An unplugged instructional approach to integrate science and computational thinking. *Journal of Science Education and Technology, 31*, 428–441.

Ratcliffe, M., & Grace, M. (2003). *Science education and citizenship: Teaching socio-scientific issues*. Open University Press.

Sadler, T. D. (2004). Moral and ethical dimensions of socioscientific decision-making as integral components of scientific literacy. *The Science Educator, 13*, 39–48.

Sadler, T. D., & Donnelly, L. A. (2006). Socioscientific argumentation: The effects of content knowledge and morality. *International Journal of Science Education, 28*, 1463–1488.

Sadler, T. D., Barab, S., & Scott, B. (2007). What do students gain by engaging in socioscientific inquiry? *Research in Science Education, 37*(4), 371–391.

Schwarz, C. V., Reiser, B. J., Davis, E. A., Kenyon, L., Achér, A., Fortus, D., Shwartz, Y., Hug, B., & Krajcik, J. (2009). Developing a learning progression for scientific modeling: Making scientific modeling accessible and meaningful for learners. *Journal of Research in Science Teaching, 46*(6), 632–654.

Schwarz, C. V., Ke, L., Salgado, M., & Man, E. (2022). Beyond assessing modeling knowledge: Moving towards expansive, equitable, and meaningful modeling practice. *Journal of Research in Science Teaching, 59*(6), 1086–1096.

Selby, C., & Woollard, J. (2013). *Computational thinking: The developing definition*. Retrieved from https://eprints.soton.ac.uk/356481/

Taylor, P., Lee, S. H., & Tal, T. (2006). Toward socio-scientific participation: Changing culture in the science classroom and more: Setting the Stage. *Cultural Studies of Science Education, 1*, 645–656.

Williams, M. A., Friedrichsen, P. J., Sadler, T. D., & Brown, P. J. (2018). Modeling the emergence of antibiotic resistance in bacterial populations. *American Biology Teacher, 80*, 214–220.

Windschitl, M., Thompson, J., & Braaten, M. (2008). Beyond the scientific method: Model-based inquiry as a new paradigm of preference for school science investigations. *Science Education, 92*(5), 941–967.

Yager, R. E. (1991). Science/technology/society as a major reform in science education: Its importance for teacher education. *Teaching Education, 3*(2), 91–100. https://doi.org/10.1080/1047621910030209

Zangori, L., & Forbes, C. T. (2015). Exploring third-grade student model-based explanations about plant relationships within an ecosystem. *International Journal of Science Education, 37*, 2942–2964.

Zangori, L., Peel, A., Kinslow, A., Friedrichsen, P., & Sadler, T. D. (2017). Student development of model-based reasoning about carbon cycling and climate change in a socio-scientific issues unit. *Journal of Research in Science Teaching, 54*(10), 1249–1273.

Zeidler, D. L. (2003). *The role of moral reasoning on socioscientific issues and discourse in science education*. Springer.

Zeidler, D. L., Walker, K. A., Ackett, W. A., & Simmons, M. L. (2002). Tangled up in views: Beliefs in the nature of science and responses to socioscientific dilemmas. *Science Education*. https://doi.org/10.1002/sce.10025

Zeidler, D. L., Sadler, T. D., Simmons, M. L., & Howes, E. V. (2005). Beyond STS: A research-based framework for socioscientific issues education. *Science Education, 89*(3), 357–377.

Afterward

Benjamin C. Herman

Lessons from History and the Human Condition: The Need for SSI Scholarship and Instruction

I feel very honored Dana Zeidler asked me to write concluding remarks to this collection of admirable work advanced by a diverse group of notable scholars. I have known Dana since 2010, meeting him first at an ASTE conference. I was on the job market and my first (informal) interview with him was in the gym—I think I left a favorable impression. Succeeding that, I worked with Dana at the University of South Florida where I was a new and shiny Assistant Professor of Science Education whose work largely focused on nature of science (NOS) teaching. Interested from a young age in how to help people effectively engage complex environmental issues, I found my discussions with Dana about socioscientific issues (SSI) scholarship to be invigorating and informed my thinking regarding how NOS relates to socioscientific decision-making. Not surprisingly, most of those discussions occurred in the gym or over long lunches—we ate a lot. Those discussions and the work that Dana and I shared were formative for my career. More importantly, those activities were part of a friendship that has persisted since. A friendship that I am deeply grateful for.

I turn my attention to the other authors in this book to express my humble appreciation for their contribution. I find it encouraging and exciting that there are so many bright scholars seeking to move the SSI field forward. I will use this space to briefly advocate why the collection of work appearing in this book must be seriously considered by educators, researchers, policy makers, and anyone else holding an interest in preparing people to effectively resolve SSI.

B. C. Herman (✉)
Texas A&M University, College Station, TX, USA

D. L. Zeidler (ed.), *A Moral Inquiry into Epistemic Insights in Science Education*, Contemporary Trends and Issues in Science Education 61,
https://doi.org/10.1007/978-3-031-63382-9

Diversity of Perspectives

A foremost goal of SSI efforts is promoting the kind of perspective taking and moral consideration necessary for civic and just SSI resolution at personal to local to global scales (Zeidler & Sadler, 2023; Herman et al., 2021; Kahn & Zeidler, 2019). This initiative is not only evident in the diversity of issues and topics addressed in this book's chapters, but also reflected in the diverse representation of this book's authors who seek to improve the human and environmental condition through their SSI scholarship and teaching. Contributors to this book include 40 scholars at varying stages of their careers from 37 institutions representing 16 countries across six continents. While each chapter largely situates SSI scholarship within its country of origin, the overlapping conversation across the chapters can be interpreted as what it means to promote people to develop a *functional scientific literacy* as local and global citizens that demonstrate civic character and values in their interactions and actions. The SSI scholarship and pedagogy featured in this book serves an important role toward achieving this kind of literacy which is crucial for: (1) preventing the recommitting of past science associated mistakes and atrocities and (2) resolving current and future SSI in a manner that conscientiously weighs science-based evidence, moral justification, sociocultural factors, and other facets relevant to those issues. Diverse voices are important for advancing varied perspectives and avenues for reaching these two significant goals and engaging SSI in a humane and compassionate fashion. Historical episodes where science has been wrongly used buttress this point.

Lessons from History

History is rife with horrific examples of what can occur when people eschew ethical and compassionate perspective-taking and action, and instead engage in ideologically fueled self-interest, tribalism, and othering. Science has been coopted and at times corrupted to advance such efforts. For instance, consider the eugenics movement instigated in the nineteenth century and informed by Samuel Morton's work on racial ranking of intelligence through skull measurements. Morton's work, which ranked Whites first and then Native American and Black peoples, was so apparently biased Stephen Jay Gould (1978, p. 509) described it as "an a priori conviction about racial ranking so powerful that it directed his tabulations along preestablished lines." Francis Galton, who performed similar work, formally introduced the term eugenics and framed it as "the study of agencies under social control that may improve or impair the racial qualities of future generations". The ideologically driven science gained popularity in many countries including those with social democratic governments (Weigmann, 2001, p. 871).

Eugenics was promoted among many respected scientists and used to justify ideologically fueled "othering" and human injustices and atrocities. This occurred most notably in Germany where a form of eugenics with the goal of "racial hygiene" and the protection of "German blood and honor" was put into action. Prominent scientists including those associated with the Kaiser Wilhelm Society/Institutes (the predecessor of the Max Planck Society[1]) afforded the Nazi initiative expert justification (Weigmann, 2001). Making clear the ideological motivation underpinning the use of science to justify racial and ethnic atrocities Hitler's deputy Rudolph Hess said in 1934 that "National Socialism is nothing but applied biology" (Lifton, 1986, p. 31). In 1935 the Nuremburg Laws were implemented and socially adopted, which classified all Germans as 'Aryans', 'Full' 'Half' or 'Quarter Jew' and prohibited marriage and sexual relationships between Jews and German citizens (Fig. 1). The eugenics informed narrative was promoted through Germany's education system, news media, and even children's books (Feldman, 2022); and its implementation was supported by scientists and resulted in horrific atrocities. For instance, Wetzell (2017, p. 148) writes:

> Germany's physicians and biomedical scientists—including leading academics in the related fields of anthropology, eugenics, human genetics, and Rassenforschung (racial science)—were complicit in the eugenic and racial policies of the Nazi regime that culminated in the Holocaust.

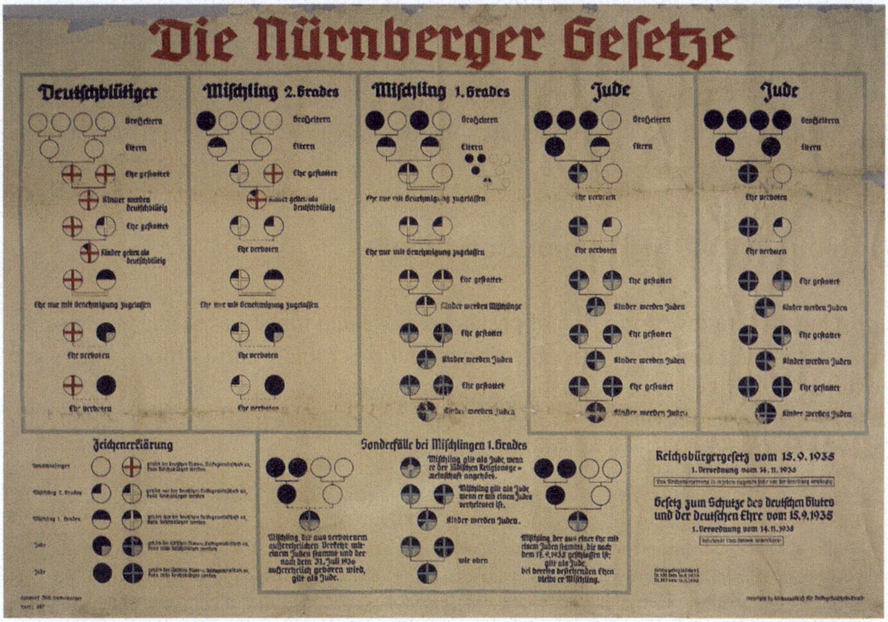

Fig. 1 German Nuremberg Law teaching chart distributed in 1936 that distinguishes hierarchal differences among German-blooded people, Jewish people, and those who comprise hybrid-variations (United States Holocaust Museum)

---

[1] Acknowledgement of and an apology for the history of the Kaiser Wilhelm Society under National Socialism provided by the Max Planck Society can be found at: https://www.mpg.de/history/kws-under-national-socialism

The preceding example focuses on a highly profiled instance of science being coopted and corrupted in a manner that induced extreme human suffering. History is littered with such instances that occurred and were allowed to persist in diverse cultures. For instance, published works of the American eugenics movements and California sterilization laws (which persisted until 1979, see Stern et al., 2017) informed Nazi initiatives (Farber, 2008). Eugenics was also present in many other "advanced" countries to include Great Britain, Italy, Mexico, and Canada (National Institute of Health, 2023; United States Holocaust Memorial Museum, 2020); and in Sweden 63,000 people (including most resident gypsies) were legally sterilized between 1934 and 1975, mainly because of "antisocial behavior" (Weigmann, 2001). Lysenkoism was a Lamarckian agricultural approach that denied the existence of genes and advocated the inheritance of desired plant characteristics could be acquired through exposure to environmental factors. This practice, particularly "vernalization", was adopted in the Union of Soviet Socialist Republics (USSR) in the 1930's to 1950's because of its perceived consistency with Marxist-Leninist ideologies and contributed to the famines that killed millions of Soviets (Borinskaya et al., 2019; Kolchinsky et al., 2017). The United States Public Health Service, collaborating with the Tuskegee Institute in Macon, AL, conducted the Tuskegee study from 1932 to 1972 that recorded the effects of untreated syphilis in Black males. The study entailed unethical practices such as failing to obtain participants' informed consent and deceptively informing participants that they were being treated for "bad blood" (Alsan & Wanamaker, 2018; CDC, 2022). These are but a handful of historical examples where science entangled with ideology resulted in terrible outcomes.

A Reminder of the Human Condition

I provide the lengthy discussion of past science related injustices as a grim reminder that humans are capable of terrible things and when committing those things can genuinely feel like they are on the side of righteousness. Aleksandr Solzhenitsyn's sentiments presented in *The Gulag Archipelago* (2007, p. 174, 746) eloquently describe this vexing aspect of the human condition.

> Ideology—that is what gives evildoing its long-sought justification and gives the evildoer the necessary steadfastness and determination. That is the social theory which helps to make his acts seem good instead of bad in his own and others' eyes, so that he won't hear reproaches and curses but will receive praise and honors…The line separating good and evil passes not through states, nor between classes, nor between political parties either—but right through every human heart—and through all human hearts.

Part of the human condition is responding to issues, particularly those that are perceived to be socially and morally contentious, first intuitively and emotively and then rationally selecting evidence and information that justifies that initial response and preserves identity and sociocultural group membership (Haidt, 2001, 2013; Herman et al., 2024a, b). Science, our most powerful way of understanding and predicting occurrences in the natural world, like all human endeavors, can be entangled with good and evil agendas that are ideologically driven and hold irreversible consequences for the environment and human health and well-being.

Ideologically driven, polarized, and polarizing decision-making have occurred with more contemporary SSI, particularly those that exhibit high degrees of uncertainty and risk (e.g., climate change and the COVID-19 pandemic), which has significantly impacted the resolution of those issues (Bolsen & Shapiro, 2018; Herman et al., 2022; Herman et al., 2024a, b; Pietrocola et al., 2021). This is exacerbated by the post-truth era where science mis/disinformation is rampant and spreads rapidly and, unfortunately, siloed science, technology, engineering, and mathematics (STEM) education movements that neglect the humanities and lessons from history (Au et al., 2022; Barzilai & Chinn, 2020; Shahi et al., 2021; Valladares, 2022; Zeidler, 2016; Zeidler et al., 2016). Siloed approaches to science education fail to promote among students critical mis/disinformation detection skills (e.g., understanding methods of perspective taking and persuasion) and a sense of character, humility, and ethical sensitivity. Science educators must recognize this and seek to engage in scholarship, teaching, and communication with the public that considers the myriad factors associated with helping people become participants in a global society whose quality of existence depends on informed, compassionate, and civic SSI resolution. Encouragingly, those in the STEM fields have also been called to share this responsibility by organizations such as the National Academies of Sciences, Engineering, and Medicine (NASEM, 2017) and the United Nations Educational, Scientific, and Cultural Organization (Mede, 2023).

In Conclusion

I hope the sobering conclusion readers glean from this text is that there is a crucial need for education efforts to promote among the public not only a deep understanding of science, but also morality and compassionate perspective-taking—particularly that which can be enacted during socioscientific decision-making. SSI scholarship and instruction can contribute to this effort—and it is crucial that this comes from a diverse group of global contributors (Mede, 2023). That this book advances many diverse voices in a global conversation regarding epistemic insights in science education and how to prepare people to engage SSI through the "negotiation of multiple complex and often disconcerting dimensions that include scientific evidence, compassion, ethical sensitivity, NOS, willingness to act, and the unequal impacts on people and the environment (Herman et al., 2021, p. 3)" is refreshing and sorely needed. Furthermore, that many diverse voices collectively advanced SSI work occurring in their respective nations indicates two things. First, all cultures are not immune to using science or resolving SSI in ways that hold the potential to cause human and environmental harm. Equally, all cultures have the potential to work within and across communities and leverage science, sociocultural capital, and viewpoint diversity to find optimal SSI resolutions that are utilitarian in the sense that they best minimize negative and best maximize positive human and environmental outcomes locally and globally. Second, and more succinctly, the diversity of voices in this book indicate that that SSI are not a "you" or "them" problem—remember history and the catastrophic consequences of ideologically fueled othering and the abdication of moral responsibility. They are *US* problems, *YOU* the reader included, with widespread and long-term implications. Educators, researchers, policy makers, and others impacted by SSI, locally and globally, must take note of the diverse perspectives represented by the authors in this book that appear to coalesce into a global collective consciousness in the interest of preparing people to optimally resolve these current and future grand challenges, we call SSI.

References

Alsan, M., & Wanamaker, M. (2018). Tuskegee and the health of black men. *Quarterly Journal of Economics, 133*(1), 407–455. https://doi.org/10.1093/qje/qjx029

Au, C. H., Ho, K. K. W., & Chiu, D. K. (2022). The role of online misinformation and fake news in ideological polarization: Barriers, catalysts, and implications. *Information Systems Frontiers, 24*, 1331–1354. https://doi.org/10.1007/s10796-021-10133-9

Barzilai, S., & Chinn, C. A. (2020). A review of educational responses to the "post-truth" condition: Four lenses on "post-truth" problems. *Educational Psychologist, 55*(3), 107–119. https://doi.org/10.1080/00461520.2020.1786388

Bolsen, T., & Shapiro, M. A. (2018). The US news media, polarization on climate change, and pathways to effective communication. *Environmental Communication, 12*(2), 149–163. https://doi.org/10.1080/17524032.2017.1397039

Borinskaya, S. A., Ermolaev, A. I., & Kolchinsky, E. I. (2019). Lysenkoism against genetics: The meeting of the Lenin all-union academy of agricultural sciences of august 1948, its background, causes, and aftermath. *Genetics, 212*(1), 1–12. https://doi.org/10.1534/genetics.118.301413

Centers for Disease Control and Prevention. (2022). *The untreated syphilis study at Tuskegee timeline*. Retrieved on April 18, 2024, from https://www.cdc.gov/tuskegee/timeline.htm.

Farber, S. A. (2008). US scientists' role in the eugenics movement (1907–1939): A contemporary biologist's perspective. *Zebrafish, 5*(4), 243–245. https://doi.org/10.1089/zeb.2008.0576

Feldman, D. (2022). Reading poison: Science and story in children's propaganda. *Children's Literature in Education, 53*(2), 199–220.

Gould, S. J. (1978). Morton's ranking of races by cranial capacity: Unconscious manipulation of data may be a scientific norm. *Science, 200*, 503–509.

Haidt, J. (2001). The emotional dog and its rational tail: A social intuitionist approach to moral judgment. *Psychological Review, 108*(4), 814–834. https://doi.org/10.1037/0033-295X.108.4.814

Haidt, J. (2013). *The righteous mind*. Penguin Books.

Herman, B. C., Newton, M. H., & Zeidler, D. L. (2021). Impact of place-based socioscientific issues instruction on students' contextualization of socioscientific orientations. *Science Education, 105*, 587–627. https://doi.org/10.1002/sce.21618

Herman, B. C., Clough, M. P., & Rao, A. (2022). Socioscientific issues thinking and action in the midst of science-in-the-making. *Science and Education, 31*, 1105–1139. https://doi.org/10.1007/s11191-021-00306-y

Herman, B. C., Poor, S., Clough, M. P., Rao, A., Kidd, A., De Jesús, D., & Varghese, D. (2024a). It's not just a science thing: Educating future STEM professionals through mis/disinformation responsive instruction. *Journal of Research in Science Teaching*, 1–50. https://doi.org/10.1002/tea.21934

Herman, B. C., Clough, M. P., & Sobotka, A. (2024b). The influence of COVID-19 science views, risk perceptions, and group membership on socioscientific decisions. *International Journal of Science Education*, 1–28. https://doi.org/10.1080/09500693.2023.2300378

Kahn, S., & Zeidler, D. L. (2019). A conceptual analysis of perspective taking in support of socioscientific reasoning. *Science & Education, 28*(6–7), 605–638. https://doi.org/10.1007/s11191-019-00044-2

Kolchinsky, E. I., Kutschera, U., Hossfeld, U., & Levit, G. S. (2017). Russia's new Lysenkoism. *Current Biology, 27*, R1042–R1047. https://doi.org/10.1016/j.cub.2017.07.045

Lifton, R. J. (1986). *The Nazi doctors: Medical killing and the psychology of genocide*. Basic Books.

Mede, N. G. (2023). Rethink science communication for the post-pandemic era. *UNESCO Inclusive Policy Lab*. Retrieved on April 15, 2024, from https://en.unesco.org/inclusivepolicylab/analytics/rethink-science-communication-post-pandemic-era

National Academies of Sciences, Engineering, and Medicine. (2017). *Communicating science effectively: A research agenda*. The National Academies Press. https://doi.org/10.17226/23674.

National Institute of Health. (2023). *Eugenics and scientific racism.* Retrieved on January 15, 2023 from: https://www.genome.gov/about-genomics/fact-sheets/Eugenics-and-Scientific-Racism

Pietrocola, M., Rodrigues, E., Bercot, F., & Schnorr, S. (2021). Risk society and science education: Lessons from the COVID-19 pandemic. *Science & Education, 30*(2), 209–233. https://doi.org/10.1007/s11191-020-00176-w

Shahi, G., Dirkson, A., & Majchrzak, T. (2021). An exploratory study of COVID-19 misinformation on twitter. *Online Soc Netw Media, 22*, 1–16. https://doi.org/10.1016/j.osnem.2020.100104

Solzhenitsyn, A. I. (2007). *The Gulag Archipelago: An experiment in literary investigation, Volumes 1 & 2,* reissue editions. Harper Perennial.

Stern, A. M., Novak, N. L., Lira, N., O'Connor, K., Harlow, S., & Kardia, S. (2017). California's sterilization survivors: An estimate and call for redress. *American Journal of Public Health, 107*(1), 50–54.

United States Holocaust Memorial Museum. (2020). *Eugenics: Theories of eugenics, or "racial hygiene" in the German context, shaped many of Nazi Germany's persecutory policies.* Retrieved from: https://encyclopedia.ushmm.org/content/en/article/eugenics on 02/20/2024.

Valladares, L. (2022). Post-truth and education. *Science & Education, 31*(5), 1311–1337. https://doi.org/10.1007/s11191-021-00293-0

Weigmann, K. (2001). In the name of science. The role of biologists in Nazi atrocities: Lessons for today's scientists. *EMBO Reports, 2*, 871–875.

Wetzell, R. F. (2017). Eugenics, racial science, and Nazi biopolitics: Was there a genesis of the "final solution" from the spirit of science? In D. O. Pendas et al. (Eds.), *Beyond the racial state: Rethinking Nazi Germany* (pp. 147–175). Cambridge University Press.

Zeidler, D. L. (2016). STEM education: A deficit framework for the twenty-first century? A sociocultural socioscientific response. *Cultural Studies of Science Education, 11*(1), 11–26. https://doi.org/10.1007/s11422-014-9578-z

Zeidler, D. L., & Sadler, T. D. (2023). Exploring and expanding the frontiers of socioscientific issues: Crossroads and future directions. In N. G. Lederman, D. L. Zeidler, & J. S. Lederman (Eds.), *Handbook of research on science education* (Vol. III, pp. 899–929). Routledge.

Zeidler, D. L., Herman, B. C., Clough, M. P., Olson, J. K., Kahn, S., & Newton, M. (2016). Humanitas emptor: Reconsidering recent trends and policy in science teacher education. *Journal of Science Teacher Education, 27*(5), 465–476. https://doi.org/10.1007/s10972-016-9481-4

The manufacturer's authorised representative in the EU is Springer Nature Customer Service Centre GmbH, Europaplatz 3, 69115 Heidelberg, Germany. If you have any concerns regarding our products, please contact ProductSafety@springernature.com

Printed and bound by CPI Group (UK) Ltd, Croydon, CR0 4YY
01/10/2026
02225859-0016